Boundary Value Problems for Systems of Differential, Difference and Fractional Equations

Boundary Value Problems for Systems of Differential, Difference and Fractional Equations

Positive Solutions

Johnny Henderson and *Rodica Luca*

ELSEVIER

AMSTERDAM • BOSTON • HEIDELBERG • LONDON • NEW YORK • OXFORD
PARIS • SAN DIEGO • SAN FRANCISCO • SINGAPORE • SYDNEY • TOKYO

Elsevier
Radarweg 29, PO Box 211, 1000 AE Amsterdam, Netherlands
The Boulevard, Langford Lane, Kidlington, Oxford OX5 1GB, UK
225 Wyman Street, Waltham, MA 02451, USA

ISBN: 978-0-12-803652-5

British Library Cataloguing in Publication Data
A catalogue record for this book is available from the British Library

Library of Congress Control Number: 2015939629

Johnny Henderson dedicates this book to his siblings, Monty, Madonna, Jana, and Chrissie, and to the memory of his parents, Ernest and Madora. Rodica Luca dedicates this book to her husband, Mihai Tudorache, and her son, Alexandru-Gabriel Tudorache, and to the memory of her parents, Viorica and Constantin Luca.

Contents

Preface

In recent decades, nonlocal boundary value problems for ordinary differential equations, difference equations, or fractional differential equations have become a rapidly growing area of research. The study of these types of problems is driven not only by a theoretical interest, but also by the fact that several phenomena in engineering, physics, and the life sciences can be modeled in this way. Boundary value problems with positive solutions describe many phenomena in the applied sciences such as the nonlinear diffusion generated by nonlinear sources, thermal ignition of gases, and concentration in chemical or biological problems. Various problems arising in heat conduction, underground water flow, thermoelasticity, and plasma physics can be reduced to nonlinear differential problems with integral boundary conditions. Fractional differential equations describe many phenomena in several fields of engineering and scientific disciplines such as physics, biophysics, chemistry, biology (such as blood flow phenomena), economics, control theory, signal and image processing, aerodynamics, viscoelasticity, and electromagnetics.

Hundreds of researchers are working on boundary value problems for differential equations, difference equations, and fractional equations, and at the heart of the community are researchers whose interest is in positive solutions. The authors of this monograph occupy a niche in the center of that group. The monograph contains many of their results related to these topics obtained in recent years.

In Chapter 1, questions are addressed on the existence, multiplicity, and nonexistence of positive solutions for some classes of systems of nonlinear second-order ordinary differential equations with parameters or without parameters, subject to Riemann–Stieltjes boundary conditions, and for which the nonlinearities are nonsingular or singular functions. Chapter 2 is devoted to the existence, multiplicity, and nonexistence of positive solutions for some classes of systems of nonlinear higher-order ordinary differential equations with parameters or without parameters, subject to multipoint boundary conditions, and for which the nonlinearities are nonsingular or singular functions. A system of higher-order differential equations with sign-changing nonlinearities and Riemann–Stieltjes integral boundary conditions is also investigated. Chapter 3 deals with the existence, multiplicity, and nonexistence of positive solutions for some classes of systems of nonlinear second-order difference equations, also with or without parameters, subject to multipoint boundary conditions.

Chapter 4 is concerned with the existence, multiplicity, and nonexistence of positive solutions for some classes of systems of nonlinear Riemann–Liouville fractional differential equations with parameters or without parameters, subject to uncoupled Riemann–Stieltjes integral boundary conditions, and for which the nonlinearities are nonsingular or singular functions. A system of fractional equations with sign-changing nonlinearities and integral boundary conditions is also investigated. Chapter 5 is focused on the existence, multiplicity, and nonexistence of positive solutions for some classes of systems of nonlinear Riemann–Liouville fractional differential equations with parameters or without parameters, subject to coupled Riemann–Stieltjes integral boundary conditions, and for which the nonlinearities are nonsingular or singular functions. A system of fractional equations with sign-changing nonsingular or singular nonlinearities and integral boundary conditions is also investigated. In each chapter, various examples are presented which support the main results.

Central to the results of each chapter are applications of the Guo–Krasnosel'skii fixed point theorem for nonexpansive and noncontractive operators on a cone (Theorem 1.1.1). Unique to applications of the fixed point theorem is the novel representation of the Green's functions, which ultimately provide almost a checklist for determining conditions for which positive solutions exist relative to given nonlinearities. In the proof of many of the main results, applications are also made of the Schauder fixed-point theorem (Theorem 1.6.1), the nonlinear alternative of Leray–Schauder type (Theorem 2.5.1), and some theorems from the fixed point index theory (Theorems 1.3.1–1.3.3).

There have been other books in the past on positive solutions for boundary value problems, but in spite of the area receiving much attention, there have been no new books recently. This monograph provides a springboard for other researchers to emulate the authors' methods. The audience for this book includes the family of mathematical and scientific researchers in boundary value problems for which positive solutions are important, and in addition, the monograph can serve as a great source for topics to be studied in graduate seminars.

Johnny Henderson
Rodica Luca

About the authors

Johnny Henderson is a distinguished professor of Mathematics at the Baylor University, Waco, Texas, USA. He has also held faculty positions at the Auburn University and the Missouri University of Science and Technology. His published research is primarily in the areas of boundary value problems for ordinary differential equations, finite difference equations, functional differential equations, and dynamic equations on time scales. He is an Inaugural Fellow of the American Mathematical Society.

Rodica Luca is a professor of Mathematics at the "Gheorghe Asachi" Technical University of Iasi, Romania. She obtained her PhD degree in mathematics from "Alexandru Ioan Cuza" University of Iasi. Her research interests are boundary value problems for nonlinear systems of ordinary differential equations, finite difference equations, and fractional differential equations, and initial-boundary value problems for nonlinear hyperbolic systems of partial differential equations.

Acknowledgments

We are grateful to all the anonymous referees for carefully reviewing early drafts of the manuscript. We also express our thanks to Glyn Jones, the Mathematics publisher at Elsevier, and to Steven Mathews, the Editorial project manager for Mathematics and Statistics at Elsevier, for their support and encouragement of us during the preparation of this book; and to Poulouse Joseph, the project manager for book production at Elsevier, for all of his work on our book.

The work of Rodica Luca was supported by a grant of the Romanian National Authority for Scientific Research, CNCS-UEFISCDI, project number PN-II-ID-PCE-2011-3-0557.

Systems of second-order ordinary differential equations with integral boundary conditions

1

1.1 Existence of positive solutions for systems with parameters

Boundary value problems with positive solutions describe many phenomena in the applied sciences, such as the nonlinear diffusion generated by nonlinear sources, thermal ignition of gases, and concentration in chemical or biological problems (see Boucherif and Henderson, 2006; Cac et al., 1997; de Figueiredo et al., 1982; Guo and Lakshmikantham, 1988a,b; Joseph and Sparrow, 1970; Keller and Cohen, 1967). Integral boundary conditions arise in thermal conduction, semiconductor, and hydrodynamic problems (e.g., Cannon, 1964; Chegis, 1984; Ionkin, 1977; Samarskii, 1980). In recent decades, many authors have investigated scalar problems with integral boundary conditions (e.g., Ahmad et al., 2008; Boucherif, 2009; Jankowski, 2013; Jia and Wang, 2012; Karakostas and Tsamatos, 2002; Ma and An, 2009; Webb and Infante, 2008; Yang, 2006). We also mention references (Cui and Sun, 2012; Goodrich, 2012; Hao et al., 2012; Infante et al., 2012; Infante and Pietramala, 2009a,b; Kang and Wei, 2009; Lan, 2011; Song and Gao, 2011; Yang, 2005; Yang and O'Regan, 2005; Yang and Zhang, 2012), where the authors studied the existence of positive solutions for some systems of differential equations with integral boundary conditions.

1.1.1 Presentation of the problem

In this section, we consider the system of nonlinear second-order ordinary differential equations

$$\begin{cases} (a(t)u'(t))' - b(t)u(t) + \lambda p(t)f(t, u(t), v(t)) = 0, & 0 < t < 1, \\ (c(t)v'(t))' - d(t)v(t) + \mu q(t)g(t, u(t), v(t)) = 0, & 0 < t < 1, \end{cases} \tag{S}$$

with the integral boundary conditions

$$\begin{cases} \alpha u(0) - \beta a(0)u'(0) = \int_0^1 u(s)\,\mathrm{d}H_1(s), & \gamma u(1) + \delta a(1)u'(1) = \int_0^1 u(s)\,\mathrm{d}H_2(s), \\ \tilde{\alpha} v(0) - \tilde{\beta} c(0)v'(0) = \int_0^1 v(s)\,\mathrm{d}K_1(s), & \tilde{\gamma} v(1) + \tilde{\delta} c(1)v'(1) = \int_0^1 v(s)\,\mathrm{d}K_2(s), \end{cases} \tag{BC}$$

Boundary Value Problems for Systems of Differential, Difference and Fractional Equations.
http://dx.doi.org/10.1016/B978-0-12-803652-5.00001-6

where the above integrals are Riemann–Stieltjes integrals. The boundary conditions above include multipoint and integral boundary conditions and the sum of these in a single framework.

We give sufficient conditions on f and g and on the parameters λ and μ such that positive solutions of problem (S)–(BC) exist. By a positive solution of problem (S)–(BC) we mean a pair of functions $(u, v) \in C^2([0, 1]) \times C^2([0, 1])$ satisfying (S) and (BC) with $u(t) \geq 0$, $v(t) \geq 0$ for all $t \in [0, 1]$ and $(u, v) \neq (0, 0)$. The case in which the functions H_1, H_2, K_1, and K_2 are step functions—that is, the boundary conditions (BC) become multipoint boundary conditions

$$\begin{cases} \alpha u(0) - \beta a(0)u'(0) = \sum_{i=1}^{m} a_i u(\xi_i), \quad \gamma u(1) + \delta a(1)u'(1) = \sum_{i=1}^{n} b_i u(\eta_i), \\ \tilde{\alpha} v(0) - \tilde{\beta} c(0)v'(0) = \sum_{i=1}^{r} c_i v(\zeta_i), \quad \tilde{\gamma} v(1) + \tilde{\delta} c(1)v'(1) = \sum_{i=1}^{l} d_i v(\rho_i), \end{cases} \tag{BC$_1$}$$

where $m, n, r, l \in \mathbb{N}$, ($\mathbb{N} = \{1, 2, \ldots\}$)—was studied in Henderson and Luca (2013g). System (S) with $a(t) = 1$, $c(t) = 1$, $b(t) = 0$, and $d(t) = 0$ for all $t \in [0, 1]$, $f(t, u, v) = \tilde{f}(u, v)$, and $g(t, u, v) = \tilde{g}(u, v)$ (denoted by (S$_1$)) and (BC$_1$) was investigated in Henderson and Luca (2014e). Some particular cases of the problem from Henderson and Luca (2014e) were studied in Henderson and Luca (2012e) (where in (BC$_1$), $a_i = 0$ for all $i = 1, \ldots, m$, $c_i = 0$ for all $i = 1, \ldots, r$, $\gamma = \tilde{\gamma} = 1$, and $\delta = \tilde{\delta} = 0$—denoted by (BC$_2$)), in Luca (2011) (where in (S$_1$), $\tilde{f}(u, v) = \tilde{f}(v)$ and $\tilde{g}(u, v) = \tilde{g}(u)$—denoted by (S$_2$)—and in (BC$_2$) we have $n = l$, $b_i = d_i$, and $\eta_i = \rho_i$ for all $i = 1, \ldots, n$, $\alpha = \tilde{\alpha}$, and $\beta = \tilde{\beta}$—denoted by (BC$_3$)), in Henderson et al. (2008b) (problem (S$_2$)–(BC$_3$) with $\alpha = \tilde{\alpha} = 1$ and $\beta = \tilde{\beta} = 0$), and in Henderson and Ntouyas (2008b) and Henderson et al. (2008a) (system (S$_2$) with the boundary conditions $u(0) = 0$, $u(1) = \alpha u(\eta)$, $v(0) = 0$, $v(1) = \alpha v(\eta)$, $\eta \in (0, 1)$, and $0 < \alpha < 1/\eta$, or $u(0) = \beta u(\eta)$, $u(1) = \alpha u(\eta)$, $v(0) = \beta v(\eta)$, and $v(1) = \alpha v(\eta)$). In Henderson and Ntouyas (2008a), the authors investigated system (S$_2$) with the boundary conditions $\alpha u(0) - \beta u'(0) = 0$, $\gamma u(1) + \delta u'(1) = 0$, $\alpha v(0) - \beta v'(0) = 0$, and $\gamma v(1) + \delta v'(1) = 0$, where $\alpha, \beta, \gamma, \delta \geq 0$ and $\alpha + \beta + \gamma + \delta > 0$.

In the proof of our main results, we shall use the Guo–Krasnosel'skii fixed point theorem (see Guo and Lakshmikantham, 1988a), which we present now:

Theorem 1.1.1. *Let X be a Banach space and let $C \subset X$ be a cone in X. Assume Ω_1 and Ω_2 are bounded open subsets of X with $0 \in \Omega_1 \subset \overline{\Omega_1} \subset \Omega_2$, and let $\mathcal{A} : C \cap (\overline{\Omega_2} \setminus \Omega_1) \to C$ be a completely continuous operator (continuous, and compact—that is, it maps bounded sets into relatively compact sets) such that either*

(1) $\|\mathcal{A}u\| \leq \|u\|, \quad u \in C \cap \partial\Omega_1$, *and* $\|\mathcal{A}u\| \geq \|u\|, \quad u \in C \cap \partial\Omega_2$, *or*
(2) $\|\mathcal{A}u\| \geq \|u\|, \quad u \in C \cap \partial\Omega_1$, *and* $\|\mathcal{A}u\| \leq \|u\|, \quad u \in C \cap \partial\Omega_2$.

Then, $\mathcal{A}$ has a fixed point in $C \cap (\overline{\Omega_2} \setminus \Omega_1)$.

1.1.2 Auxiliary results

In this section, we present some auxiliary results related to the following second-order differential equation with integral boundary conditions:

$$(a(t)u'(t))' - b(t)u(t) + y(t) = 0, \quad t \in (0,1), \tag{1.1}$$

$$\alpha u(0) - \beta a(0)u'(0) = \int_0^1 u(s)\,\mathrm{d}H_1(s), \quad \gamma u(1) + \delta a(1)u'(1) = \int_0^1 u(s)\,\mathrm{d}H_2(s). \tag{1.2}$$

For $a \in C^1([0,1],(0,\infty))$, $b \in C([0,1],[0,\infty))$, $\alpha,\beta,\gamma,\delta \in \mathbb{R}$, $|\alpha|+|\beta| \neq 0$, $|\gamma|+|\delta| \neq 0$, we denote by ψ and ϕ the solutions of the following linear problems:

$$\begin{cases} (a(t)\psi'(t))' - b(t)\psi(t) = 0, & 0 < t < 1, \\ \psi(0) = \beta, & a(0)\psi'(0) = \alpha, \end{cases} \tag{1.3}$$

and

$$\begin{cases} (a(t)\phi'(t))' - b(t)\phi(t) = 0, & 0 < t < 1, \\ \phi(1) = \delta, & a(1)\phi'(1) = -\gamma, \end{cases} \tag{1.4}$$

respectively.

We denote by θ_1 the function $\theta_1(t) = a(t)(\phi(t)\psi'(t) - \phi'(t)\psi(t))$ for $t \in [0,1]$. By using (1.3) and (1.4), we deduce that $\theta_1'(t) = 0$—that is, $\theta_1(t) = \text{const.}$ for all $t \in [0,1]$. We denote this constant by τ_1. Then $\theta_1(t) = \tau_1$ for all $t \in [0,1]$, and so $\tau_1 = \theta_1(0) = a(0)(\phi(0)\psi'(0) - \phi'(0)\psi(0)) = \alpha\phi(0) - \beta a(0)\phi'(0)$ and $\tau_1 = \theta_1(1) = a(1)(\phi(1)\psi'(1) - \phi'(1)\psi(1)) = \delta a(1)\psi'(1) + \gamma\psi(1)$.

Lemma 1.1.1. *We assume that* $a \in C^1([0,1],(0,\infty))$, $b \in C([0,1],[0,\infty))$, $\alpha,\beta,\gamma,\delta \in \mathbb{R}$, $|\alpha|+|\beta| \neq 0$, $|\gamma|+|\delta| \neq 0$, *and* $H_1, H_2 : [0,1] \to \mathbb{R}$ *are functions of bounded variation. If* $\tau_1 \neq 0$, $\Delta_1 = \left(\tau_1 - \int_0^1 \psi(s)\,\mathrm{d}H_2(s)\right)\left(\tau_1 - \int_0^1 \phi(s)\,\mathrm{d}H_1(s)\right) - \left(\int_0^1 \psi(s)\,\mathrm{d}H_1(s)\right)\left(\int_0^1 \phi(s)\,\mathrm{d}H_2(s)\right) \neq 0$, *and* $y \in C(0,1) \cap L^1(0,1)$, *then the unique solution of* (1.1)–(1.2) *is given by* $u(t) = \int_0^1 G_1(t,s)y(s)\,\mathrm{d}s$, *where the Green's function* G_1 *is defined by*

$$\begin{aligned} G_1(t,s) = g_1(t,s) &+ \frac{1}{\Delta_1}\left[\psi(t)\left(\int_0^1 \phi(s)\,\mathrm{d}H_2(s)\right) + \phi(t)\left(\tau_1 - \int_0^1 \psi(s)\,\mathrm{d}H_2(s)\right)\right] \\ &\times \int_0^1 g_1(\tau,s)\,\mathrm{d}H_1(\tau) + \frac{1}{\Delta_1}\left[\psi(t)\left(\tau_1 - \int_0^1 \phi(s)\,\mathrm{d}H_1(s)\right)\right. \\ &\left. + \phi(t)\left(\int_0^1 \psi(s)\,\mathrm{d}H_1(s)\right)\right]\int_0^1 g_1(\tau,s)\,\mathrm{d}H_2(\tau), \end{aligned} \tag{1.5}$$

for all $(t,s) \in [0,1] \times [0,1]$, *with*

$$g_1(t,s) = \frac{1}{\tau_1} \begin{cases} \phi(t)\psi(s), & 0 \le s \le t \le 1, \\ \phi(s)\psi(t), & 0 \le t \le s \le 1, \end{cases} \tag{1.6}$$

and ψ *and* ϕ *are the functions defined by* (1.3) *and* (1.4), *respectively.*

Proof. Because $\tau_1 \neq 0$, the functions ψ and ϕ are two linearly independent solutions of the equation $(a(t)u'(t))' - b(t)u(t) = 0$. Then the general solution of (1.1) is $u(t) = A\psi(t) + B\phi(t) + u_0(t)$, with $A, B \in \mathbb{R}$, and u_0 is a particular solution of (1.1). We shall determine u_0 by the method of variation of constants—namely, we shall look for two functions $C(t)$ and $D(t)$ such that $u_0(t) = C(t)\psi(t) + D(t)\phi(t)$ is a solution of (1.1). The derivatives of $C(t)$ and $D(t)$ satisfy the system

$$\begin{cases} C'(t)\psi(t) + D'(t)\phi(t) = 0, & t \in (0,1), \\ C'(t)a(t)\psi'(t) + D'(t)a(t)\phi'(t) = -y(t), & t \in (0,1). \end{cases}$$

The above system has the determinant $d_0 = -\tau_1 \neq 0$, and the solution of the above system is $C'(t) = -\frac{1}{\tau_1}\phi(t)y(t)$ and $D'(t) = \frac{1}{\tau_1}\psi(t)y(t)$. Then we choose $C(t) = \frac{1}{\tau_1}\int_t^1 \phi(s)y(s)\,\mathrm{d}s$ and $D(t) = \frac{1}{\tau_1}\int_0^t \psi(s)y(s)\,\mathrm{d}s$. We deduce that the general solution of (1.1) is

$$u(t) = \frac{\psi(t)}{\tau_1}\int_t^1 \phi(s)y(s)\,\mathrm{d}s + \frac{\phi(t)}{\tau_1}\int_0^t \psi(s)y(s)\,\mathrm{d}s + A\psi(t) + B\phi(t).$$

Then we obtain

$$u(t) = \int_0^1 g_1(t,s)y(s)\,\mathrm{d}s + A\psi(t) + B\phi(t),$$

where g_1 is defined in (1.6).

By using condition (1.2), we conclude

$$\begin{cases} \alpha\left(\frac{1}{\tau_1}\int_0^1 \psi(0)\phi(s)y(s)\,\mathrm{d}s + A\psi(0) + B\phi(0)\right) - \beta a(0)\left(\frac{1}{\tau_1}\int_0^1 \phi(s)\psi'(0)y(s)\,\mathrm{d}s \right. \\ \left. +A\psi'(0) + B\phi'(0)\right) = \int_0^1 \left(\int_0^1 g_1(s,\tau)y(\tau)\,\mathrm{d}\tau + A\psi(s) + B\phi(s)\right)\mathrm{d}H_1(s), \\ \gamma\left(\frac{1}{\tau_1}\int_0^1 \phi(1)\psi(s)y(s)\,\mathrm{d}s + A\psi(1) + B\phi(1)\right) + \delta a(1)\left(\frac{1}{\tau_1}\int_0^1 \phi'(1)\psi(s)y(s)\,\mathrm{d}s \right. \\ \left. +A\psi'(1) + B\phi'(1)\right) = \int_0^1 \left(\int_0^1 g_1(s,\tau)y(\tau)\,\mathrm{d}\tau + A\psi(s) + B\phi(s)\right)\mathrm{d}H_2(s), \end{cases}$$

or

$$
\begin{cases}
A\left(-\alpha\psi(0)+\beta a(0)\psi'(0)+\displaystyle\int_0^1 \psi(s)\,\mathrm{d}H_1(s)\right) \\
+B\left(-\alpha\phi(0)+\beta a(0)\phi'(0)+\displaystyle\int_0^1 \phi(s)\,\mathrm{d}H_1(s)\right) \\
=\dfrac{\alpha}{\tau_1}\displaystyle\int_0^1 \phi(s)\psi(0)y(s)\,\mathrm{d}s-\dfrac{\beta a(0)}{\tau_1}\displaystyle\int_0^1 \phi(s)\psi'(0)y(s)\,\mathrm{d}s \\
-\displaystyle\int_0^1\left(\int_0^1 g_1(s,\tau)y(\tau)\,\mathrm{d}\tau\right)\mathrm{d}H_1(s), \\
A\left(\gamma\psi(1)+\delta a(1)\psi'(1)-\displaystyle\int_0^1 \psi(s)\,\mathrm{d}H_2(s)\right) \\
+B\left(\gamma\phi(1)+\delta a(1)\phi'(1)-\displaystyle\int_0^1 \phi(s)\,\mathrm{d}H_2(s)\right) \\
=-\dfrac{\gamma}{\tau_1}\displaystyle\int_0^1 \phi(1)\psi(s)y(s)\,\mathrm{d}s-\dfrac{\delta a(1)}{\tau_1}\displaystyle\int_0^1 \phi'(1)\psi(s)y(s)\,\mathrm{d}s \\
+\displaystyle\int_0^1\left(\int_0^1 g_1(s,\tau)y(\tau)\,\mathrm{d}\tau\right)\mathrm{d}H_2(s).
\end{cases}
$$

Therefore, we obtain

$$
\begin{cases}
A\left(\displaystyle\int_0^1 \psi(s)\,\mathrm{d}H_1(s)\right)+B\left(-\tau_1+\displaystyle\int_0^1 \phi(s)\,\mathrm{d}H_1(s)\right) \\
=-\displaystyle\int_0^1\left(\int_0^1 g_1(s,\tau)y(\tau)\,\mathrm{d}\tau\right)\mathrm{d}H_1(s), \\
A\left(\tau_1-\displaystyle\int_0^1 \psi(s)\,\mathrm{d}H_2(s)\right)+B\left(-\displaystyle\int_0^1 \phi(s)\,\mathrm{d}H_2(s)\right) \\
=\displaystyle\int_0^1\left(\int_0^1 g_1(s,\tau)y(\tau)\,\mathrm{d}\tau\right)\mathrm{d}H_2(s).
\end{cases}
\tag{1.7}
$$

The above system with the unknowns A and B has the determinant

$$
\Delta_1=\left(\tau_1-\int_0^1 \psi(s)\,\mathrm{d}H_2(s)\right)\left(\tau_1-\int_0^1 \phi(s)\,\mathrm{d}H_1(s)\right) -\left(\int_0^1 \psi(s)\,\mathrm{d}H_1(s)\right)\left(\int_0^1 \phi(s)\,\mathrm{d}H_2(s)\right).
$$

By using the assumptions of this lemma, we have $\Delta_1\neq 0$. Hence, system (1.7) has a unique solution—namely,

$$A = \frac{1}{\Delta_1}\left[\left(\int_0^1 \phi(s)\,\mathrm{d}H_2(s)\right)\left(\int_0^1 \left(\int_0^1 g_1(s,\tau)y(\tau)\,\mathrm{d}\tau\right)\mathrm{d}H_1(s)\right)\right.$$
$$\left. + \left(\tau_1 - \int_0^1 \phi(s)\,\mathrm{d}H_1(s)\right)\left(\int_0^1 \left(\int_0^1 g_1(s,\tau)y(\tau)\,\mathrm{d}\tau\right)\mathrm{d}H_2(s)\right)\right],$$
$$B = \frac{1}{\Delta_1}\left[\left(\int_0^1 \psi(s)\,\mathrm{d}H_1(s)\right)\left(\int_0^1 \left(\int_0^1 g_1(s,\tau)y(\tau)\,\mathrm{d}\tau\right)\mathrm{d}H_2(s)\right)\right.$$
$$\left. + \left(\tau_1 - \int_0^1 \psi(s)\,\mathrm{d}H_2(s)\right)\left(\int_0^1 \left(\int_0^1 g_1(s,\tau)y(\tau)\,\mathrm{d}\tau\right)\mathrm{d}H_1(s)\right)\right].$$

Then the solution of problem (1.1)–(1.2) is

$$u(t) = \int_0^1 g_1(t,s)y(s)\,\mathrm{d}s$$
$$+ \frac{\psi(t)}{\Delta_1}\left[\left(\int_0^1 \phi(s)\,\mathrm{d}H_2(s)\right)\left(\int_0^1 \left(\int_0^1 g_1(s,\tau)y(\tau)\,\mathrm{d}\tau\right)\mathrm{d}H_1(s)\right)\right.$$
$$\left. + \left(\tau_1 - \int_0^1 \phi(s)\,\mathrm{d}H_1(s)\right)\left(\int_0^1 \left(\int_0^1 g_1(s,\tau)y(\tau)\,\mathrm{d}\tau\right)\mathrm{d}H_2(s)\right)\right]$$
$$+ \frac{\phi(t)}{\Delta_1}\left[\left(\int_0^1 \psi(s)\,\mathrm{d}H_1(s)\right)\left(\int_0^1 \left(\int_0^1 g_1(s,\tau)y(\tau)\,\mathrm{d}\tau\right)\mathrm{d}H_2(s)\right)\right.$$
$$\left. + \left(\tau_1 - \int_0^1 \psi(s)\,\mathrm{d}H_2(s)\right)\left(\int_0^1 \left(\int_0^1 g_1(s,\tau)y(\tau)\,\mathrm{d}\tau\right)\mathrm{d}H_1(s)\right)\right].$$

Therefore, we deduce

$$u(t) = \int_0^1 g_1(t,s)y(s)\,\mathrm{d}s$$
$$+ \frac{1}{\Delta_1}\left[\psi(t)\left(\int_0^1 \phi(s)\,\mathrm{d}H_2(s)\right) + \phi(t)\left(\tau_1 - \int_0^1 \psi(s)\,\mathrm{d}H_2(s)\right)\right]$$
$$\times \int_0^1 \left(\int_0^1 g_1(s,\tau)\,\mathrm{d}H_1(s)\right)y(\tau)\,\mathrm{d}\tau + \frac{1}{\Delta_1}\left[\psi(t)\left(\tau_1 - \int_0^1 \phi(s)\,\mathrm{d}H_1(s)\right)\right.$$
$$\left. + \phi(t)\int_0^1 \psi(s)\,\mathrm{d}H_1(s)\right]\int_0^1 \left(\int_0^1 g_1(s,\tau)\,\mathrm{d}H_2(s)\right)y(\tau)\,\mathrm{d}\tau$$
$$= \int_0^1 g_1(t,s)y(s)\,\mathrm{d}s$$
$$+ \frac{1}{\Delta_1}\left[\psi(t)\left(\int_0^1 \phi(s)\,\mathrm{d}H_2(s)\right) + \phi(t)\left(\tau_1 - \int_0^1 \psi(s)\,\mathrm{d}H_2(s)\right)\right]$$
$$\times \int_0^1 \left(\int_0^1 g_1(\tau,s)\,\mathrm{d}H_1(\tau)\right)y(s)\,\mathrm{d}s + \frac{1}{\Delta_1}\left[\psi(t)\left(\tau_1 - \int_0^1 \phi(s)\,\mathrm{d}H_1(s)\right)\right.$$
$$\left. + \phi(t)\int_0^1 \psi(s)\,\mathrm{d}H_1(s)\right]\int_0^1 \left(\int_0^1 g_1(\tau,s)\,\mathrm{d}H_2(\tau)\right)y(s)\,\mathrm{d}s.$$

So, the solution u of (1.1)–(1.2) is $u(t) = \int_0^1 G_1(t,s)y(s)\,\mathrm{d}s$, $t \in [0,1]$, where G_1 is given in (1.5). □

Now, we introduce the following assumptions:

(A1) $a \in C^1([0,1],(0,\infty))$, $b \in C([0,1],[0,\infty))$.
(A2) $\alpha,\beta,\gamma,\delta \in [0,\infty)$, with $\alpha+\beta>0$ and $\gamma+\delta>0$.
(A3) If $b(t)\equiv 0$, then $\alpha+\gamma>0$.
(A4) $H_1, H_2 : [0,1] \to \mathbb{R}$ are nondecreasing functions.
(A5) $\tau_1 - \int_0^1 \phi(s)\,\mathrm{d}H_1(s) > 0$, $\tau_1 - \int_0^1 \psi(s)\,\mathrm{d}H_2(s) > 0$, and $\Delta_1 > 0$.

Lemma 1.1.2 (Atici and Guseinov, 2002). *Let (A1) and (A2) hold. Then*

(a) *the function ψ is nondecreasing on $[0,1]$, $\psi(t) \geq 0$ for all $t \in [0,1]$ and $\psi(t) > 0$ on $(0,1]$;*
(b) *the function ϕ is nonincreasing on $[0,1]$, $\phi(t) \geq 0$ for all $t \in [0,1]$ and $\phi(t) > 0$ on $[0,1)$.*

Lemma 1.1.3 (Atici and Guseinov, 2002). *Let (A1) and (A2) hold.*

(a) *If $b(t)$ is not identically zero, then $\tau_1 > 0$.*
(b) *If $b(t)$ is identically zero, then $\tau_1 > 0$ if and only if $\alpha+\gamma>0$.*

Lemma 1.1.4. *Let (A1)–(A3) hold. Then the function g_1 given by* (1.6) *has the following properties:*

(a) *g_1 is a continuous function on $[0,1]\times[0,1]$.*
(b) *$g_1(t,s) \geq 0$ for all $t,s \in [0,1]$, and $g_1(t,s) > 0$ for all $t,s \in (0,1)$.*
(c) *$g_1(t,s) \leq g_1(s,s)$ for all $t,s \in [0,1]$.*
(d) *For any $\sigma \in (0,1/2)$, we have $\min_{t\in[\sigma,1-\sigma]} g_1(t,s) \geq \nu_1 g_1(s,s)$ for all $s \in [0,1]$, where $\nu_1 = \min\left\{\frac{\phi(1-\sigma)}{\phi(0)}, \frac{\psi(\sigma)}{\psi(1)}\right\}$.*

For the proof of Lemma 1.1.4 (a) and (b), see Atici and Guseinov (2002), and for the proof of Lemma 1.1.4 (c) and (d), see Ma and Thompson (2004).

Lemma 1.1.5. *Let (A1)–(A5) hold. Then the Green's function G_1 of problem* (1.1)–(1.2) *is continuous on $[0,1]\times[0,1]$ and satisfies $G_1(t,s) \geq 0$ for all $(t,s) \in [0,1]\times[0,1]$. Moreover, if $y \in C(0,1)\cap L^1(0,1)$ satisfies $y(t) \geq 0$ for all $t \in (0,1)$, then the solution u of problem* (1.1)–(1.2) *satisfies $u(t) \geq 0$ for all $t \in [0,1]$.*

Proof. By using the assumptions of this lemma, we deduce $G_1(t,s) \geq 0$ for all $(t,s) \in [0,1]\times[0,1]$, and so $u(t) \geq 0$ for all $t \in [0,1]$. □

Lemma 1.1.6. *Assume that (A1)–(A5) hold. Then the Green's function G_1 of problem* (1.1)–(1.2) *satisfies the following inequalities:*

(a) *$G_1(t,s) \leq J_1(s)$, $\forall\,(t,s) \in [0,1]\times[0,1]$, where*

$$J_1(s) = g_1(s,s)$$
$$+\frac{1}{\Delta_1}\left[\psi(1)\left(\int_0^1 \phi(s)\,\mathrm{d}H_2(s)\right)+\phi(0)\left(\tau_1-\int_0^1 \psi(s)\,\mathrm{d}H_2(s)\right)\right]\int_0^1 g_1(\tau,s)\,\mathrm{d}H_1(\tau)$$
$$+\frac{1}{\Delta_1}\left[\psi(1)\left(\tau_1-\int_0^1 \phi(s)\,\mathrm{d}H_1(s)\right)+\phi(0)\left(\int_0^1 \psi(s)\,\mathrm{d}H_1(s)\right)\right]\int_0^1 g_1(\tau,s)\,\mathrm{d}H_2(\tau).$$

(b) *For every $\sigma \in (0,1/2)$ we have*

$$\min_{t\in[\sigma,1-\sigma]} G_1(t,s) \geq \nu_1 J_1(s) \geq \nu_1 G_1(t',s), \quad \forall t',s\in[0,1],$$

where ν_1 is given in Lemma 1.1.4.

Proof. The first inequality, (a), is evident. For the second inequality, (b), for $\sigma \in (0,1/2)$ and $t\in[\sigma,1-\sigma]$, $s\in[0,1]$ we conclude

$$\begin{aligned}
G_1(t,s) \geq& \nu_1 g_1(s,s) \\
&+ \frac{1}{\Delta_1}\left[\psi(\sigma)\left(\int_0^1 \phi(s)\,dH_2(s)\right) + \phi(1-\sigma)\left(\tau_1 - \int_0^1 \psi(s)\,dH_2(s)\right)\right] \\
&\times \int_0^1 g_1(\tau,s)\,dH_1(\tau) + \frac{1}{\Delta_1}\left[\psi(\sigma)\left(\tau_1 - \int_0^1 \phi(s)\,dH_1(s)\right)\right. \\
&\left.+ \phi(1-\sigma)\left(\int_0^1 \psi(s)\,dH_1(s)\right)\right]\int_0^1 g_1(\tau,s)\,dH_2(\tau) \\
&= \nu_1 g_1(s,s) + \frac{1}{\Delta_1}\left[\frac{\psi(\sigma)}{\psi(1)}\psi(1)\left(\int_0^1 \phi(s)\,dH_2(s)\right) + \frac{\phi(1-\sigma)}{\phi(0)}\phi(0)\right. \\
&\left.\times\left(\tau_1 - \int_0^1 \psi(s)\,dH_2(s)\right)\right]\int_0^1 g_1(\tau,s)\,dH_1(\tau) \\
&+ \frac{1}{\Delta_1}\left[\frac{\psi(\sigma)}{\psi(1)}\psi(1)\left(\tau_1 - \int_0^1 \phi(s)\,dH_1(s)\right)\right. \\
&\left.+ \frac{\phi(1-\sigma)}{\phi(0)}\phi(0)\left(\int_0^1 \psi(s)\,dH_1(s)\right)\right]\int_0^1 g_1(\tau,s)\,dH_2(\tau) \geq \nu_1 J_1(s).
\end{aligned}$$

□

Lemma 1.1.7. *Assume that (A1)–(A5) hold and let $\sigma \in (0,1/2)$. If $y \in C(0,1) \cap L^1(0,1)$, $y(t) \geq 0$ for all $t \in (0,1)$, then the solution $u(t)$, $t \in [0,1]$, of problem (1.1)–(1.2) satisfies the inequality* $\inf_{t\in[\sigma,1-\sigma]} u(t) \geq \nu_1 \sup_{t'\in[0,1]} u(t')$.

Proof. For $\sigma \in (0,1/2)$, $t \in [\sigma,1-\sigma]$, and $t' \in [0,1]$ we have

$$u(t) = \int_0^1 G_1(t,s)y(s)\,ds \geq \nu_1 \int_0^1 J_1(s)y(s)\,ds \geq \nu_1 \int_0^1 G_1(t',s)y(s)\,ds = \nu_1 u(t'),$$

and so $\inf_{t\in[\sigma,1-\sigma]} u(t) \geq \nu_1 \sup_{t'\in[0,1]} u(t')$. □

We can also formulate results similar to those in Lemmas 1.1.1–1.1.7 for the boundary value problem

$$(c(t)v'(t))' - d(t)v(t) + h(t) = 0, \quad 0 < t < 1, \tag{1.8}$$

$$\tilde{\alpha}v(0) - \tilde{\beta}c(0)v'(0) = \int_0^1 v(s)\,\mathrm{d}K_1(s), \tilde{\gamma}v(1) + \tilde{\delta}c(1)v'(1) = \int_0^1 v(s)\,\mathrm{d}K_2(s), \tag{1.9}$$

under assumptions similar to (A1)–(A5) and $h \in C(0,1) \cap L^1(0,1)$. We denote by $\tilde{\psi}, \tilde{\phi}, \theta_2, \tau_2, \Delta_2, g_2, G_2, \nu_2$, and J_2 the corresponding constants and functions for problem (1.8)–(1.9) defined in a similar manner as $\psi, \phi, \theta_1, \tau_1, \Delta_1, g_1, G_1, \nu_1$, and J_1, respectively.

1.1.3 Main existence results

In this section, we give sufficient conditions on λ, μ, f, and g such that positive solutions with respect to a cone for our problem (S)–(BC) exist.

We present the assumptions that we shall use in the sequel:

(I1) The functions $a, c \in C^1([0,1], (0,\infty))$ and $b, d \in C([0,1], [0,\infty))$.
(I2) $\alpha, \beta, \gamma, \delta, \tilde{\alpha}, \tilde{\beta}, \tilde{\gamma}, \tilde{\delta} \in [0,\infty)$ with $\alpha + \beta > 0$, $\gamma + \delta > 0$, $\tilde{\alpha} + \tilde{\beta} > 0$, and $\tilde{\gamma} + \tilde{\delta} > 0$; if $b \equiv 0$, then $\alpha + \gamma > 0$; if $d \equiv 0$, then $\tilde{\alpha} + \tilde{\gamma} > 0$.
(I3) $H_1, H_2, K_1, K_2 : [0,1] \to \mathbb{R}$ are nondecreasing functions.
(I4) $\tau_1 - \int_0^1 \phi(s)\,\mathrm{d}H_1(s) > 0$, $\tau_1 - \int_0^1 \psi(s)\,\mathrm{d}H_2(s) > 0$, $\tau_2 - \int_0^1 \tilde{\phi}(s)\,\mathrm{d}K_1(s) > 0$, $\tau_2 - \int_0^1 \tilde{\psi}(s)\,\mathrm{d}K_2(s) > 0$, and $\Delta_1 > 0$, $\Delta_2 > 0$, where τ_1, τ_2, Δ_1, and Δ_2 are defined in Section 1.1.2.
(I5) The functions $p, q \in C([0,1], [0,\infty))$, and there exist $t_1, t_2 \in (0,1)$ such that $p(t_1) > 0$, $q(t_2) > 0$.
(I6) The functions $f, g \in C([0,1] \times [0,\infty) \times [0,\infty), [0,\infty))$.

From assumption (I5), there exists $\sigma \in (0, 1/2)$ such that $t_1, t_2 \in (\sigma, 1-\sigma)$. We shall work in this section and in Section 1.2 with this number σ. This implies that

$$\int_\sigma^{1-\sigma} g_1(t,s)p(s)\,\mathrm{d}s > 0, \quad \int_\sigma^{1-\sigma} g_2(t,s)q(s)\,\mathrm{d}s > 0,$$

$$\int_\sigma^{1-\sigma} J_1(s)p(s)\,\mathrm{d}s > 0, \quad \int_\sigma^{1-\sigma} J_2(s)q(s)\,\mathrm{d}s > 0,$$

for all $t \in [0,1]$, where g_1, g_2, J_1, and J_2 are defined in Section 1.1.2 (Lemmas 1.1.1 and 1.1.6).

For σ defined above, we introduce the following extreme limits:

$$f_0^{\mathrm{s}} = \limsup_{u+v\to 0^+} \max_{t\in[0,1]} \frac{f(t,u,v)}{u+v}, \quad g_0^{\mathrm{s}} = \limsup_{u+v\to 0^+} \max_{t\in[0,1]} \frac{g(t,u,v)}{u+v},$$

$$f_0^{\mathrm{i}} = \liminf_{u+v\to 0^+} \min_{t\in[\sigma,1-\sigma]} \frac{f(t,u,v)}{u+v}, \quad g_0^{\mathrm{i}} = \liminf_{u+v\to 0^+} \min_{t\in[\sigma,1-\sigma]} \frac{g(t,u,v)}{u+v},$$

$$f_\infty^{\rm s} = \limsup_{u+v\to\infty} \max_{t\in[0,1]} \frac{f(t,u,v)}{u+v}, \quad g_\infty^{\rm s} = \limsup_{u+v\to\infty} \max_{t\in[0,1]} \frac{g(t,u,v)}{u+v},$$

$$f_\infty^{\rm i} = \liminf_{u+v\to\infty} \min_{t\in[\sigma,1-\sigma]} \frac{f(t,u,v)}{u+v}, \quad g_\infty^{\rm i} = \liminf_{u+v\to\infty} \min_{t\in[\sigma,1-\sigma]} \frac{g(t,u,v)}{u+v}.$$

In the definitions of the extreme limits above, the variables u and v are nonnegative.

By using the Green's functions G_1 and G_2 from Section 1.1.2 (Lemma 1.1.1), we can write our problem (S)–(BC) equivalently as the following nonlinear system of integral equations:

$$\begin{cases} u(t) = \lambda \int_0^1 G_1(t,s)p(s)f(s,u(s),v(s))\,{\rm d}s, & 0 \le t \le 1, \\ v(t) = \mu \int_0^1 G_2(t,s)q(s)g(s,u(s),v(s))\,{\rm d}s, & 0 \le t \le 1. \end{cases}$$

We consider the Banach space $X = C([0,1])$ with the supremum norm $\|u\| = \sup_{t\in[0,1]} |u(t)|$ and the Banach space $Y = X \times X$ with the norm $\|(u,v)\|_Y = \|u\| + \|v\|$. We define the cone $P \subset Y$ by

$$P = \Big\{ (u,v) \in Y; u(t) \ge 0, v(t) \ge 0, \forall\, t \in [0,1] \text{ and } \inf_{t\in[\sigma,1-\sigma]} (u(t)+v(t)) \ge \nu \|(u,v)\|_Y \Big\},$$

where $\nu = \min\{\nu_1, \nu_2\}$, and ν_1 and ν_2 are the constants defined in Section 1.1.2 (Lemma 1.1.4) with respect to the above σ.

For $\lambda, \mu > 0$, we introduce the operators $Q_1, Q_2 : Y \to X$ and $Q : Y \to Y$ defined by

$$Q_1(u,v)(t) = \lambda \int_0^1 G_1(t,s)p(s)f(s,u(s),v(s))\,{\rm d}s, \quad 0 \le t \le 1,$$

$$Q_2(u,v)(t) = \mu \int_0^1 G_2(t,s)q(s)g(s,u(s),v(s))\,{\rm d}s, \quad 0 \le t \le 1,$$

and $Q(u,v) = (Q_1(u,v), Q_2(u,v))$, $(u,v) \in Y$. The solutions of our problem (S)–(BC) coincide with the fixed points of the operator Q.

Lemma 1.1.8. *If (I1)–(I6) hold, then $Q : P \to P$ is a completely continuous operator.*

Proof. Let $(u,v) \in P$ be an arbitrary element. Because $Q_1(u,v)$ and $Q_2(u,v)$ satisfy problem (1.1)–(1.2) for $y(t) = \lambda p(t)f(t,u(t),v(t))$, $t \in [0,1]$, and problem (1.8)–(1.9) for $h(t) = \mu q(t)g(t,u(t),v(t))$, $t \in [0,1]$, respectively, then by Lemma 1.1.7, we obtain

$$\inf_{t\in[\sigma,1-\sigma]} Q_1(u,v)(t) \geq \nu_1 \sup_{t'\in[0,1]} Q_1(u,v)(t') = \nu_1 \|Q_1(u,v)\|,$$

$$\inf_{t\in[\sigma,1-\sigma]} Q_2(u,v)(t) \geq \nu_2 \sup_{t'\in[0,1]} Q_2(u,v)(t') = \nu_2 \|Q_2(u,v)\|.$$

Hence, we conclude

$$\begin{aligned}\inf_{t\in[\sigma,1-\sigma]}[Q_1(u,v)(t)+Q_2(u,v)(t)] &\geq \inf_{t\in[\sigma,1-\sigma]} Q_1(u,v)(t) + \inf_{t\in[\sigma,1-\sigma]} Q_2(u,v)(t)\\ &\geq \nu_1\|Q_1(u,v)\| + \nu_2\|Q_2(u,v)\|\\ &\geq \nu\|(Q_1(u,v),Q_2(u,v))\|_Y = \nu\|\mathcal{Q}(u,v)\|_Y.\end{aligned}$$

By (I1)–(I6) and Lemma 1.1.5, we obtain $Q_1(u,v)(t) \geq 0$, $Q_2(u,v)(t) \geq 0$ for all $t \in [0,1]$, and so we deduce that $\mathcal{Q}(u,v) \in P$. Hence, we get $\mathcal{Q}(P) \subset P$. By using standard arguments, we can easily show that Q_1 and Q_2 are completely continuous, and then $\mathcal{Q}$ is a completely continuous operator. □

We denote $A = \int_\sigma^{1-\sigma} J_1(s)p(s)\,ds$, $B = \int_0^1 J_1(s)p(s)\,ds$, $C = \int_\sigma^{1-\sigma} J_2(s)q(s)\,ds$, and $D = \int_0^1 J_2(s)q(s)\,ds$.

First, for $f_0^s, g_0^s, f_\infty^i, g_\infty^i \in (0,\infty)$ and numbers $\alpha_1, \alpha_2 \geq 0$ and $\tilde{\alpha}_1, \tilde{\alpha}_2 > 0$ such that $\alpha_1 + \alpha_2 = 1$ and $\tilde{\alpha}_1 + \tilde{\alpha}_2 = 1$, we define the numbers L_1, L_2, L_3, L_4, L_2', and L_4' by

$$L_1 = \frac{\alpha_1}{\nu\nu_1 f_\infty^i A}, \quad L_2 = \frac{\tilde{\alpha}_1}{f_0^s B}, \quad L_3 = \frac{\alpha_2}{\nu\nu_2 g_\infty^i C}, \quad L_4 = \frac{\tilde{\alpha}_2}{g_0^s D},$$
$$L_2' = \frac{1}{f_0^s B}, \quad L_4' = \frac{1}{g_0^s D}.$$

Theorem 1.1.2. *Assume that (I1)–(I6) hold, and $\alpha_1, \alpha_2 \geq 0$ and $\tilde{\alpha}_1, \tilde{\alpha}_2 > 0$ such that $\alpha_1 + \alpha_2 = 1$ and $\tilde{\alpha}_1 + \tilde{\alpha}_2 = 1$.*

(1) If $f_0^s, g_0^s, f_\infty^i, g_\infty^i \in (0,\infty)$, $L_1 < L_2$, and $L_3 < L_4$, then for each $\lambda \in (L_1, L_2)$ and $\mu \in (L_3, L_4)$ there exists a positive solution $(u(t), v(t)), t \in [0,1]$ for (S)–(BC).

(2) If $f_0^s = 0$, $g_0^s, f_\infty^i, g_\infty^i \in (0,\infty)$, and $L_3 < L_4'$, then for each $\lambda \in (L_1, \infty)$ and $\mu \in (L_3, L_4')$ there exists a positive solution $(u(t), v(t)), t \in [0,1]$ for (S)–(BC).

(3) If $g_0^s = 0$, $f_0^s, f_\infty^i, g_\infty^i \in (0,\infty)$, and $L_1 < L_2'$, then for each $\lambda \in (L_1, L_2')$ and $\mu \in (L_3, \infty)$ there exists a positive solution $(u(t), v(t)), t \in [0,1]$ for (S)–(BC).

(4) If $f_0^s = g_0^s = 0$ and $f_\infty^i, g_\infty^i \in (0,\infty)$, then for each $\lambda \in (L_1, \infty)$ and $\mu \in (L_3, \infty)$ there exists a positive solution $(u(t), v(t)), t \in [0,1]$ for (S)–(BC).

(5) If $\{f_0^s, g_0^s, f_\infty^i \in (0,\infty), g_\infty^i = \infty\}$ or $\{f_0^s, g_0^s, g_\infty^i \in (0,\infty), f_\infty^i = \infty\}$ or $\{f_0^s, g_0^s \in (0,\infty), f_\infty^i = g_\infty^i = \infty\}$, then for each $\lambda \in (0, L_2)$ and $\mu \in (0, L_4)$ there exists a positive solution $(u(t), v(t)), t \in [0,1]$ for (S)–(BC).

(6) If $\{f_0^s = 0, g_0^s, f_\infty^i \in (0,\infty), g_\infty^i = \infty\}$ or $\{f_0^s = 0, f_\infty^i = \infty, g_0^s, g_\infty^i \in (0,\infty)\}$ or $\{f_0^s = 0, g_0^s \in (0,\infty), f_\infty^i = g_\infty^i = \infty\}$, then for each $\lambda \in (0,\infty)$ and $\mu \in (0, L_4')$ there exists a positive solution $(u(t), v(t)), t \in [0,1]$ for (S)–(BC).

(7) If $\{f_0^s, f_\infty^i \in (0,\infty), g_0^s = 0, g_\infty^i = \infty\}$ or $\{f_0^s, g_\infty^i \in (0,\infty), g_0^s = 0, f_\infty^i = \infty\}$ or $\{f_0^s \in (0,\infty), g_0^s = 0, f_\infty^i = g_\infty^i = \infty\}$, then for each $\lambda \in (0, L_2')$ and $\mu \in (0,\infty)$ there exists a positive solution $(u(t), v(t)), t \in [0,1]$ for (S)–(BC).

(8) If $\{f_0^s = g_0^s = 0, f_\infty^i \in (0,\infty), g_\infty^i = \infty\}$ or $\{f_0^s = g_0^s = 0, f_\infty^i = \infty, g_\infty^i \in (0,\infty)\}$ or $\{f_0^s = g_0^s = 0, f_\infty^i = g_\infty^i = \infty\}$, then for each $\lambda \in (0,\infty)$ and $\mu \in (0,\infty)$ there exists a positive solution $(u(t), v(t)), t \in [0,1]$ for (S)–(BC).

Proof. We consider the above cone $P \subset Y$ and the operators $\mathcal{Q}_1$, $\mathcal{Q}_2$, and $\mathcal{Q}$. Because the proofs of the above cases are similar, in what follows we shall prove one of them—namely, case (2). So, we suppose $f_0^s = 0$, $g_0^s, f_\infty^i, g_\infty^i \in (0, \infty)$, and $L_3 < L_4'$. Let $\lambda \in (L_1, \infty)$ and $\mu \in (L_3, L_4')$—that is, $\lambda \in \left(\frac{\alpha_1}{\nu\nu_1 f_\infty^i A}, \infty\right)$ and $\mu \in \left(\frac{\alpha_2}{\nu\nu_2 g_\infty^i C}, \frac{1}{g_0^s D}\right)$. We choose $\tilde{\alpha}_2' \in (\mu g_0^s D, 1)$. Let $\tilde{\alpha}_1' = 1 - \tilde{\alpha}_2'$ and let $\varepsilon > 0$ be a positive number such that $\varepsilon < \min\{f_\infty^i, g_\infty^i\}$ and

$$\frac{\alpha_1}{\nu\nu_1(f_\infty^i - \varepsilon)A} \le \lambda, \quad \frac{\alpha_2}{\nu\nu_2(g_\infty^i - \varepsilon)C} \le \mu, \quad \frac{\tilde{\alpha}_1'}{\varepsilon B} \ge \lambda, \quad \frac{\tilde{\alpha}_2'}{(g_0^s + \varepsilon)D} \ge \mu.$$

By using (I6) and the definitions of f_0^s and g_0^s, we deduce that there exists $R_1 > 0$ such that $f(t, u, v) \le \varepsilon(u + v)$ and $g(t, u, v) \le (g_0^s + \varepsilon)(u + v)$ for all $t \in [0, 1]$ and $u, v \ge 0$, with $0 \le u + v \le R_1$. We define the set $\Omega_1 = \{(u, v) \in Y, \|(u, v)\|_Y < R_1\}$. Now let $(u, v) \in P \cap \partial\Omega_1$—that is, $(u, v) \in P$ with $\|(u, v)\|_Y = R_1$ or equivalently $\|u\| + \|v\| = R_1$. Then $u(t) + v(t) \le R_1$ for all $t \in [0, 1]$, and by Lemma 1.1.6, we obtain

$$\begin{aligned}
\mathcal{Q}_1(u, v)(t) &\le \lambda \int_0^1 J_1(s)p(s)f(s, u(s), v(s))\, ds \\
&\le \lambda \int_0^1 J_1(s)p(s)\varepsilon(u(s) + v(s))\, ds \\
&\le \lambda\varepsilon \int_0^1 J_1(s)p(s)(\|u\| + \|v\|)\, ds \\
&= \lambda\varepsilon B\|(u, v)\|_Y \le \tilde{\alpha}_1'\|(u, v)\|_Y, \quad \forall t \in [0, 1].
\end{aligned}$$

Therefore, $\|\mathcal{Q}_1(u, v)\| \le \tilde{\alpha}_1'\|(u, v)\|_Y$. In a similar manner, we conclude

$$\begin{aligned}
\mathcal{Q}_2(u, v)(t) &\le \mu \int_0^1 J_2(s)q(s)g(s, u(s), v(s))\, ds \\
&\le \mu \int_0^1 J_2(s)q(s)(g_0^s + \varepsilon)(u(s) + v(s))\, ds \\
&\le \mu(g_0^s + \varepsilon) \int_0^1 J_2(s)q(s)(\|u\| + \|v\|)\, ds \\
&= \mu(g_0^s + \varepsilon)D\|(u, v)\|_Y \le \tilde{\alpha}_2'\|(u, v)\|_Y, \quad \forall t \in [0, 1].
\end{aligned}$$

Therefore, $\|\mathcal{Q}_2(u, v)\| \le \tilde{\alpha}_2'\|(u, v)\|_Y$.

Then for $(u, v) \in P \cap \partial\Omega_1$, we deduce

$$\|\mathcal{Q}(u, v)\|_Y = \|\mathcal{Q}_1(u, v)\| + \|\mathcal{Q}_2(u, v)\| \le \tilde{\alpha}_1'\|(u, v)\|_Y + \tilde{\alpha}_2'\|(u, v)\|_Y = \|(u, v)\|_Y. \tag{1.10}$$

By the definitions of f_∞^i and g_∞^i, there exists $\bar{R}_2 > 0$ such that $f(t, u, v) \ge (f_\infty^i - \varepsilon)(u + v)$ and $g(t, u, v) \ge (g_\infty^i - \varepsilon)(u + v)$ for all $u, v \ge 0$, with $u + v \ge \bar{R}_2$, and $t \in [\sigma, 1 - \sigma]$. We consider $R_2 = \max\{2R_1, \bar{R}_2/\nu\}$, and we define $\Omega_2 = \{(u, v) \in Y, \|(u, v)\|_Y < R_2\}$. Then for $(u, v) \in P$ with $\|(u, v)\|_Y = R_2$, we obtain

$$u(t)+v(t)\geq \inf_{t\in[\sigma,1-\sigma]}(u(t)+v(t))\geq \nu\|(u,v)\|_Y=\nu R_2\geq \bar{R}_2$$

for all $t\in[\sigma,1-\sigma]$.

Then, by Lemma 1.1.6, we conclude

$$\begin{aligned}Q_1(u,v)(\sigma)&\geq \lambda\nu_1\int_0^1 J_1(s)p(s)f(s,u(s),v(s))\,ds\\&\geq \lambda\nu_1\int_\sigma^{1-\sigma} J_1(s)p(s)f(s,u(s),v(s))\,ds\\&\geq \lambda\nu_1\int_\sigma^{1-\sigma} J_1(s)p(s)(f_\infty^{\rm i}-\varepsilon)(u(s)+v(s))\,ds\\&\geq \lambda\nu_1(f_\infty^{\rm i}-\varepsilon)A\nu\|(u,v)\|_Y\geq \alpha_1\|(u,v)\|_Y.\end{aligned}$$

So, $\|Q_1(u,v)\|\geq Q_1(u,v)(\sigma)\geq \alpha_1\|(u,v)\|_Y$.

In a similar manner, we deduce

$$\begin{aligned}Q_2(u,v)(\sigma)&\geq \mu\nu_2\int_0^1 J_2(s)q(s)g(s,u(s),v(s))\,ds\\&\geq \mu\nu_2\int_\sigma^{1-\sigma} J_2(s)q(s)g(s,u(s),v(s))\,ds\\&\geq \mu\nu_2\int_\sigma^{1-\sigma} J_2(s)q(s)(g_\infty^{\rm i}-\varepsilon)(u(s)+v(s))\,ds\\&\geq \mu\nu_2(g_\infty^{\rm i}-\varepsilon)C\nu\|(u,v)\|_Y\geq \alpha_2\|(u,v)\|_Y.\end{aligned}$$

So, $\|Q_2(u,v)\|\geq Q_2(u,v)(\sigma)\geq \alpha_2\|(u,v)\|_Y$.

Hence, for $(u,v)\in P\cap\partial\Omega_2$ we obtain

$$\|Q(u,v)\|_Y=\|Q_1(u,v)\|+\|Q_2(u,v)\|\geq(\alpha_1+\alpha_2)\|(u,v)\|_Y=\|(u,v)\|_Y. \tag{1.11}$$

By using (1.10), (1.11), Lemma 1.1.8, and Theorem 1.1.1 (1), we conclude that Q has a fixed point $(u,v)\in P\cap(\bar{\Omega}_2\setminus\Omega_1)$ such that $R_1\leq\|u\|+\|v\|\leq R_2$. □

Remark 1.1.1. We mention that in Theorem 1.1.2 we have the possibility to choose $\alpha_1=0$ or $\alpha_2=0$. Therefore, each of the first four cases contains three subcases. For example, in the second case $f_0^{\rm s}=0, g_0^{\rm s}, f_\infty^{\rm i}, g_\infty^{\rm i}\in(0,\infty)$, we have the following situations:

(a) If $\alpha_1,\alpha_2\in(0,1)$, $\alpha_1+\alpha_2=1$, and $L_3<L_4'$, then $\lambda\in(L_1,\infty)$ and $\mu\in(L_3,L_4')$.
(b) If $\alpha_1=1$ and $\alpha_2=0$, then $\lambda\in(L_1',\infty)$ and $\mu\in(0,L_4')$, where $L_1'=\frac{1}{\nu\nu_1 f_\infty^{\rm i}A}$.
(c) If $\alpha_1=0$, $\alpha_2=1$, and $L_3'<L_4'$, then $\lambda\in(0,\infty)$ and $\mu\in(L_3',L_4')$, where $L_3'=\frac{1}{\nu\nu_2 g_\infty^{\rm i}C}$.

In what follows, for $f_0^{\rm i}, g_0^{\rm i}, f_\infty^{\rm s}, g_\infty^{\rm s}\in(0,\infty)$ and numbers $\alpha_1,\alpha_2\geq 0$ and $\tilde{\alpha}_1,\tilde{\alpha}_2>0$ such that $\alpha_1+\alpha_2=1$ and $\tilde{\alpha}_1+\tilde{\alpha}_2=1$, we define the numbers $\tilde{L}_1$, $\tilde{L}_2$, $\tilde{L}_3$, $\tilde{L}_4$, $\tilde{L}_2'$, and $\tilde{L}_4'$ by

$$\tilde{L}_1 = \frac{\alpha_1}{\nu\nu_1 f_0^{\rm i} A}, \quad \tilde{L}_2 = \frac{\tilde{\alpha}_1}{f_\infty^{\rm s} B}, \quad \tilde{L}_3 = \frac{\alpha_2}{\nu\nu_2 g_0^{\rm i} C}, \quad \tilde{L}_4 = \frac{\tilde{\alpha}_2}{g_\infty^{\rm s} D},$$

$$\tilde{L}_2' = \frac{1}{f_\infty^{\rm s} B}, \quad \tilde{L}_4' = \frac{1}{g_\infty^{\rm s} D}.$$

Theorem 1.1.3. *Assume that (I1)–(I6) hold, and $\alpha_1, \alpha_2 \geq 0$ and $\tilde{\alpha}_1, \tilde{\alpha}_2 > 0$ such that $\alpha_1 + \alpha_2 = 1$ and $\tilde{\alpha}_1 + \tilde{\alpha}_2 = 1$.*

(1) If $f_0^{\rm i}, g_0^{\rm i}, f_\infty^{\rm s}, g_\infty^{\rm s} \in (0,\infty)$, $\tilde{L}_1 < \tilde{L}_2$, and $\tilde{L}_3 < \tilde{L}_4$, then for each $\lambda \in (\tilde{L}_1, \tilde{L}_2)$ and $\mu \in (\tilde{L}_3, \tilde{L}_4)$ there exists a positive solution $(u(t), v(t)), t \in [0,1]$ for (S)–(BC).

(2) If $f_0^{\rm i}, g_0^{\rm i}, f_\infty^{\rm s} \in (0,\infty)$, $g_\infty^{\rm s} = 0$, and $\tilde{L}_1 < \tilde{L}_2'$, then for each $\lambda \in (\tilde{L}_1, \tilde{L}_2')$ and $\mu \in (\tilde{L}_3, \infty)$ there exists a positive solution $(u(t), v(t)), t \in [0,1]$ for (S)–(BC).

(3) If $f_0^{\rm i}, g_0^{\rm i}, g_\infty^{\rm s} \in (0,\infty)$, $f_\infty^{\rm s} = 0$, and $\tilde{L}_3 < \tilde{L}_4'$, then for each $\lambda \in (\tilde{L}_1, \infty)$ and $\mu \in (\tilde{L}_3, \tilde{L}_4')$ there exists a positive solution $(u(t), v(t)), t \in [0,1]$ for (S)–(BC).

(4) If $f_0^{\rm i}, g_0^{\rm i} \in (0,\infty)$ and $f_\infty^{\rm s} = g_\infty^{\rm s} = 0$, then for each $\lambda \in (\tilde{L}_1, \infty)$ and $\mu \in (\tilde{L}_3, \infty)$ there exists a positive solution $(u(t), v(t)), t \in [0,1]$ for (S)–(BC).

(5) If $\{f_0^{\rm i} = \infty,\ g_0^{\rm i}, f_\infty^{\rm s}, g_\infty^{\rm s} \in (0,\infty)\}$ or $\{f_0^{\rm i}, f_\infty^{\rm s}, g_\infty^{\rm s} \in (0,\infty),\ g_0^{\rm i} = \infty\}$ or $\{f_0^{\rm i} = g_0^{\rm i} = \infty,\ f_\infty^{\rm s}, g_\infty^{\rm s} \in (0,\infty)\}$, then for each $\lambda \in (0, \tilde{L}_2)$ and $\mu \in (0, \tilde{L}_4)$ there exists a positive solution $(u(t), v(t)), t \in [0,1]$ for (S)–(BC).

(6) If $\{f_0^{\rm i} = \infty,\ g_0^{\rm i}, f_\infty^{\rm s} \in (0,\infty),\ g_\infty^{\rm s} = 0\}$ or $\{f_0^{\rm i}, f_\infty^{\rm s} \in (0,\infty),\ g_0^{\rm i} = \infty,\ g_\infty^{\rm s} = 0\}$ or $\{f_0^{\rm i} = g_0^{\rm i} = \infty,\ f_\infty^{\rm s} \in (0,\infty),\ g_\infty^{\rm s} = 0\}$, then for each $\lambda \in (0, \tilde{L}_2')$ and $\mu \in (0,\infty)$ there exists a positive solution $(u(t), v(t)), t \in [0,1]$ for (S)–(BC).

(7) If $\{f_0^{\rm i} = \infty,\ g_0^{\rm i}, g_\infty^{\rm s} \in (0,\infty),\ f_\infty^{\rm s} = 0\}$ or $\{f_0^{\rm i}, g_\infty^{\rm s} \in (0,\infty),\ g_0^{\rm i} = \infty,\ f_\infty^{\rm s} = 0\}$ or $\{f_0^{\rm i} = g_0^{\rm i} = \infty,\ f_\infty^{\rm s} = 0,\ g_\infty^{\rm s} \in (0,\infty)\}$, then for each $\lambda \in (0,\infty)$ and $\mu \in (0, \tilde{L}_4')$ there exists a positive solution $(u(t), v(t)), t \in [0,1]$ for (S)–(BC).

(8) If $\{f_0^{\rm i} = \infty,\ g_0^{\rm i} \in (0,\infty),\ f_\infty^{\rm s} = g_\infty^{\rm s} = 0\}$ or $\{f_0^{\rm i} \in (0,\infty),\ g_0^{\rm i} = \infty,\ f_\infty^{\rm s} = g_\infty^{\rm s} = 0\}$ or $\{f_0^{\rm i} = g_0^{\rm i} = \infty,\ f_\infty^{\rm s} = g_\infty^{\rm s} = 0\}$, then for each $\lambda \in (0,\infty)$ and $\mu \in (0,\infty)$ there exists a positive solution $(u(t), v(t)), t \in [0,1]$ for (S)–(BC).

Proof. We consider the above cone $P \subset Y$ and the operators Q_1, Q_2, and $\mathcal{Q}$. Because the proofs of the above cases are similar, in what follows we shall prove one of them—namely, the second case of (6). So, we suppose $f_0^{\rm i}, f_\infty^{\rm s} \in (0,\infty)$, $g_0^{\rm i} = \infty$, and $g_\infty^{\rm s} = 0$. Let $\lambda \in (0, \tilde{L}_2')$—that is, $\lambda \in \left(0, \frac{1}{f_\infty^{\rm s} B}\right)$—and $\mu \in (0,\infty)$. We choose $\alpha_1' > 0$, $\alpha_1' < \min\{\lambda\nu\nu_1 f_0^{\rm i} A, 1\}$, and $\tilde{\alpha}_1' \in (\lambda f_\infty^{\rm s} B, 1)$. Let $\alpha_2' = 1 - \alpha_1'$ and $\tilde{\alpha}_2' = 1 - \tilde{\alpha}_1'$, and let $\varepsilon > 0$ be a positive number such that $\varepsilon < f_0^{\rm i}$ and

$$\frac{\alpha_1'}{\nu\nu_1(f_0^{\rm i} - \varepsilon)A} \leq \lambda, \quad \frac{\alpha_2'\varepsilon}{\nu\nu_2 C} \leq \mu, \quad \frac{\tilde{\alpha}_1'}{(f_\infty^{\rm s} + \varepsilon)B} \geq \lambda, \quad \frac{\tilde{\alpha}_2'}{\varepsilon D} \geq \mu.$$

By using (I6) and the definitions of $f_0^{\rm i}$ and $g_0^{\rm i}$, we deduce that there exists $R_3 > 0$ such that $f(t,u,v) \geq (f_0^{\rm i} - \varepsilon)(u+v)$ and $g(t,u,v) \geq \frac{1}{\varepsilon}(u+v)$ for all $u, v \geq 0$, with $0 \leq u + v \leq R_3$, and $t \in [\sigma, 1-\sigma]$. We denote $\Omega_3 = \{(u,v) \in Y, \|(u,v)\|_Y < R_3\}$. Let $(u,v) \in P$ with $\|(u,v)\|_Y = R_3$—that is, $\|u\| + \|v\| = R_3$. Because $u(t) + v(t) \leq \|u\| + \|v\| = R_3$ for all $t \in [0,1]$, then, by using Lemma 1.1.6, we obtain

$$
\begin{aligned}
Q_1(u,v)(\sigma) &\geq \lambda\nu_1 \int_{\sigma}^{1-\sigma} J_1(s)p(s)f(s,u(s),v(s))\,\mathrm{d}s \\
&\geq \lambda\nu_1 \int_{\sigma}^{1-\sigma} J_1(s)p(s)(f_0^{\mathrm{i}}-\varepsilon)(u(s)+v(s))\,\mathrm{d}s \\
&\geq \lambda\nu\nu_1(f_0^{\mathrm{i}}-\varepsilon) \int_{\sigma}^{1-\sigma} J_1(s)p(s)(\|u\|+\|v\|)\,\mathrm{d}s \\
&= \lambda\nu\nu_1(f_0^{\mathrm{i}}-\varepsilon)A\|(u,v)\|_Y \geq \alpha_1'\|(u,v)\|_Y.
\end{aligned}
$$

Therefore, $\|Q_1(u,v)\| \geq Q_1(u,v)(\sigma) \geq \alpha_1'\|(u,v)\|_Y$. In a similar manner, we conclude

$$
\begin{aligned}
Q_2(u,v)(\sigma) &\geq \mu\nu_2 \int_{\sigma}^{1-\sigma} J_2(s)q(s)g(s,u(s),v(s))\,\mathrm{d}s \\
&\geq \mu\nu_2 \int_{\sigma}^{1-\sigma} J_2(s)q(s)\frac{1}{\varepsilon}(u(s)+v(s))\,\mathrm{d}s \\
&\geq \mu\nu\nu_2\frac{1}{\varepsilon} \int_{\sigma}^{1-\sigma} J_2(s)q(s)(\|u\|+\|v\|)\,\mathrm{d}s \\
&= \mu\nu\nu_2\frac{1}{\varepsilon}C\|(u,v)\|_Y \geq \alpha_2'\|(u,v)\|_Y.
\end{aligned}
$$

So, $\|Q_2(u,v)\| \geq Q_2(u,v)(\sigma) \geq \alpha_2'\|(u,v)\|_Y$.

Thus, for an arbitrary element $(u,v) \in P \cap \partial\Omega_3$ we deduce

$$
\|\mathcal{Q}(u,v)\|_Y \geq (\alpha_1'+\alpha_2')\|(u,v)\|_Y = \|(u,v)\|_Y. \tag{1.12}
$$

Now, we define the functions $f^*, g^* : [0,1]\times\mathbb{R}_+ \to \mathbb{R}_+$, $f^*(t,x) = \max_{0\leq u+v\leq x} f(t,u,v)$, $g^*(t,x) = \max_{0\leq u+v\leq x} g(t,u,v)$, $t\in[0,1]$, $x\in\mathbb{R}_+$. Then $f(t,u,v) \leq f^*(t,x)$ and $g(t,u,v)\leq g^*(t,x)$ for all $t\in[0,1]$, $u\geq 0$, $v\geq 0$, and $u+v\leq x$. The functions $f^*(t,\cdot)$ and $g^*(t,\cdot)$ are nondecreasing for every $t\in[0,1]$, and they satisfy the conditions

$$
\limsup_{x\to\infty}\max_{t\in[0,1]}\frac{f^*(t,x)}{x} \leq f_\infty^{\mathrm{s}}, \quad \lim_{x\to\infty}\max_{t\in[0,1]}\frac{g^*(t,x)}{x} = 0.
$$

Therefore, for $\varepsilon>0$ there exists $\bar{R}_4>0$ such that for all $x\geq\bar{R}_4$ and $t\in[0,1]$ we have

$$
\begin{aligned}
\frac{f^*(t,x)}{x} &\leq \limsup_{x\to\infty}\max_{t\in[0,1]}\frac{f^*(t,x)}{x}+\varepsilon \leq f_\infty^{\mathrm{s}}+\varepsilon, \\
\frac{g^*(t,x)}{x} &\leq \lim_{x\to\infty}\max_{t\in[0,1]}\frac{g^*(t,x)}{x}+\varepsilon = \varepsilon,
\end{aligned}
$$

and so $f^*(t,x)\leq(f_\infty^{\mathrm{s}}+\varepsilon)x$ and $g^*(t,x)\leq\varepsilon x$.

We consider $R_4 = \max\{2R_3, \bar{R}_4\}$, and we denote $\Omega_4 = \{(u,v) \in Y, \|(u,v)\|_Y < R_4\}$. Let $(u,v) \in P \cap \partial\Omega_4$. By the definitions of f^* and g^*, we conclude

$$f(t,u(t),v(t)) \le f^*(t,\|(u,v)\|_Y), \quad g(t,u(t),v(t)) \le g^*(t,\|(u,v)\|_Y), \quad \forall t \in [0,1].$$

Then for all $t \in [0,1]$, we obtain

$$\begin{aligned}
Q_1(u,v)(t) &\le \lambda \int_0^1 J_1(s)p(s)f(s,u(s),v(s))\,\mathrm{d}s \le \lambda \int_0^1 J_1(s)p(s)f^*(s,\|(u,v)\|_Y)\,\mathrm{d}s \\
&\le \lambda(f_\infty^{\mathrm{s}} + \varepsilon)\int_0^1 J_1(s)p(s)\|(u,v)\|_Y\,\mathrm{d}s \\
&= \lambda(f_\infty^{\mathrm{s}} + \varepsilon)B\|(u,v)\|_Y \le \tilde{\alpha}_1'\|(u,v)\|_Y,
\end{aligned}$$

and so $\|Q_1(u,v)\| \le \tilde{\alpha}_1'\|(u,v)\|_Y$.

In a similar manner, we deduce

$$\begin{aligned}
Q_2(u,v)(t) &\le \mu \int_0^1 J_2(s)q(s)g(s,u(s),v(s))\,\mathrm{d}s \le \mu \int_0^1 J_2(s)q(s)g^*(s,\|(u,v)\|_Y)\,\mathrm{d}s \\
&\le \mu\varepsilon \int_0^1 J_2(s)q(s)\|(u,v)\|_Y\,\mathrm{d}s = \mu\varepsilon D\|(u,v)\|_Y \le \tilde{\alpha}_2'\|(u,v)\|_Y,
\end{aligned}$$

and so $\|Q_2(u,v)\| \le \tilde{\alpha}_2'\|(u,v)\|_Y$.

Therefore, for $(u,v) \in P \cap \partial\Omega_4$ it follows that

$$\|\mathcal{Q}(u,v)\|_Y \le (\tilde{\alpha}_1' + \tilde{\alpha}_2')\|(u,v)\|_Y = \|(u,v)\|_Y. \tag{1.13}$$

By using (1.12), (1.13), Lemma 1.1.8, and Theorem 1.1.1 (2), we conclude that $\mathcal{Q}$ has a fixed point $(u,v) \in P \cap (\bar{\Omega}_4 \setminus \Omega_3)$ such that $R_3 \le \|(u,v)\|_Y \le R_4$. □

1.1.4 Examples

Let $a(t) = 1$, $b(t) = 4$, $c(t) = 1$, $d(t) = 1$, $p(t) = 1$, and $q(t) = 1$ for all $t \in [0,1]$, $\alpha = 1$, $\beta = 3$, $\gamma = 1$, $\delta = 1$, $\tilde{\alpha} = 3$, $\tilde{\beta} = 2$, $\tilde{\gamma} = 1$, $\tilde{\delta} = 3/2$,

$$H_1(t) = t^2, \quad H_2(t) = \begin{cases} 0, & t \in [0,1/3), \\ 7/2, & t \in [1/3,2/3), \\ 11/2, & t \in [2/3,1], \end{cases}$$

$$K_1(t) = \begin{cases} 0, & t \in [0,1/2), \\ 4/3, & t \in [1/2,1], \end{cases} \quad K_2(t) = t^3.$$

Then $\int_0^1 u(s)\,\mathrm{d}H_2(s) = \frac{7}{2}u\left(\frac{1}{3}\right) + 2u\left(\frac{2}{3}\right)$, $\int_0^1 u(s)\,\mathrm{d}K_1(s) = \frac{4}{3}u\left(\frac{1}{2}\right)$, $\int_0^1 u(s)\,\mathrm{d}H_1(s) = 2\int_0^1 su(s)\,\mathrm{d}s$, and $\int_0^1 u(s)\,\mathrm{d}K_2(s) = 3\int_0^1 s^2u(s)\,\mathrm{d}s$.

We consider the second-order differential system

$$\begin{cases} u''(t) - 4u(t) + \lambda f(t,u(t),v(t)) = 0, & t \in (0,1), \\ v''(t) - v(t) + \mu g(t,u(t),v(t)) = 0, & t \in (0,1), \end{cases} \tag{$\bar{\mathrm{S}}$}$$

with the boundary conditions

$$\left\{\begin{array}{ll} u(0) - 3u'(0) = 2\displaystyle\int_0^1 su(s)\,\mathrm{d}s, & u(1) + u'(1) = \dfrac{7}{2}u\left(\dfrac{1}{3}\right) + 2u\left(\dfrac{2}{3}\right), \\ 3v(0) - 2v'(0) = \dfrac{4}{3}v\left(\dfrac{1}{2}\right), & v(1) + \dfrac{3}{2}v'(1) = 3\displaystyle\int_0^1 s^2 v(s)\,\mathrm{d}s. \end{array}\right. \tag{$\overline{\mathrm{BC}}$}$$

The functions ψ and ϕ from Section 1.1.2 are the solutions of the following problems:

$$\left\{\begin{array}{ll} \psi''(t) - 4\psi(t) = 0, & 0 < t < 1, \\ \psi(0) = 3, & \psi'(0) = 1, \end{array}\right. \qquad \left\{\begin{array}{ll} \phi''(t) - 4\phi(t) = 0, & 0 < t < 1, \\ \phi(1) = 1, & \phi'(1) = -1. \end{array}\right.$$

We obtain $\psi(t) = \frac{7\,\mathrm{e}^{4t}+5}{4\,\mathrm{e}^{2t}}$ and $\phi(t) = \frac{1+3\,\mathrm{e}^{4-4t}}{4\,\mathrm{e}^{2-2t}}$ for all $t \in [0,1]$, $\tau_1 = \frac{21\,\mathrm{e}^4 - 5}{4\,\mathrm{e}^2}$,

$$\Lambda_1 := \tau_1 - \int_0^1 \psi(s)\,\mathrm{d}H_2(s) = \tau_1 - \left(\frac{7}{2}\psi\left(\frac{1}{3}\right) + 2\psi\left(\frac{2}{3}\right)\right)$$

$$= (42\,\mathrm{e}^4 - 28\,\mathrm{e}^{10/3} - 49\,\mathrm{e}^{8/3} - 35\,\mathrm{e}^{4/3} - 20\,\mathrm{e}^{2/3} - 10)/(8\,\mathrm{e}^2) \approx 10.51047404 > 0,$$

$$\Lambda_2 := \tau_1 - \int_0^1 \phi(s)\,\mathrm{d}H_1(s) = \tau_1 - 2\int_0^1 s\phi(s)\,\mathrm{d}s$$

$$= (21\,\mathrm{e}^4 - 5)/(4\,\mathrm{e}^2) - 2\int_0^1 s(1 + 3\,\mathrm{e}^{4-4s})/(4\,\mathrm{e}^{2-2s})\,\mathrm{d}s \approx 36.83556247 > 0,$$

$$\Lambda_3 := \int_0^1 \psi(s)\,\mathrm{d}H_1(s) = 2\int_0^1 s\psi(s)\,\mathrm{d}s = 2\int_0^1 s(7\,\mathrm{e}^{4s} + 5)/(4\,\mathrm{e}^{2s})\,\mathrm{d}s \approx 7.71167043,$$

$$\Lambda_4 := \int_0^1 \phi(s)\,\mathrm{d}H_2(s) = \frac{7}{2}\phi\left(\frac{1}{3}\right) + 2\phi\left(\frac{2}{3}\right)$$

$$= 7(1 + 3\,\mathrm{e}^{8/3})/(8\,\mathrm{e}^{4/3}) + 2(1 + 3\,\mathrm{e}^{4/3})/(4\,\mathrm{e}^{2/3}) \approx 13.36733534,$$

$$\Delta_1 = \Lambda_1\Lambda_2 - \Lambda_3\Lambda_4 \approx 284.07473844 > 0.$$

The functions g_1 and J_1 are given by

$$g_1(t,s) = \frac{1}{\tau_1}\left\{\begin{array}{ll} \phi(t)\psi(s), & 0 \le s \le t \le 1, \\ \phi(s)\psi(t), & 0 \le t \le s \le 1, \end{array}\right. \quad \text{or}$$

$$g_1(t,s) = \frac{4\,\mathrm{e}^2}{21\,\mathrm{e}^4 - 5}\left\{\begin{array}{ll} \dfrac{(1 + 3\,\mathrm{e}^{4-4t})(7\,\mathrm{e}^{4s} + 5)}{16\,\mathrm{e}^{2-2t+2s}}, & 0 \le s \le t \le 1, \\ \dfrac{(1 + 3\,\mathrm{e}^{4-4s})(7\,\mathrm{e}^{4t} + 5)}{16\,\mathrm{e}^{2-2s+2t}}, & 0 \le t \le s \le 1, \end{array}\right.$$

$$J_1(s) = g_1(s,s) + \frac{1}{\Delta_1}(\Lambda_4\psi(1) + \Lambda_1\phi(0)) \times 2\int_0^1 \tau g_1(\tau,s)\,d\tau$$

$$+ \frac{1}{\Delta_1}(\Lambda_2\psi(1) + \Lambda_3\phi(0))\left[\frac{7}{2}g_1\left(\frac{1}{3},s\right) + 2g_1\left(\frac{2}{3},s\right)\right]$$

$$= g_1(s,s) + \frac{1}{\Delta_1}(\Lambda_4\psi(1) + \Lambda_1\phi(0)) \times 2\left(\int_0^s \tau g_1(\tau,s)\,d\tau + \int_s^1 \tau g_1(\tau,s)\,d\tau\right)$$

$$+ \frac{1}{\Delta_1}(\Lambda_2\psi(1) + \Lambda_3\phi(0))\left[\frac{7}{2}g_1\left(\frac{1}{3},s\right) + 2g_1\left(\frac{2}{3},s\right)\right]$$

$$= \begin{cases} \frac{1}{\tau_1}\left\{\phi(s)\psi(s) + \frac{1}{\Delta_1}(\Lambda_4\psi(1)+\Lambda_1\phi(0)) \times 2\left(\phi(s)\int_0^s \tau\psi(\tau)\,d\tau + \psi(s)\int_s^1 \tau\phi(\tau)\,d\tau\right)\right. \\ \left. + \frac{1}{\Delta_1}(\Lambda_2\psi(1)+\Lambda_3\phi(0))\left(\frac{7}{2}\phi\left(\frac{1}{3}\right)\psi(s) + 2\phi\left(\frac{2}{3}\right)\psi(s)\right)\right\}, \quad 0 \le s < \frac{1}{3}, \\ \frac{1}{\tau_1}\left\{\phi(s)\psi(s) + \frac{1}{\Delta_1}(\Lambda_4\psi(1)+\Lambda_1\phi(0)) \times 2\left(\phi(s)\int_0^s \tau\psi(\tau)\,d\tau + \psi(s)\int_s^1 \tau\phi(\tau)\,d\tau\right)\right. \\ \left. + \frac{1}{\Delta_1}(\Lambda_2\psi(1) + \Lambda_3\phi(0))\left(\frac{7}{2}\phi(s)\psi\left(\frac{1}{3}\right) + 2\phi\left(\frac{2}{3}\right)\psi(s)\right)\right\}, \quad \frac{1}{3} \le s < \frac{2}{3}, \\ \frac{1}{\tau_1}\left\{\phi(s)\psi(s) + \frac{1}{\Delta_1}(\Lambda_4\psi(1)+\Lambda_1\phi(0)) \times 2\left(\phi(s)\int_0^s \tau\psi(\tau)\,d\tau + \psi(s)\int_s^1 \tau\phi(\tau)\,d\tau\right)\right. \\ \left. + \frac{1}{\Delta_1}(\Lambda_2\psi(1) + \Lambda_3\phi(0))\left(\frac{7}{2}\phi(s)\psi\left(\frac{1}{3}\right) + 2\phi(s)\psi\left(\frac{2}{3}\right)\right)\right\}, \quad \frac{2}{3} \le s \le 1. \end{cases}$$

The functions $\tilde{\psi}$ and $\tilde{\phi}$ from Section 1.1.2 are the solutions of the following problems:

$$\begin{cases} \tilde{\psi}''(t) - \tilde{\psi}(t) = 0, & 0 < t < 1, \\ \tilde{\psi}(0) = 2, & \tilde{\psi}'(0) = 3, \end{cases} \qquad \begin{cases} \tilde{\phi}''(t) - \tilde{\phi}(t) = 0, & 0 < t < 1, \\ \tilde{\phi}(1) = \frac{3}{2}, & \tilde{\phi}'(1) = -1. \end{cases}$$

We obtain $\tilde{\psi}(t) = \frac{5e^{2t}-1}{2e^t}$ and $\tilde{\phi}(t) = \frac{1+5e^{2-2t}}{4e^{1-t}}$ for all $t \in [0,1]$, $\tau_2 = \frac{25e^2+1}{4e}$,

$$\tilde{\Lambda}_1 := \tau_2 - \int_0^1 \tilde{\psi}(s)\,dK_2(s) = \tau_2 - 3\int_0^1 s^2\tilde{\psi}(s)\,ds$$

$$= (25e^2+1)/(4e) - 3\int_0^1 s^2(5e^{2s}-1)/(2e^s)\,ds \approx 11.93502177 > 0,$$

$$\tilde{\Lambda}_2 := \tau_2 - \int_0^1 \tilde{\phi}(s)\,dK_1(s) = \tau_2 - \frac{4}{3}\tilde{\phi}\left(\frac{1}{2}\right)$$
$$= (25e^2+1)/(4e) - (1+5e)/(3e^{1/2}) \approx 14.13118562 > 0,$$

$$\tilde{\Lambda}_3 := \int_0^1 \tilde{\psi}(s)\,dK_1(s) = \frac{4}{3}\tilde{\psi}\left(\frac{1}{2}\right) = (10e-2)/(3e^{1/2}) \approx 5.09138379,$$

$$\tilde{\Lambda}_4 := \int_0^1 \tilde{\phi}(s)\,\mathrm{d}K_2(s) = 3\int_0^1 s^2\tilde{\phi}(s)\,\mathrm{d}s = 3\int_0^1 s^2(1+5\,\mathrm{e}^{2-2s})/(4\,\mathrm{e}^{1-s})\,\mathrm{d}s \approx 1.83529455,$$

$$\Delta_2 = \tilde{\Lambda}_1\tilde{\Lambda}_2 - \tilde{\Lambda}_3\tilde{\Lambda}_4 \approx 159.31181898 > 0.$$

The functions g_2 and J_2 are given by

$$g_2(t,s) = \frac{1}{\tau_2}\begin{cases} \tilde{\phi}(t)\tilde{\psi}(s), & 0 \le s \le t \le 1, \\ \tilde{\phi}(s)\tilde{\psi}(t), & 0 \le t \le s \le 1, \end{cases} \quad \text{or}$$

$$g_2(t,s) = \frac{4\,\mathrm{e}}{25\,\mathrm{e}^2+1}\begin{cases} \dfrac{(1+5\,\mathrm{e}^{2-2t})(5\,\mathrm{e}^{2s}-1)}{8\,\mathrm{e}^{1-t+s}}, & 0 \le s \le t \le 1, \\ \dfrac{(1+5\,\mathrm{e}^{2-2s})(5\,\mathrm{e}^{2t}-1)}{8\,\mathrm{e}^{1-s+t}}, & 0 \le t \le s \le 1, \end{cases}$$

$$\begin{aligned}
J_2(s) &= g_2(s,s) + \frac{1}{\Delta_2}(\tilde{\Lambda}_4\tilde{\psi}(1) + \tilde{\Lambda}_1\tilde{\phi}(0))\frac{4}{3}g_2\left(\frac{1}{2},s\right) \\
&\quad + \frac{1}{\Delta_2}(\tilde{\Lambda}_2\tilde{\psi}(1) + \tilde{\Lambda}_3\tilde{\phi}(0)) \times 3\int_0^1 \tau^2 g_2(\tau,s)\,\mathrm{d}\tau \\
&= g_2(s,s) + \frac{1}{\Delta_2}(\tilde{\Lambda}_4\tilde{\psi}(1) + \tilde{\Lambda}_1\tilde{\phi}(0))\frac{4}{3}g_2\left(\frac{1}{2},s\right) \\
&\quad + \frac{1}{\Delta_2}(\tilde{\Lambda}_2\tilde{\psi}(1) + \tilde{\Lambda}_3\tilde{\phi}(0)) \times 3\left(\int_0^s \tau^2 g_2(\tau,s)\,\mathrm{d}\tau + \int_s^1 \tau^2 g_2(\tau,s)\,\mathrm{d}\tau\right)
\end{aligned}$$

$$= \begin{cases}
\dfrac{1}{\tau_2}\Bigg\{\tilde{\phi}(s)\tilde{\psi}(s) + \dfrac{1}{\Delta_2}\left(\tilde{\Lambda}_4\tilde{\psi}(1) + \tilde{\Lambda}_1\tilde{\phi}(0)\right) \times \dfrac{4}{3}\tilde{\phi}\left(\dfrac{1}{2}\right)\tilde{\psi}(s) \\
+ \dfrac{1}{\Delta_2}\left(\tilde{\Lambda}_2\tilde{\psi}(1) + \tilde{\Lambda}_3\tilde{\phi}(0)\right) \\
\times 3\left(\tilde{\phi}(s)\int_0^s \tau^2\tilde{\psi}(\tau)\,\mathrm{d}\tau + \tilde{\psi}(s)\int_s^1 \tau^2\tilde{\phi}(\tau)\,\mathrm{d}\tau\right)\Bigg\}, & 0 \le s < \frac{1}{2}, \\
\dfrac{1}{\tau_2}\Bigg\{\tilde{\phi}(s)\tilde{\psi}(s) + \dfrac{1}{\Delta_2}\left(\tilde{\Lambda}_4\tilde{\psi}(1) + \tilde{\Lambda}_1\tilde{\phi}(0)\right) \times \dfrac{4}{3}\tilde{\phi}(s)\tilde{\psi}\left(\dfrac{1}{2}\right) \\
+ \dfrac{1}{\Delta_2}\left(\tilde{\Lambda}_2\tilde{\psi}(1) + \tilde{\Lambda}_3\tilde{\phi}(0)\right) \\
\times 3\left(\tilde{\phi}(s)\displaystyle\int_0^s \tau^2\tilde{\psi}(\tau)\,\mathrm{d}\tau + \tilde{\psi}(s)\int_s^1 \tau^2\tilde{\phi}(\tau)\,\mathrm{d}\tau\right)\Bigg\}, & \dfrac{1}{2} \le s \le 1.
\end{cases}$$

For $\sigma = 1/4$, we have $\nu = \nu_1 = (1+3\,\mathrm{e})\,\mathrm{e}^{3/2}/(1+3\,\mathrm{e}^4)$ and $\nu_2 = (5\,\mathrm{e}^{1/2} - 1)\,\mathrm{e}^{3/4}/(5\,\mathrm{e}^2 - 1)$. We also deduce $A = \int_{1/4}^{3/4} J_1(s)\,\mathrm{d}s \approx 1.35977188$, $B = \int_0^1 J_1(s)\,\mathrm{d}s \approx 2.51890379$, $C = \int_{1/4}^{3/4} J_2(s)\,\mathrm{d}s \approx 0.48198213$, and $D = \int_0^1 J_2(s)\,\mathrm{d}s \approx 0.93192847$.

Example 1.1.1. First, we consider the functions

$$f(t,u,v) = \frac{\mathrm{e}^t[p_1(u+v)+1](u+v)(q_1+\sin v)}{u+v+1},$$

$$g(t,u,v) = \frac{\mathrm{e}^{-t}[p_2(u+v)+1](u+v)(q_2+\cos u)}{u+v+1}$$

for all $t \in [0,1]$ and $u, v \in [0,\infty)$, with $p_1, p_2 > 0$ and $q_1, q_2 > 1$.

We deduce $f_0^{\mathrm{s}} = eq_1$, $g_0^{\mathrm{s}} = q_2 + 1$, $f_\infty^{\mathrm{i}} = \mathrm{e}^{1/4} p_1(q_1 - 1)$, and $g_\infty^{\mathrm{i}} = \mathrm{e}^{-3/4} p_2(q_2 - 1)$. For $\alpha_1, \alpha_2 > 0$ with $\alpha_1 + \alpha_2 = 1$, we consider $\tilde{\alpha}_1 = \alpha_1$ and $\tilde{\alpha}_2 = \alpha_2$. Then we obtain

$$L_1 = \frac{\alpha_1(1+3\,\mathrm{e}^4)^2}{\mathrm{e}^{13/4}(1+3\,\mathrm{e})^2 p_1(q_1-1)A}, \quad L_2 = \frac{\alpha_1}{eq_1 B},$$

$$L_3 = \frac{\alpha_2(1+3\,\mathrm{e}^4)(5\,\mathrm{e}^2-1)}{\mathrm{e}^{3/2}(1+3\,\mathrm{e})(5\,\mathrm{e}^{1/2}-1)p_2(q_2-1)C}, \quad L_4 = \frac{\alpha_2}{(q_2+1)D}.$$

The conditions $L_1 < L_2$ and $L_3 < L_4$ become

$$\frac{p_1(q_1-1)}{q_1} > \frac{(1+3\,\mathrm{e}^4)^2\,\mathrm{e}B}{(1+3\,\mathrm{e})^2\,\mathrm{e}^{13/4}A}, \quad \frac{p_2(q_2-1)}{q_2+1} > \frac{(1+3\,\mathrm{e}^4)(5\,\mathrm{e}^2-1)D}{(1+3\,\mathrm{e})(5\,\mathrm{e}^{1/2}-1)\,\mathrm{e}^{3/2}C}.$$

If $p_1(q_1-1)/q_1 \ge 64$ and $p_2(q_2-1)/(q_2+1) \ge 39$, then the above conditions are satisfied. For example, if $\alpha_1 = \alpha_2 = 1/2$, $p_1 = 128$, $q_1 = 2$, $p_2 = 117$, and $q_2 = 2$, we obtain $L_1 \approx 0.03609275$, $L_2 \approx 0.03651185$, $L_3 \approx 0.17672178$, and $L_4 \approx 0.17884062$. Therefore, by Theorem 1.1.2 (1), for each $\lambda \in (L_1, L_2)$ and $\mu \in (L_3, L_4)$ there exists a positive solution $(u(t), v(t))$, $t \in [0,1]$ for problem $(\overline{\mathrm{S}})$–$(\overline{\mathrm{BC}})$.

Example 1.1.2. We consider the functions

$$f(t,u,v) = (u+v)^{\beta_1}, \qquad g(t,u,v) = (u+v)^{\beta_2}, \quad \forall t \in [0,1], \ \ u,v \in [0,\infty),$$

with $\beta_1, \beta_2 > 1$. Then $f_0^{\mathrm{s}} = g_0^{\mathrm{s}} = 0$ and $f_\infty^{\mathrm{i}} = g_\infty^{\mathrm{i}} = \infty$. By Theorem 1.1.2 (8), we deduce that for each $\lambda \in (0,\infty)$ and $\mu \in (0,\infty)$ there exists a positive solution $(u(t), v(t))$, $t \in [0,1]$ for problem $(\overline{\mathrm{S}})$–$(\overline{\mathrm{BC}})$.

Example 1.1.3. We consider the functions

$$f(t,u,v) = (u+v)^{\gamma_1}, \qquad g(t,u,v) = (u+v)^{\gamma_2}, \quad \forall t \in [0,1], \ \ u,v \in [0,\infty),$$

with $\gamma_1, \gamma_2 \in (0,1)$. Then $f_0^{\mathrm{i}} = g_0^{\mathrm{i}} = \infty$ and $f_\infty^{\mathrm{s}} = g_\infty^{\mathrm{s}} = 0$. By Theorem 1.1.3 (8), we deduce that for each $\lambda \in (0,\infty)$ and $\mu \in (0,\infty)$ there exists a positive solution $(u(t), v(t))$, $t \in [0,1]$ for problem $(\overline{\mathrm{S}})$–$(\overline{\mathrm{BC}})$.

1.2 Nonexistence of positive solutions

We consider again problem (S)–(BC) from Section 1.1 with the auxiliary results from Section 1.1.2 and assumptions (I1)–(I6) from Section 1.1.3. In this section, we determine intervals for λ and μ for which there exists no positive solution (in the sense of the definition from Section 1.1.1—that is, $u(t) \geq 0$, $v(t) \geq 0$ for all $t \in [0,1]$ and $(u,v) \neq (0,0)$) of problem (S)–(BC).

1.2.1 Main nonexistence results

Our main results related to the nonexistence of positive solutions for problem (S)–(BC) are as follows:

Theorem 1.2.1. *Assume that (I1)–(I6) hold. If* $f_0^s, f_\infty^s, g_0^s, g_\infty^s < \infty$, *then there exist positive constants* λ_0 *and* μ_0 *such that for every* $\lambda \in (0, \lambda_0)$ *and* $\mu \in (0, \mu_0)$ *the boundary value problem (S)–(BC) has no positive solution.*

Proof. Since $f_0^s, f_\infty^s < \infty$, we deduce that there exist $M_1', M_1'', r_1, r_1' > 0$, $r_1 < r_1'$ such that

$$f(t,u,v) \leq M_1'(u+v), \quad \forall u,v \geq 0, \quad u+v \in [0, r_1], \quad t \in [0,1],$$

$$f(t,u,v) \leq M_1''(u+v), \quad \forall u,v \geq 0, \quad u+v \in [r_1', \infty), \quad t \in [0,1].$$

We consider $M_1 = \max\left\{M_1', M_1'', \max_{r_1 \leq u+v \leq r_1'} \max_{t \in [0,1]} \frac{f(t,u,v)}{u+v}\right\} > 0$. Then we obtain

$$f(t,u,v) \leq M_1(u+v), \quad \forall u,v \geq 0, \ t \in [0,1].$$

Since $g_0^s, g_\infty^s < \infty$, we deduce that there exist $M_2', M_2'', r_2, r_2' > 0$, $r_2 < r_2'$ such that

$$g(t,u,v) \leq M_2'(u+v), \quad \forall u,v \geq 0, \quad u+v \in [0, r_2], \quad t \in [0,1],$$

$$g(t,u,v) \leq M_2''(u+v), \quad \forall u,v \geq 0, \quad u+v \in [r_2', \infty), \quad t \in [0,1].$$

We consider $M_2 = \max\left\{M_2', M_2'', \max_{r_2 \leq u+v \leq r_2'} \max_{t \in [0,1]} \frac{g(t,u,v)}{u+v}\right\} > 0$. Then we obtain

$$g(t,u,v) \leq M_2(u+v), \quad \forall u,v \geq 0, \quad t \in [0,1].$$

We define $\lambda_0 = \frac{1}{2M_1 B}$ and $\mu_0 = \frac{1}{2M_2 D}$, where $B = \int_0^1 J_1(s)p(s)\,ds$ and $D = \int_0^1 J_2(s)q(s)\,ds$. We shall show that for every $\lambda \in (0, \lambda_0)$ and $\mu \in (0, \mu_0)$ problem (S)–(BC) has no positive solution.

Let $\lambda \in (0, \lambda_0)$ and $\mu \in (0, \mu_0)$. We suppose that (S)–(BC) has a positive solution $(u(t), v(t))$, $t \in [0,1]$. Then we have

$$u(t) = Q_1(u,v)(t) = \lambda \int_0^1 G_1(t,s)p(s)f(s,u(s),v(s))\,\mathrm{d}s$$
$$\le \lambda \int_0^1 J_1(s)p(s)f(s,u(s),v(s))\,\mathrm{d}s \le \lambda M_1 \int_0^1 J_1(s)p(s)(u(s)+v(s))\,\mathrm{d}s$$
$$\le \lambda M_1(\|u\|+\|v\|)\int_0^1 J_1(s)p(s)\,\mathrm{d}s = \lambda M_1 B\|(u,v)\|_Y, \quad \forall t\in[0,1].$$

Therefore, we conclude

$$\|u\| \le \lambda M_1 B\|(u,v)\|_Y < \lambda_0 M_1 B\|(u,v)\|_Y = \frac{1}{2}\|(u,v)\|_Y.$$

In a similar manner, we have

$$v(t) = Q_2(u,v)(t) = \mu \int_0^1 G_2(t,s)q(s)g(s,u(s),v(s))\,\mathrm{d}s$$
$$\le \mu \int_0^1 J_2(s)q(s)g(s,u(s),v(s))\,\mathrm{d}s \le \mu M_2 \int_0^1 J_2(s)q(s)(u(s)+v(s))\,\mathrm{d}s$$
$$\le \mu M_2(\|u\|+\|v\|)\int_0^1 J_2(s)q(s)\,\mathrm{d}s = \mu M_2 D\|(u,v)\|_Y, \quad \forall t\in[0,1].$$

Therefore, we conclude

$$\|v\| \le \mu M_2 D\|(u,v)\|_Y < \mu_0 M_2 D\|(u,v)\|_Y = \frac{1}{2}\|(u,v)\|_Y.$$

Hence, $\|(u,v)\|_Y = \|u\|+\|v\| < \frac{1}{2}\|(u,v)\|_Y + \frac{1}{2}\|(u,v)\|_Y = \|(u,v)\|_Y$, which is a contradiction. So, the boundary value problem (S)–(BC) has no positive solution. □

Theorem 1.2.2. *Assume that (I1)–(I6) hold. If $f_0^{\mathrm{i}}, f_\infty^{\mathrm{i}} > 0$ and $f(t,u,v) > 0$ for all $t\in[\sigma,1-\sigma], u\ge 0, v\ge 0$, and $u+v>0$, then there exists a positive constant $\tilde{\lambda}_0$ such that for every $\lambda > \tilde{\lambda}_0$ and $\mu > 0$ the boundary value problem (S)–(BC) has no positive solution.*

Proof. By the assumptions of this theorem, we deduce that there exist $m_1', m_1'', r_3, r_3' > 0$, $r_3 < r_3'$ such that

$$f(t,u,v) \ge m_1'(u+v), \quad \forall u,v\ge 0, \;\; u+v\in[0,r_3], \;\; t\in[\sigma,1-\sigma],$$
$$f(t,u,v) \ge m_1''(u+v), \quad \forall u,v\ge 0, \;\; u+v\in[r_3',\infty), \;\; t\in[\sigma,1-\sigma].$$

We introduce $m_1 = \min\left\{m_1', m_1'', \min_{u+v\in[r_3,r_3']}\min_{t\in[\sigma,1-\sigma]}\frac{f(t,u,v)}{u+v}\right\} > 0$. Then we obtain

$$f(t,u,v) \ge m_1(u+v), \quad \forall u,v\ge 0, \;\; t\in[\sigma,1-\sigma].$$

We define $\tilde{\lambda}_0 = \frac{1}{\nu\nu_1 m_1 A} > 0$, where $A = \int_\sigma^{1-\sigma} J_1(s)p(s)\,\mathrm{d}s$. We shall show that for every $\lambda > \tilde{\lambda}_0$ and $\mu > 0$ problem (S)–(BC) has no positive solution.

Let $\lambda > \tilde{\lambda}_0$ and $\mu > 0$. We suppose that (S)–(BC) has a positive solution $(u(t),v(t))$, $t\in[0,1]$. Then we obtain

$$\begin{aligned} u(\sigma) = Q_1(u,v)(\sigma) &= \lambda \int_0^1 G_1(\sigma,s)p(s)f(s,u(s),v(s))\,ds \\ &\geq \lambda \int_\sigma^{1-\sigma} G_1(\sigma,s)p(s)f(s,u(s),v(s))\,ds \\ &\geq \lambda m_1 \int_\sigma^{1-\sigma} G_1(\sigma,s)p(s)(u(s)+v(s))\,ds \\ &\geq \lambda m_1 \nu_1 \int_\sigma^{1-\sigma} J_1(s)p(s)\nu(\|u\|+\|v\|)\,ds = \lambda m_1 \nu \nu_1 A \|(u,v)\|_Y. \end{aligned}$$

Therefore, we deduce

$$\|u\| \geq u(\sigma) \geq \lambda m_1 \nu \nu_1 A \|(u,v)\|_Y > \tilde{\lambda}_0 m_1 \nu \nu_1 A \|(u,v)\|_Y = \|(u,v)\|_Y,$$

and so $\|(u,v)\|_Y = \|u\|+\|v\| \geq \|u\| > \|(u,v)\|_Y$, which is a contradiction. Therefore, the boundary value problem (S)–(BC) has no positive solution. □

Theorem 1.2.3. *Assume that (I1)–(I6) hold. If $g_0^{\rm i}, g_\infty^{\rm i} > 0$ and $g(t,u,v) > 0$ for all $t \in [\sigma, 1-\sigma]$, $u \geq 0$, $v \geq 0$, and $u+v > 0$, then there exists a positive constant $\tilde{\mu}_0$ such that for every $\mu > \tilde{\mu}_0$ and $\lambda > 0$ the boundary value problem (S)–(BC) has no positive solution.*

Proof. By the assumptions of this theorem, we deduce that there exist $m_2', m_2'', r_4, r_4' > 0$, $r_4 < r_4'$ such that

$$g(t,u,v) \geq m_2'(u+v), \quad \forall\, u,v \geq 0, \quad u+v \in [0,r_4], \quad t \in [\sigma,1-\sigma],$$

$$g(t,u,v) \geq m_2''(u+v), \quad \forall\, u,v \geq 0, \quad u+v \in [r_4',\infty), \quad t \in [\sigma,1-\sigma].$$

We introduce $m_2 = \min\left\{m_2', m_2'', \min_{u+v\in[r_4,r_4']} \min_{t\in[\sigma,1-\sigma]} \frac{g(t,u,v)}{u+v}\right\} > 0$. Then we obtain

$$g(t,u,v) \geq m_2(u+v), \quad \forall\, u,v \geq 0, \quad t \in [\sigma,1-\sigma].$$

We define $\tilde{\mu}_0 = \frac{1}{\nu\nu_2 m_2 C} > 0$, where $C = \int_\sigma^{1-\sigma} J_2(s)q(s)\,ds$. We shall show that for every $\mu > \tilde{\mu}_0$ and $\lambda > 0$ problem (S)–(BC) has no positive solution.

Let $\mu > \tilde{\mu}_0$ and $\lambda > 0$. We suppose that (S)–(BC) has a positive solution $(u(t), v(t))$, $t \in [0,1]$. Then we obtain

$$\begin{aligned} v(\sigma) = Q_2(u,v)(\sigma) &= \mu \int_0^1 G_2(\sigma,s)q(s)g(s,u(s),v(s))\,ds \\ &\geq \mu \int_\sigma^{1-\sigma} G_2(\sigma,s)q(s)g(s,u(s),v(s))\,ds \\ &\geq \mu m_2 \int_\sigma^{1-\sigma} G_2(\sigma,s)q(s)(u(s)+v(s))\,ds \\ &\geq \mu m_2 \nu_2 \int_\sigma^{1-\sigma} J_2(s)q(s)\nu(\|u\|+\|v\|)\,ds = \mu m_2 \nu \nu_2 C \|(u,v)\|_Y. \end{aligned}$$

Therefore, we deduce

$$\|v\| \geq v(\sigma) \geq \mu m_2 \nu\nu_2 C\|(u,v)\|_Y > \tilde{\mu}_0 m_2 \nu\nu_2 C\|(u,v)\|_Y = \|(u,v)\|_Y,$$

and so $\|(u,v)\|_Y = \|u\|+\|v\| \geq \|v\| > \|(u,v)\|_Y$, which is a contradiction. Therefore, the boundary value problem (S)–(BC) has no positive solution. □

Theorem 1.2.4. *Assume that (I1)–(I6) hold. If $f_0^i, f_\infty^i, g_0^i, g_\infty^i > 0$, $f(t,u,v) > 0$, and $g(t,u,v) > 0$ for all $t \in [\sigma, 1-\sigma]$, $u \geq 0$, $v \geq 0$, and $u+v > 0$, then there exist positive constants $\hat{\lambda}_0$ and $\hat{\mu}_0$ such that for every $\lambda > \hat{\lambda}_0$ and $\mu > \hat{\mu}_0$ the boundary value problem (S)–(BC) has no positive solution.*

Proof. By the assumptions of this theorem, we deduce as above that there exist $m_1, m_2 > 0$ such that

$$f(t,u,v) \geq m_1(u+v), \qquad g(t,u,v) \geq m_2(u+v), \quad \forall\, u,v \geq 0, \quad t \in [\sigma, 1-\sigma].$$

We define $\hat{\lambda}_0 = \frac{1}{2\nu\nu_1 m_1 A}\left(= \frac{\tilde{\lambda}_0}{2}\right)$ and $\hat{\mu}_0 = \frac{1}{2\nu\nu_2 m_2 C}\left(= \frac{\tilde{\mu}_0}{2}\right)$. Then for every $\lambda > \hat{\lambda}_0$ and $\mu > \hat{\mu}_0$, problem (S)–(BC) has no positive solution. Let $\lambda > \hat{\lambda}_0$ and $\mu > \hat{\mu}_0$. We suppose that (S)–(BC) has a positive solution $(u(t), v(t)), t \in [0,1]$. Then in a manner similar to that above, we deduce

$$\|u\| \geq \lambda m_1 \nu\nu_1 A\|(u,v)\|_Y, \quad \|v\| \geq \mu m_2 \nu\nu_2 C\|(u,v)\|_Y,$$

and so

$$\begin{aligned}\|(u,v)\|_Y &= \|u\| + \|v\| \geq \lambda m_1 \nu\nu_1 A\|(u,v)\|_Y + \mu m_2 \nu\nu_2 C\|(u,v)\|_Y \\ &> \hat{\lambda}_0 m_1 \nu\nu_1 A\|(u,v)\|_Y + \hat{\mu}_0 m_2 \nu\nu_2 C\|(u,v)\|_Y \\ &= \frac{1}{2}\|(u,v)\|_Y + \frac{1}{2}\|(u,v)\|_Y = \|(u,v)\|_Y,\end{aligned}$$

which is a contradiction. Therefore, the boundary value problem (S)–(BC) has no positive solution. □

1.2.2 An example

Example 1.2.1. We consider the first example from Section 1.1.4—that is, problem $(\overline{S})$–$(\overline{BC})$ with f and g given in Example 1.1.1. Because $f_0^s = eq_1$, $f_\infty^s = ep_1(q_1+1)$, $g_0^s = q_2+1$, and $g_\infty^s = p_2(q_2+1)$ are finite, we can apply Theorem 1.2.1. Using the same values for p_1, q_1, p_2, and q_2 as in Example 1.1.1 from Section 1.1.4—that is, $p_1 = 128$, $q_1 = 2$, $p_2 = 117$, and $q_2 = 2$, we deduce

$$M_1 = \sup_{u,v\geq 0} \max_{t\in[0,1]} \frac{f(t,u,v)}{u+v} = e \sup_{u,v\geq 0} \frac{[p_1(u+v)+1](q_1+\sin v)}{u+v+1} \approx 1043.82022212,$$

$$M_2 = \sup_{u,v\geq 0} \max_{t\in[0,1]} \frac{g(t,u,v)}{u+v} = \sup_{u,v\geq 0} \frac{[p_2(u+v)+1](q_2+\cos u)}{u+v+1} = 351.$$

Then we obtain $\lambda_0 = \frac{1}{2M_1B} \approx 0.00019016$ and $\mu_0 = \frac{1}{2M_2D} \approx 0.00152855$. Therefore, by Theorem 1.2.1, we conclude that for every $\lambda \in (0, \lambda_0)$ and $\mu \in (0, \mu_0)$ problem $(\overline{\text{S}})$–$(\overline{\text{BC}})$ has no positive solution.

Remark 1.2.1. The results presented in Sections 1.1 and 1.2 were published in Henderson and Luca (2013j).

1.3 Existence and multiplicity of positive solutions for systems without parameters

In this section, we investigate the existence and multiplicity of positive solutions for problem (S)–(BC) from Section 1.1 with $\lambda = \mu = 1$, $p(t) = 1$, and $q(t) = 1$ for all $t \in [0, 1]$, and where f and g are dependent only on t and v, and t and u, respectively.

1.3.1 Presentation of the problem

We consider the system of nonlinear second-order ordinary differential equations

$$\begin{cases} (a(t)u'(t))' - b(t)u(t) + f(t, v(t)) = 0, & 0 < t < 1, \\ (c(t)v'(t))' - d(t)v(t) + g(t, u(t)) = 0, & 0 < t < 1, \end{cases} \tag{S$'$}$$

with the integral boundary conditions

$$\begin{cases} \alpha u(0) - \beta a(0)u'(0) = \int_0^1 u(s)\, \mathrm{d}H_1(s), & \gamma u(1) + \delta a(1)u'(1) = \int_0^1 u(s)\, \mathrm{d}H_2(s), \\ \tilde{\alpha} v(0) - \tilde{\beta} c(0)v'(0) = \int_0^1 v(s)\, \mathrm{d}K_1(s), & \tilde{\gamma} v(1) + \tilde{\delta} c(1)v'(1) = \int_0^1 v(s)\, \mathrm{d}K_2(s), \end{cases} \tag{BC}$$

where the above integrals are Riemann–Stieltjes integrals.

By applying some theorems from the fixed point index theory, we prove the existence and multiplicity of positive solutions of problem (S′)–(BC) when f and g satisfy various assumptions. By a positive solution of (S′)–(BC) we mean a pair of functions $(u, v) \in C^2([0, 1]) \times C^2([0, 1])$ satisfying (S′) and (BC) with $u(t) \geq 0$ and $v(t) \geq 0$ for all $t \in [0, 1]$ and $\sup_{t\in[0,1]} u(t) > 0$ and $\sup_{t\in[0,1]} v(t) > 0$. This problem is a generalization of the problem studied in Henderson and Luca (2012a), where $a(t) = c(t) = 1$ and $b(t) = d(t) = 0$ for all $t \in (0, 1)$ in system (S′) (denoted by $(\tilde{\text{S}}')$), $\alpha = \tilde{\alpha} = 1$, $\beta = \tilde{\beta} = 0$, $\gamma = \tilde{\gamma} = 1$, $\delta = \tilde{\delta} = 0$, H_1 and K_1 are constant functions, and H_2 and K_2 are step functions (i.e., boundary conditions (BC) become multipoint boundary conditions). In Zhou and Xu (2006), the authors investigated the existence and multiplicity of positive solutions for system $(\tilde{\text{S}}')$ with the boundary conditions $u(0) = 0$, $u(1) = \alpha u(\eta)$, $v(0) = 0$, $v(1) = \alpha v(\eta)$, $\eta \in (0, 1)$, and $0 < \alpha\eta < 1$.

We recall now some theorems concerning the fixed point index theory. Let E be a real Banach space, $P \subset E$ a cone, "$\leq$" the partial ordering defined by P, and θ the zero element in E. For $\varrho > 0$, let $B_\varrho = \{u \in E, \|u\| < \varrho\}$ be the open ball of radius ϱ centered at 0, and let its boundary be $\partial B_\varrho = \{u \in E, \|u\| = \varrho\}$. The proofs of our results are based on the following fixed point index theorems:

Theorem 1.3.1 (Amann, 1976). *Let $A : \bar{B}_\varrho \cap P \to P$ be a completely continuous operator which has no fixed point on $\partial B_\varrho \cap P$. If $\|Au\| \leq \|u\|$ for all $u \in \partial B_\varrho \cap P$, then $i(A, B_\varrho \cap P, P) = 1$.*

Theorem 1.3.2 (Amann, 1976). *Let $A : \bar{B}_\varrho \cap P \to P$ be a completely continuous operator. If there exists $u_0 \in P \setminus \{\theta\}$ such that $u - Au \neq \lambda u_0$ for all $\lambda \geq 0$ and $u \in \partial B_\varrho \cap P$, then $i(A, B_\varrho \cap P, P) = 0$.*

Theorem 1.3.3 (Zhou and Xu, 2006). *Let $A : \bar{B}_\varrho \cap P \to P$ be a completely continuous operator which has no fixed point on $\partial B_\varrho \cap P$. If there exists a linear operator $L : P \to P$ and $u_0 \in P \setminus \{\theta\}$ such that*

$$(1)\ u_0 \leq Lu_0 \quad \text{and} \quad (2)\ Lu \leq Au, \quad \forall u \in \partial B_\varrho \cap P,$$

then $i(A, B_\varrho \cap P, P) = 0$.

We shall suppose that assumptions (A1)–(A5) from Section 1.1.2 hold, and we shall use in our main results from this section Lemmas 1.1.1–1.1.7 from Section 1.1.2.

1.3.2 Main results

We investigate the existence and multiplicity of positive solutions for our problem (S′)–(BC), under various assumptions on f and g.

We present the basic assumptions that we shall use in the sequel:

(H1) The functions $a, c \in C^1([0,1], (0,\infty))$ and $b, d \in C([0,1], [0,\infty))$.

(H2) $\alpha, \beta, \gamma, \delta, \tilde{\alpha}, \tilde{\beta}, \tilde{\gamma}, \tilde{\delta} \in [0,\infty)$ with $\alpha + \beta > 0$, $\gamma + \delta > 0$, $\tilde{\alpha} + \tilde{\beta} > 0$, and $\tilde{\gamma} + \tilde{\delta} > 0$; if $b \equiv 0$, then $\alpha + \gamma > 0$; if $d \equiv 0$, then $\tilde{\alpha} + \tilde{\gamma} > 0$.

(H3) $H_1, H_2, K_1, K_2 : [0,1] \to \mathbb{R}$ are nondecreasing functions.

(H4) $\tau_1 - \int_0^1 \phi(s)\, dH_1(s) > 0$, $\tau_1 - \int_0^1 \psi(s)\, dH_2(s) > 0$, $\tau_2 - \int_0^1 \tilde{\phi}(s)\, dK_1(s) > 0$, $\tau_2 - \int_0^1 \tilde{\psi}(s)\, dK_2(s) > 0$, $\Delta_1 > 0$, and $\Delta_2 > 0$, where τ_1, τ_2, Δ_1, and Δ_2 are defined in Section 1.1.2.

(H5) The functions $f, g \in C([0,1] \times [0,\infty), [0,\infty))$ and $f(t,0) = 0$, $g(t,0) = 0$ for all $t \in [0,1]$.

The pair of functions $(u, v) \in C^2([0,1]) \times C^2([0,1])$ is a solution for our problem (S′)–(BC) if and only if $(u, v) \in C([0,1]) \times C([0,1])$ is a solution for the nonlinear integral system

$$\begin{cases} u(t) = \displaystyle\int_0^1 G_1(t,s) f(s, v(s))\, ds, & t \in [0,1], \\ v(t) = \displaystyle\int_0^1 G_2(t,s) g(s, u(s))\, ds, & t \in [0,1]. \end{cases} \tag{1.14}$$

Besides, system (1.14) can be written as the nonlinear integral system

$$\begin{cases} u(t) = \int_0^1 G_1(t,s) f\left(s, \int_0^1 G_2(s,\tau) g(\tau, u(\tau))\,d\tau\right) ds, & t \in [0,1], \\ v(t) = \int_0^1 G_2(t,s) g(s, u(s))\,ds, & t \in [0,1]. \end{cases}$$

We consider the Banach space $X = C([0,1])$ with the supremum norm $\|\cdot\|$, and define the cone $P' \subset X$ by $P' = \{u \in X, u(t) \geq 0, \forall t \in [0,1]\}$.

We also define the operator $\mathcal{A} : P' \to X$ by

$$(\mathcal{A}u)(t) = \int_0^1 G_1(t,s) f\left(s, \int_0^1 G_2(s,\tau) g(\tau, u(\tau))\,d\tau\right) ds, \quad t \in [0,1],$$

and the operators $\mathcal{B} : P' \to X$ and $\mathcal{C} : P' \to X$ by

$$(\mathcal{B}u)(t) = \int_0^1 G_1(t,s) u(s)\,ds, \qquad (\mathcal{C}u)(t) = \int_0^1 G_2(t,s) u(s)\,ds, \quad t \in [0,1].$$

Under assumptions (H1)–(H5), using also Lemma 1.1.5, we find that $\mathcal{A}$, $\mathcal{B}$, and $\mathcal{C}$ are completely continuous from P' to P'. Thus, the existence and multiplicity of positive solutions of problem (S′)–(BC) are equivalent to the existence and multiplicity of fixed points of the operator $\mathcal{A}$.

Theorem 1.3.4. *Assume that (H1)–(H5) hold. If the functions f and g also satisfy the following conditions (H6) and (H7), then problem* (S′)–(BC) *has at least one positive solution* $(u(t), v(t))$, $t \in [0,1]$*:*

(H6) *There exist $\sigma \in (0, 1/2)$ and $p \in (0,1]$ such that*

$$(1)\ \tilde{f}_\infty^{\mathrm{i}} = \liminf_{u\to\infty} \inf_{t\in[\sigma,1-\sigma]} \frac{f(t,u)}{u^p} \in (0,\infty] \ \text{and} \ (2)\ \tilde{g}_\infty^{\mathrm{i}} = \liminf_{u\to\infty} \inf_{t\in[\sigma,1-\sigma]} \frac{g(t,u)}{u^{1/p}} = \infty.$$

(H7) *There exist $q_1, q_2 > 0$ with $q_1 q_2 \geq 1$ such that*

$$(1)\ \tilde{f}_0^{\mathrm{s}} = \limsup_{u\to 0+} \sup_{t\in[0,1]} \frac{f(t,u)}{u^{q_1}} \in [0,\infty) \quad \text{and} \quad (2)\ \tilde{g}_0^{\mathrm{s}} = \limsup_{u\to 0+} \sup_{t\in[0,1]} \frac{g(t,u)}{u^{q_2}} = 0.$$

Proof. For σ from (H6), we define the cone $P_0 = \{u \in P'; \inf_{t\in[\sigma,1-\sigma]} u(t) \geq \nu\|u\|\}$, where $\nu = \min\{\nu_1, \nu_2\}$, and ν_1 and ν_2 are defined in Section 1.1.2. From our assumptions and Lemma 1.1.7, we obtain $\mathcal{B}(P') \subset P_0$ and $\mathcal{C}(P') \subset P_0$. Now we consider the function $u_0(t) = \int_0^1 G_1(t,s)\,ds = (\mathcal{B}y_0)(t)$, $t \in [0,1]$, where $y_0(t) = 1$ for all $t \in [0,1]$, and the set

$$M = \{u \in P'; \quad \text{there exists} \quad \lambda \geq 0 \quad \text{such that} \quad u = \mathcal{A}u + \lambda u_0\}.$$

We shall show that $M \subset P_0$ and that M is a bounded subset of X. If $u \in M$, then there exists $\lambda \geq 0$ such that $u(t) = (\mathcal{A}u)(t) + \lambda u_0(t), t \in [0,1]$. Hence, we have

$$u(t) = (\mathcal{A}u)(t) + \lambda(\mathcal{B}y_0)(t) = \mathcal{B}(Fu(t)) + \lambda(\mathcal{B}y_0)(t) = \mathcal{B}(Fu(t) + \lambda y_0(t)) \in P_0,$$

where F: $P' \to P'$ is defined by $(Fu)(t) = f\left(t, \int_0^1 G_2(t,s) g(s,u(s))\,ds\right)$. Therefore, $M \subset P_0$, and

$$\|u\| \le \frac{1}{\nu} \inf_{t\in[\sigma,1-\sigma]} u(t), \quad \forall\, u \in M. \tag{1.15}$$

From (1) of assumption (H6), we deduce that there exist $C_1, C_2 > 0$ such that

$$f(t,u) \ge C_1 u^p - C_2, \quad \forall\,(t,u) \in [\sigma, 1-\sigma] \times [0,\infty). \tag{1.16}$$

If $\tilde{f}^{\rm i}_\infty \in (0,\infty)$, then we obtain

$$\forall\, \varepsilon > 0\ \exists\, \delta_\varepsilon > 0 \text{ such that } \forall\, u \ge \delta_\varepsilon \text{ we have } \frac{f(t,u)}{u^p} \ge \tilde{f}^{\rm i}_\infty - \varepsilon,\ \forall\, t \in [\sigma, 1-\sigma].$$

We choose $\varepsilon \in (0, \tilde{f}^{\rm i}_\infty)$, and we denote $C_1 = \tilde{f}^{\rm i}_\infty - \varepsilon > 0$. Then from the above relation, we deduce that there exists $\delta_0 > 0$ such that $f(t,u) \ge C_1 u^p$ for all $u \ge \delta_0$ and $t \in [\sigma, 1-\sigma]$. Because $f(t,u) \ge 0 \ge C_1 u^p - C_2$ for all $u \in [0, \delta_0]$ and $t \in [\sigma, 1-\sigma]$, with $C_2 = C_1 \delta_0^p$, then we obtain relation (1.16).

If $\tilde{f}^{\rm i}_\infty = \infty$—that is, $\lim_{u\to\infty} \inf_{t\in[\sigma,1-\sigma]} \frac{f(t,u)}{u^p} = \infty$—then we obtain

$$\forall\, \varepsilon > 0 \quad \exists\, \delta_\varepsilon > 0 \quad \text{such that} \quad \forall u \ge \delta_\varepsilon \quad \text{we have} \quad \frac{f(t,u)}{u^p} \ge \varepsilon, \quad \forall\, t \in [\sigma, 1-\sigma].$$

For $\varepsilon = 1$, we deduce that there exists $\tilde{\delta}_0 > 0$ such that $f(t,u) \ge u^p$ for all $u \ge \tilde{\delta}_0$ and $t \in [\sigma, 1-\sigma]$. Because $f(t,u) \ge 0 \ge u^p - C_2$ for all $u \in [0, \tilde{\delta}_0]$ and $t \in [\sigma, 1-\sigma]$, with $C_2 = \tilde{\delta}_0^p$, then we obtain relation (1.16) with $C_1 = 1$. Therefore, in both cases, we obtain inequality (1.16).

From (2) of assumption (H6), we have $\lim_{u\to\infty} \inf_{t\in[\sigma,1-\sigma]} \frac{g(t,u)}{u^{1/p}} = \infty$—that is,

$$\forall\, \varepsilon > 0 \quad \exists\, u_\varepsilon > 0 \quad \text{such that} \quad \forall\, u \ge u_\varepsilon \quad \text{we have} \quad \frac{g(t,u)}{u^{1/p}} \ge \varepsilon, \quad \forall\, t \in [\sigma, 1-\sigma].$$

We consider $\varepsilon_0 = 2/(C_1 \nu_1 \nu_2^p m_1 m_2) > 0$, where $m_1 = \int_\sigma^{1-\sigma} J_1(\tau)\, d\tau > 0$ and $m_2 = \int_\sigma^{1-\sigma} (J_2(\tau))^p\, d\tau > 0$. Then from the above relation, we deduce that there exists $u_0' > 0$ such that $(g(t,u))^p \ge \varepsilon_0 u$ for all $(t,u) \in [\sigma, 1-\sigma] \times [u_0', \infty)$. For $C_3 = \varepsilon_0 u_0' > 0$ we obtain

$$(g(t,u))^p \ge \varepsilon_0 u - C_3, \quad \forall\,(t,u) \in [\sigma, 1-\sigma] \times [0,\infty). \tag{1.17}$$

Then for any $u \in P'$, by using (1.16), Lemma 1.1.6, and the reverse form of Hölder's inequality, we have for $p \in (0,1)$

$$\begin{aligned}
(\mathcal{A}u)(t) &= \int_0^1 G_1(t,s) f\left(s, \int_0^1 G_2(s,\tau) g(\tau, u(\tau))\, d\tau\right) ds \\
&\ge \int_\sigma^{1-\sigma} G_1(t,s) f\left(s, \int_0^1 G_2(s,\tau) g(\tau, u(\tau))\, d\tau\right) ds \\
&\ge \int_\sigma^{1-\sigma} G_1(t,s) \left[C_1 \left(\int_0^1 G_2(s,\tau) g(\tau, u(\tau))\, d\tau \right)^p - C_2 \right] ds
\end{aligned}$$

$$\geq \int_\sigma^{1-\sigma} G_1(t,s)\left[C_1\int_0^1 (G_2(s,\tau)g(\tau,u(\tau)))^p\,d\tau\left(\int_0^1 d\tau\right)^{p/q_0}\right]ds - C_2\int_0^1 J_1(s)\,ds$$
$$\geq C_1\int_\sigma^{1-\sigma} G_1(t,s)\left(\int_\sigma^{1-\sigma}(G_2(s,\tau))^p(g(\tau,u(\tau)))^p\,d\tau\right)ds - C_4,\quad \forall t\in[0,1],$$

where $q_0 = p/(p-1)$ and $C_4 = C_2\int_0^1 J_1(s)\,ds$.

The above inequality is also valid for $p = 1$. Therefore, for $u \in P'$ and $p \in (0,1]$ we have

$$(\mathcal{A}u)(t)\geq C_1\int_\sigma^{1-\sigma} G_1(t,s)\left(\int_\sigma^{1-\sigma}(G_2(s,\tau))^p(g(\tau,u(\tau)))^p\,d\tau\right)ds - C_4,\quad \forall t\in[0,1]. \tag{1.18}$$

Next, for $u \in M$ and $t \in [\sigma, 1-\sigma]$, by using Lemma 1.1.6 and relations (1.17) and (1.18), we obtain

$$\begin{aligned}
u(t) &= (\mathcal{A}u)(t) + \lambda u_0(t) \geq (\mathcal{A}u)(t)\\
&\geq C_1\int_\sigma^{1-\sigma} G_1(t,s)\left(\int_\sigma^{1-\sigma}(G_2(s,\tau))^p(g(\tau,u(\tau)))^p\,d\tau\right)ds - C_4\\
&\geq C_1\nu_1\nu_2^p\left(\int_\sigma^{1-\sigma}J_1(s)\,ds\right)\left(\int_\sigma^{1-\sigma}(J_2(\tau))^p\,(\varepsilon_0 u(\tau) - C_3)\,d\tau\right) - C_4\\
&\geq C_1\nu_1\nu_2^p\varepsilon_0\left(\int_\sigma^{1-\sigma}J_1(s)\,ds\right)\left(\int_\sigma^{1-\sigma}(J_2(\tau))^p\,d\tau\right)\inf_{\tau\in[\sigma,1-\sigma]}u(\tau) - C_5\\
&= 2\inf_{\tau\in[\sigma,1-\sigma]}u(\tau) - C_5,
\end{aligned}$$

where $C_5 = C_4 + C_3C_1\nu_1\nu_2^p m_1 m_2 > 0$.

Hence, $\inf_{t\in[\sigma,1-\sigma]} u(t) \geq 2\inf_{t\in[\sigma,1-\sigma]} u(t) - C_5$, and so

$$\inf_{t\in[\sigma,1-\sigma]} u(t) \leq C_5,\quad \forall u \in M. \tag{1.19}$$

From relations (1.15) and (1.19) we obtain $\|u\| \leq C_5/\nu$ for all $u \in M$—that is, M is a bounded subset of X.

Moreover, there exists a sufficiently large R_1 ($R_1 \geq 1$) such that

$$u \neq \mathcal{A}u + \lambda u_0,\quad \forall u \in \partial B_{R_1}\cap P',\ \ \forall \lambda \geq 0.$$

From Theorem 1.3.2 we deduce that the fixed point index of the operator $\mathcal{A}$ is

$$i(\mathcal{A}, B_{R_1}\cap P', P') = 0. \tag{1.20}$$

Next, from (1) of assumption (H7) we have

$$\forall \varepsilon > 0\quad \exists u_\varepsilon > 0\quad \text{such that}\quad \frac{f(t,u)}{u^{q_1}} \leq \tilde{f}_0^s + \varepsilon,\quad \forall (t,u)\in[0,1]\times(0,u_\varepsilon].$$

For $\varepsilon = 1$ there exists $\tilde{u}_0 > 0$ such that $\frac{f(t,u)}{u^{q_1}} \leq (\tilde{f}_0^s + 1)$ for all $(t,u) \in [0,1]\times(0,\tilde{u}_0]$. If $\tilde{u}_0 \geq 1$, then the last inequality is true for every $u \in (0,1]$. If $\tilde{u}_0 < 1$, then $u^{q_1} > \tilde{u}_0^{q_1}$

and $f(t,u) \le \tilde{m}$ for all $u \in (\tilde{u}_0, 1]$ and $t \in [0,1]$, with $\tilde{m} > 0$. Therefore, for $u \in (\tilde{u}_0, 1]$ and $t \in [0,1]$ we obtain $\frac{f(t,u)}{u^{q_1}} \le \frac{\tilde{m}}{\tilde{u}_0^{q_1}}$. So, we deduce

$$\frac{f(t,u)}{u^{q_1}} \le M_0, \quad \forall\,(t,u) \in [0,1] \times (0,1], \quad \text{where } M_0 = \max\left\{\tilde{f}_0^{\rm s} + 1, \frac{\tilde{m}}{\tilde{u}_0^{q_1}}\right\} > 0,$$

and using also (H5), we obtain

$$f(t,u) \le M_0 u^{q_1}, \quad \forall\,(t,u) \in [0,1] \times [0,1]. \tag{1.21}$$

From (2) of assumption (H7) and from (H5) we have

$$\forall\,\varepsilon > 0 \quad \exists\,\delta_\varepsilon > 0 \quad \text{such that} \quad g(t,u) \le \varepsilon u^{q_2}, \quad \forall\,(t,u) \in [0,1] \times [0,\delta_\varepsilon].$$

We consider $\varepsilon_1 = \min\left\{1/M_2, (1/(2M_0M_1M_2^{q_1}))^{1/q_1}\right\} > 0$, where $M_1 = \int_0^1 J_1(s)\,{\rm d}s > 0$ and $M_2 = \int_0^1 J_2(s)\,{\rm d}s > 0$. Then we deduce that there exists $r_1 \in (0,1)$ such that

$$g(t,u) \le \varepsilon_1 u^{q_2}, \quad \forall\,(t,u) \in [0,1] \times [0,r_1]. \tag{1.22}$$

Hence, by (1.22), for any $u \in \bar{B}_{r_1} \cap P'$ and $t \in [0,1]$ we obtain

$$\int_0^1 G_2(t,s)g(s,u(s))\,{\rm d}s \le \varepsilon_1 \int_0^1 J_2(s)(u(s))^{q_2}\,{\rm d}s \le \varepsilon_1 M_2 \|u\|^{q_2} \le 1. \tag{1.23}$$

Therefore, by (1.21) and (1.23), we deduce that for any $u \in \bar{B}_{r_1} \cap P'$ and $t \in [0,1]$

$$\begin{aligned}(\mathcal{A}u)(t) &\le M_0 \int_0^1 G_1(t,s)\left(\int_0^1 G_2(s,\tau)g(\tau,u(\tau))\,{\rm d}\tau\right)^{q_1} {\rm d}s\\ &\le M_0\varepsilon_1^{q_1}M_1M_2^{q_1}\|u\|^{q_1q_2} \le M_0\varepsilon_1^{q_1}M_1M_2^{q_1}\|u\| \le \|u\|/2.\end{aligned}$$

This gives us $\|\mathcal{A}u\| \le \|u\|/2$ for all $u \in \partial B_{r_1} \cap P'$. From Theorem 1.3.1 we conclude that the fixed point index of $\mathcal{A}$ is

$$i(\mathcal{A}, B_{r_1} \cap P', P') = 1. \tag{1.24}$$

Combining (1.20) and (1.24), we obtain

$$i(\mathcal{A}, (B_{R_1} \setminus \bar{B}_{r_1}) \cap P', P') = i(\mathcal{A}, B_{R_1} \cap P', P') - i(\mathcal{A}, B_{r_1} \cap P', P') = -1.$$

Hence, we deduce that $\mathcal{A}$ has at least one fixed point $u_1 \in (B_{R_1} \setminus \bar{B}_{r_1}) \cap P'$—that is, $r_1 < \|u_1\| < R_1$. Let $v_1(t) = \int_0^1 G_2(t,s)g(s,u_1(s))\,{\rm d}s$. Then $(u_1, v_1) \in P' \times P'$ is a solution of (S′)–(BC). By using (H5), we also have $\|v_1\| > 0$. If we suppose that $v_1(t) = 0$ for all $t \in [0,1]$, then by using (H5), we have $f(s, v_1(s)) = f(s,0) = 0$ for all $s \in [0,1]$. This implies $u_1(t) = 0$ for all $t \in [0,1]$, which contradicts $\|u_1\| > 0$. The proof of Theorem 1.3.4 is completed. □

Theorem 1.3.5. *Assume that (H1)–(H5) hold. If the functions f and g also satisfy the following conditions (H8) and (H9), then problem* (S′)–(BC) *has at least one positive solution* $(u(t), v(t))$, $t \in [0,1]$*:*

(H8) *There exist* $\alpha_1, \alpha_2 > 0$ *with* $\alpha_1\alpha_2 \le 1$ *such that*

$$(1)\ \tilde{f}^{\mathrm{s}}_{\infty} = \limsup_{u\to\infty} \sup_{t\in[0,1]} \frac{f(t,u)}{u^{\alpha_1}} \in [0,\infty) \quad \textit{and} \quad (2)\ \tilde{g}^{\mathrm{s}}_{\infty} = \limsup_{u\to\infty} \sup_{t\in[0,1]} \frac{g(t,u)}{u^{\alpha_2}} = 0.$$

(H9) *There exists* $\sigma \in (0,1/2)$ *such that*

$$(1)\ \tilde{f}^{\mathrm{i}}_{0} = \liminf_{u\to0^+} \inf_{t\in[\sigma,1-\sigma]} \frac{f(t,u)}{u} \in (0,\infty] \quad \textit{and} \quad (2)\ \tilde{g}^{\mathrm{i}}_{0} = \liminf_{u\to0^+} \inf_{t\in[\sigma,1-\sigma]} \frac{g(t,u)}{u} = \infty.$$

Proof. From assumption (H8) (1), we deduce that there exist $C_6, C_7 > 0$ such that

$$f(t,u) \le C_6 u^{\alpha_1} + C_7, \quad \forall\,(t,u) \in [0,1]\times[0,\infty). \tag{1.25}$$

If $\tilde{f}^{\mathrm{s}}_{\infty} \in [0,\infty)$, then we obtain

$$\forall\,\varepsilon > 0 \quad \exists\, u_\varepsilon > 0 \quad \text{such that} \quad \frac{f(t,u)}{u^{\alpha_1}} \le \tilde{f}^{\mathrm{s}}_{\infty} + \varepsilon, \quad \forall\,(t,u) \in [0,1]\times[u_\varepsilon,\infty).$$

For $\varepsilon = 1$ there exists $\overline{u}_0 > 0$ such that $f(t,u) \le (\tilde{f}^{\mathrm{s}}_{\infty} + 1)u^{\alpha_1}$ for all $(t,u) \in [0,1]\times[\overline{u}_0,\infty)$. Because for $u \in [0,\overline{u}_0]$ and $t \in [0,1]$ there exists $C_7 > 0$ such that $f(t,u) \le C_7$, we obtain relation (1.25) with $C_6 = \tilde{f}^{\mathrm{s}}_{\infty} + 1 > 0$.

From (H8) (2) we have $\lim_{u\to\infty} \sup_{t\in[0,1]} \frac{g(t,u)}{u^{\alpha_2}} = 0$—that is,

$$\forall\,\varepsilon > 0 \quad \exists\, u_\varepsilon > 0 \quad \text{such that} \quad \frac{g(t,u)}{u^{\alpha_2}} \le \varepsilon, \quad \forall\,(t,u) \in [0,1]\times[u_\varepsilon,\infty).$$

We consider $\varepsilon_2 = (1/(2^{\alpha_1+1}C_6M_1M_2^{\alpha_1}))^{1/\alpha_1}$, where M_1 and M_2 are defined in the proof of Theorem 1.3.4. Then we deduce that there exists $\overline{\overline{u}}_0 > 0$ such that $g(t,u) \le \varepsilon_2 u^{\alpha_2}$ for all $(t,u) \in [0,1]\times[\overline{\overline{u}}_0,\infty)$. For $u \in [0,\overline{\overline{u}}_0]$ there exists $C_8 > 0$ such that $g(t,u) \le C_8$ for all $t \in [0,1]$. Therefore,

$$g(t,u) \le \varepsilon_2 u^{\alpha_2} + C_8, \quad \forall\,(t,u) \in [0,1]\times[0,\infty). \tag{1.26}$$

Hence, for $u \in P'$, by using (1.25) and (1.26), we obtain for all $t \in [0,1]$

$$\begin{aligned}
(\mathcal{A}u)(t) &\le \int_0^1 G_1(t,s)\left[C_6\left(\int_0^1 G_2(s,\tau)g(\tau,u(\tau))\,\mathrm{d}\tau\right)^{\alpha_1} + C_7\right]\mathrm{d}s\\
&\le C_6\int_0^1 G_1(t,s)\left[\int_0^1 G_2(s,\tau)\left(\varepsilon_2(u(\tau))^{\alpha_2} + C_8\right)\mathrm{d}\tau\right]^{\alpha_1}\mathrm{d}s + M_1C_7\\
&\le C_6\int_0^1 J_1(s)\left[\int_0^1 J_2(\tau)\left(\varepsilon_2\|u\|^{\alpha_2} + C_8\right)\mathrm{d}\tau\right]^{\alpha_1}\mathrm{d}s + M_1C_7\\
&= C_6\left(\varepsilon_2\|u\|^{\alpha_2} + C_8\right)^{\alpha_1}\left(\int_0^1 J_1(s)\,\mathrm{d}s\right)\left(\int_0^1 J_2(\tau)\,\mathrm{d}\tau\right)^{\alpha_1} + M_1C_7.
\end{aligned}$$

Therefore, we have

$$\begin{aligned}
(\mathcal{A}u)(t) &\le C_6M_1M_2^{\alpha_1}\left(\varepsilon_2\|u\|^{\alpha_2} + C_8\right)^{\alpha_1} + M_1C_7\\
&\le 2^{\alpha_1}C_6M_1M_2^{\alpha_1}\varepsilon_2^{\alpha_1}\|u\|^{\alpha_1\alpha_2} + 2^{\alpha_1}C_6M_1M_2^{\alpha_1}C_8^{\alpha_1} + M_1C_7, \quad \forall\,t \in [0,1].
\end{aligned}$$

For $u \in P'$ with $\|u\| \geq 1$ we obtain

$$(\mathcal{A}u)(t) \leq 2^{\alpha_1} C_6 M_1 M_2^{\alpha_1} \varepsilon_2^{\alpha_1} \|u\| + 2^{\alpha_1} C_6 M_1 M_2^{\alpha_1} C_8^{\alpha_1} + M_1 C_7 =: \Lambda(\|u\|), \quad \forall t \in [0,1]. \tag{1.27}$$

Because $\lim_{\|u\|\to\infty} \Lambda(\|u\|)/\|u\| = 1/2$, there exists a large R_2 ($R_2 \geq 1$) such that

$$\Lambda(\|u\|) \leq \frac{3}{4}\|u\|, \quad \forall\, u \in P', \quad \|u\| \geq R_2. \tag{1.28}$$

Hence, from (1.27) and (1.28) we deduce $\|\mathcal{A}u\| \leq \frac{3}{4}\|u\|$ for all $u \in \partial B_{R_2} \cap P'$, and from Theorem 1.3.1 we have that the fixed point index of the operator $\mathcal{A}$ is

$$i(\mathcal{A}, B_{R_2} \cap P', P') = 1. \tag{1.29}$$

On the other hand, from (H9) (1) we deduce that there exist positive constants $C_9 > 0$ and $\tilde{u}_1 > 0$ such that

$$f(t,u) \geq C_9 u, \quad \forall\, (t,u) \in [\sigma, 1-\sigma] \times [0, \tilde{u}_1]. \tag{1.30}$$

If $\tilde{f}_0^{\mathrm{i}} \in (0,\infty)$, then we obtain

$$\forall\, \varepsilon > 0 \quad \exists\, u_\varepsilon > 0 \quad \text{such that} \quad \frac{f(t,u)}{u} \geq \tilde{f}_0^{\mathrm{i}} - \varepsilon, \quad \forall\, (t,u) \in [\sigma, 1-\sigma] \times (0, u_\varepsilon].$$

We choose $\varepsilon > 0$ such that $\tilde{f}_0^{\mathrm{i}} - \varepsilon > 0$; we denote $C_9 = \tilde{f}_0^{\mathrm{i}} - \varepsilon$. Then there exists $\tilde{u}_2 > 0$ such that $f(t,u) \geq C_9 u$ for all $(t,u) \in [\sigma, 1-\sigma] \times [0, \tilde{u}_2]$. We obtain relation (1.30) with $\tilde{u}_1 = \tilde{u}_2$.

If $\tilde{f}_0^{\mathrm{i}} = \infty$, then $\lim_{u\to 0^+} \inf_{t\in[\sigma,1-\sigma]} \frac{f(t,u)}{u} = \infty$, and we have

$$\forall\, \varepsilon > 0 \quad \exists\, u_\varepsilon > 0 \quad \text{such that} \quad \frac{f(t,u)}{u} \geq \varepsilon, \quad \forall\, (t,u) \in [\sigma, 1-\sigma] \times (0, u_\varepsilon].$$

For $\varepsilon = 1$ there exists $\hat{u}_2$ such that $f(t,u) \geq u$ for all $(t,u) \in [\sigma, 1-\sigma] \times [0, \hat{u}_2]$. We obtain relation (1.30) with $C_9 = 1$ and $\tilde{u}_1 = \hat{u}_2$.

From (H9) (2) we have $\tilde{g}_0^{\mathrm{i}} = \lim_{u\to 0+} \inf_{t\in[\sigma,1-\sigma]} \frac{g(t,u)}{u} = \infty$, and so

$$\forall\, \varepsilon > 0 \quad \exists\, u_\varepsilon > 0 \quad \text{such that} \quad \frac{g(t,u)}{u} \geq \varepsilon, \quad \forall\, (t,u) \in [\sigma, 1-\sigma] \times (0, u_\varepsilon].$$

For $\varepsilon = C_0/C_9$ with $C_0 = 1/(\nu_1\nu_2 m_1 m_3) > 0$ and $m_3 = \int_\sigma^{1-\sigma} J_2(\tau)\,\mathrm{d}\tau$, we deduce that there exists $\hat{u}_1 > 0$ such that $g(t,u) \geq \frac{C_0}{C_9} u$ for all $(t,u) \in [\sigma, 1-\sigma] \times [0, \hat{u}_1]$.

We consider $\delta_1 = \min\{\tilde{u}_1, \hat{u}_1\}$, and then

$$f(t,u) \geq C_9 u, \qquad g(t,u) \geq \frac{C_0}{C_9} u, \quad \forall\, (t,u) \in [\sigma, 1-\sigma] \times [0, \delta_1]. \tag{1.31}$$

Because $g(t,0) = 0$ for all $t \in [0,1]$, and g is continuous, it can be shown that there exists a sufficiently small $r_2 \in (0, \min\{\delta_1, 1\})$ such that $g(t,u) \leq \delta_1/m_3$ for all $(t,u) \in [\sigma, 1-\sigma] \times [0, r_2]$. Hence, for any $u \in \bar{B}_{r_2} \cap P'$ we obtain

$$\int_\sigma^{1-\sigma} G_2(s,\tau) g(\tau, u(\tau))\,\mathrm{d}\tau \leq \int_\sigma^{1-\sigma} J_2(\tau) g(\tau, u(\tau))\,\mathrm{d}\tau \leq \delta_1, \quad \forall\, s \in [\sigma, 1-\sigma]. \tag{1.32}$$

From (1.31), (1.32), and Lemma 1.1.6, we deduce that for any $u \in \bar{B}_{r_2} \cap P'$ we have

$$\begin{aligned}(\mathcal{A}u)(t) &\geq \int_\sigma^{1-\sigma} G_1(t,s)f\left(s, \int_\sigma^{1-\sigma} G_2(s,\tau)g(\tau,u(\tau))\,\mathrm{d}\tau\right)\mathrm{d}s\\ &\geq C_9 \int_\sigma^{1-\sigma} G_1(t,s)\left(\int_\sigma^{1-\sigma} G_2(s,\tau)g(\tau,u(\tau))\,\mathrm{d}\tau\right)\mathrm{d}s\\ &\geq C_0 \int_\sigma^{1-\sigma} G_1(t,s)\left(\int_\sigma^{1-\sigma} G_2(s,\tau)u(\tau)\,\mathrm{d}\tau\right)\mathrm{d}s\\ &\geq C_0\nu_2 \int_\sigma^{1-\sigma} G_1(t,s)\left(\int_\sigma^{1-\sigma} J_2(\tau)u(\tau)\,\mathrm{d}\tau\right)\mathrm{d}s =: (\mathcal{L}u)(t), \quad t \in [0,1],\end{aligned}$$

where the linear operator $\mathcal{L} : P' \to P'$ is defined by

$$(\mathcal{L}u)(t) = C_0\nu_2 \left(\int_\sigma^{1-\sigma} J_2(\tau)u(\tau)\,\mathrm{d}\tau\right)\left(\int_\sigma^{1-\sigma} G_1(t,s)\,\mathrm{d}s\right), \quad t \in [0,1].$$

Hence, we obtain

$$\mathcal{A}u \geq \mathcal{L}u, \quad \forall\, u \in \partial B_{r_2} \cap P'. \tag{1.33}$$

For $w_0(t) = \int_\sigma^{1-\sigma} G_1(t,s)\,\mathrm{d}s, t \in [0,1]$ we have $w_0 \in P'$ and $w_0 \neq 0$. Besides, for all $t \in [0,1]$ we deduce

$$\begin{aligned}(\mathcal{L}w_0)(t) &= C_0\nu_2 \left[\int_\sigma^{1-\sigma} J_2(\tau)\left(\int_\sigma^{1-\sigma} G_1(\tau,s)\,\mathrm{d}s\right)\mathrm{d}\tau\right]\left(\int_\sigma^{1-\sigma} G_1(t,s)\,\mathrm{d}s\right)\\ &\geq C_0\nu_1\nu_2 \left(\int_\sigma^{1-\sigma} J_2(\tau)\,\mathrm{d}\tau\right)\left(\int_\sigma^{1-\sigma} J_1(\tau)\,\mathrm{d}\tau\right)\left(\int_\sigma^{1-\sigma} G_1(t,s)\,\mathrm{d}s\right)\\ &= C_0\nu_1\nu_2 m_1 m_3 \int_\sigma^{1-\sigma} G_1(t,s)\,\mathrm{d}s = \int_\sigma^{1-\sigma} G_1(t,s)\,\mathrm{d}s = w_0(t).\end{aligned}$$

Therefore,

$$\mathcal{L}w_0 \geq w_0. \tag{1.34}$$

We may suppose that $\mathcal{A}$ has no fixed point on $\partial B_{r_2} \cap P'$ (otherwise the proof is finished). From (1.33), (1.34), and Theorem 1.3.3, we conclude that the fixed point index of $\mathcal{A}$ is

$$i(\mathcal{A}, B_{r_2} \cap P', P') = 0. \tag{1.35}$$

Therefore, from (1.29) and (1.35) we have

$$i(\mathcal{A}, (B_{R_2} \setminus \bar{B}_{r_2}) \cap P', P') = i(\mathcal{A}, B_{R_2} \cap P', P') - i(\mathcal{A}, B_{r_2} \cap P', P') = 1.$$

Then $\mathcal{A}$ has at least one fixed point u_1 in $(B_{R_2} \setminus \bar{B}_{r_2}) \cap P'$—that is, $r_2 < \|u_1\| < R_2$. Let $v_1(t) = \int_0^1 G_2(t,s)g(s,u_1(s))\,\mathrm{d}s$. Then $(u_1, v_1) \in P' \times P'$ is a solution of (S′)–(BC). By (H5), we also deduce that $\|v_1\| > 0$. This completes the proof of Theorem 1.3.5. □

Theorem 1.3.6. *Assume that (H1)–(H5) hold. If the functions f and g also satisfy conditions (H6) and (H9) and the following condition (H10), then problem* (S′)–(BC) *has at least two positive solutions $(u_1(t), v_1(t))$, $(u_2(t), v_2(t))$, $t \in [0, 1]$:*

(H10) *For each $t \in [0, 1]$, $f(t, u)$ and $g(t, u)$ are nondecreasing with respect to u, and there exists a constant $N > 0$ such that*

$$f\left(t, m_0 \int_0^1 g(s, N)\, \mathrm{d}s\right) < \frac{N}{m_0}, \quad \forall t \in [0, 1],$$

where $m_0 = \max\{\tilde{K}_1, \tilde{K}_2\}$, $\tilde{K}_1 = \max_{s\in[0,1]} J_1(s)$, $\tilde{K}_2 = \max_{s\in[0,1]} J_2(s)$, and J_1 and J_2 are defined in Section 1.1.2.

Proof. From Section 1.1.2, we have $0 \le G_1(t, s) \le J_1(s) \le \tilde{K}_1$ and $G_2(t, s) \le J_2(s) \le \tilde{K}_2$ for all $(t, s) \in [0, 1] \times [0, 1]$. By using (H10), for any $u \in \partial B_N \cap P'$, we obtain

$$(\mathcal{A}u)(t) \le \int_0^1 G_1(t, s) f\left(s, \tilde{K}_2 \int_0^1 g(\tau, u(\tau))\, \mathrm{d}\tau\right) \mathrm{d}s$$
$$\le \int_0^1 J_1(s) f\left(s, m_0 \int_0^1 g(\tau, N)\, \mathrm{d}\tau\right) \mathrm{d}s, \quad \forall t \in [0, 1].$$

Then

$$\|\mathcal{A}u\| \le \int_0^1 J_1(s) f\left(s, m_0 \int_0^1 g(\tau, N)\, \mathrm{d}\tau\right) \mathrm{d}s < \frac{N}{m_0} \int_0^1 J_1(s)\, \mathrm{d}s \le \frac{N\tilde{K}_1}{m_0} \le N.$$

So, $\|\mathcal{A}u\| < \|u\|$ for all $u \in \partial B_N \cap P'$.

By Theorem 1.3.1, we conclude that the fixed point index of the operator $\mathcal{A}$ is

$$i(\mathcal{A}, B_N \cap P', P') = 1. \tag{1.36}$$

On the other hand, from (H6), (H9), and the proofs of Theorems 1.3.4 and 1.3.5, we know that there exists a sufficiently large $R_1 > N$ and a sufficiently small r_2 with $0 < r_2 < N$ such that

$$i(\mathcal{A}, B_{R_1} \cap P', P') = 0, \quad i(\mathcal{A}, B_{r_2} \cap P', P') = 0. \tag{1.37}$$

From relations (1.36) and (1.37) we obtain

$$i(\mathcal{A}, (B_{R_1} \setminus \bar{B}_N) \cap P', P') = i(\mathcal{A}, B_{R_1} \cap P', P') - i(\mathcal{A}, B_N \cap P', P') = -1,$$
$$i(\mathcal{A}, (B_N \setminus \bar{B}_{r_2}) \cap P', P') = i(\mathcal{A}, B_N \cap P', P') - i(\mathcal{A}, B_{r_2} \cap P', P') = 1.$$

Then $\mathcal{A}$ has at least one fixed point u_1 in $(B_{R_1} \setminus \bar{B}_N) \cap P'$ and has at least one fixed point u_2 in $(B_N \setminus \bar{B}_{r_2}) \cap P'$. If in Theorem 1.3.5 the operator $\mathcal{A}$ has at least one fixed point on $\partial B_{r_2} \cap P'$, then by using the first relation from formula above, we deduce that $\mathcal{A}$ has at least one fixed point u_1 in $(B_{R_1} \setminus \bar{B}_N) \cap P'$ and has at least one fixed point u_2 on $\partial B_{r_2} \cap P'$. Therefore, problem (S′)–(BC) has two distinct positive solutions $(u_1, v_1), (u_2, v_2) \in P' \times P'$, where $v_i(t) = \int_0^1 G_2(t,s)g(s,u_i(s))\,ds$, $i = 1,2$, with $u_i(t) \geq 0$, $v_i(t) \geq 0$ for all $t \in [0,1]$ and $\|u_i\| > 0$, $\|v_i\| > 0$, $i = 1,2$. The proof of Theorem 1.3.6 is completed. □

1.3.3 Examples

In this section, we shall present some examples which illustrate our results.

Example 1.3.1. Let $f(t,u) = (at+1)u^{\theta}$ and $g(t,u) = (bt+1)u^{\zeta}$ for all $t \in [0,1]$ and $u \in [0,\infty)$, with $\theta > 1/2$ and $\zeta > 2$, and $p = 1/2$, $q_1 = 1/2$, $q_2 = 2$, $a,b \geq 0$, and $\sigma = 1/4$. Then assumptions (H6) and (H7) are satisfied; we have $\tilde{f}^{\rm i}_{\infty} = \infty$, $\tilde{g}^{\rm i}_{\infty} = \infty$, $\tilde{f}^{\rm s}_0 = 0$, and $\tilde{g}^{\rm s}_0 = 0$. Under assumptions (H1)–(H4), by Theorem 1.3.4, we deduce that problem (S′)–(BC) has at least one positive solution. Here, g is a superlinear function in variable u, and f may be linear ($\theta = 1$), superlinear ($\theta > 1$), or sublinear ($\theta < 1$) in variable u.

Example 1.3.2. Let $f(t,u) = u^{1/2}(2+\sin u)$ and $g(t,u) = u^{\rho}(3+\cos u)$ for all $t \in [0,1]$ and $u \in [0,\infty)$, with $\rho > 2$, and $p = 1/2$, $q_1 = 1/2$, $q_2 = 2$, and $\sigma = 1/4$. Then assumptions (H6) and (H7) are satisfied; we have $\tilde{f}^{\rm i}_{\infty} = 1$, $\tilde{g}^{\rm i}_{\infty} = \infty$, $\tilde{f}^{\rm s}_0 = 2$, and $\tilde{g}^{\rm s}_0 = 0$. Under assumptions (H1)–(H4), by Theorem 1.3.4, we deduce that problem (S′)–(BC) has at least one positive solution.

Example 1.3.3. Let $f(t,u) = u^{1/2}$ and $g(t,u) = u^{\varrho}$ for all $t \in [0,1]$ and $u \in [0,\infty)$, with $\varrho < 1$, and $\alpha_1 = 1/2$ and $\alpha_2 = 2$. Then assumptions (H8) and (H9) are satisfied; we have $\tilde{f}^{\rm s}_{\infty} = 1$, $\tilde{g}^{\rm s}_{\infty} = 0$, $\tilde{f}^{\rm i}_0 = \infty$, and $\tilde{g}^{\rm i}_0 = \infty$. Under assumptions (H1)–(H4), by Theorem 1.3.5, we deduce that problem (S′)–(BC) has at least one positive solution. Here, f and g are both sublinear.

Example 1.3.4. Let $a(t) = 1$, $b(t) = 4$, $c(t) = 1$, and $d(t) = 1$ for all $t \in [0,1]$, $\alpha = 1$, $\beta = 3$, $\gamma = 1$, $\delta = 1$, $\tilde{\alpha} = 3$, $\tilde{\beta} = 2$, $\tilde{\gamma} = 1$, $\tilde{\delta} = 3/2$,

$$H_1(t) = t^2, \quad H_2(t) = \begin{cases} 0, & t \in [0,1/3), \\ 7/2, & t \in [1/3,2/3), \\ 11/2, & t \in [2/3,1], \end{cases} \qquad K_1(t) = \begin{cases} 0, & t \in [0,1/2), \\ 4/3, & t \in [1/2,1], \end{cases}$$

$K_2(t) = t^3$, $f(t,x) = a(x^{\hat{\alpha}} + x^{\hat{\beta}})$, and $g(t,x) = b(x^{\hat{\gamma}} + x^{\hat{\delta}})$ for all $t \in [0,1]$, $x \in [0,\infty)$, with $a,b > 0$, $\hat{\alpha} > 1$, $\hat{\beta} < 1$, $\hat{\gamma} > 2$, and $\hat{\delta} < 1$. Then $\int_0^1 u(s)\,dH_2(s) = \frac{7}{2}u\left(\frac{1}{3}\right) + 2u\left(\frac{2}{3}\right)$, $\int_0^1 u(s)\,dK_1(s) = \frac{4}{3}u\left(\frac{1}{2}\right)$, $\int_0^1 u(s)\,dH_1(s) = 2\int_0^1 su(s)\,ds$, and $\int_0^1 u(s)\,dK_2(s) = 3\int_0^1 s^2u(s)\,ds$.

We consider the second-order differential system

$$\begin{cases} u''(t) - 4u(t) + a(v^{\hat{\alpha}}(t) + v^{\hat{\beta}}(t)) = 0, & t \in (0,1), \\ v''(t) - v(t) + b(u^{\hat{\gamma}}(t) + u^{\hat{\delta}}(t)) = 0, & t \in (0,1), \end{cases} \qquad (\overline{S'})$$

with the boundary conditions

$$\begin{cases} u(0) - 3u'(0) = 2\int_0^1 su(s)\,ds, & u(1) + u'(1) = \frac{7}{2}u\left(\frac{1}{3}\right) + 2u\left(\frac{2}{3}\right), \\ 3v(0) - 2v'(0) = \frac{4}{3}v\left(\frac{1}{2}\right), & v(1) + \frac{3}{2}v'(1) = 3\int_0^1 s^2 v(s)\,ds. \end{cases} \qquad (\overline{\text{BC}'})$$

By using the expressions for J_1 and J_2 given in Section 1.1.4, we obtain $\tilde{K}_1 = \max_{s\in[0,1]} J_1(s) \approx 3.04247954$ and $\tilde{K}_2 = \max_{s\in[0,1]} J_2(s) \approx 1.05002303$. Then $m_0 = \max\{\tilde{K}_1, \tilde{K}_2\} = \tilde{K}_1$. The functions $f(t,u)$ and $g(t,u)$ are nondecreasing with respect to u for any $t \in [0,1]$, and for $p = 1/2$ and $\sigma \in (0, 1/2)$ fixed, assumptions (H6) and (H9) are satisfied; we have $\tilde{f}^{\text{i}}_{\infty} = \infty$, $\tilde{g}^{\text{i}}_{\infty} = \infty$, $\tilde{f}^{\text{i}}_0 = \infty$, and $\tilde{g}^{\text{i}}_0 = \infty$.

We take $N = 1$, and then $\int_0^1 g(s,1)\,ds = 2b$ and $f(t, 2bm_0) = a[(2bm_0)^{\hat{\alpha}} + (2bm_0)^{\hat{\beta}}]$. If $a[m_0^{\hat{\alpha}+1}(2b)^{\hat{\alpha}} + m_0^{\hat{\beta}+1}(2b)^{\hat{\beta}}] < 1$, assumption (H10) is satisfied. For example, if $\hat{\alpha} = 3/2$, $\hat{\beta} = 1/2$, $b = 1/2$, and $a < 1/(m_0^{5/2} + m_0^{3/2})$ (e.g., $a \le 0.04$), then the above inequality is satisfied. By Theorem 1.3.6, we deduce that problem ($\overline{\text{S}'}$)–($\overline{\text{BC}'}$) has at least two positive solutions.

1.4 Systems with singular nonlinearities

In this section, we study the existence of positive solutions for problem (S′)–(BC) from Section 1.3, where the nonlinearities f and g do not possess any sublinear or superlinear growth conditions and may be singular at $t = 0$ and/or $t = 1$.

1.4.1 Presentation of the problem

We consider the system of nonlinear second-order ordinary differential equations

$$\begin{cases} (a(t)u'(t))' - b(t)u(t) + f(t, v(t)) = 0, & 0 < t < 1, \\ (c(t)v'(t))' - d(t)v(t) + g(t, u(t)) = 0, & 0 < t < 1, \end{cases} \qquad (\text{S}')$$

with the integral boundary conditions

$$\begin{cases} \alpha u(0) - \beta a(0)u'(0) = \int_0^1 u(s)\,dH_1(s), & \gamma u(1) + \delta a(1)u'(1) = \int_0^1 u(s)\,dH_2(s), \\ \tilde{\alpha} v(0) - \tilde{\beta} c(0)v'(0) = \int_0^1 v(s)\,dK_1(s), & \tilde{\gamma} v(1) + \tilde{\delta} c(1)v'(1) = \int_0^1 v(s)\,dK_2(s), \end{cases} \qquad (\text{BC})$$

where the above integrals are Riemann–Stieltjes integrals, and the functions f and g may be singular at $t = 0$ and/or $t = 1$.

By a positive solution of (S′)–(BC), we mean a pair of functions $(u, v) \in (C([0,1], \mathbb{R}_+) \cap C^2(0,1))^2$ satisfying (S′) and (BC) with $\sup_{t\in[0,1]} u(t) > 0$, $\sup_{t\in[0,1]} v(t) > 0$.

This problem is a generalization of the problem studied in Henderson and Luca (2013i), where in (S′) we have $a(t) = 1$, $c(t) = 1$, $b(t) = 0$, and $d(t) = 0$ for all $t \in (0, 1)$ (denoted by $(\tilde{\text{S}}')$), and $\alpha = \tilde{\alpha} = 1$, $\beta = \tilde{\beta} = 0$, $\gamma = \tilde{\gamma} = 1$, $\delta = \tilde{\delta} = 0$, H_1 and K_1 are constant functions, and H_2 and K_2 are step functions. Problem $(\tilde{\text{S}}')$–(BC) also generalizes the problem investigated in Liu et al. (2007), where the authors studied the existence of positive solutions for system $(\tilde{\text{S}}')$ with the boundary conditions $u(0) = 0$, $u(1) = \alpha u(\eta)$, $v(0) = 0$, and $v(1) = \alpha v(\eta)$ with $\eta \in (0, 1), 0 < \alpha\eta < 1$.

We shall suppose that assumptions (A1)–(A5) from Section 1.1.2 hold, and we shall use in our main results from this section Lemmas 1.1.1–1.1.7 from Section 1.1.2.

1.4.2 Main results

We investigate the existence of positive solutions for problem (S′)–(BC) under various assumptions on singular functions f and g.

We present the basic assumptions that we shall use in the sequel:

(L1) The functions $a, c \in C^1([0, 1], (0, \infty))$ and $b, d \in C([0, 1], [0, \infty))$.
(L2) $\alpha, \beta, \gamma, \delta, \tilde{\alpha}, \tilde{\beta}, \tilde{\gamma}, \tilde{\delta} \in [0, \infty)$ with $\alpha + \beta > 0$, $\gamma + \delta > 0$, $\tilde{\alpha} + \tilde{\beta} > 0$, and $\tilde{\gamma} + \tilde{\delta} > 0$; if $b \equiv 0$, then $\alpha + \gamma > 0$; if $d \equiv 0$, then $\tilde{\alpha} + \tilde{\gamma} > 0$.
(L3) $H_1, H_2, K_1, K_2 : [0, 1] \to \mathbb{R}$ are nondecreasing functions.
(L4) $\tau_1 - \int_0^1 \phi(s)\, dH_1(s) > 0$, $\tau_1 - \int_0^1 \psi(s)\, dH_2(s) > 0$, $\tau_2 - \int_0^1 \tilde{\phi}(s)\, dK_1(s) > 0$, $\tau_2 - \int_0^1 \tilde{\psi}(s)\, dK_2(s) > 0$, $\Delta_1 > 0$, and $\Delta_2 > 0$, where τ_1, τ_2, Δ_1, and Δ_2 are defined in Section 1.1.2.
(L5) The functions $f, g \in C((0, 1) \times \mathbb{R}_+, \mathbb{R}_+)$, and there exist $p_i \in C((0, 1), \mathbb{R}_+)$ and $q_i \in C(\mathbb{R}_+, \mathbb{R}_+)$, $i = 1, 2$, with $0 < \int_0^1 p_i(t)\, dt < \infty$, $i = 1, 2$, $q_1(0) = 0$, $q_2(0) = 0$ such that

$$f(t, x) \le p_1(t)q_1(x), \qquad g(t, x) \le p_2(t)q_2(x), \ \ \forall\, t \in (0, 1), \ \ x \in \mathbb{R}_+.$$

The pair of functions $(u, v) \in (C([0, 1]) \cap C^2(0, 1))^2$ is a solution for our problem (S′)–(BC) if and only if $(u, v) \in (C([0, 1]))^2$ is a solution for the nonlinear integral equations

$$\begin{cases} u(t) = \displaystyle\int_0^1 G_1(t, s) f\left(s, \int_0^1 G_2(s, \tau) g(\tau, u(\tau))\, d\tau\right) ds, & t \in [0, 1], \\ v(t) = \displaystyle\int_0^1 G_2(t, s) g(s, u(s))\, ds, & t \in [0, 1]. \end{cases}$$

We consider again the Banach space $X = C([0, 1])$ with the supremum norm $\|\cdot\|$ and the cone $P' \subset X$ by $P' = \{u \in X,\ u(t) \ge 0,\ \forall\, t \in [0, 1]\}$.

We also define the operator $\mathcal{D}$: $P' \to X$ by

$$\mathcal{D}(u)(t) = \int_0^1 G_1(t, s) f\left(s, \int_0^1 G_2(s, \tau) g(\tau, u(\tau))\, d\tau\right) ds.$$

Lemma 1.4.1. *Assume that (L1)–(L5) hold. Then $\mathcal{D}$: $P' \to P'$ is completely continuous.*

Proof. We denote $\alpha_0 = \int_0^1 J_1(s)p_1(s)\,\mathrm{d}s$ and $\beta_0 = \int_0^1 J_2(s)p_2(s)\,\mathrm{d}s$. Using (L5), we deduce that $0 < \alpha_0 < \infty$ and $0 < \beta_0 < \infty$. By Lemma 1.1.5, we deduce that $\mathcal{D}$ maps P' into P'. We shall prove that $\mathcal{D}$ maps bounded sets into relatively compact sets. Suppose $E \subset P'$ is an arbitrary bounded set. First, we prove that $\mathcal{D}(E)$ is a bounded set. Because E is bounded, then there exists $M_1 > 0$ such that $\|u\| \le M_1$ for all $u \in E$. By the continuity of q_2, there exists $M_2 > 0$ such that $M_2 = \sup_{x\in[0,M_1]} q_2(x)$. By using Lemma 1.1.6, for any $u \in E$ and $s \in [0,1]$, we obtain

$$\int_0^1 G_2(s,\tau)g(\tau,u(\tau))\,\mathrm{d}\tau \le \int_0^1 G_2(s,\tau)p_2(\tau)q_2(u(\tau))\,\mathrm{d}\tau \le \beta_0 M_2. \tag{1.38}$$

Because q_1 is continuous, there exists $M_3 > 0$ such that $M_3 = \sup_{x\in[0,\beta_0 M_2]} q_1(x)$. Therefore, from (1.38), (L5), and Lemma 1.1.6, we deduce

$$\begin{aligned}(\mathcal{D}u)(t) &\le \int_0^1 G_1(t,s)p_1(s)q_1\left(\int_0^1 G_2(s,\tau)g(\tau,u(\tau))\,\mathrm{d}\tau\right)\mathrm{d}s\\ &\le M_3\int_0^1 J_1(s)p_1(s)\,\mathrm{d}s = \alpha_0 M_3, \quad \forall t\in[0,1].\end{aligned} \tag{1.39}$$

So, $\|\mathcal{D}u\| \le \alpha_0 M_3$ for all $u \in E$. Therefore, $\mathcal{D}(E)$ is a bounded set.

In what follows, we shall prove that $\mathcal{D}(E)$ is equicontinuous. By using (1.5) from Lemma 1.1.1, we have for all $t \in [0,1]$

$$\begin{aligned}(\mathcal{D}u)(t) &= \int_0^1 G_1(t,s)f\left(s,\int_0^1 G_2(s,\tau)g(\tau,u(\tau))\,\mathrm{d}\tau\right)\mathrm{d}s\\
&= \int_0^1 \left\{g_1(t,s) + \frac{1}{\Delta_1}\left[\psi(t)\left(\int_0^1 \phi(\tau)\,\mathrm{d}H_2(\tau)\right) + \phi(t)\left(\tau_1 - \int_0^1 \psi(\tau)\,\mathrm{d}H_2(\tau)\right)\right]\right.\\
&\quad\times\left(\int_0^1 g_1(\tau,s)\,\mathrm{d}H_1(\tau)\right) + \frac{1}{\Delta_1}\left[\psi(t)\left(\tau_1 - \int_0^1 \phi(\tau)\,\mathrm{d}H_1(\tau)\right)\right.\\
&\quad\left.\left. + \phi(t)\left(\int_0^1 \psi(\tau)\,\mathrm{d}H_1(\tau)\right)\right]\left(\int_0^1 g_1(\tau,s)\,\mathrm{d}H_2(\tau)\right)\right\}\\
&\quad\times f\left(s,\int_0^1 G_2(s,\tau)g(\tau,u(\tau))\,\mathrm{d}\tau\right)\mathrm{d}s\\
&= \int_0^t \frac{1}{\tau_1}\phi(t)\psi(s)f\left(s,\int_0^1 G_2(s,\tau)g(\tau,u(\tau))\,\mathrm{d}\tau\right)\mathrm{d}s\\
&\quad+\int_t^1 \frac{1}{\tau_1}\phi(s)\psi(t)f\left(s,\int_0^1 G_2(s,\tau)g(\tau,u(\tau))\,\mathrm{d}\tau\right)\mathrm{d}s\\
&\quad+\frac{1}{\Delta_1}\int_0^1 \left[\psi(t)\left(\int_0^1 \phi(\tau)\,\mathrm{d}H_2(\tau)\right) + \phi(t)\left(\tau_1 - \int_0^1 \psi(\tau)\,\mathrm{d}H_2(\tau)\right)\right]\\
&\quad\times\left(\int_0^1 g_1(\tau,s)\,\mathrm{d}H_1(\tau)\right)f\left(s,\int_0^1 G_2(s,\tau)g(\tau,u(\tau))\,\mathrm{d}\tau\right)\mathrm{d}s\\
&\quad+\frac{1}{\Delta_1}\int_0^1 \left[\psi(t)\left(\tau_1 - \int_0^1 \phi(\tau)\,\mathrm{d}H_1(\tau)\right) + \phi(t)\left(\int_0^1 \psi(\tau)\,\mathrm{d}H_1(\tau)\right)\right]\\
&\quad\times\left(\int_0^1 g_1(\tau,s)\,\mathrm{d}H_2(\tau)\right)f\left(s,\int_0^1 G_2(s,\tau)g(\tau,u(\tau))\,\mathrm{d}\tau\right)\mathrm{d}s.\end{aligned}$$

Therefore, we obtain for any $t \in (0,1)$

$$
\begin{aligned}
(\mathcal{D}u)'(t) = \;& \frac{1}{\tau_1}\phi(t)\psi(t)f\left(t, \int_0^1 G_2(t,\tau)g(\tau,u(\tau))\,\mathrm{d}\tau\right) \\
& + \frac{1}{\tau_1}\int_0^t \phi'(t)\psi(s)f\left(s, \int_0^1 G_2(s,\tau)g(\tau,u(\tau))\,\mathrm{d}\tau\right)\mathrm{d}s \\
& - \frac{1}{\tau_1}\phi(t)\psi(t)f\left(t, \int_0^1 G_2(t,\tau)g(\tau,u(\tau))\,\mathrm{d}\tau\right) \\
& + \frac{1}{\tau_1}\int_t^1 \psi'(t)\phi(s)f\left(s, \int_0^1 G_2(s,\tau)g(\tau,u(\tau))\,\mathrm{d}\tau\right)\mathrm{d}s \\
& + \frac{1}{\Delta_1}\int_0^1 \left[\psi'(t)\left(\int_0^1 \phi(\tau)\,\mathrm{d}H_2(\tau)\right) + \phi'(t)\left(\tau_1 - \int_0^1 \psi(\tau)\,\mathrm{d}H_2(\tau)\right)\right] \\
& \times \left(\int_0^1 g_1(\tau,s)\,\mathrm{d}H_1(\tau)\right) f\left(s, \int_0^1 G_2(s,\tau)g(\tau,u(\tau))\,\mathrm{d}\tau\right)\mathrm{d}s \\
& + \frac{1}{\Delta_1}\int_0^1 \left[\psi'(t)\left(\tau_1 - \int_0^1 \phi(\tau)\,\mathrm{d}H_1(\tau)\right) + \phi'(t)\left(\int_0^1 \psi(\tau)\,\mathrm{d}H_1(\tau)\right)\right] \\
& \times \left(\int_0^1 g_1(\tau,s)\,\mathrm{d}H_2(\tau)\right) f\left(s, \int_0^1 G_2(s,\tau)g(\tau,u(\tau))\,\mathrm{d}\tau\right)\mathrm{d}s.
\end{aligned}
$$

So, for any $t \in (0,1)$ we deduce

$$
\begin{aligned}
|(\mathcal{D}u)'(t)| \le \;& \frac{1}{\tau_1}\int_0^t |\phi'(t)\psi(s)| p_1(s) q_1\left(\int_0^1 G_2(s,\tau)g(\tau,u(\tau))\,\mathrm{d}\tau\right)\mathrm{d}s \\
& + \frac{1}{\tau_1}\int_t^1 |\psi'(t)\phi(s)| p_1(s) q_1\left(\int_0^1 G_2(s,\tau)g(\tau,u(\tau))\,\mathrm{d}\tau\right)\mathrm{d}s \\
& + \frac{1}{\Delta_1}\int_0^1 \left[|\psi'(t)|\left(\int_0^1 \phi(\tau)\,\mathrm{d}H_2(\tau)\right) + |\phi'(t)|\left(\tau_1 - \int_0^1 \psi(\tau)\,\mathrm{d}H_2(\tau)\right)\right] \\
& \times \left(\int_0^1 g_1(\tau,s)\,\mathrm{d}H_1(\tau)\right) p_1(s) q_1\left(\int_0^1 G_2(s,\tau)g(\tau,u(\tau))\,\mathrm{d}\tau\right)\mathrm{d}s \\
& + \frac{1}{\Delta_1}\int_0^1 \left[|\psi'(t)|\left(\tau_1 - \int_0^1 \phi(\tau)\,\mathrm{d}H_1(\tau)\right) + |\phi'(t)|\left(\int_0^1 \psi(\tau)\,\mathrm{d}H_1(\tau)\right)\right] \\
& \times \left(\int_0^1 g_1(\tau,s)\,\mathrm{d}H_2(\tau)\right) p_1(s) q_1\left(\int_0^1 G_2(s,\tau)g(\tau,u(\tau))\,\mathrm{d}\tau\right)\mathrm{d}s \\
& \le M_3\left\{-\frac{1}{\tau_1}\int_0^t \phi'(t)\psi(s)p_1(s)\,\mathrm{d}s + \frac{1}{\tau_1}\int_t^1 \psi'(t)\phi(s)p_1(s)\,\mathrm{d}s\right. \\
& + \frac{1}{\Delta_1}\int_0^1 \left[\psi'(t)\left(\int_0^1 \phi(\tau)\,\mathrm{d}H_2(\tau)\right) - \phi'(t)\left(\tau_1 - \int_0^1 \psi(\tau)\,\mathrm{d}H_2(\tau)\right)\right] \\
& \times \left(\int_0^1 g_1(\tau,s)\,\mathrm{d}H_1(\tau)\right) p_1(s)\,\mathrm{d}s + \frac{1}{\Delta_1}\int_0^1 \left[\psi'(t)\left(\tau_1 - \int_0^1 \phi(\tau)\,\mathrm{d}H_1(\tau)\right)\right. \\
& \left. - \phi'(t)\left(\int_0^1 \psi(\tau)\,\mathrm{d}H_1(\tau)\right)\right]\left(\int_0^1 g_1(\tau,s)\,\mathrm{d}H_2(\tau)\right) p_1(s)\,\mathrm{d}s\Bigg\}.
\end{aligned}
\tag{1.40}
$$

We denote

$$h(t) = -\frac{1}{\tau_1}\int_0^t \phi'(t)\psi(s)p_1(s)\,\mathrm{d}s + \frac{1}{\tau_1}\int_t^1 \psi'(t)\phi(s)p_1(s)\,\mathrm{d}s, \quad t \in (0,1),$$

$$\begin{aligned}\mu(t) = h(t) &+ \frac{1}{\Delta_1}\int_0^1 \left[\psi'(t)\left(\int_0^1 \phi(\tau)\,\mathrm{d}H_2(\tau)\right) - \phi'(t)\left(\tau_1 - \int_0^1 \psi(\tau)\,\mathrm{d}H_2(\tau)\right)\right] \\ &\times \left(\int_0^1 g_1(\tau,s)\,\mathrm{d}H_1(\tau)\right)p_1(s)\,\mathrm{d}s + \frac{1}{\Delta_1}\int_0^1 \left[\psi'(t)\left(\tau_1 - \int_0^1 \phi(\tau)\,\mathrm{d}H_1(\tau)\right)\right. \\ &\left. - \phi'(t)\left(\int_0^1 \psi(\tau)\,\mathrm{d}H_1(\tau)\right)\right]\left(\int_0^1 g_1(\tau,s)\,\mathrm{d}H_2(\tau)\right)p_1(s)\,\mathrm{d}s, \quad t \in (0,1).\end{aligned}$$

For the integral of the function h, by exchanging the order of integration, we obtain

$$\begin{aligned}\int_0^1 h(t)\,\mathrm{d}t &= \frac{1}{\tau_1}\int_0^1\left(\int_0^t (-\phi'(t))\psi(s)p_1(s)\,\mathrm{d}s\right)\mathrm{d}t + \frac{1}{\tau_1}\int_0^1\left(\int_t^1 \psi'(t)\phi(s)p_1(s)\,\mathrm{d}s\right)\mathrm{d}t \\ &= \frac{1}{\tau_1}\int_0^1\left(\int_s^1 (-\phi'(t))\psi(s)p_1(s)\,\mathrm{d}t\right)\mathrm{d}s + \frac{1}{\tau_1}\int_0^1\left(\int_0^s \psi'(t)\phi(s)p_1(s)\,\mathrm{d}t\right)\mathrm{d}s \\ &= \frac{1}{\tau_1}\int_0^1 \psi(s)(\phi(s)-\phi(1))p_1(s)\,\mathrm{d}s + \frac{1}{\tau_1}\int_0^1 \phi(s)(\psi(s)-\psi(0))p_1(s)\,\mathrm{d}s \\ &\le \frac{1}{\tau_1}[\psi(1)(\phi(0)-\phi(1))+\phi(0)(\psi(1)-\psi(0))]\int_0^1 p_1(s)\,\mathrm{d}s = \tilde{M}_0\int_0^1 p_1(s)\,\mathrm{d}s < \infty,\end{aligned}$$

where $\tilde{M}_0 = \frac{1}{\tau_1}[\psi(1)(\phi(0)-\phi(1)) + \phi(0)(\psi(1)-\psi(0))]$.

For the integral of the function μ, we have

$$\begin{aligned}\int_0^1 \mu(t)\,\mathrm{d}t &\le \tilde{M}_0\int_0^1 p_1(s)\,\mathrm{d}s + \frac{1}{\Delta_1}\left[\left(\int_0^1 \psi'(t)\,\mathrm{d}t\right)\left(\int_0^1 \phi(\tau)\,\mathrm{d}H_2(\tau)\right)\right. \\ &\left. - \left(\int_0^1 \phi'(t)\,\mathrm{d}t\right)\left(\tau_1 - \int_0^1 \psi(\tau)\,\mathrm{d}H_2(\tau)\right)\right]\left(\int_0^1\left(\int_0^1 g_1(\tau,s)\,\mathrm{d}H_1(\tau)\right)p_1(s)\,\mathrm{d}s\right) \\ &+ \frac{1}{\Delta_1}\left[\left(\int_0^1 \psi'(t)\,\mathrm{d}t\right)\left(\tau_1 - \int_0^1 \phi(\tau)\,\mathrm{d}H_1(\tau)\right) - \left(\int_0^1 \phi'(t)\,\mathrm{d}t\right)\left(\int_0^1 \psi(\tau)\,\mathrm{d}H_1(\tau)\right)\right] \\ &\times \left(\int_0^1\left(\int_0^1 g_1(\tau,s)\,\mathrm{d}H_2(\tau)\right)p_1(s)\,\mathrm{d}s\right) \\ &\le \tilde{M}_0\int_0^1 p_1(s)\,\mathrm{d}s + \frac{1}{\Delta_1}\left[(\psi(1)-\psi(0))\left(\int_0^1 \phi(\tau)\,\mathrm{d}H_2(\tau)\right)\left(\int_0^1 \mathrm{d}H_1(\tau)\right)\right. \\ &+ (\phi(0)-\phi(1))\left(\tau_1 - \int_0^1 \psi(\tau)\,\mathrm{d}H_2(\tau)\right)\left(\int_0^1 \mathrm{d}H_1(\tau)\right) \\ &+ (\psi(1)-\psi(0))\left(\tau_1 - \int_0^1 \phi(\tau)\,\mathrm{d}H_1(\tau)\right)\left(\int_0^1 \mathrm{d}H_2(\tau)\right) \\ &\left. + (\phi(0)-\phi(1))\left(\int_0^1 \psi(\tau)\,\mathrm{d}H_1(\tau)\right)\left(\int_0^1 \mathrm{d}H_2(\tau)\right)\right]\left(\int_0^1 g_1(s,s)p_1(s)\,\mathrm{d}s\right)\end{aligned}$$

$$
\begin{aligned}
&\le \tilde{M}_0 \int_0^1 p_1(s)\,\mathrm{d}s + \frac{1}{\tau_1\Delta_1}\phi(0)\psi(1)\Bigg[(\psi(1)-\psi(0))(H_1(1)-H_1(0))\left(\int_0^1 \phi(\tau)\,\mathrm{d}H_2(\tau)\right)\\
&+(\phi(0)-\phi(1))(H_1(1)-H_1(0))\left(\tau_1-\int_0^1 \psi(\tau)\,\mathrm{d}H_2(\tau)\right)\\
&+(\psi(1)-\psi(0))(H_2(1)-H_2(0))\left(\tau_1-\int_0^1 \phi(\tau)\,\mathrm{d}H_1(\tau)\right)\\
&+(\phi(0)-\phi(1))(H_2(1)-H_2(0))\left(\int_0^1 \psi(\tau)\,\mathrm{d}H_1(\tau)\right)\Bigg]\int_0^1 p_1(s)\,\mathrm{d}s<\infty.
\end{aligned}
\tag{1.41}
$$

We deduce that $\mu \in L^1(0,1)$. Thus, for any given $t_1, t_2 \in [0,1]$ with $t_1 \le t_2$ and $u \in E$, by (1.40), we obtain

$$
|(\mathcal{D}u)(t_1)-(\mathcal{D}u)(t_2)| = \left|\int_{t_1}^{t_2} (\mathcal{D}u)'(t)\,\mathrm{d}t\right| \le M_3\int_{t_1}^{t_2} \mu(t)\,\mathrm{d}t. \tag{1.42}
$$

From (1.41), (1.42), and the absolute continuity of the integral function, we find that $\mathcal{D}(E)$ is equicontinuous. This conclusion, together with (1.39) and the Ascoli–Arzelà theorem, yields that $\mathcal{D}(E)$ is relatively compact. Therefore, $\mathcal{D}$ is a compact operator.

By using arguments similar to those used in the proof of Lemma 2.4 from Liu et al. (2007), we show that $\mathcal{D}$ is continuous. Suppose that $u_p, u \in E$ for $p \in \mathbb{N}$, and $\|u_p - u\| \to 0$ as $p \to \infty$. Then there exists $M_4 > 0$ such that $\|u_p\| \le M_4$ and $\|u\| \le M_4$. From the first part of this proof we know that $\{\mathcal{D}u_p, p \in \mathbb{N}\}$ is relatively compact. We shall prove that $\|\mathcal{D}u_p - \mathcal{D}u\| \to 0$ as $p \to \infty$. If we suppose that this is not true, then there exists $\varepsilon_0 > 0$ and a subsequence $(u_{p_k})_k \subset (u_k)_k$ such that $\|\mathcal{D}u_{p_k} - \mathcal{D}u\| \ge \varepsilon_0, k = 1,2,\ldots$. Since $\{\mathcal{D}u_{p_k}, k = 1,2,\ldots\}$ is relatively compact, there exists a subsequence of $(\mathcal{D}u_{p_k})_k$ which converges in P' to some $u^* \in P'$. Without loss of generality, we assume that $(\mathcal{D}u_{p_k})_k$ itself converges to u^*—that is, $\lim_{k\to\infty} \|\mathcal{D}u_{p_k} - u^*\| = 0$. From the above relation, we deduce that $(\mathcal{D}u_{p_k})(t) \to u^*(t)$, as $k \to \infty$ for all $t \in [0,1]$. By (L5) and Lemma 1.1.6, we obtain

$$
G_2(s,\tau)g(\tau,u_{p_k}(\tau)) \le J_2(\tau)p_2(\tau)q_2(u_{p_k}(\tau)) \le M_5J_2(\tau)p_2(\tau)
$$

for all $s,\tau \in [0,1]$, where $M_5 = \sup_{x\in[0,M_4]} q_2(x) < \infty$. Therefore, we obtain

$$
\begin{aligned}
G_1(t,s)f\left(s,\int_0^1 G_2(s,\tau)g(\tau,u_{p_k}(\tau))\,\mathrm{d}\tau\right) &\le J_1(s)p_1(s)q_1\left(\int_0^1 G_2(s,\tau)g(\tau,u_{p_k}(\tau))\,\mathrm{d}\tau\right)\\
&\le M_6J_1(s)p_1(s),
\end{aligned}
\tag{1.43}
$$

where $M_6 = \sup_{x\in[0,\beta_0M_5]} q_1(x)$.

By (L5), (1.43), and Lebesgue's dominated convergence theorem, we obtain

$$
u^*(t) = \lim_{k\to\infty} (\mathcal{D}u_{p_k})(t) = (\mathcal{D}u)(t), \quad \forall t \in [0,1];
$$

that is, $u^* = \mathcal{D}u$. This relation contradicts the inequality $\|\mathcal{D}u_{p_k} - u^*\| \ge \varepsilon_0, k = 1,2,\ldots$. Therefore, $\mathcal{D}$ is continuous in u, and in general on P'. Lemma 1.4.1 is completely proved. □

For $\sigma \in (0, 1/2)$ we define again the cone

$$P_0 = \left\{ u \in X, \quad u(t) \geq 0, \quad \forall t \in [0,1], \quad \inf_{t \in [\sigma, 1-\sigma]} u(t) \geq \nu \|u\| \right\} \subset P',$$

where $\nu = \min\{\nu_1, \nu_2\}$, and ν_1 and ν_2 are defined in Section 1.1.2 (Lemma 1.1.4). Under assumptions (L1)–(L5), we have $\mathcal{D}(P') \subset P_0$. For $u \in P'$, let $v = \mathcal{D}(u)$. By Lemma 1.1.7, we have $\inf_{t \in [\sigma, 1-\sigma]} v(t) \geq \nu_1 \|v\| \geq \nu \|v\|$—that is, $v \in P_0$.

Theorem 1.4.1. *Assume that (L1)–(L5) hold. If the functions f and g also satisfy the following conditions (L6) and (L7), then problem* (S′)–(BC) *has at least one positive solution $(u(t), v(t)), t \in [0,1]$:*

(L6) There exist $r_1, r_2 \in (0, \infty)$ with $r_1 r_2 \geq 1$ such that

$$(1)\ q_{10}^{\rm s} = \limsup_{x \to 0^+} \frac{q_1(x)}{x^{r_1}} \in [0, \infty) \quad \textit{and} \quad (2)\ q_{20}^{\rm s} = \limsup_{x \to 0^+} \frac{q_2(x)}{x^{r_2}} = 0.$$

(L7) There exist $l_1, l_2 \in (0, \infty)$ with $l_1 l_2 \geq 1$ and $\sigma \in (0, 1/2)$ such that

$$(1)\ \hat{f}_\infty^{\rm i} = \liminf_{x \to \infty} \inf_{t \in [\sigma, 1-\sigma]} \frac{f(t,x)}{x^{l_1}} \in (0, \infty] \quad \textit{and} \quad (2)\ \hat{g}_\infty^{\rm i} = \liminf_{x \to \infty} \inf_{t \in [\sigma, 1-\sigma]} \frac{g(t,x)}{x^{l_2}} = \infty.$$

Proof. We consider the cone P_0 with σ given in (L7). From (L6) (1) and (L5), we deduce that there exists $C_1 > 0$ such that

$$q_1(x) \leq C_1 x^{r_1}, \quad \forall x \in [0,1]. \tag{1.44}$$

By using (L6) (2) and (L5), for $C_2 = \min\{(1/(C_1 \alpha_0 \beta_0^{r_1}))^{1/r_1}, 1/\beta_0\} > 0$ with α_0, β_0 defined in the proof of Lemma 1.4.1, we conclude that there exists $\delta_1 \in (0,1)$ such that

$$q_2(x) \leq C_2 x^{r_2}, \quad \forall x \in [0, \delta_1]. \tag{1.45}$$

From (1.45), (L5), and Lemma 1.1.6, for any $u \in \partial B_{\delta_1} \cap P_0$ and $s \in [0,1]$ we obtain

$$\int_0^1 G_2(s,\tau) g(\tau, u(\tau))\, d\tau \leq C_2 \int_0^1 J_2(\tau) p_2(\tau)\, d\tau \cdot \|u\|^{r_2} = C_2 \beta_0 \delta_1^{r_2} \leq \delta_1^{r_2} < 1. \tag{1.46}$$

Now by using (1.44)–(1.46) and (L5), for any $u \in \partial B_{\delta_1} \cap P_0$ and $t \in [0,1]$, we deduce

$$\begin{aligned}
(\mathcal{D}u)(t) &\leq C_1 \int_0^1 G_1(t,s) p_1(s) \left(\int_0^1 G_2(s,\tau) g(\tau, u(\tau))\, d\tau \right)^{r_1} ds \\
&\leq C_1 \int_0^1 G_1(t,s) p_1(s) \left(C_2 \int_0^1 G_2(s,\tau) p_2(\tau) (u(\tau))^{r_2}\, d\tau \right)^{r_1} ds \\
&\leq C_1 \int_0^1 J_1(s) p_1(s)\, ds \cdot \left(C_2 \int_0^1 J_2(\tau) p_2(\tau)\, d\tau \right)^{r_1} \cdot \|u\|^{r_1 r_2} \leq \|u\|.
\end{aligned}$$

Therefore,

$$\|\mathcal{D}u\| \le \|u\|, \quad \forall u \in \partial B_{\delta_1} \cap P_0. \tag{1.47}$$

From (L7) (1), we conclude that there exist $C_3 > 0$ and $x_1 > 0$ such that

$$f(t,x) \ge C_3 x^{l_1}, \quad \forall x \ge x_1, \ \forall t \in [\sigma, 1-\sigma]. \tag{1.48}$$

We consider now $C_4 = \max\{(\nu_2 \nu^{l_2}\theta_2)^{-1}, (C_3 \nu_1 \nu_2^{l_1} \nu^{l_1 l_2} \theta_1 \theta_2^{l_1})^{-1/l_1}\} > 0$, where $\theta_1 = \int_\sigma^{1-\sigma} J_1(s)\,ds > 0$ and $\theta_2 = \int_\sigma^{1-\sigma} J_2(s)\,ds > 0$. From (L7) (2), we deduce that there exists $x_2 \ge 1$ such that

$$g(t,x) \ge C_4 x^{l_2}, \quad \forall x \ge x_2, \ \forall t \in [\sigma, 1-\sigma]. \tag{1.49}$$

We choose $R_0 = \max\{x_1, x_2\}$ and $R > \max\{R_0/\nu, R_0^{1/l_2}\}$. Then for any $u \in \partial B_R \cap P_0$, we have $\inf_{t\in[\sigma,1-\sigma]} u(t) \ge \nu\|u\| = \nu R > R_0$.

By using (1.48) and (1.49), for any $u \in \partial B_R \cap P_0$ and $s \in [\sigma, 1-\sigma]$, we obtain

$$\begin{aligned}\int_0^1 G_2(s,\tau) g(\tau, u(\tau))\,d\tau &\ge \nu_2 C_4 \int_\sigma^{1-\sigma} J_2(\tau)(u(\tau))^{l_2}\,d\tau \\ &\ge \nu_2 C_4 \nu^{l_2} \int_\sigma^{1-\sigma} J_2(\tau)\,d\tau \cdot \|u\|^{l_2} \ge \|u\|^{l_2} = R^{l_2} > R_0.\end{aligned}$$

Then for any $u \in \partial B_R \cap P_0$ and $t \in [\sigma, 1-\sigma]$, we have

$$\begin{aligned}(\mathcal{D}u)(t) &\ge \int_\sigma^{1-\sigma} G_1(t,s) f\left(s, \int_0^1 G_2(s,\tau) g(\tau, u(\tau))\,d\tau\right) ds \\ &\ge C_3 \int_\sigma^{1-\sigma} G_1(t,s)\left(\nu_2 \int_\sigma^{1-\sigma} J_2(\tau) C_4 (u(\tau))^{l_2}\,d\tau\right)^{l_1} ds \\ &\ge C_3 C_4^{l_1} \nu_2^{l_1} \int_\sigma^{1-\sigma} G_1(t,s) \nu^{l_1 l_2} \|u\|^{l_1 l_2} \left(\int_\sigma^{1-\sigma} J_2(\tau)\,d\tau\right)^{l_1} ds \\ &\ge C_3 C_4^{l_1} \nu_2^{l_1} \nu_1 \nu^{l_1 l_2} \left(\int_\sigma^{1-\sigma} J_1(s)\,ds\right)\left(\int_\sigma^{1-\sigma} J_2(\tau)\,d\tau\right)^{l_1} \|u\|^{l_1 l_2} \ge \|u\|.\end{aligned}$$

Therefore, we obtain

$$\|\mathcal{D}u\| \ge \|u\|, \quad \forall u \in \partial B_R \cap P_0. \tag{1.50}$$

By (1.47), (1.50), Lemma 1.4.1, and Theorem 1.1.1 (1), we conclude that $\mathcal{D}$ has a fixed point $u_1 \in (\bar{B}_R \setminus B_{\delta_1}) \cap P_0$—that is, $\delta_1 \le \|u_1\| \le R$. Let $v_1(t) = \int_0^1 G_2(t,s) g(s, u_1(s))\,ds$. Then $(u_1, v_1) \in P_0 \times P_0$ is a solution of (S′)–(BC). In addition, $\|v_1\| > 0$. If we suppose that $v_1(t) = 0$ for all $t \in [0,1]$, then by using (L5), we have $f(s, v_1(s)) = f(s,0) = 0$ for all $s \in [0,1]$. This implies $u_1(t) = 0$ for all $t \in [0,1]$, which contradicts $\|u_1\| > 0$. The proof of Theorem 1.4.1 is completed. □

Theorem 1.4.2. *Assume that (L1)–(L5) hold. If the functions f and g also satisfy the following conditions (L8) and (L9), then problem* (S′)–(BC) *has at least one positive solution $(u(t), v(t)), t \in [0,1]$:*

(L8) There exist $\alpha_1, \alpha_2 \in (0,\infty)$ with $\alpha_1\alpha_2 \le 1$ such that

$$(1)\ q_{1\infty}^{s} = \limsup_{x\to\infty} \frac{q_1(x)}{x^{\alpha_1}} \in [0,\infty) \quad and \quad (2)\ q_{2\infty}^{s} = \limsup_{x\to\infty} \frac{q_2(x)}{x^{\alpha_2}} = 0.$$

(L9) There exist $\beta_1, \beta_2 \in (0,\infty)$ with $\beta_1\beta_2 \le 1$ and $\sigma \in (0, 1/2)$ such that

$$(1)\ \hat{f}_0^{i} = \liminf_{x\to 0+} \inf_{t\in[\sigma,1-\sigma]} \frac{f(t,x)}{x^{\beta_1}} \in (0,\infty] \quad and \quad (2)\ \hat{g}_0^{i} = \liminf_{x\to 0+} \inf_{t\in[\sigma,1-\sigma]} \frac{g(t,x)}{x^{\beta_2}} = \infty.$$

Proof. We consider the cone P_0 with σ given in (L9). By (L8) (1), we deduce that there exist $C_5 > 0$ and $C_6 > 0$ such that

$$q_1(x) \le C_5 x^{\alpha_1} + C_6, \quad \forall x \in [0,\infty). \tag{1.51}$$

From (L8) (2), for $\varepsilon_0 > 0$, $\varepsilon_0 < (2^{\alpha_1} C_5 \alpha_0 \beta_0^{\alpha_1})^{-1/\alpha_1}$, we conclude that there exists $C_7 > 0$ such that

$$q_2(x) \le \varepsilon_0 x^{\alpha_2} + C_7, \quad \forall x \in [0,\infty). \tag{1.52}$$

By using (1.51), (1.52), and (L5), for any $u \in P_0$, we obtain

$$\begin{aligned}
(\mathcal{D}u)(t) &\le \int_0^1 G_1(t,s)p_1(s)q_1\left(\int_0^1 G_2(s,\tau)g(\tau,u(\tau))\,d\tau\right) ds\\
&\le C_5 \int_0^1 G_1(t,s)p_1(s)\left(\int_0^1 G_2(s,\tau)g(\tau,u(\tau))\,d\tau\right)^{\alpha_1} ds + C_6\int_0^1 J_1(s)p_1(s)\,ds\\
&\le C_5 \int_0^1 J_1(s)p_1(s)\,ds \left(\int_0^1 J_2(\tau)p_2(\tau)\,d\tau\right)^{\alpha_1} (\varepsilon_0\|u\|^{\alpha_2} + C_7)^{\alpha_1} + \alpha_0 C_6\\
&\le C_5 2^{\alpha_1}\varepsilon_0^{\alpha_1}\alpha_0\beta_0^{\alpha_1}\|u\|^{\alpha_1\alpha_2} + C_5 2^{\alpha_1}\alpha_0\beta_0^{\alpha_1}C_7^{\alpha_1} + \alpha_0 C_6, \quad \forall t \in [0,1].
\end{aligned}$$

By the definition of ε_0, we can choose sufficiently large $R_1 > 0$ such that

$$\|\mathcal{D}u\| \le \|u\|, \quad \forall u \in \partial B_{R_1} \cap P_0. \tag{1.53}$$

From (L9) (1), we deduce that there exist positive constants $C_8 > 0$ and $x_3 > 0$ such that $f(t,x) \ge C_8 x^{\beta_1}$ for all $x \in [0, x_3]$ and $t \in [\sigma, 1-\sigma]$. From (L9) (2), for $\varepsilon_1 = (C_8 \nu_1 \nu_2^{\beta_1} \nu^{\beta_1\beta_2}\theta_1\theta_2^{\beta_1})^{-1/\beta_1} > 0$, we conclude that there exists $x_4 > 0$ such that $g(t,x) \ge \varepsilon_1 x^{\beta_2}$ for all $x \in [0, x_4]$ and $t \in [\sigma, 1-\sigma]$.
We consider $x_5 = \min\{x_3, x_4\}$. So we obtain

$$f(t,x) \ge C_8 x^{\beta_1}, \quad g(t,x) \ge \varepsilon_1 x^{\beta_2}, \quad \forall (t,x) \in [\sigma, 1-\sigma] \times [0, x_5]. \tag{1.54}$$

From the assumption $q_2(0) = 0$ and the continuity of q_2, we deduce that there exists sufficiently small $\varepsilon_2 \in (0, \min\{x_5, 1\})$ such that $q_2(x) \le \beta_0^{-1} x_5$ for all $x \in [0, \varepsilon_2]$.

Therefore, for any $u \in \partial B_{\varepsilon_2} \cap P_0$ and $s \in [0,1]$ we have

$$\int_0^1 G_2(s,\tau)g(\tau,u(\tau))\,d\tau \le \beta_0^{-1} x_5 \int_0^1 J_2(\tau)p_2(\tau)\,d\tau = x_5. \tag{1.55}$$

By (1.54), (1.55), Lemma 1.1.6, and Lemma 1.1.7, for any $t \in [\sigma, 1-\sigma]$ we obtain

$$\begin{aligned}(\mathcal{D}u)(t) &\geq C_8 \int_\sigma^{1-\sigma} G_1(t,s) \left(\int_\sigma^{1-\sigma} G_2(s,\tau) g(\tau, u(\tau))\, d\tau \right)^{\beta_1} ds \\ &\geq C_8 \nu_1 \int_\sigma^{1-\sigma} J_1(s) \left[(\varepsilon_1 \nu_2)^{\beta_1} \left(\int_\sigma^{1-\sigma} J_2(\tau)(u(\tau))^{\beta_2}\, d\tau \right)^{\beta_1} \right] ds \\ &\geq C_8 \nu_1 \nu_2^{\beta_1} \varepsilon_1^{\beta_1} \nu^{\beta_1\beta_2} \theta_1 \theta_2^{\beta_1} \|u\|^{\beta_1\beta_2} \geq \|u\|.\end{aligned}$$

Therefore,

$$\|\mathcal{D}u\| \geq \|u\|, \quad \forall u \in \partial B_{\varepsilon_2} \cap P_0. \tag{1.56}$$

By (1.53), (1.56), Lemma 1.4.1, and Theorem 1.1.1 (2), we deduce that $\mathcal{D}$ has at least one fixed point $u_2 \in (\bar{B}_{R_1} \setminus B_{\varepsilon_2}) \cap P_0$. Then our problem (S′)–(BC) has at least one positive solution $(u_2, v_2) \in P_0 \times P_0$, where $v_2(t) = \int_0^1 G_2(t,s) g(s, u_2(s))\, ds$. This completes the proof of Theorem 1.4.2. □

1.4.3 Examples

In this section, we shall present two examples which illustrate our results above.

Example 1.4.1. Let

$$f(t,x) = \frac{x^a}{t^{\gamma_1}(1-t)^{\delta_1}}, \quad g(t,x) = \frac{x^b}{t^{\gamma_2}(1-t)^{\delta_2}}, \quad \forall t \in (0,1), \ x \in [0,\infty),$$

with $a, b > 1$ and $\gamma_1, \delta_1, \gamma_2, \delta_2 \in (0,1)$. Here $f(t,x) = p_1(t) q_1(x)$ and $g(t,x) = p_2(t) q_2(x)$, where

$$p_1(t) = \frac{1}{t^{\gamma_1}(1-t)^{\delta_1}}, \quad p_2(t) = \frac{1}{t^{\gamma_2}(1-t)^{\delta_2}}, \quad q_1(x) = x^a, \quad q_2(x) = x^b.$$

We have $0 < \int_0^1 p_1(s)\, ds < \infty$ and $0 < \int_0^1 p_2(s)\, ds < \infty$.
In (L6), for $r_1 < a$, $r_2 < b$, and $r_1 r_2 \geq 1$ we have

$$\limsup_{x\to 0^+} \frac{q_1(x)}{x^{r_1}} = \lim_{x\to 0^+} x^{a-r_1} = 0, \quad \limsup_{x\to 0^+} \frac{q_2(x)}{x^{r_2}} = \lim_{x\to 0^+} x^{b-r_2} = 0.$$

In (L7), for $l_1 < a$, $l_2 < b$, $l_1 l_2 \geq 1$, and $\sigma \in \left(0, \frac{1}{2}\right)$ we have

$$\begin{aligned}\liminf_{x\to\infty} \inf_{t\in[\sigma,1-\sigma]} \frac{f(t,x)}{x^{l_1}} &= \liminf_{x\to\infty} \inf_{t\in[\sigma,1-\sigma]} \frac{x^{a-l_1}}{t^{\gamma_1}(1-t)^{\delta_1}} \\ &= \left(\max\left\{ \frac{\gamma_1^{\gamma_1}\delta_1^{\delta_1}}{(\gamma_1+\delta_1)^{\gamma_1+\delta_1}}, \sigma^{\gamma_1}(1-\sigma)^{\delta_1}, \sigma^{\delta_1}(1-\sigma)^{\gamma_1} \right\} \right)^{-1} \\ &\quad \times \lim_{x\to\infty} x^{a-l_1} = \infty.\end{aligned}$$

In a similar manner, we have

$$\liminf_{x\to\infty} \inf_{t\in[\sigma,1-\sigma]} \frac{g(t,x)}{x^{l_2}} = \infty.$$

For example, if $a = 2$, $b = 3/2$, $r_1 = 1$, $r_2 = 4/3$, $l_1 = 3/2$, and $l_2 = 1$, the above conditions are satisfied. Under assumptions (L1)–(L4), by Theorem 1.4.1, we deduce that problem (S′)–(BC) has at least one positive solution.

Example 1.4.2. Let

$$f(t,x) = \frac{x^a(2+\cos x)}{t^{\gamma_1}}, \quad g(t,x) = \frac{x^b(1+\sin x)}{(1-t)^{\delta_1}}, \quad \forall t \in (0,1), \ x \in [0,\infty),$$

with $a, b \in (0,1)$ and $\gamma_1, \delta_1 \in (0,1)$. Here $f(t,x) = p_1(t)q_1(x)$ and $g(t,x) = p_2(t)q_2(x)$, where

$$p_1(t) = \frac{1}{t^{\gamma_1}}, \quad p_2(t) = \frac{1}{(1-t)^{\delta_1}}, \quad q_1(x) = x^a(2+\cos x), \quad q_2(x) = x^b(1+\sin x).$$

We have $0 < \int_0^1 p_1(s)\,ds < \infty$ and $0 < \int_0^1 p_2(s)\,ds < \infty$.
In (L8), for $\alpha_1 = a$, $\alpha_2 > b$, and $\alpha_1\alpha_2 \le 1$ we have

$$\limsup_{x\to\infty} \frac{q_1(x)}{x^{\alpha_1}} = \limsup_{x\to\infty} \frac{x^a(2+\cos x)}{x^{\alpha_1}} = 3, \quad \limsup_{x\to\infty} \frac{q_2(x)}{x^{\alpha_2}} = \limsup_{x\to\infty} \frac{x^b(1+\sin x)}{x^{\alpha_2}} = 0.$$

In (L9), for $\beta_1 = a$, $\beta_2 > b$, $\beta_1\beta_2 \le 1$, and $\sigma \in \left(0, \frac{1}{2}\right)$ we have

$$\liminf_{x\to0^+} \inf_{t\in[\sigma,1-\sigma]} \frac{f(t,x)}{x^{\beta_1}} = \liminf_{x\to0^+} \inf_{t\in[\sigma,1-\sigma]} \frac{x^a(2+\cos x)}{t^{\gamma_1}x^{\beta_1}} = \frac{3}{(1-\sigma)^{\gamma_1}} > 0,$$

$$\liminf_{x\to0^+} \inf_{t\in[\sigma,1-\sigma]} \frac{g(t,x)}{x^{\beta_2}} = \liminf_{x\to0^+} \inf_{t\in[\sigma,1-\sigma]} \frac{x^b(1+\sin x)}{(1-t)^{\delta_1}x^{\beta_2}} = \frac{1}{(1-\sigma)^{\delta_1}} \lim_{x\to0^+} x^{b-\beta_2} = \infty.$$

For example, if $a = 1/3$, $b = 1/2$, $\alpha_1 = 1/3$, $\alpha_2 = 1$, $\beta_1 = 1/3$, and $\beta_2 = 1$, the above conditions are satisfied. Under assumptions (L1)–(L4), by Theorem 1.4.2, we deduce that problem (S′)–(BC) has at least one positive solution.

1.5 Remarks on some particular cases

In this section, we present briefly some existence and multiplicity results for the positive solutions for some particular cases of problem (S)–(BC) from Section 1.1 and of problem (S′)–(BC) from Sections 1.3 and 1.4.

1.5.1 Systems with parameters

We investigate in this section problem (S)–(BC) from Section 1.1 with $a(t) = c(t) = 1$, $b(t) = d(t) = 0$ for all $t \in [0,1]$, $\gamma = \tilde{\gamma} = 1$, $\delta = \tilde{\delta} = 0$, H_1 and K_1 are constant functions, and H_2 and K_2 are step functions—that is, the integral boundary conditions become multipoint boundary conditions.

Namely, we consider here the system of nonlinear second-order ordinary differential equations

$$\begin{cases} u''(t) + \lambda c(t) f(u(t), v(t)) = 0, & t \in (0,T), \\ v''(t) + \mu d(t) g(u(t), v(t)) = 0, & t \in (0,T), \end{cases} \tag{S_0}$$

with the multipoint boundary conditions

$$\begin{cases} \alpha u(0) - \beta u'(0) = 0, \quad u(T) = \sum_{i=1}^{m} a_i u(\xi_i), \\ \gamma v(0) - \delta v'(0) = 0, \quad v(T) = \sum_{i=1}^{n} b_i v(\eta_i), \end{cases} \tag{BC_0}$$

where $m, n \in \mathbb{N}$, $\alpha, \beta, \gamma, \delta, a_i, \xi_i \in \mathbb{R}$ for all $i = 1, \ldots, m$; $b_i, \eta_i \in \mathbb{R}$ for all $i = 1, \ldots, n$; $0 < \xi_1 < \cdots < \xi_m < T$, $0 < \eta_1 < \cdots < \eta_n < T$.

By a positive solution of problem (S_0)–(BC_0) we mean a pair of functions $(u, v) \in (C^2([0, T]))^2$ satisfying (S_0) and (BC_0) with $u(t) \geq 0$, $v(t) \geq 0$ for all $t \in [0, T]$, and $u(t) > 0$ for all $t \in (0, T]$ or $v(t) > 0$ for all $t \in (0, T]$.

First, we consider the following second-order differential equation with $m+2$-point boundary conditions:

$$u''(t) + y(t) = 0, \quad t \in (0, T), \tag{1.57}$$

$$\alpha u(0) - \beta u'(0) = 0, \quad u(T) = \sum_{i=1}^{m} a_i u(\xi_i), \tag{1.58}$$

where $m \in \mathbb{N}$, $\alpha, \beta, a_i, \xi_i \in \mathbb{R}$ for all $i = 1, \ldots, m$, $0 < \xi_1 < \cdots < \xi_m < T$.

Lemma 1.5.1 (Li and Sun, 2006; Luca, 2011). *If* $\alpha, \beta, a_i, \xi_i \in \mathbb{R}$ *for all* $i = 1, \ldots, m$, $0 < \xi_1 < \cdots < \xi_m < T$, $d = \alpha\left(T - \sum_{i=1}^{m} a_i \xi_i\right) + \beta\left(1 - \sum_{i=1}^{m} a_i\right) \neq 0$, *and* $y \in C(0, T) \cap L^1(0, T)$, *then the unique solution of problem* (1.57)–(1.58) *is given by*

$$u(t) = \frac{\alpha t + \beta}{d} \int_0^T (T - s) y(s)\, ds - \frac{\alpha t + \beta}{d} \sum_{i=1}^{m} a_i \int_0^{\xi_i} (\xi_i - s) y(s)\, ds - \int_0^t (t - s) y(s)\, ds, \quad 0 \leq t \leq T.$$

Lemma 1.5.2 (Luca, 2011). *Under the assumptions of Lemma 1.5.1, the Green's function for the boundary value problem* (1.57)–(1.58) *is given by*

$$G_1(t, s) = \begin{cases} \dfrac{\alpha t + \beta}{d} \left[(T - s) - \sum_{j=i}^{m} a_j (\xi_j - s) \right] - (t - s), & \text{if } \xi_{i-1} \leq s < \xi_i, \quad s \leq t, \\ & i = 1, \ldots, m, \ (\xi_0 = 0), \\ \dfrac{\alpha t + \beta}{d} \left[(T - s) - \sum_{j=i}^{m} a_j (\xi_j - s) \right], & \text{if } \xi_{i-1} \leq s < \xi_i, \quad s \geq t, \\ & i = 1, \ldots, m, \\ \dfrac{\alpha t + \beta}{d} (T - s) - (t - s), & \text{if } \xi_m \leq s \leq T, \quad s \leq t, \\ \dfrac{\alpha t + \beta}{d} (T - s), & \text{if } \xi_m \leq s \leq T, \quad s \geq t. \end{cases}$$

Using the above Green's function, we can express the solution of problem (1.57)–(1.58) as $u(t) = \int_0^T G_1(t,s)y(s)\,ds$.

Lemma 1.5.3 (Li and Sun, 2006; Luca, 2011). *If $\alpha \geq 0, \beta \geq 0$, $a_i \geq 0$ for all $i = 1,\dots,m$, $0 < \xi_1 < \cdots < \xi_m < T$, $T \geq \sum_{i=1}^m a_i\xi_i$, $d > 0$, and $y \in C(0,T) \cap L^1(0,T)$, $y(t) \geq 0$ for all $t \in (0,T)$, then the solution u of problem* (1.57)–(1.58) *satisfies $u(t) \geq 0$ for all $t \in [0,T]$.*

Lemma 1.5.4 (Henderson and Luca, 2012e). *If $\alpha \geq 0, \beta \geq 0$, $a_i \geq 0$ for all $i = 1,\dots,m$, $0 < \xi_1 < \cdots < \xi_m < T$, $T \geq \sum_{i=1}^m a_i\xi_i$, $d > 0$, and $y \in C(0,T) \cap L^1(0,T)$, $y(t) \geq 0$ for all $t \in (0,T)$, then the solution of problem* (1.57)–(1.58) *satisfies*

$$u(t) \leq \frac{\alpha T + \beta}{d} \int_0^T (T-s)y(s)\,ds, \quad 0 \leq t \leq T,$$

$$u(\xi_j) \geq \frac{\alpha\xi_j + \beta}{d} \int_{\xi_m}^T (T-s)y(s)\,ds, \quad \forall j = 1,\dots,m.$$

Lemma 1.5.5 (Li and Sun, 2006). *Assume that $\alpha \geq 0, \beta \geq 0$, $a_i \geq 0$ for all $i = 1,\dots,m$, $0 < \xi_1 < \cdots < \xi_m < T$, $T > \sum_{i=1}^m a_i\xi_i > 0$, $d > 0$, and $y \in C(0,T) \cap L^1(0,T)$, $y(t) \geq 0$ for all $t \in (0,T)$. Then the solution of problem* (1.57)–(1.58) *satisfies* $\inf_{t\in[\xi_1,T]} u(t) \geq r_1 \sup_{t'\in[0,T]} u(t')$, *where*

$$r_1 = \min\left\{\frac{\xi_1}{T}, \frac{\sum_{i=1}^m a_i(T-\xi_i)}{T - \sum_{i=1}^m a_i\xi_i}, \frac{\sum_{i=1}^m a_i\xi_i}{T}, \frac{\sum_{i=1}^{s-1} a_i\xi_i + \sum_{i=s}^m a_i(T-\xi_i)}{T - \sum_{i=s}^m a_i\xi_i}, \quad s = 2,\dots,m\right\}.$$

We can also formulate results similar to those of Lemmas 1.5.1–1.5.5 for the boundary value problem

$$v''(t) + h(t) = 0, \quad t \in (0,T), \tag{1.59}$$

$$\gamma v(0) - \delta v'(0) = 0, \quad v(T) = \sum_{i=1}^n b_i v(\eta_i), \tag{1.60}$$

where $n \in \mathbb{N}$, $\gamma, \delta, b_i, \eta_i \in \mathbb{R}$ for all $i = 1,\dots,n$, $0 < \eta_1 < \cdots < \eta_n < T$.

If $e = \gamma\left(T - \sum_{i=1}^n b_i\eta_i\right) + \delta\left(1 - \sum_{i=1}^n b_i\right) \neq 0$, we denote by G_2 the Green's function corresponding to problem (1.59)–(1.60)—that is,

$$G_2(t,s) = \begin{cases} \dfrac{\gamma t + \delta}{e}\left[(T-s) - \displaystyle\sum_{j=i}^n b_j(\eta_j - s)\right] - (t-s), & \text{if } \eta_{i-1} \leq s < \eta_i, \quad s \leq t, \\ & i = 1,\dots,n, \ (\eta_0 = 0), \\ \dfrac{\gamma t + \delta}{e}\left[(T-s) - \displaystyle\sum_{j=i}^n b_j(\eta_j - s)\right], & \text{if } \eta_{i-1} \leq s < \eta_i, \quad s \geq t, \\ & i = 1,\dots,n, \\ \dfrac{\gamma t + \delta}{e}(T-s) - (t-s), & \text{if } \eta_n \leq s \leq T, \quad s \leq t, \\ \dfrac{\gamma t + \delta}{e}(T-s), & \text{if } \eta_n \leq s \leq T, \quad s \geq t. \end{cases}$$

Under assumptions similar to those from Lemma 1.5.5, we have the inequality $\inf_{t\in[\eta_1,T]} v(t) \geq r_2 \sup_{t'\in[0,T]} v(t')$, where v is the solution of problem (1.59)–(1.60) and r_2 is given by

$$r_2 = \min\left\{\frac{\eta_1}{T}, \frac{\sum_{i=1}^n b_i(T-\eta_i)}{T-\sum_{i=1}^n b_i\eta_i}, \frac{\sum_{i=1}^n b_i\eta_i}{T}, \frac{\sum_{i=1}^{s-1} b_i\eta_i + \sum_{i=s}^n b_i(T-\eta_i)}{T-\sum_{i=s}^n b_i\eta_i}, \quad s=2,\ldots,n\right\}.$$

We present the assumptions that we shall use in the sequel:

($\tilde{\mathrm{I}}1$) $\alpha \geq 0$, $\beta \geq 0$, $\gamma \geq 0$, $\delta \geq 0$, $a_i \geq 0, i = 1,\ldots,m$, $b_i \geq 0, i = 1,\ldots,n$, $0 < \xi_1 < \cdots < \xi_m < T$, $0 < \eta_1 < \cdots < \eta_n < T$, $T > \sum_{i=1}^m a_i\xi_i > 0$, $T > \sum_{i=1}^n b_i\eta_i > 0$, $d = \alpha\left(T - \sum_{i=1}^m a_i\xi_i\right) + \beta\left(1 - \sum_{i=1}^m a_i\right) > 0$, $e = \gamma\left(T - \sum_{i=1}^n b_i\eta_i\right) + \delta\left(1 - \sum_{i=1}^n b_i\right) > 0$, $\xi_m \geq \eta_1$, $\eta_n \geq \xi_1$.

($\tilde{\mathrm{I}}2$) The functions $c, d : [0,T] \to [0,\infty)$ are continuous, and there exists $t_1 \in [\xi_m, T]$ and $t_2 \in [\eta_n, T]$ such that $c(t_1) > 0$ and $d(t_2) > 0$.

($\tilde{\mathrm{I}}3$) The functions $f, g : [0,\infty) \times [0,\infty) \to [0,\infty)$ are continuous.

We introduce the following extreme limits:

$$f_0^{\mathrm{s}} = \limsup_{u+v\to 0^+} \frac{f(u,v)}{u+v}, \quad g_0^{\mathrm{s}} = \limsup_{u+v\to 0^+} \frac{g(u,v)}{u+v}, \quad f_0^{\mathrm{i}} = \liminf_{u+v\to 0^+} \frac{f(u,v)}{u+v}, \quad g_0^{\mathrm{i}} = \liminf_{u+v\to 0^+} \frac{g(u,v)}{u+v},$$

$$f_\infty^{\mathrm{s}} = \limsup_{u+v\to \infty} \frac{f(u,v)}{u+v}, \quad g_\infty^{\mathrm{s}} = \limsup_{u+v\to \infty} \frac{g(u,v)}{u+v}, \quad f_\infty^{\mathrm{i}} = \liminf_{u+v\to \infty} \frac{f(u,v)}{u+v}, \quad g_\infty^{\mathrm{i}} = \liminf_{u+v\to \infty} \frac{g(u,v)}{u+v}.$$

In the definitions of the extreme limits above, the variables u and v are nonnegative.

We consider the Banach space $X = C([0,T])$ with the supremum norm $\|\cdot\|$ and the Banach space $Y = X \times X$ with the norm $\|(u,v)\|_Y = \|u\| + \|v\|$.

We define the cone $C \subset Y$ by

$$C = \left\{(u,v) \in Y;\quad u(t) \geq 0, v(t) \geq 0, \forall t \in [0,T] \quad \text{and} \quad \inf_{t\in[\theta_0,T]}(u(t)+v(t)) \geq r\|(u,v)\|_Y\right\},$$

where $\theta_0 = \max\{\xi_1, \eta_1\}$, $r = \min\{r_1, r_2\}$, r_1, and r_2 are defined above.

Let $A_1, A_2 : Y \to X$ and $\mathcal{P} : Y \to Y$ be the operators defined by

$$A_1(u,v)(t) = \lambda \int_0^T G_1(t,s)c(s)f(u(s),v(s))\,\mathrm{d}s, \quad t \in [0,T],$$

$$A_2(u,v)(t) = \mu \int_0^T G_2(t,s)d(s)g(u(s),v(s))\,\mathrm{d}s, \quad t \in [0,T],$$

and $\mathcal{P}(u,v) = (A_1(u,v), A_2(u,v))$, $(u,v) \in Y$, where G_1 and G_2 are the Green's functions defined above.

The solutions of problem (S_0)–(BC_0) coincide with the fixed points of the operator $\mathcal{P}$. In addition, the operator $\mathcal{P} : C \to C$ is a completely continuous operator.

We suppose first that $f_0^{\mathrm{s}}, g_0^{\mathrm{s}}, f_\infty^{\mathrm{i}}, g_\infty^{\mathrm{i}} \in (0,\infty)$ and for positive numbers $\alpha_1, \alpha_2 > 0$ such that $\alpha_1 + \alpha_2 = 1$. We define the positive numbers L_1, L_2, L_3, and L_4 by

$$L_1 = \alpha_1\left(\frac{r(\alpha\xi_m+\beta)}{d}\int_{\xi_m}^T (T-s)c(s)f_\infty^{\mathrm{i}}\,\mathrm{d}s\right)^{-1}, \quad L_2 = \alpha_1\left(\frac{\alpha T+\beta}{d}\int_0^T (T-s)c(s)f_0^{\mathrm{s}}\,\mathrm{d}s\right)^{-1},$$

$$L_3 = \alpha_2\left(\frac{r(\gamma\eta_n+\delta)}{e}\int_{\eta_n}^T (T-s)d(s)g_\infty^{\mathrm{i}}\,\mathrm{d}s\right)^{-1}, \quad L_4 = \alpha_2\left(\frac{\gamma T+\delta}{e}\int_0^T (T-s)d(s)g_0^{\mathrm{s}}\,\mathrm{d}s\right)^{-1}.$$

By using Lemmas 1.5.1–1.5.5, the operator $\mathcal{P}$, Theorem 1.1.1, and arguments similar to those used in the proofs of Theorems 1.1.2 and 1.1.3, we obtain the following results:

Theorem 1.5.1. *Assume that* $(\widetilde{I1})$–$(\widetilde{I3})$ *hold,* $f_0^s, g_0^s, f_\infty^i, g_\infty^i \in (0,\infty)$, $\alpha_1, \alpha_2 > 0$ *with* $\alpha_1+\alpha_2 = 1$, $L_1 < L_2$, *and* $L_3 < L_4$. *Then for each* $\lambda \in (L_1, L_2)$ *and* $\mu \in (L_3, L_4)$ *there exists a positive solution* $(u(t), v(t)), t \in [0,T]$ *for* (S_0)–(BC_0).

Theorem 1.5.2. *Assume that* $(\widetilde{I1})$–$(\widetilde{I3})$ *hold,* $f_0^s = g_0^s = 0, f_\infty^i, g_\infty^i \in (0,\infty)$, *and* $\alpha_1, \alpha_2 > 0$ *with* $\alpha_1+\alpha_2 = 1$. *Then for each* $\lambda \in (L_1,\infty)$ *and* $\mu \in (L_3,\infty)$ *there exists a positive solution* $(u(t), v(t)), t \in [0,T]$ *for* (S_0)–(BC_0).

Theorem 1.5.3. *Assume that* $(\widetilde{I1})$–$(\widetilde{I3})$ *hold,* $f_0^s, g_0^s \in (0,\infty)$, $f_\infty^i = g_\infty^i = \infty$, *and* $\alpha_1, \alpha_2 > 0$ *with* $\alpha_1+\alpha_2 = 1$. *Then for each* $\lambda \in (0, L_2)$ *and* $\mu \in (0, L_4)$ *there exists a positive solution* $(u(t), v(t)), t \in [0,T]$ *for* (S_0)–(BC_0).

Theorem 1.5.4. *Assume that* $(\widetilde{I1})$–$(\widetilde{I3})$ *hold,* $f_0^s = g_0^s = 0$, *and* $f_\infty^i = g_\infty^i = \infty$. *Then for each* $\lambda \in (0,\infty)$ *and* $\mu \in (0,\infty)$ *there exists a positive solution* $(u(t), v(t))$, $t \in [0,T]$ *for* (S_0)–(BC_0).

We suppose now that $f_0^i, g_0^i, f_\infty^s, g_\infty^s \in (0,\infty)$ and for positive numbers $\alpha_1, \alpha_2 > 0$ such that $\alpha_1+\alpha_2 = 1$ we define the positive numbers $\tilde{L}_1, \tilde{L}_2, \tilde{L}_3$, and $\tilde{L}_4$ by

$$\tilde{L}_1 = \alpha_1\left(\frac{r(\alpha\xi_m+\beta)}{d}\int_{\xi_m}^T (T-s)c(s)f_0^i\,ds\right)^{-1},\quad \tilde{L}_2 = \alpha_1\left(\frac{\alpha T+\beta}{d}\int_0^T (T-s)c(s)f_\infty^s\,ds\right)^{-1},$$

$$\tilde{L}_3 = \alpha_2\left(\frac{r(\gamma\eta_n+\delta)}{e}\int_{\eta_n}^T (T-s)d(s)g_0^i\,ds\right)^{-1},\quad \tilde{L}_4 = \alpha_2\left(\frac{\gamma T+\delta}{e}\int_0^T (T-s)d(s)g_\infty^s\,ds\right)^{-1}.$$

Theorem 1.5.5. *Assume that* $(\widetilde{I1})$–$(\widetilde{I3})$ *hold,* $f_0^i, g_0^i, f_\infty^s, g_\infty^s \in (0,\infty)$, $\alpha_1, \alpha_2 > 0$ *with* $\alpha_1+\alpha_2 = 1$, $\tilde{L}_1 < \tilde{L}_2$, *and* $\tilde{L}_3 < \tilde{L}_4$. *Then for each* $\lambda \in (\tilde{L}_1, \tilde{L}_2)$ *and* $\mu \in (\tilde{L}_3, \tilde{L}_4)$ *there exists a positive solution* $(u(t), v(t)), t \in [0,T]$ *for* (S_0)–(BC_0).

Theorem 1.5.6. *Assume that* $(\widetilde{I1})$–$(\widetilde{I3})$ *hold,* $f_\infty^s = g_\infty^s = 0$, $f_0^i, g_0^i \in (0,\infty)$, *and* $\alpha_1, \alpha_2 > 0$ *with* $\alpha_1+\alpha_2 = 1$. *Then for each* $\lambda \in (\tilde{L}_1,\infty)$ *and* $\mu \in (\tilde{L}_3,\infty)$ *there exists a positive solution* $(u(t), v(t)), t \in [0,T]$ *for* (S_0)–(BC_0).

Theorem 1.5.7. *Assume that* $(\widetilde{I1})$–$(\widetilde{I3})$ *hold,* $f_\infty^s, g_\infty^s \in (0,\infty)$, $f_0^i = g_0^i = \infty$, *and* $\alpha_1, \alpha_2 > 0$ *with* $\alpha_1+\alpha_2 = 1$. *Then for each* $\lambda \in (0, \tilde{L}_2)$ *and* $\mu \in (0, \tilde{L}_4)$ *there exists a positive solution* $(u(t), v(t)), t \in [0,T]$ *for* (S_0)–(BC_0).

Theorem 1.5.8. *Assume that* $(\widetilde{I1})$–$(\widetilde{I3})$ *hold,* $f_\infty^s = g_\infty^s = 0$, *and* $f_0^i = g_0^i = \infty$. *Then for each* $\lambda \in (0,\infty)$ *and* $\mu \in (0,\infty)$ *there exists a positive solution* $(u(t), v(t))$, $t \in [0,T]$ *for* (S_0)–(BC_0).

In Theorems 1.5.1–1.5.8, we find that the operator $\mathcal{P}$ has at least one fixed point $(u,v) \in C$ such that $K_1 \le \|(u,v)\|_Y \le K_2$, with some $K_1, K_2 > 0$, $K_1 < K_2$. The last inequalities imply that $\|u\| > 0$ or $\|v\| > 0$. If $\|u\| > 0$, then $u(t) > 0$ for all $t \in (0,T]$. Because $u''(t) = -\lambda c(t)f(u(t), v(t)) \le 0$ for all $t \in (0,T)$, we deduce that u is concave on $(0,T)$. This, together with the fact that $u(T) \ge \inf_{t\in[\xi_1,T]} u(t) \ge r_1\|u\| > 0$, implies that $u(t) > 0$ for all $t \in (0,T]$. In a similar manner, if $\|v\| > 0$, then $v(t) > 0$ for all $t \in (0,T]$.

Remark 1.5.1. The results presented in this section were published in Henderson and Luca (2012e).

1.5.2 Systems without parameters and nonsingular nonlinearities

We investigate in this section problem (S′)–(BC) from Section 1.3 with $a(t) = c(t) = 1$ and $b(t) = d(t) = 0$ for all $t \in [0, 1]$, $\alpha = \tilde{\alpha} = 1$, $\beta = \tilde{\beta} = 0$, $\gamma = \tilde{\gamma} = 1$, $\delta = \tilde{\delta} = 0$, H_1 and K_1 are constant functions, and H_2 and K_2 are step functions—that is, the integral boundary conditions become multipoint boundary conditions.

Namely, we consider here the system of nonlinear second-order ordinary differential equations

$$\begin{cases} u''(t) + f(t, v(t)) = 0, & t \in (0, T), \\ v''(t) + g(t, u(t)) = 0, & t \in (0, T), \end{cases} \tag{S'_0}$$

with the multipoint boundary conditions

$$\begin{cases} u(0) = 0, \quad u(T) = \sum_{i=1}^{m-2} b_i u(\xi_i), \\ v(0) = 0, \quad v(T) = \sum_{i=1}^{n-2} c_i v(\eta_i), \end{cases} \tag{BC'_0}$$

where $m, n \in \mathbb{N}$, $m, n \geq 3$, $b_i, \xi_i \in \mathbb{R}$ for all $i = 1, \ldots, m-2$; $c_i, \eta_i \in \mathbb{R}$ for all $i = 1, \ldots, n-2$; $0 < \xi_1 < \cdots < \xi_{m-2} < T$, $0 < \eta_1 < \cdots < \eta_{n-2} < T$.

By a positive solution of (S'_0)–(BC'_0) we mean a pair of functions $(u, v) \in (C^2([0, T]))^2$ which satisfy (S'_0) and (BC'_0), and $u(t) > 0$, $v(t) > 0$ for all $t \in (0, T]$.

We consider first the following second-order differential equation with m-point boundary conditions:

$$u''(t) + y(t) = 0, \quad t \in (0, T), \tag{1.61}$$

$$u(0) = 0, \quad u(T) = \sum_{i=1}^{m-2} b_i u(\xi_i), \tag{1.62}$$

where $m \in \mathbb{N}$, $m \geq 3$, $b_i, \xi_i \in \mathbb{R}$ for all $i = 1, \ldots, m-2$, $0 < \xi_1 < \cdots < \xi_{m-2} < T$.

Lemma 1.5.6 (Li and Sun, 2006; Luca, 2011). *If $b_i \in \mathbb{R}$ for all $i = 1, \ldots, m-2$, $0 < \xi_1 < \cdots < \xi_{m-2} < T$, $\tilde{d} = T - \sum_{i=1}^{m-2} b_i \xi_i \neq 0$, and $y \in C(0, T) \cap L^1(0, T)$, then the unique solution of problem* (1.61)–(1.62) *is given by*

$$u(t) = \frac{t}{\tilde{d}} \int_0^T (T - s) y(s)\, ds - \frac{t}{\tilde{d}} \sum_{i=1}^{m-2} b_i \int_0^{\xi_i} (\xi_i - s) y(s)\, ds - \int_0^t (t - s) y(s)\, ds, \quad 0 \leq t \leq T.$$

Lemma 1.5.7 (Luca, 2011). *Under the assumptions of Lemma 1.5.6, the Green's function for the boundary value problem* (1.61)–(1.62) *is given by*

$$\tilde{G}_1(t,s)=\begin{cases}\dfrac{t}{\tilde{d}}\left[(T-s)-\displaystyle\sum_{j=i}^{m-2}b_j(\xi_j-s)\right]-(t-s), & \text{if } \xi_{i-1}\le s<\xi_i, \quad s\le t,\\ & i=1,\ldots,m-2,\ (\xi_0=0),\\ \dfrac{t}{\tilde{d}}\left[(T-s)-\displaystyle\sum_{j=i}^{m-2}b_j(\xi_j-s)\right], & \text{if } \xi_{i-1}\le s<\xi_i, \quad s\ge t,\\ & i=1,\ldots,m-2,\\ \dfrac{t}{\tilde{d}}(T-s)-(t-s), & \text{if } \xi_{m-2}\le s\le T, \quad s\le t,\\ \dfrac{t}{\tilde{d}}(T-s), & \text{if } \xi_{m-2}\le s\le T, \quad s\ge t.\end{cases}$$

Using the above Green's function, we can express the solution of problem (1.61)–(1.62) as $u(t)=\int_0^T \tilde{G}_1(t,s)y(s)\,ds$, $t\in[0,T]$.

Lemma 1.5.8 (Li and Sun, 2006; Luca, 2011). *Assume that $b_i\ge 0$ for all $i=1,\ldots,m-2$, $0<\xi_1<\cdots<\xi_{m-2}<T$, $\tilde{d}>0$, and $y\in C(0,T)\cap L^1(0,T)$, $y(t)\ge 0$ for all $t\in(0,T)$. Then the solution u of problem* (1.61)–(1.62) *satisfies $u(t)\ge 0$ for all $t\in[0,T]$ and $\tilde{G}_1(t,s)\ge 0$ for all $(t,s)\in[0,T]\times[0,T]$.*

Lemma 1.5.9 (Li and Sun, 2006). *Assume that $b_i\ge 0$ for all $i=1,\ldots,m-2$, $0<\xi_1<\cdots<\xi_{m-2}<T$, $\sum_{i=1}^{m-2}b_i\xi_i>0$, $\tilde{d}>0$, and $y\in C(0,T)\cap L^1(0,T)$, $y(t)\ge 0$ for all $t\in(0,T)$. Then the solution of problem* (1.61)–(1.62) *satisfies $\inf_{t\in[\xi_1,T]}u(t)\ge \gamma_1\sup_{t'\in[0,T]}u(t')$, where*

$$\gamma_1=\min\left\{\frac{\xi_1}{T},\ \frac{\sum_{i=1}^{m-2}b_i(T-\xi_i)}{T-\sum_{i=1}^{m-2}b_i\xi_i},\ \frac{\sum_{i=1}^{m-2}b_i\xi_i}{T},\ \frac{\sum_{i=1}^{s-1}b_i\xi_i+\sum_{i=s}^{m-2}b_i(T-\xi_i)}{T-\sum_{i=s}^{m-2}b_i\xi_i},\right.$$
$$\left. s=2,\ldots,m-2\right\}.$$

Lemma 1.5.10. *If $b_i\ge 0$ for all $i=1,\ldots,m-2$, $0<\xi_1<\cdots<\xi_{m-2}<T$, $\tilde{d}>0$, then the Green's function $\tilde{G}_1$ of problem* (1.61)–(1.62) *satisfies the following inequalities:*

$$\tilde{G}_1(t,s)\le a_1 s(T-s),\quad \forall\,(t,s)\in[0,T]\times[0,T],$$
$$\tilde{G}_1(t,s)\ge a_2 s(T-s),\quad \forall\,(t,s)\in[a,b]\times[0,T],\quad a\in(0,T),\quad b\in[\xi_{m-2},T],\quad a<b,$$

where

$$a_1=\max\left\{\frac{1}{\tilde{d}(T-\xi_1)}\sum_{j=1}^{m-2}b_j(T-\xi_j),\ \frac{1}{\tilde{d}},\ \frac{1}{\tilde{d}\xi_{m-2}}\sum_{j=1}^{m-2}b_j\xi_j,\right.$$
$$\frac{1}{\tilde{d}\xi_{i-1}(T-\xi_i)}\left[\xi_{i-1}\left(T\sum_{j=i}^{m-2}b_j-\sum_{j=1}^{m-2}b_j\xi_j\right)+T\sum_{j=1}^{i-1}b_j\xi_j\right],\quad i=2,\ldots,m-2,$$
$$\left.\frac{T}{\tilde{d}\xi_{i-1}(T-\xi_i)}\sum_{j=1}^{i-1}b_j\xi_j,\quad i=2,\ldots,m-2\right\}, \tag{1.63}$$

$$a_2 = \min\left\{\frac{1}{T}, \frac{1}{\tilde{d}T}\sum_{j=1}^{m-2} b_j\xi_j, \frac{a}{\tilde{d}T}, \frac{1}{\tilde{d}(T-\xi_{i-1})}\sum_{j=i}^{m-2}(b-\xi_j)b_j, \quad i=1,\ldots,m-2,\right.$$
$$\frac{a}{\tilde{d}}\left[T - \sum_{j=i}^{m-2} b_j\xi_j + \left(\sum_{j=i}^{m-2} b_j - 1\right)\xi_{i-1}\right]\cdot\frac{4}{T^2}, \quad i=1,\ldots,m-2,$$
$$\left.\frac{a}{\tilde{d}}\left[T - \sum_{j=i}^{m-2} b_j\xi_j + \left(\sum_{j=i}^{m-2} b_j - 1\right)\xi_i\right]\cdot\frac{4}{T^2}, \quad i=1,\ldots,m-2\right\}. \tag{1.64}$$

Proof. **(a)** For *the upper bound* of $\tilde{G}_1$ we have the following cases:
(I) If $\xi_{i-1} \le s < \xi_i, i = 1,\ldots,m-2$ and $s \le t$, we obtain

$$\begin{aligned}\tilde{G}_1(t,s) &= \frac{t}{\tilde{d}}[(T-s) - \sum_{j=i}^{m-2} b_j(\xi_j - s)] - (t-s)\\ &= s + \frac{t}{\tilde{d}}\left[T - s - \sum_{j=i}^{m-2} b_j\xi_j + s\sum_{j=i}^{m-2} b_j - T + \sum_{j=1}^{m-2} b_j\xi_j\right]\\ &= s + \frac{t}{\tilde{d}}\left[\underbrace{\left(\sum_{j=i}^{m-2} b_j - 1\right)s + \sum_{j=1}^{i-1} b_j\xi_j}_{A}\right]\end{aligned} \tag{1.65}$$

(we consider that $\sum_{j=1}^{0} d_i = 0$).

(I$_1$) If $A \ge 0$, then the maximum of $\tilde{G}_1$ is attained in $t = T$. Then we obtain

$$\begin{aligned}\tilde{G}_1(t,s) &\le s + \frac{T}{\tilde{d}}\left[\left(\sum_{j=i}^{m-2} b_j - 1\right)s + \sum_{j=1}^{i-1} b_j\xi_j\right]\\ &= \frac{1}{\tilde{d}}\left(sT - s\sum_{j=1}^{m-2} b_j\xi_j + Ts\sum_{j=i}^{m-2} b_j - Ts + T\sum_{j=1}^{i-1} b_j\xi_j\right)\\ &= \frac{1}{\tilde{d}}\left[s\left(T\sum_{j=i}^{m-2} b_j - \sum_{j=1}^{m-2} b_j\xi_j\right) + T\sum_{j=1}^{i-1} b_j\xi_j\right].\end{aligned} \tag{1.66}$$

(I$_{11}$) If $i = 1$, then for $0 \le s < \xi_1$ and $s \le t$, we obtain from (1.66)

$$\begin{aligned}\tilde{G}_1(t,s) &\le \frac{s}{\tilde{d}}\left(T\sum_{j=1}^{m-2} b_j - \sum_{j=1}^{m-2} b_j\xi_j\right)\\ &< \frac{s}{\tilde{d}}\left(T\sum_{j=1}^{m-2} b_j - \sum_{j=1}^{m-2} b_j\xi_j\right)\cdot\frac{T-s}{T-\xi_1} = \frac{\sum_{j=1}^{m-2} b_j(T-\xi_j)}{\tilde{d}(T-\xi_1)}s(T-s).\end{aligned}$$

(I_{12}) If $2 \le i \le m-2$, then for $\xi_{i-1} \le s < \xi_i$, we have from (1.66) the following cases:

(I_{121}) If $T\sum_{j=i}^{m-2} b_j - \sum_{j=1}^{m-2} b_j\xi_j > 0$, then

$$\begin{aligned}\tilde{G}_1(t,s) &\le \frac{1}{\tilde{d}}\left(T\sum_{j=i}^{m-2} b_j - \sum_{j=1}^{m-2} b_j\xi_j\right)\left(s + \underbrace{\frac{T\sum_{j=1}^{i-1} b_j\xi_j}{T\sum_{j=i}^{m-2} b_j - \sum_{j=1}^{m-2} b_j\xi_j}}_{B}\right)\\ &\le \frac{1}{\tilde{d}}\left(T\sum_{j=i}^{m-2} b_j - \sum_{j=1}^{m-2} b_j\xi_j\right)\left(1+\frac{B}{\xi_{i-1}}\right)s\\ &\le \frac{1}{\tilde{d}}\left(T\sum_{j=i}^{m-2} b_j - \sum_{j=1}^{m-2} b_j\xi_j\right)\left(1+\frac{B}{\xi_{i-1}}\right)s\cdot\frac{T-s}{T-\xi_i}\\ &= \frac{1}{\tilde{d}\xi_{i-1}(T-\xi_i)}\left[\xi_{i-1}\left(T\sum_{j=i}^{m-2} b_j - \sum_{j=1}^{m-2} b_j\xi_j\right) + T\sum_{j=1}^{i-1} b_j\xi_j\right]s(T-s).\end{aligned}$$

(I_{122}) If $T\sum_{j=i}^{m-2} b_j - \sum_{j=1}^{m-2} b_j\xi_j \le 0$, then

$$\tilde{G}_1(t,s) \le \frac{T}{\tilde{d}}\sum_{j=1}^{i-1} b_j\xi_j \le \frac{T}{\tilde{d}}\sum_{j=1}^{i-1} b_j\xi_j\cdot\frac{s}{\xi_{i-1}} \le \frac{T}{\tilde{d}\xi_{i-1}(T-\xi_i)}\sum_{j=1}^{i-1} b_j\xi_j\cdot s(T-s).$$

(I_2) If $A < 0$, then the maximum of $\tilde{G}_1$ in (1.65) appears for $t = s$, and then we obtain

$$\begin{aligned}\tilde{G}_1(t,s) &\le s + \frac{s}{\tilde{d}}\left[\left(\sum_{j=i}^{m-2} b_j - 1\right)s + \sum_{j=1}^{i-1} b_j\xi_j\right]\\ &= \frac{s}{\tilde{d}}\left[T - \sum_{j=1}^{m-2} b_j\xi_j + \left(\sum_{j=i}^{m-2} b_j - 1\right)s + \sum_{j=1}^{i-1} b_j\xi_j\right]\\ &= \frac{s}{\tilde{d}}\left[T - \sum_{j=i}^{m-2} b_j\xi_j + \left(\sum_{j=i}^{m-2} b_j - 1\right)s\right]\\ &= \frac{s}{\tilde{d}}\left(T - s - \sum_{j=i}^{m-2} b_j\xi_j + s\sum_{j=i}^{m-2} b_j\right) \le \frac{1}{\tilde{d}}s(T-s).\end{aligned}$$

(II) For $\xi_{i-1} \le s < \xi_i$ and $s \ge t$, for $i = 1, \ldots, m-2$, we have
$$\tilde{G}_1(t,s) \le \frac{1}{\tilde{d}}t(T-s) \le \frac{1}{\tilde{d}}s(T-s).$$

(III) For $\xi_{m-2} \le s \le T$ and $s \le t$, we have

$$\begin{aligned}\tilde{G}_1(t,s) &= \frac{t}{\tilde{d}}(T-s) - (t-s) = s + \frac{t}{\tilde{d}}\left(T - s - T + \sum_{j=1}^{m-2} b_j\xi_j\right)\\ &= s + \frac{t}{\tilde{d}}\left(\sum_{j=1}^{m-2} b_j\xi_j - s\right).\end{aligned} \tag{1.67}$$

(III$_1$) If $s \le \sum_{j=1}^{m-2} b_j\xi_j$, then $\tilde{G}_1$ is a nondecreasing function in variable t, so the maximum occurs when $t = T$. Then we have

$$\tilde{G}_1(t,s) \le s + \frac{T}{\tilde{d}}\left(\sum_{j=1}^{m-2} b_j\xi_j - s\right) = \frac{T-s}{\tilde{d}}\sum_{j=1}^{m-2} b_j\xi_j$$

$$= \frac{1}{\tilde{d}\xi_{m-2}}(T-s)\xi_{m-2}\sum_{j=1}^{m-2} b_j\xi_j \le \frac{1}{\tilde{d}\xi_{m-2}}\sum_{j=1}^{m-2} b_j\xi_j\, s(T-s).$$

(III$_2$) If $s > \sum_{j=1}^{m-2} b_j\xi_j$, then $\tilde{G}_1$ is a decreasing function in variable t, so the maximum occurs when $t = s$. Then we obtain

$$\tilde{G}_1(t,s) \le s + \frac{s}{\tilde{d}}\left(\sum_{j=1}^{m-2} b_j\xi_j - s\right) = \frac{1}{\tilde{d}}s(T-s).$$

(IV) If $\xi_{m-2} \le s \le T$ and $s \ge t$, then we have

$$\tilde{G}_1(t,s) = \frac{t}{\tilde{d}}(T-s) \le \frac{1}{\tilde{d}}s(T-s).$$

Therefore, we obtain $\tilde{G}_1(t,s) \le a_1 s(T-s)$ for all $(t,s) \in [0,T] \times [0,T]$, where a_1 is given by (1.63).

(b) For the *lower bound* of $\tilde{G}_1$ we have the following cases:

($\tilde{\text{I}}$) If $\xi_{i-1} \le s < \xi_i$ and $s \le t$, for $i = 1, \dots, m-2$, $\tilde{G}_1$ has the form given in (1.65), and we consider the following cases:

($\tilde{\text{I}}_1$) If $A \ge 0$, then $\tilde{G}_1$ is nondecreasing in t, so the minimum of $\tilde{G}_1$ is attained when $t = s$ and

$$\tilde{G}_1(t,s) \ge s \ge \frac{1}{T}s(T-s).$$

($\tilde{\text{I}}_2$) If $A < 0$, then $\tilde{G}_1$ is decreasing in t and the minimum is attained when $t = b \in [\xi_{m-2}, T]$. Therefore, we obtain

$$\tilde{G}_1(t,s) \ge s + \frac{b}{\tilde{d}}\left[\left(\sum_{j=i}^{m-2} b_j - 1\right)s + \sum_{j=1}^{i-1} b_j\xi_j\right]$$

$$= \frac{1}{\tilde{d}}\left(sT - s\sum_{j=1}^{m-2} b_j\xi_j + bs\sum_{j=i}^{m-2} b_j - bs + b\sum_{j=1}^{i-1} b_j\xi_j\right)$$

$$= \frac{1}{\tilde{d}}\left[(T-b)s + (b-s)\sum_{j=1}^{i-1} b_j\xi_j + s\sum_{j=i}^{m-2}(b-\xi_j)b_j\right]$$

$$\ge \frac{s}{\tilde{d}}\sum_{j=i}^{m-2}(b-\xi_j)b_j \ge \frac{1}{\tilde{d}(T-\xi_{i-1})}\sum_{j=i}^{m-2}(b-\xi_j)b_j s(T-s).$$

($\tilde{\text{II}}$) If $\xi_{i-1} \le s < \xi_i, i = 1, \dots, m-2$, and $s \ge t \ge a$, then for $\tilde{G}_1$ written as

$$\tilde{G}_1(t,s) = \frac{t}{\tilde{d}}\left[T - \sum_{j=i}^{m-2} b_j\xi_j + \left(\sum_{j=i}^{m-2} b_j - 1\right)s\right],$$

we have the following cases:

($\widetilde{\mathbf{II}}_1$) If $\sum_{j=i}^{m-2} b_j - 1 \geq 0$, then

$$\tilde{G}_1(t,s) \geq \frac{a}{\tilde{d}}\left[T - \sum_{j=i}^{m-2} b_j\xi_j + \left(\sum_{j=i}^{m-2} b_j - 1\right)\xi_{i-1}\right]$$
$$\geq \frac{a}{\tilde{d}}\left[T - \sum_{j=i}^{m-2} b_j\xi_j + \left(\sum_{j=i}^{m-2} b_j - 1\right)\xi_{i-1}\right] \cdot \frac{4}{T^2}s(T-s).$$

($\widetilde{\mathbf{II}}_2$) If $\sum_{j=i}^{m-2} b_j - 1 < 0$, then $\sum_{j=i}^{m-2} b_j\xi_j \leq \sum_{j=i}^{m-2} b_j\xi_{m-2} < T$ and

$$T - \sum_{j=i}^{m-2} b_j\xi_j + \left(\sum_{j=i}^{m-2} b_j - 1\right)s \geq T - \sum_{j=i}^{m-2} b_j\xi_j + \left(\sum_{j=i}^{m-2} b_j - 1\right)\xi_i$$
$$= T - \xi_i - \sum_{j=i}^{m-2} b_j(\xi_j - \xi_i)$$
$$\geq T - \xi_i - \sum_{j=i}^{m-2} b_j(\xi_{m-2} - \xi_i) > 0.$$

Therefore, $\tilde{G}_1$ attains its minimum when $t = a$, and

$$\tilde{G}_1(t,s) \geq \frac{a}{\tilde{d}}\left[T - \sum_{j=i}^{m-2} b_j\xi_j + \left(\sum_{j=i}^{m-2} b_j - 1\right)s\right]$$
$$\geq \frac{a}{\tilde{d}}\left[T - \sum_{j=i}^{m-2} b_j\xi_j + \left(\sum_{j=i}^{m-2} b_j - 1\right)\xi_i\right]$$
$$\geq \frac{a}{\tilde{d}}\left[T - \sum_{j=i}^{m-2} b_j\xi_j + \left(\sum_{j=i}^{m-2} b_j - 1\right)\xi_i\right] \cdot \frac{4}{T^2}s(T-s).$$

($\widetilde{\mathbf{III}}$) If $\xi_{m-2} \leq s \leq T$ and $s \leq t$, then $\tilde{G}_1$ has the form given in relation (1.67), and we have the following cases:

($\widetilde{\mathbf{III}}_1$) If $s \geq \sum_{j=1}^{m-2} b_j\xi_j$, then $\tilde{G}_1$ is nonincreasing in t and the minimum occurs when $t = T$. So

$$\tilde{G}_1(t,s) \geq s + \frac{T}{\tilde{d}}\left(\sum_{j=1}^{m-2} b_j\xi_j - s\right) = \frac{1}{\tilde{d}}\sum_{j=1}^{m-2} b_j\xi_j(T-s)$$
$$= \frac{1}{\tilde{d}T}\sum_{j=1}^{m-2} b_j\xi_j(T-s)T \geq \frac{1}{\tilde{d}T}\sum_{j=1}^{m-2} b_j\xi_j s(T-s).$$

($\widetilde{\mathbf{III}}_2$) If $s < \sum_{j=1}^{m-2} b_j\xi_j$, then $\tilde{G}_1$ is increasing in t and the minimum occurs when $t = s$. Then we have

$$\tilde{G}_1(t,s) \geq s + \frac{s}{\tilde{d}}\left(\sum_{j=1}^{m-2} b_j\xi_j - s\right) = \frac{1}{\tilde{d}}s(T-s).$$

($\widetilde{\mathbf{IV}}$) If $\xi_{m-2} \leq s \leq T$ and $s \geq t$, then we have

$$\tilde{G}_1(t,s) = \frac{t}{\tilde{d}}(T-s) \geq \frac{a}{\tilde{d}T}T(T-s) \geq \frac{a}{\tilde{d}T}s(T-s).$$

Therefore, we obtain $\tilde{G}_1(t,s) \geq a_2 s(T-s)$ for all $(t,s) \in [a,b] \times [0,T], a \in (0,T), b \in [\xi_{m-2}, T], a < b$, where a_2 is given by (1.64). □

We observe that under the assumptions of Lemma 1.5.10, the coefficient $a_1 > 0$. If, in addition, $b > \xi_{m-2}$ and there exists $j_0 \in \{1, 2, \ldots, m-2\}$ such that $b_{j_0} > 0$ (equivalently $\sum_{j=1}^{m-2} b_j\xi_j > 0$), then $a_2 > 0$. For every $i \in \{1, \ldots, m-2\}$, we have two cases:

(1) If $\sum_{j=i}^{m-2} b_j \geq 1$, then $T - \sum_{j=i}^{m-2} b_j\xi_j + \left(\sum_{j=i}^{m-2} b_j - 1\right)s > 0$, for $s = \xi_{i-1}, \xi_i$.

(2) If $\sum_{j=i}^{m-2} b_j < 1$, then

$$\begin{aligned} T - \sum_{j=i}^{m-2} b_j\xi_j + \left(\sum_{j=i}^{m-2} b_j - 1\right)s &\geq T - \sum_{j=i}^{m-2} b_j\xi_j + \left(\sum_{j=i}^{m-2} b_j - 1\right)\xi_i \\ &= T - \xi_i - \sum_{j=i}^{m-2} b_j(\xi_j - \xi_i) \\ &\geq T - \xi_i - \sum_{j=i}^{m-2} b_j(T - \xi_i) \\ &= (T-\xi_i)\left(1 - \sum_{j=i}^{m-2} b_j\right) > 0, \end{aligned}$$

for $s = \xi_{i-1}, \xi_i$.

We can also formulate results similar to those in Lemmas 1.5.6–1.5.10 for the boundary value problem

$$v''(t) + h(t) = 0, \quad t \in (0,T), \tag{1.68}$$

$$v(0) = 0, \quad v(T) = \sum_{i=1}^{n-2} c_i v(\eta_i), \tag{1.69}$$

where $n \in \mathbb{N}, n \geq 3, c_i, \eta_i \in \mathbb{R}$ for all $i = 1, \ldots, n-2, 0 < \eta_1 < \cdots < \eta_{n-2} < T$.

If $\tilde{e} = T - \sum_{i=1}^{n-2} c_i\eta_i \neq 0$, we denote by $\tilde{G}_2$ the Green's function corresponding to problem (1.68)–(1.69)—that is,

$$
\tilde{G}_2(t,s) = \begin{cases} \dfrac{t}{\tilde{e}}\left[(T-s) - \displaystyle\sum_{j=i}^{n-2} c_j(\eta_j - s)\right] - (t-s), & \text{if } \eta_{i-1} \le s < \eta_i, \quad s \le t, \\ & i = 1,\dots,n-2, (\eta_0 = 0), \\ \dfrac{t}{\tilde{e}}\left[(T-s) - \displaystyle\sum_{j=i}^{n-2} c_j(\eta_j - s)\right], & \text{if } \eta_{i-1} \le s < \eta_i, \quad s \ge t, \\ & i = 1,\dots,n-2, \\ \dfrac{t}{\tilde{e}}(T-s) - (t-s), & \text{if } \eta_{n-2} \le s \le T, \quad s \le t, \\ \dfrac{t}{\tilde{e}}(T-s), & \text{if } \eta_{n-2} \le s \le T, \quad s \ge t. \end{cases}
$$

Under assumptions similar to those from Lemma 1.5.9, we have the inequality $\inf_{t\in[\eta_1,T]} v(t) \ge \gamma_2 \sup_{t'\in[0,T]} v(t')$, where v is the solution of problem (1.68)–(1.69), and γ_2 is given by

$$
\gamma_2 = \min\left\{ \frac{\eta_1}{T}, \frac{\sum_{i=1}^{n-2} c_i(T-\eta_i)}{T - \sum_{i=1}^{n-2} c_i\eta_i}, \frac{\sum_{i=1}^{n-2} c_i\eta_i}{T}, \right.
$$

$$
\left. \frac{\sum_{i=1}^{s-1} c_i\eta_i + \sum_{i=s}^{n-2} c_i(T-\eta_i)}{T - \sum_{i=s}^{n-2} c_i\eta_i}, \quad s = 2,\dots,n-2 \right\}.
$$

Moreover, under assumptions similar to those from Lemma 1.5.10, we have the inequalities

$$
\tilde{G}_2(t,s) \le \tilde{a}_1 s(T-s), \quad \forall\,(t,s) \in [0,T]\times[0,T],
$$

$$
\tilde{G}_2(t,s) \ge \tilde{a}_2 s(T-s), \quad \forall\,(t,s) \in [a,b]\times[0,T], a\in(0,T), b \in [\eta_{n-2},T], a<b,
$$

where $\tilde{a}_1$ and $\tilde{a}_2$ are defined as a_1 and a_2 in (1.63)–(1.64), with m, b_i, and ξ_i replaced by n, c_i, and η_i, respectively.

In the rest of this section, we shall denote by a_1 and a_2 the constants from Lemma 1.5.10 for $\tilde{G}_1$ with $a = \xi_1$ and $b = T$, and by $\tilde{a}_1$ and $\tilde{a}_2$ the corresponding constants for $\tilde{G}_2$ with $a = \eta_1$ and $b = T$.

We present the assumptions that we shall use in the sequel:

($\widetilde{\mathrm{H1}}$) $0 < \xi_1 < \cdots < \xi_{m-2} < T$, $0 < \eta_1 < \cdots < \eta_{n-2} < T$, $b_i \ge 0, i = 1,\dots,m-2$, $c_i \ge 0, i = 1,\dots,n-2$, $\tilde{d} = T - \sum_{i=1}^{m-2} b_i\xi_i > 0$, $\tilde{e} = T - \sum_{i=1}^{n-2} c_i\eta_i > 0$, $\sum_{i=1}^{m-2} b_i\xi_i > 0$, $\sum_{i=1}^{n-2} c_i\eta_i > 0$.

($\widetilde{\mathrm{H2}}$) The functions $f, g \in C([0,T]\times[0,\infty),[0,\infty))$ and $f(t,0) = 0, g(t,0) = 0$, for all $t \in [0,T]$.

The pair of functions $(u,v) \in (C^2([0,T];[0,\infty)))^2$ is a solution for our problem (S_0')–(BC_0') if and only if $(u,v) \in (C([0,T];[0,\infty)))^2$ is a solution for the nonlinear integral system

$$\begin{cases} u(t) = \int_0^T \tilde{G}_1(t,s) f\left(s, \int_0^T \tilde{G}_2(s,\tau) g(\tau, u(\tau))\,d\tau\right) ds, & t \in [0,T], \\ v(t) = \int_0^T \tilde{G}_2(t,s) g(s, u(s))\,ds, & t \in [0,T]. \end{cases}$$

We consider the Banach space $X = C([0,T])$ with the supremum norm $\|\cdot\|$, and define the cone $C' \subset X$ by $C' = \{u \in X, u(t) \geq 0, \forall t \in [0,T]\}$.

We also define the operator $\mathcal{A} : C' \to X$ by

$$(\mathcal{A}u)(t) = \int_0^T \tilde{G}_1(t,s) f\left(s, \int_0^T \tilde{G}_2(s,\tau) g(\tau, u(\tau))\,d\tau\right) ds, \quad t \in [0,T],$$

and the operators $\mathcal{B} : C' \to X$ and $\mathcal{C} : C' \to X$ by

$$(\mathcal{B}u)(t) = \int_0^T \tilde{G}_1(t,s) u(s)\,ds, \quad (\mathcal{C}u)(t) = \int_0^T \tilde{G}_2(t,s) u(s)\,ds, \quad t \in [0,T].$$

Under assumptions $(\widetilde{\mathrm{H}1})$ and $(\widetilde{\mathrm{H}2})$, using also Lemma 1.5.8, we can easily see that $\mathcal{A}$, $\mathcal{B}$, and $\mathcal{C}$ are completely continuous from C' to C'. Thus, the existence and multiplicity of positive solutions of system (S_0')–(BC_0') are equivalent to the existence and multiplicity of fixed points of the operator $\mathcal{A}$.

We also define the cone $C_0 = \{u \in C'; \inf_{t\in[\theta_0,T]} u(t) \geq \gamma\|u\|\}$, where $\theta_0 = \max\{\xi_1, \eta_1\}$, $\gamma = \min\{\gamma_1, \gamma_2\}$, and $\gamma_1, \gamma_2 > 0$ are defined above. From our assumptions and Lemma 1.5.9, we deduce that $\mathcal{B}(C') \subset C_0$ and $\mathcal{C}(C') \subset C_0$.

By using Lemmas 1.5.6–1.5.10, Theorems 1.3.1–1.3.3, and arguments similar to those used in the proofs of Theorems 1.3.4–1.3.6, we obtain the following results:

Theorem 1.5.9. *Assume that* $(\widetilde{\mathrm{H}1})$ *and* $(\widetilde{\mathrm{H}2})$ *hold. If the functions f and g also satisfy the following conditions* $(\widetilde{\mathrm{H}3})$ *and* $(\widetilde{\mathrm{H}4})$, *then problem* (S_0')–(BC_0') *has at least one positive solution* $(u(t), v(t))$, $t \in [0,T]$:

$(\widetilde{\mathrm{H}3})$ *There exists a positive constant* $p \in (0,1]$ *such that*

$$(1)\ \tilde{f}_\infty^{\mathrm{i}} = \liminf_{u\to\infty} \inf_{t\in[0,T]} \frac{f(t,u)}{u^p} \in (0,\infty] \quad \text{and} \quad (2)\ \tilde{g}_\infty^{\mathrm{i}} = \liminf_{u\to\infty} \inf_{t\in[0,T]} \frac{g(t,u)}{u^{1/p}} = \infty.$$

$(\widetilde{\mathrm{H}4})$ *There exists a positive constant* $q \in (0,\infty)$ *such that*

$$(1)\ \tilde{f}_0^{\mathrm{s}} = \limsup_{u\to0^+} \sup_{t\in[0,T]} \frac{f(t,u)}{u^q} \in [0,\infty) \quad \text{and} \quad (2)\ \tilde{g}_0^{\mathrm{s}} = \limsup_{u\to0^+} \sup_{t\in[0,T]} \frac{g(t,u)}{u^{1/q}} = 0.$$

Theorem 1.5.10. *Assume that* $(\widetilde{\mathrm{H}1})$ *and* $(\widetilde{\mathrm{H}2})$ *hold. If the functions f and g also satisfy the following conditions* $(\widetilde{\mathrm{H}5})$ *and* $(\widetilde{\mathrm{H}6})$, *then problem* (S_0')–(BC_0') *has at least one positive solution* $(u(t), v(t))$, $t \in [0,T]$:

$(\widetilde{\mathrm{H}5})$ *There exists a positive constant* $r \in (0,\infty)$ *such that*

$$(1)\ \tilde{f}_\infty^{\mathrm{s}} = \limsup_{u\to\infty} \sup_{t\in[0,T]} \frac{f(t,u)}{u^r} \in [0,\infty) \quad \text{and} \quad (2)\ \tilde{g}_\infty^{\mathrm{s}} = \limsup_{u\to\infty} \sup_{t\in[0,T]} \frac{g(t,u)}{u^{1/r}} = 0.$$

$(\widetilde{\mathrm{H}6})$ *The following conditions are satisfied:*

$$(1)\ \tilde{f}_0^{\mathrm{i}} = \liminf_{u\to0^+} \inf_{t\in[0,T]} \frac{f(t,u)}{u} \in (0,\infty] \quad \text{and} \quad (2)\ \tilde{g}_0^{\mathrm{i}} = \liminf_{u\to0^+} \inf_{t\in[0,T]} \frac{g(t,u)}{u} = \infty.$$

Theorem 1.5.11. *Assume that* ($\widetilde{\text{H1}}$) *and* ($\widetilde{\text{H2}}$) *hold. If the functions f and g also satisfy conditions* ($\widetilde{\text{H3}}$) *and* ($\widetilde{\text{H6}}$) *and the following condition* ($\widetilde{\text{H7}}$), *problem* (S_0')–(BC_0') *has at least two positive solutions* $(u_1(t), v_1(t))$, $(u_2(t), v_2(t))$, $t \in [0, T]$:

($\widetilde{\text{H7}}$) *For each $t \in [0, T]$, $f(t, u)$ and $g(t, u)$ are nondecreasing with respect to u, and there exists a constant $N > 0$ such that*

$$f\left(t, m_0 \int_0^T g(s, N)\,ds\right) < \frac{N}{m_0}, \quad \forall t \in [0, T],$$

where $m_0 = \frac{T^2}{4} \max\{a_1 T, \tilde{a}_1\}$, *and a_1 and $\tilde{a}_1$ are defined above.*

In Theorems 1.5.9–1.5.11, we find that the operator $\mathcal{A}$ has at least one fixed point $u_1 \in (B_L \setminus \bar{B}_{\delta_1}) \cap C'$ (with some $\delta_1, L > 0$, $\delta_1 < L$)—that is, $\delta_1 < \|u_1\| < L$ (or at least two fixed points). Let $v_1(t) = \int_0^T \tilde{G}_2(t,s) g(s, u_1(s))\,ds$. Then $(u_1, v_1) \in C' \times C'$ is a solution of (S_0')–(BC_0'). In addition, $u_1(t) > 0, v_1(t) > 0, \forall t \in (0, T]$. Because $u_1''(t) = -f(t, v_1(t)) \leq 0$, we deduce that u_1 is concave on $(0, T)$. Combining this information with $u_1(T) \geq \inf_{t \in [\xi_1, T]} u_1(t) \geq \gamma_1 \|u_1\| > 0$, we obtain $u_1(t) > 0$ for all $t \in (0, T]$. Because $u_1(t) = \int_0^T \tilde{G}_1(t,s) f(s, v_1(s))\,ds$ and $\|u_1\| > 0$, we deduce that $\|v_1\| > 0$. If we suppose that $v_1(t) = 0$ for all $t \in [0, T]$, then by using ($\widetilde{\text{H2}}$), we have $f(s, v_1(s)) = f(s, 0) = 0$ for all $s \in [0, T]$. This implies $u_1(t) = 0$ for all $t \in [0, T]$, which contradicts $\|u_1\| > 0$. In a manner similar to that above, we obtain $v_1(t) > 0$ for all $t \in (0, T]$.

Remark 1.5.2. The results presented in this section were published in Henderson and Luca (2012a).

1.5.3 Systems without parameters and singular nonlinearities

We investigate in this section system (S′)–(BC) from Section 1.3 with $a(t) = c(t) = 1$ and $b(t) = d(t) = 0$ for all $t \in [0, 1]$, $\alpha = \tilde{\alpha} = 1$, $\beta = \tilde{\beta} = 0$, $\gamma = \tilde{\gamma} = 1$, $\delta = \tilde{\delta} = 0$, H_1 and K_1 are constant functions, and H_2 and K_2 are step functions—that is, the integral boundary conditions become multipoint boundary conditions, and the nonlinearities from the system may be singular at $t = 0$ and/or $t = 1$.

Namely, we consider the system of nonlinear second-order ordinary differential equations

$$\begin{cases} u''(t) + f(t, v(t)) = 0, & t \in (0, T), \\ v''(t) + g(t, u(t)) = 0, & t \in (0, T), \end{cases} \tag{S_0'}$$

with the multipoint boundary conditions

$$\begin{cases} u(0) = 0, \quad u(T) = \displaystyle\sum_{i=1}^{m-2} b_i u(\xi_i), \\ v(0) = 0, \quad v(T) = \displaystyle\sum_{i=1}^{n-2} c_i v(\eta_i), \end{cases} \tag{BC_0'}$$

where $m, n \in \mathbb{N}$, $m, n \geq 3$, $b_i, \xi_i \in \mathbb{R}$ for all $i = 1, \ldots, m-2$; $c_i, \eta_i \in \mathbb{R}$ for all $i = 1, \ldots, n-2$; $0 < \xi_1 < \cdots < \xi_{m-2} < T$, $0 < \eta_1 < \cdots < \eta_{n-2} < T$, and f and g may be singular at $t = 0$ and/or $t = T$.

By a positive solution of (S_0')–(BC_0') we mean a pair of functions $(u, v) \in (C([0,T]) \cap C^2(0,T))^2$ which satisfies (S_0') and (BC_0'), and $u(t) > 0$, $v(t) > 0$ for all $t \in (0, T]$.

We consider first problem (1.61)–(1.62) with the auxiliary results in Lemmas 1.5.6–1.5.10 from Section 1.5.2. In addition, we have the following lemma:

Lemma 1.5.11. *Under the assumptions of Lemma 1.5.6, the Green's function for the boundary value problem* (1.61)–(1.62) *can be expressed as*

$$\tilde{G}_1(t,s) = k_0(t,s) + \frac{t}{\tilde{d}} \sum_{j=1}^{m-2} b_j k_0(\xi_j, s), \tag{1.70}$$

where

$$k_0(t,s) = \begin{cases} \dfrac{t(T-s)}{T}, & 0 \leq t \leq s \leq T, \\ \dfrac{s(T-t)}{T}, & 0 \leq s \leq t \leq T. \end{cases}$$

Proof. If $s \leq t$ and $\xi_{i-1} \leq s < \xi_i$, for $i \in \{1, 2, \ldots, m-2\}$, we have

$$\begin{aligned}
\tilde{G}_1(t,s) &= \frac{t(T-s)}{\tilde{d}} - \frac{t}{\tilde{d}} \sum_{j=i}^{m-2} b_j(\xi_j - s) - \frac{t-s}{\tilde{d}} \left(T - \sum_{j=1}^{m-2} b_j \xi_j \right) \\
&= \frac{1}{\tilde{d}} \left(sT - ts - t \sum_{j=i}^{m-2} b_j(\xi_j - s) + (t-s) \sum_{j=1}^{m-2} b_j \xi_j \right) \\
&= \frac{1}{\tilde{d}} \left[\frac{s(T-t)}{T} \left(T - \sum_{j=1}^{m-2} b_j \xi_j \right) + \frac{s(T-t)}{T} \sum_{j=1}^{m-2} b_j \xi_j - t \sum_{j=i}^{m-2} b_j(\xi_j - s) + (t-s) \sum_{j=1}^{m-2} b_j \xi_j \right] \\
&= \frac{s(T-t)}{T} + \frac{1}{\tilde{d}} \left(\frac{t(T-s)}{T} \sum_{j=1}^{m-2} b_j \xi_j - t \sum_{j=i}^{m-2} b_j \xi_j + ts \sum_{j=i}^{m-2} b_j \right) \\
&= \frac{s(T-t)}{T} + \frac{ts}{\tilde{d}T} \sum_{j=i}^{m-2} b_j (T - \xi_j) + \frac{1}{\tilde{d}} \left(\frac{t(s-T)}{T} \sum_{j=i}^{m-2} b_j \xi_j + \frac{t(T-s)}{T} \sum_{j=1}^{m-2} b_j \xi_j \right) \\
&= \frac{s(T-t)}{T} + \frac{ts}{\tilde{d}T} \sum_{j=i}^{m-2} b_j (T - \xi_j) + \frac{t(T-s)}{\tilde{d}T} \sum_{j=1}^{i-1} b_j \xi_j \\
&= \frac{s(T-t)}{T} + \frac{t}{\tilde{d}} \sum_{j=1}^{i-1} b_j \frac{\xi_j(T-s)}{T} + \frac{t}{\tilde{d}} \sum_{j=i}^{m-2} b_j \frac{s(T-\xi_j)}{T} \\
&= k_0(t,s) + \frac{t}{\tilde{d}} \sum_{j=1}^{m-2} b_j k_0(\xi_j, s).
\end{aligned}$$

Here, we consider that $\sum_{i=1}^{0} z_i = 0$.

If $s \geq t$ and $\xi_{i-1} \leq s < \xi_i$, for $i \in \{1, 2, \dots, m-2\}$, we obtain

$$\begin{aligned}
\tilde{G}_1(t,s) &= \frac{t(T-s)}{\tilde{d}} - \frac{t}{\tilde{d}} \sum_{j=i}^{m-2} b_j(\xi_j - s) \\
&= \frac{t(T-s)}{T} + \frac{1}{\tilde{d}T} \left(t(T-s) \sum_{j=1}^{m-2} b_j \xi_j - tT \sum_{j=i}^{m-2} b_j(\xi_j - s) \right) \\
&= \frac{t(T-s)}{T} + \frac{1}{\tilde{d}T} \left(t(T-s) \sum_{j=1}^{i-1} b_j \xi_j + t(T-s) \sum_{j=i}^{m-2} b_j \xi_j - tT \sum_{j=i}^{m-2} b_j \xi_j + tsT \sum_{j=i}^{m-2} b_j \right) \\
&= \frac{t(T-s)}{T} + \frac{t(T-s)}{\tilde{d}T} \sum_{j=1}^{i-1} b_j \xi_j + \frac{ts}{\tilde{d}T} \sum_{j=i}^{m-2} b_j (T - \xi_j) \\
&= \frac{t(T-s)}{T} + \frac{t}{\tilde{d}} \sum_{j=1}^{i-1} b_j \frac{\xi_j (T-s)}{T} + \frac{t}{\tilde{d}} \sum_{j=i}^{m-2} b_j \frac{s(T-\xi_j)}{T} \\
&= k_0(t,s) + \frac{t}{\tilde{d}} \sum_{j=1}^{m-2} b_j k_0(\xi_j, s).
\end{aligned}$$

If $s \leq t$ and $\xi_{m-2} \leq s \leq T$, we have

$$\begin{aligned}
\tilde{G}_1(t,s) &= \frac{t(T-s)}{T} - (t-s) \\
&= \frac{s(T-t)}{T} + \frac{1}{\tilde{d}} \left[\frac{st}{T} \left(T - \sum_{j=1}^{m-2} b_j \xi_j \right) - t \left(T - \sum_{j=1}^{m-2} b_j \xi_j \right) + tT - st \right] \\
&= \frac{s(T-t)}{T} + \frac{1}{\tilde{d}} \left(-\frac{st}{T} \sum_{j=1}^{m-2} b_j \xi_j + t \sum_{j=1}^{m-2} b_j \xi_j \right) \\
&= \frac{s(T-t)}{T} + \frac{t}{\tilde{d}} \sum_{j=1}^{m-2} b_j \frac{\xi_j (T-s)}{T} = k_0(t,s) + \frac{t}{\tilde{d}} \sum_{j=1}^{m-2} b_j k_0(\xi_j, s).
\end{aligned}$$

If $s \geq t$ and $\xi_{m-2} \leq s \leq T$, we obtain

$$\tilde{G}_1(t,s) = \frac{t(T-s)}{\tilde{d}} = \frac{t(T-s)}{T} + \frac{t}{\tilde{d}} \sum_{j=1}^{m-2} b_j \frac{\xi_j (T-s)}{T} = k_0(t,s) + \frac{t}{\tilde{d}} \sum_{j=1}^{m-2} b_j k_0(\xi_j, s).$$

Therefore, in every case from above, we obtain relation (1.70). □

We can also formulate a result similar to Lemma 1.5.11 for problem (1.68)–(1.69).

We present now the assumptions that we shall use in the sequel:

($\widetilde{\text{L}1}$) $0 < \xi_1 < \cdots < \xi_{m-2} < T$, $0 < \eta_1 < \cdots < \eta_{n-2} < T$, $b_i \geq 0, i = 1, \ldots, m-2$, $c_i \geq 0, i = 1, \ldots, n-2$, $\tilde{d} = T - \sum_{i=1}^{m-2} b_i \xi_i > 0$, $\tilde{e} = T - \sum_{i=1}^{n-2} c_i \eta_i > 0$, $\sum_{i=1}^{m-2} b_i \xi_i > 0$, $\sum_{i=1}^{n-2} c_i \eta_i > 0$.

($\widetilde{\text{L}2}$) The functions $f, g \in C((0,T) \times \mathbb{R}_+, \mathbb{R}_+)$, and there exists $p_i \in C((0,T), \mathbb{R}_+)$ and $q_i \in C(\mathbb{R}_+, \mathbb{R}_+)$, $i = 1, 2$, $q_1(0) = 0$, $q_2(0) = 0$ such that

$$f(t,x) \leq p_1(t) q_1(x), \quad g(t,x) \leq p_2(t) q_2(x), \quad \forall t \in (0,T), \quad x \in \mathbb{R}_+,$$

and $\alpha = \int_0^T s(T-s) p_1(s)\, ds < \infty$, $\beta = \int_0^T s(T-s) p_2(s)\, ds < \infty$.

The pair of functions $(u, v) \in (C([0,T]) \cap C^2(0,T))^2$ is a solution for our problem (S_0')–(BC_0') if and only if $(u, v) \in (C([0,T]))^2$ is a solution for the nonlinear integral equations

$$\begin{cases} u(t) = \displaystyle\int_0^T \tilde{G}_1(t,s) f\left(s, \int_0^T \tilde{G}_2(s,\tau) g(\tau, u(\tau))\, d\tau\right) ds, & t \in [0,T]. \\ v(t) = \displaystyle\int_0^T \tilde{G}_2(t,s) g(s, u(s))\, ds, & t \in [0,T]. \end{cases}$$

We consider the Banach space X and the cones C' and C_0 from Section 1.5.2. We also define the operator $\mathcal{D} : C' \to X$ by

$$(\mathcal{D}u)(t) = \int_0^T \tilde{G}_1(t,s) f\left(s, \int_0^T \tilde{G}_2(s,\tau) g(\tau, u(\tau))\, d\tau\right) ds, \ t \in [0,T].$$

Under assumptions ($\widetilde{\text{L}1}$) and ($\widetilde{\text{L}2}$), by using arguments similar to those used in the proof of Lemma 1.4.1 (see Henderson and Luca, 2013i), we deduce that the operator $\mathcal{D} : C' \to C'$ is completely continuous and $\mathcal{D}(C') \subset C_0$.

By using Lemmas 1.5.6–1.5.11, Theorem 1.1.1, and methods similar to those used in the proofs of Theorems 1.4.1 and 1.4.2, we obtain the following results:

Theorem 1.5.12. *Assume that* ($\widetilde{\text{L}1}$) *and* ($\widetilde{\text{L}2}$) *hold. If the functions f and g also satisfy the following conditions* ($\widetilde{\text{L}3}$) *and* ($\widetilde{\text{L}4}$), *then problem* (S_0')–(BC_0') *has at least one positive solution* $(u(t), v(t)), t \in [0,T]$*:*

($\widetilde{\text{L}3}$) *There exist* $r_1, r_2 \in (0,\infty)$ *with* $r_1 r_2 \geq 1$ *such that*

$$(1)\ q_{10}^{\text{s}} = \limsup_{x \to 0^+} \frac{q_1(x)}{x^{r_1}} \in [0,\infty) \quad \textit{and} \quad (2)\ q_{20}^{\text{s}} = \limsup_{x \to 0^+} \frac{q_2(x)}{x^{r_2}} = 0.$$

($\widetilde{\text{L}4}$) *There exist* $l_1, l_2 \in (0,\infty)$ *with* $l_1 l_2 \geq 1$ *such that*

$$(1)\ \hat{f}_\infty^{\text{i}} = \liminf_{x \to \infty} \inf_{t \in [\theta_0, T]} \frac{f(t,x)}{x^{l_1}} \in (0,\infty] \quad \textit{and} \quad (2)\ \hat{g}_\infty^{\text{i}} = \liminf_{x \to \infty} \inf_{t \in [\theta_0, T]} \frac{g(t,x)}{x^{l_2}} = \infty.$$

Theorem 1.5.13. *Assume that* ($\widetilde{\text{L}1}$) *and* ($\widetilde{\text{L}2}$) *hold. If the functions f and g also satisfy the following conditions* ($\widetilde{\text{L}5}$) *and* ($\widetilde{\text{L}6}$), *then problem* (S_0')–(BC_0') *has at least one positive solution* $(u(t), v(t)), t \in [0,T]$*:*

$(\widetilde{L5})$ *There exist* $\alpha_1, \alpha_2 \in (0, \infty)$ *with* $\alpha_1\alpha_2 \le 1$ *such that*

$$(1)\ q_{1\infty}^{s} = \limsup_{x\to\infty} \frac{q_1(x)}{x^{\alpha_1}} \in [0,\infty) \quad \text{and} \quad (2)\ q_{2\infty}^{s} = \limsup_{x\to\infty} \frac{q_2(x)}{x^{\alpha_2}} = 0.$$

$(\widetilde{L6})$ *There exist* $\beta_1, \beta_2 \in (0, \infty)$ *with* $\beta_1\beta_2 \le 1$ *such that*

$$(1)\ \hat{f}_0^{i} = \liminf_{x\to 0+} \inf_{t\in[\theta_0,T]} \frac{f(t,x)}{x^{\beta_1}} \in (0,\infty] \quad \text{and} \quad (2)\ \hat{g}_0^{i} = \liminf_{x\to 0+} \inf_{t\in[\theta_0,T]} \frac{g(t,x)}{x^{\beta_2}} = \infty.$$

In the proofs of Theorems 1.5.12 and 1.5.13, we obtain in a manner similar to that in Section 1.5.2 that $u(t) > 0$ and $v(t) > 0$ for all $t \in (0, T]$.

Remark 1.5.3. The results presented in this section were published in Henderson and Luca (2013i).

1.6 Boundary conditions with additional positive constants

In this section, we shall investigate the existence and nonexistence of positive solutions for a system of second-order ordinary differential equations with integral boundary conditions in which some positive constants appear.

1.6.1 Presentation of the problem

We consider the system of nonlinear second-order ordinary differential equations

$$\begin{cases} u''(t) + p(t)f(v(t)) = 0, & 0 < t < 1, \\ v''(t) + q(t)g(u(t)) = 0, & 0 < t < 1, \end{cases} \tag{S^0}$$

with the integral boundary conditions

$$\begin{cases} \alpha u(0) - \beta u'(0) = \int_0^1 u(s)\,dH_1(s), & \gamma u(1) + \delta u'(1) = \int_0^1 u(s)\,dH_2(s) + a_0, \\ \tilde{\alpha} v(0) - \tilde{\beta} v'(0) = \int_0^1 v(s)\,dK_1(s), & \tilde{\gamma} v(1) + \tilde{\delta} v'(1) = \int_0^1 v(s)\,dK_2(s) + b_0, \end{cases} \tag{BC^0}$$

where the above integrals are Riemann–Stieltjes integrals, and a_0 and b_0 are positive constants.

By using the Schauder fixed point theorem, we shall prove the existence of positive solutions of problem (S^0)–(BC^0). By a positive solution of (S^0)–(BC^0) we mean a pair of functions $(u, v) \in (C^2([0, 1]; \mathbb{R}_+))^2$ satisfying (S^0) and (BC^0), with $u(t) > 0$ and $v(t) > 0$ for all $t \in (0, 1]$. We shall also give sufficient conditions for the nonexistence of the positive solutions for this problem. The particular case of the above problem when $\gamma = \tilde{\gamma} = 1$, $\delta = \tilde{\delta} = 0$, the functions H_1 and K_1 are constant, and

H_2 and K_2 are step functions was investigated in Henderson and Luca (2012d). We also mention Li and Sun (2006), where the authors studied the existence and nonexistence of positive solutions for the m-point boundary value problem on time scales $u^{\Delta\nabla}(t) + a(t)f(u(t)) = 0, t \in (0, T)$, $\beta u(0) - \gamma u^{\Delta}(0) = 0$, $u(T) - \sum_{i=1}^{m-2} a_i u(\xi_i) = b$, $m \geq 3$, and $b > 0$.

We present the assumptions that we shall use in the sequel:

(J1) $\alpha, \beta, \gamma, \delta, \tilde{\alpha}, \tilde{\beta}, \tilde{\gamma}, \tilde{\delta} \in [0, \infty)$.
(J2) $H_1, H_2, K_1, K_2 : [0, 1] \to \mathbb{R}$ are nondecreasing functions.
(J3) $\alpha - \int_0^1 \mathrm{d}H_1(\tau) > 0$, $\gamma - \int_0^1 \mathrm{d}H_2(\tau) > 0$, $\tilde{\alpha} - \int_0^1 \mathrm{d}K_1(\tau) > 0$, $\tilde{\gamma} - \int_0^1 \mathrm{d}K_2(\tau) > 0$.
(J4) The functions $p, q : [0, 1] \to [0, \infty)$ are continuous, and there exist $t_1, t_2 \in (0, 1)$ such that $p(t_1) > 0, q(t_2) > 0$.
(J5) $f, g : [0, \infty) \to [0, \infty)$ are continuous functions, and there exists $c_0 > 0$ such that $f(u) < \frac{c_0}{L}$, $g(u) < \frac{c_0}{L}$ for all $u \in [0, c_0]$, where $L = \max\{\int_0^1 p(s)J_1(s)\,\mathrm{d}s, \int_0^1 q(s)J_2(s)\,\mathrm{d}s\}$, and J_1 and J_2 are defined in Section 1.1.2 (see also Section 1.6.2).
(J6) $f, g : [0, \infty) \to [0, \infty)$ are continuous functions and satisfy the conditions

$$\lim_{u\to\infty} \frac{f(u)}{u} = \infty, \quad \lim_{u\to\infty} \frac{g(u)}{u} = \infty.$$

By (J4), we deduce that $\int_0^1 p(s)J_1(s)\,\mathrm{d}s > 0$ and $\int_0^1 q(s)J_2(s)\,\mathrm{d}s > 0$—that is, the constant L from (J5) is positive. In the proof of our main result, we shall use the Schauder fixed point theorem, which we present now:

Theorem 1.6.1. *Let X be a Banach space and $Y \subset X$ be a nonempty, bounded, convex, and closed subset. If the operator $A : Y \to Y$ is completely continuous, then A has at least one fixed point.*

1.6.2 Auxiliary results

We consider the second-order differential equations with integral boundary conditions

$$u''(t) + \tilde{x}(t) = 0, \quad t \in (0, 1), \tag{1.71}$$

$$\alpha u(0) - \beta u'(0) = \int_0^1 u(s)\,\mathrm{d}H_1(s), \quad \gamma u(1) + \delta u'(1) = \int_0^1 u(s)\,\mathrm{d}H_2(s), \tag{1.72}$$

and

$$v''(t) + \tilde{y}(t) = 0, \quad t \in (0, 1), \tag{1.73}$$

$$\tilde{\alpha} v(0) - \tilde{\beta} v'(0) = \int_0^1 v(s)\,\mathrm{d}K_1(s), \quad \tilde{\gamma} v(1) + \tilde{\delta} v'(1) = \int_0^1 v(s)\,\mathrm{d}K_2(s); \tag{1.74}$$

that is, problem (1.1)–(1.2) and problem (1.8)–(1.9) from Section 1.1.2 with $a(t) = c(t) = 1$ and $b(t) = d(t) = 0$ for all $t \in [0, 1]$.

The functions ψ and ϕ defined by (1.3) and (1.4) are $\psi(t) = \alpha t + \beta$ and $\phi(t) = -\gamma t + \gamma + \delta$, and the corresponding functions $\tilde{\psi}$ and $\tilde{\phi}$ are $\tilde{\psi}(t) = \tilde{\alpha} t + \tilde{\beta}$ and $\tilde{\phi}(t) = -\tilde{\gamma} t + \tilde{\gamma} + \tilde{\delta}$. In addition, we have $\tau_1 = \alpha\gamma + \alpha\delta + \beta\gamma$ and $\tau_2 = \tilde{\alpha}\tilde{\gamma} + \tilde{\alpha}\tilde{\delta} + \tilde{\beta}\tilde{\gamma}$.

By using assumptions (J1)–(J3), we deduce that assumptions (I2)–(I4) from Section 1.1.3 are satisfied. By (J1)–(J3), we obtain $\alpha > 0$, $\gamma > 0$, $\tilde{\alpha} > 0$, and $\tilde{\gamma} > 0$, and so $\alpha+\beta > 0$, $\gamma+\delta > 0$, $\alpha+\gamma > 0$, $\tilde{\alpha}+\tilde{\beta} > 0$, $\tilde{\gamma}+\tilde{\delta} > 0$, and $\tilde{\alpha}+\tilde{\gamma} > 0$. Besides,

$$\tau_1 - \int_0^1 \phi(s)\,\mathrm{d}H_1(s) = (\gamma+\delta)\left(\alpha - \int_0^1 \mathrm{d}H_1(s)\right) + \beta\gamma + \gamma\int_0^1 s\,\mathrm{d}H_1(s) > 0,$$

$$\tau_1 - \int_0^1 \psi(s)\,\mathrm{d}H_2(s) = (\alpha+\beta)\left(\gamma - \int_0^1 \mathrm{d}H_2(s)\right) + \alpha\delta + \alpha\int_0^1 (1-s)\,\mathrm{d}H_2(s) > 0,$$

$$\tau_2 - \int_0^1 \tilde{\phi}(s)\,\mathrm{d}K_1(s) = (\tilde{\gamma}+\tilde{\delta})\left(\tilde{\alpha} - \int_0^1 \mathrm{d}K_1(s)\right) + \tilde{\beta}\tilde{\gamma} + \tilde{\gamma}\int_0^1 s\,\mathrm{d}K_1(s) > 0,$$

$$\tau_2 - \int_0^1 \tilde{\psi}(s)\,\mathrm{d}K_2(s) = (\tilde{\alpha}+\tilde{\beta})\left(\tilde{\gamma} - \int_0^1 \mathrm{d}K_2(s)\right) + \tilde{\alpha}\tilde{\delta} + \tilde{\alpha}\int_0^1 (1-s)\,\mathrm{d}K_2(s) > 0.$$

For Δ_1 and Δ_2 from Section 1.1.2, we conclude after some computations that $\Delta_1 = \tau_1\tilde{\Delta}_1 > 0$ and $\Delta_2 = \tau_2\tilde{\Delta}_2 > 0$, where

$$\begin{aligned}\tilde{\Delta}_1 &= \left(\beta + \int_0^1 s\,\mathrm{d}H_1(s)\right)\left(\gamma - \int_0^1 \mathrm{d}H_2(s)\right)\\ &\quad + \left(\alpha - \int_0^1 \mathrm{d}H_1(s)\right)\left(\gamma+\delta - \int_0^1 s\,\mathrm{d}H_2(s)\right) > 0,\\ \tilde{\Delta}_2 &= \left(\tilde{\beta} + \int_0^1 s\,\mathrm{d}K_1(s)\right)\left(\tilde{\gamma} - \int_0^1 \mathrm{d}K_2(s)\right)\\ &\quad + \left(\tilde{\alpha} - \int_0^1 \mathrm{d}K_1(s)\right)\left(\tilde{\gamma}+\tilde{\delta} - \int_0^1 s\,\mathrm{d}K_2(s)\right) > 0.\end{aligned}$$

In this way, all auxiliary results in Lemmas 1.1.1–1.1.7 from Section 1.1.2 for problem (1.71)–(1.72) and the corresponding auxiliary results for problem (1.73)–(1.74) are satisfied.

Therefore, under assumptions (J1)–(J3) and for $\tilde{x}, \tilde{y} \in C(0,1)\cap L^1(0,1)$, the solutions of problems (1.71)–(1.72) and (1.73)–(1.74) are given by $u(t) = \int_0^1 G_1(t,s)\tilde{x}(s)\,\mathrm{d}s$ and $v(t) = \int_0^1 G_2(t,s)\tilde{y}(s)\,\mathrm{d}s$, respectively, where

$$\begin{aligned}G_1(t,s) &= g_1(t,s) + \frac{1}{\Delta_1}\left[(\alpha t+\beta)\int_0^1 (-\gamma s+\gamma+\delta)\,\mathrm{d}H_2(s)\right.\\ &\quad \left. + (-\gamma t+\gamma+\delta)\left(\alpha\gamma+\alpha\delta+\beta\gamma - \int_0^1 (\alpha s+\beta)\,\mathrm{d}H_2(s)\right)\right]\int_0^1 g_1(\tau,s)\,\mathrm{d}H_1(\tau)\\ &\quad + \frac{1}{\Delta_1}\left[(\alpha t+\beta)\left(\alpha\gamma+\alpha\delta+\beta\gamma - \int_0^1 (-\gamma s+\gamma+\delta)\,\mathrm{d}H_1(s)\right)\right.\\ &\quad \left. + (-\gamma t+\gamma+\delta)\int_0^1 (\alpha s+\beta)\,\mathrm{d}H_1(s)\right]\int_0^1 g_1(\tau,s)\,\mathrm{d}H_2(\tau),\end{aligned}$$

$$
\begin{aligned}
G_2(t,s) = g_2(t,s) &+ \frac{1}{\Delta_2}\left[(\tilde{\alpha}t+\tilde{\beta})\int_0^1(-\tilde{\gamma}s+\tilde{\gamma}+\tilde{\delta})\,\mathrm{d}K_2(s)\right.\\
&\left.+(-\tilde{\gamma}t+\tilde{\gamma}+\tilde{\delta})\left(\tilde{\alpha}\tilde{\gamma}+\tilde{\alpha}\tilde{\delta}+\tilde{\beta}\tilde{\gamma}-\int_0^1(\tilde{\alpha}s+\tilde{\beta})\,\mathrm{d}K_2(s)\right)\right]\int_0^1 g_2(\tau,s)\,\mathrm{d}K_1(\tau)\\
&+\frac{1}{\Delta_2}\left[(\tilde{\alpha}t+\tilde{\beta})\left(\tilde{\alpha}\tilde{\gamma}+\tilde{\alpha}\tilde{\delta}+\tilde{\beta}\tilde{\gamma}-\int_0^1(-\tilde{\gamma}s+\tilde{\gamma}+\tilde{\delta})\,\mathrm{d}K_1(s)\right)\right.\\
&\left.+(-\tilde{\gamma}t+\tilde{\gamma}+\tilde{\delta})\int_0^1(\tilde{\alpha}s+\tilde{\beta})\,\mathrm{d}K_1(s)\right]\int_0^1 g_2(\tau,s)\,\mathrm{d}K_2(\tau),
\end{aligned}
$$

and

$$
g_1(t,s)=\frac{1}{\alpha\gamma+\alpha\delta+\beta\gamma}\begin{cases}(-\gamma t+\gamma+\delta)(\alpha s+\beta), & 0\le s\le t\le 1,\\[2mm] (-\gamma s+\gamma+\delta)(\alpha t+\beta), & 0\le t\le s\le 1,\end{cases}
$$

$$
g_2(t,s)=\frac{1}{\tilde{\alpha}\tilde{\gamma}+\tilde{\alpha}\tilde{\delta}+\tilde{\beta}\tilde{\gamma}}\begin{cases}(-\tilde{\gamma}t+\tilde{\gamma}+\tilde{\delta})(\tilde{\alpha}s+\tilde{\beta}), & 0\le s\le t\le 1,\\[2mm] (-\tilde{\gamma}s+\tilde{\gamma}+\tilde{\delta})(\tilde{\alpha}t+\tilde{\beta}), & 0\le t\le s\le 1.\end{cases}
$$

From assumption (J4), there exists $\sigma\in(0,1/2)$ such that $t_1,t_2\in(\sigma,1-\sigma)$. Besides, the constants ν_1 and ν_2 from Section 1.1.2 (Lemma 1.1.4) are $\nu_1=\min\left\{\frac{\gamma\sigma+\delta}{\gamma+\delta},\frac{\alpha\sigma+\beta}{\alpha+\beta}\right\}$ and $\nu_2=\min\left\{\frac{\tilde{\gamma}\sigma+\tilde{\delta}}{\tilde{\gamma}+\tilde{\delta}},\frac{\tilde{\alpha}\sigma+\tilde{\beta}}{\tilde{\alpha}+\tilde{\beta}}\right\}$, and the functions J_1 and J_2 from Section 1.1.2 are given by

$$
\begin{aligned}
J_1(s) = g_1(s,s) &+ \frac{1}{\Delta_1}\left[(\alpha+\beta)\int_0^1(-\gamma s+\gamma+\delta)\,\mathrm{d}H_2(s)\right.\\
&\left.+(\gamma+\delta)\left(\alpha\gamma+\alpha\delta+\beta\gamma-\int_0^1(\alpha s+\beta)\,\mathrm{d}H_2(s)\right)\right]\int_0^1 g_1(\tau,s)\,\mathrm{d}H_1(\tau)\\
&+\frac{1}{\Delta_1}\left[(\alpha+\beta)\left(\alpha\gamma+\alpha\delta+\beta\gamma-\int_0^1(-\gamma s+\gamma+\delta)\,\mathrm{d}H_1(s)\right)\right.\\
&\left.+(\gamma+\delta)\int_0^1(\alpha s+\beta)\,\mathrm{d}H_1(s)\right]\int_0^1 g_1(\tau,s)\,\mathrm{d}H_2(\tau),\\
J_2(s) = g_2(s,s) &+ \frac{1}{\Delta_2}\left[(\tilde{\alpha}+\tilde{\beta})\int_0^1(-\tilde{\gamma}s+\tilde{\gamma}+\tilde{\delta})\,\mathrm{d}K_2(s)\right.\\
&\left.+(\tilde{\gamma}+\tilde{\delta})\left(\tilde{\alpha}\tilde{\gamma}+\tilde{\alpha}\tilde{\delta}+\tilde{\beta}\tilde{\gamma}-\int_0^1(\tilde{\alpha}s+\tilde{\beta})\,\mathrm{d}K_2(s)\right)\right]\int_0^1 g_2(\tau,s)\,\mathrm{d}K_1(\tau)\\
&+\frac{1}{\Delta_2}\left[(\tilde{\alpha}+\tilde{\beta})\left(\tilde{\alpha}\tilde{\gamma}+\tilde{\alpha}\tilde{\delta}+\tilde{\beta}\tilde{\gamma}-\int_0^1(-\tilde{\gamma}s+\tilde{\gamma}+\tilde{\delta})\,\mathrm{d}K_1(s)\right)\right.\\
&\left.+(\tilde{\gamma}+\tilde{\delta})\int_0^1(\tilde{\alpha}s+\tilde{\beta})\,\mathrm{d}K_1(s)\right]\int_0^1 g_2(\tau,s)\,\mathrm{d}K_2(\tau).
\end{aligned}
$$

1.6.3 Main results

Our first theorem is the following existence result for problem (S^0)–(BC^0):

Theorem 1.6.2. *Assume that assumptions (J1)–(J5) hold. Then problem* (S^0)–(BC^0) *has at least one positive solution for* $a_0 > 0$ *and* $b_0 > 0$ *sufficiently small.*

Proof. We consider the problems

$$\begin{cases} h''(t) = 0, & t \in (0,1), \\ \alpha h(0) - \beta h'(0) = \int_0^1 h(s)\,dH_1(s), & \gamma h(1) + \delta h'(1) = \int_0^1 h(s)\,dH_2(s) + 1, \end{cases} \tag{1.75}$$

$$\begin{cases} k''(t) = 0, & t \in (0,1), \\ \tilde{\alpha} k(0) - \tilde{\beta} k'(0) = \int_0^1 k(s)\,dK_1(s), & \tilde{\gamma} k(1) + \tilde{\delta} k'(1) = \int_0^1 k(s)\,dK_2(s) + 1. \end{cases} \tag{1.76}$$

Problems (1.75) and (1.76) have the solutions

$$h(t) = \frac{\tau_1}{\Delta_1}\left[t\left(\alpha - \int_0^1 dH_1(s)\right) + \beta + \int_0^1 s\,dH_1(s)\right], \quad t \in [0,1],$$

$$k(t) = \frac{\tau_2}{\Delta_2}\left[t\left(\tilde{\alpha} - \int_0^1 dK_1(s)\right) + \tilde{\beta} + \int_0^1 s\,dK_1(s)\right], \quad t \in [0,1], \tag{1.77}$$

respectively. By assumptions (J1)–(J3), we obtain $h(t) > 0$ and $k(t) > 0$ for all $t \in (0,1]$.

We define the functions $x(t)$ and $y(t)$, $t \in [0,1]$ by

$$x(t) = u(t) - a_0 h(t), \quad y(t) = v(t) - b_0 k(t), \quad t \in [0,1],$$

where (u,v) is a solution of (S^0)–(BC^0). Then (S^0)–(BC^0) can be equivalently written as

$$\begin{cases} x''(t) + p(t) f(y(t) + b_0 k(t)) = 0, & t \in (0,1), \\ y''(t) + q(t) g(x(t) + a_0 h(t)) = 0, & t \in (0,1), \end{cases} \tag{1.78}$$

with the boundary conditions

$$\begin{cases} \alpha x(0) - \beta x'(0) = \int_0^1 x(s)\,dH_1(s), & \gamma x(1) + \delta x'(1) = \int_0^1 x(s)\,dH_2(s), \\ \tilde{\alpha} y(0) - \tilde{\beta} y'(0) = \int_0^1 y(s)\,dK_1(s), & \tilde{\gamma} y(1) + \tilde{\delta} y'(1) = \int_0^1 y(s)\,dK_2(s). \end{cases} \tag{1.79}$$

Using the Green's functions G_1 and G_2 from Section 1.6.2, we find a pair (x,y) is a solution of problem (1.78)–(1.79) if and only if (x,y) is a solution for the nonlinear integral equations

$$\begin{cases} x(t) = \int_0^1 G_1(t,s)p(s)f\left(\int_0^1 G_2(s,\tau)q(\tau)g(x(\tau)+a_0h(\tau))\,\mathrm{d}\tau + b_0k(s)\right)\mathrm{d}s, & 0 \le t \le 1, \\ y(t) = \int_0^1 G_2(t,s)q(s)g(x(s)+a_0h(s))\,\mathrm{d}s, & 0 \le t \le 1, \end{cases} \tag{1.80}$$

where $h(t)$ and $k(t), t \in [0,1]$ are given by (1.77).

We consider the Banach space $X = C([0,1])$ with the supremum norm $\|\cdot\|$, and define the set

$$E = \{x \in C([0,1]), \quad 0 \le x(t) \le c_0, \quad \forall t \in [0,1]\} \subset X.$$

We also define the operator $\mathcal{S} : E \to X$ by

$$(\mathcal{S}x)(t) = \int_0^1 G_1(t,s)p(s)f\left(\int_0^1 G_2(s,\tau)q(\tau)g(x(\tau)+a_0h(\tau))\,\mathrm{d}\tau + b_0k(s)\right)\mathrm{d}s,$$
$$0 \le t \le 1, \quad x \in E.$$

For sufficiently small $a_0 > 0$ and $b_0 > 0$, by (J5), we deduce

$$f(y(t)+b_0k(t)) \le \frac{c_0}{L}, \quad g(x(t)+a_0h(t)) \le \frac{c_0}{L}, \quad \forall t \in [0,1], \quad \forall x,y \in E.$$

Then, by using Lemma 1.1.5, we obtain $(\mathcal{S}x)(t) \ge 0$ for all $t \in [0,1]$ and $x \in E$. By Lemma 1.1.6, for all $x \in E$, we have

$$\begin{aligned} \int_0^1 G_2(s,\tau)q(\tau)g(x(\tau)+a_0h(\tau))\,\mathrm{d}\tau &\le \int_0^1 J_2(\tau)q(\tau)g(x(\tau)+a_0h(\tau))\,\mathrm{d}\tau \\ &\le \frac{c_0}{L}\int_0^1 J_2(\tau)q(\tau)\,\mathrm{d}\tau \le c_0, \quad \forall s \in [0,1], \end{aligned}$$

and

$$\begin{aligned} (\mathcal{S}x)(t) &\le \int_0^1 J_1(s)p(s)f\left(\int_0^1 G_2(s,\tau)q(\tau)g(x(\tau)+a_0h(\tau))\,\mathrm{d}\tau + b_0k(s)\right)\mathrm{d}s \\ &\le \frac{c_0}{L}\int_0^1 J_1(s)p(s)\,\mathrm{d}s \le c_0, \quad \forall t \in [0,1]. \end{aligned}$$

Therefore, $\mathcal{S}(E) \subset E$.

Using standard arguments, we deduce that $\mathcal{S}$ is completely continuous ($\mathcal{S}$ is compact, i.e., for any bounded set $B \subset E, \mathcal{S}(B) \subset E$ is relatively compact from the Ascoli–Arzèla theorem, and $\mathcal{S}$ is continuous). By Theorem 1.6.1, we conclude that $\mathcal{S}$ has a fixed point $x \in E$. This element together with y given by (1.80) represents a solution for (1.78)–(1.79). This shows that our problem (S^0)–(BC^0) has a positive solution (u,v) with $u = x + a_0h$, $v = y + b_0k$ for sufficiently small $a_0 > 0$ and $b_0 > 0$. □

In what follows, we present sufficient conditions for the nonexistence of positive solutions of (S^0)–(BC^0).

Theorem 1.6.3. *Assume that assumptions (J1)–(J4) and (J6) hold. Then problem* (S^0)–(BC^0) *has no positive solution for* a_0 *and* b_0 *sufficiently large.*

Proof. We suppose that (u, v) is a positive solution of (S^0)–(BC^0). Then (x, y) with $x = u - a_0 h$, $y = v - b_0 k$ is a solution for (1.78)–(1.79), where h and k are the solutions of problems (1.75) and (1.76), respectively, (given by (1.77)). By (J4), there exists $\sigma \in (0, 1/2)$ such that $t_1, t_2 \in (\sigma, 1-\sigma)$, and then $\int_\sigma^{1-\sigma} p(s)J_1(s)\,ds > 0$, $\int_\sigma^{1-\sigma} q(s)J_2(s)\,ds > 0$. Now by using Lemma 1.1.5, we have $x(t) \geq 0$, $y(t) \geq 0$ for all $t \in [0, 1]$, and by Lemma 1.1.7 we obtain $\inf_{t\in[\sigma,1-\sigma]} x(t) \geq \nu_1 \|x\|$ and $\inf_{t\in[\sigma,1-\sigma]} y(t) \geq \nu_2 \|y\|$.

Using now (1.77), we deduce that

$$\inf_{t\in[\sigma,1-\sigma]} h(t) = h(\sigma) = \frac{h(\sigma)}{h(1)}\|h\|, \qquad \inf_{t\in[\sigma,1-\sigma]} k(t) = k(\sigma) = \frac{k(\sigma)}{k(1)}\|k\|.$$

Therefore, we obtain

$$\inf_{t\in[\sigma,1-\sigma]} (x(t) + a_0 h(t)) \geq \nu_1 \|x\| + a_0 \frac{h(\sigma)}{h(1)}\|h\| \geq r_1(\|x\| + a_0\|h\|) \geq r_1\|x + a_0 h\|,$$

$$\inf_{t\in[\sigma,1-\sigma]} (y(t) + b_0 k(t)) \geq \nu_2 \|y\| + b_0 \frac{k(\sigma)}{k(1)}\|k\| \geq r_2(\|y\| + b_0\|k\|) \geq r_2\|y + b_0 k\|,$$

where

$$r_1 = \min\left\{\nu_1, \frac{h(\sigma)}{h(1)}\right\} = \min\left\{\nu_1, \frac{\sigma\left(\alpha - \int_0^1 dH_1(s)\right) + \beta + \int_0^1 s\,dH_1(s)}{\alpha - \int_0^1 dH_1(s) + \beta + \int_0^1 s\,dH_1(s)}\right\},$$

$$r_2 = \min\left\{\nu_2, \frac{k(\sigma)}{k(1)}\right\} = \min\left\{\nu_2, \frac{\sigma\left(\tilde{\alpha} - \int_0^1 dK_1(s)\right) + \tilde{\beta} + \int_0^1 s\,dK_1(s)}{\tilde{\alpha} - \int_0^1 dK_1(s) + \tilde{\beta} + \int_0^1 s\,dK_1(s)}\right\}.$$

We now consider $R = (\min\{\nu_2 r_1 \int_\sigma^{1-\sigma} q(s)J_2(s)\,ds, \nu_1 r_2 \int_\sigma^{1-\sigma} p(s)J_1(s)\,ds\})^{-1} > 0$.

By (J6), for R defined above, we conclude that there exists $M > 0$ such that $f(u) > 2Ru$, $g(u) > 2Ru$ for all $u \geq M$. We consider $a_0 > 0$ and $b_0 > 0$ sufficiently large such that

$$\inf_{t\in[\sigma,1-\sigma]} (x(t) + a_0 h(t)) \geq M, \qquad \inf_{t\in[\sigma,1-\sigma]} (y(t) + b_0 k(t)) \geq M.$$

By (J4), (1.78), (1.79), and the above inequalities, we deduce that $\|x\| > 0$ and $\|y\| > 0$.

Now by using Lemma 1.1.6 and the above considerations, we have

$$\begin{aligned} y(\sigma) &= \int_0^1 G_2(\sigma, s)q(s)g(x(s) + a_0 h(s))\,ds \geq \nu_2 \int_0^1 J_2(s)q(s)g(x(s) + a_0 h(s))\,ds \\ &\geq \nu_2 \int_\sigma^{1-\sigma} J_2(s)q(s)g(x(s) + a_0 h(s))\,ds \geq 2R\nu_2 \int_\sigma^{1-\sigma} J_2(s)q(s)(x(s) + a_0 h(s))\,ds \\ &\geq 2R\nu_2 \int_\sigma^{1-\sigma} J_2(s)q(s) \inf_{\tau\in[\sigma,1-\sigma]} (x(\tau) + a_0 h(\tau))\,ds \\ &\geq 2R\nu_2 r_1 \int_\sigma^{1-\sigma} J_2(s)q(s)\|x + a_0 h\|\,ds \geq 2\|x + a_0 h\| \geq 2\|x\|. \end{aligned}$$

Therefore, we obtain

$$\|x\| \le y(\sigma)/2 \le \|y\|/2. \tag{1.81}$$

In a similar manner, we deduce

$$\begin{aligned} x(\sigma) &= \int_0^1 G_1(\sigma,s)p(s)f(y(s)+b_0k(s))\,\mathrm{d}s \ge \nu_1 \int_0^1 J_1(s)p(s)f(y(s)+b_0k(s))\,\mathrm{d}s \\ &\ge \nu_1 \int_\sigma^{1-\sigma} J_1(s)p(s)f(y(s)+b_0k(s))\,\mathrm{d}s \ge 2R\nu_1 \int_\sigma^{1-\sigma} J_1(s)p(s)(y(s)+b_0k(s))\,\mathrm{d}s \\ &\ge 2R\nu_1 \int_\sigma^{1-\sigma} J_1(s)p(s) \inf_{\tau\in[\sigma,1-\sigma]}(y(\tau)+b_0k(\tau))\,\mathrm{d}s \\ &\ge 2R\nu_1 r_2 \int_\sigma^{1-\sigma} J_1(s)p(s)\|y+b_0k\|\,\mathrm{d}s \ge 2\|y+b_0k\| \ge 2\|y\|. \end{aligned}$$

So, we obtain

$$\|y\| \le x(\sigma)/2 \le \|x\|/2. \tag{1.82}$$

By (1.81) and (1.82), we conclude that $\|x\| \le \|y\|/2 \le \|x\|/4$, which is a contradiction, because $\|x\| > 0$. Then, for a_0 and b_0 sufficiently large, our problem (S^0)–(BC^0) has no positive solution. □

Results similar to those of Theorems 1.6.2 and 1.6.3 can be obtained if instead of boundary conditions (BC^0) we have

$$\begin{cases} \alpha u(0) - \beta u'(0) = \int_0^1 u(s)\,\mathrm{d}H_1(s) + a_0, & \gamma u(1) + \delta u'(1) = \int_0^1 u(s)\,\mathrm{d}H_2(s), \\ \tilde{\alpha} v(0) - \tilde{\beta} v'(0) = \int_0^1 v(s)\,\mathrm{d}K_1(s) + b_0, & \tilde{\gamma} v(1) + \tilde{\delta} v'(1) = \int_0^1 v(s)\,\mathrm{d}K_2(s), \end{cases} \tag{BC_1^0}$$

or

$$\begin{cases} \alpha u(0) - \beta u'(0) = \int_0^1 u(s)\,\mathrm{d}H_1(s) + a_0, & \gamma u(1) + \delta u'(1) = \int_0^1 u(s)\,\mathrm{d}H_2(s), \\ \tilde{\alpha} v(0) - \tilde{\beta} v'(0) = \int_0^1 v(s)\,\mathrm{d}K_1(s), & \tilde{\gamma} v(1) + \tilde{\delta} v'(1) = \int_0^1 v(s)\,\mathrm{d}K_2(s) + b_0, \end{cases} \tag{BC_2^0}$$

or

$$\begin{cases} \alpha u(0) - \beta u'(0) = \int_0^1 u(s)\,\mathrm{d}H_1(s), & \gamma u(1) + \delta u'(1) = \int_0^1 u(s)\,\mathrm{d}H_2(s) + a_0, \\ \tilde{\alpha} v(0) - \tilde{\beta} v'(0) = \int_0^1 v(s)\,\mathrm{d}K_1(s) + b_0, & \tilde{\gamma} v(1) + \tilde{\delta} v'(1) = \int_0^1 v(s)\,\mathrm{d}K_2(s), \end{cases} \tag{BC_3^0}$$

where a_0 and b_0 are positive constants.

For problem (S^0)–(BC_1^0), instead of functions h and k from the proof of Theorem 1.6.2, the solutions of problems

$$\begin{cases} h_1''(t) = 0, & t \in (0,1), \\ \alpha h_1(0) - \beta h_1'(0) = \int_0^1 h_1(s)\,\mathrm{d}H_1(s) + 1, & \gamma h_1(1) + \delta h_1'(1) = \int_0^1 h_1(s)\,\mathrm{d}H_2(s), \end{cases} \tag{1.83}$$

$$\begin{cases} k_1''(t) = 0, & t \in (0,1), \\ \tilde{\alpha} k_1(0) - \tilde{\beta} k_1'(0) = \int_0^1 k_1(s)\,\mathrm{d}K_1(s) + 1, & \tilde{\gamma} k_1(1) + \tilde{\delta} k_1'(1) = \int_0^1 k_1(s)\,\mathrm{d}K_2(s), \end{cases} \tag{1.84}$$

are

$$h_1(t) = \frac{\tau_1}{\Delta_1}\left[-t\left(\gamma - \int_0^1 \mathrm{d}H_2(s)\right) + \gamma + \delta - \int_0^1 s\,\mathrm{d}H_2(s)\right], \quad t \in [0,1],$$

$$k_1(t) = \frac{\tau_2}{\Delta_2}\left[-t\left(\tilde{\gamma} - \int_0^1 \mathrm{d}K_2(s)\right) + \tilde{\gamma} + \tilde{\delta} - \int_0^1 s\,\mathrm{d}K_2(s)\right], \quad t \in [0,1],$$

respectively. By assumptions (J1)–(J3), we obtain $h_1(t) > 0$ and $k_1(t) > 0$ for all $t \in [0,1)$.

For problem (S^0)–(BC_2^0), instead of functions h and k from the proof of Theorem 1.6.2, the solutions of problems (1.83) and (1.76) are the functions h_1 and k, respectively, which satisfy $h_1(t) > 0$ for all $t \in [0,1)$ and $k(t) > 0$ for all $t \in (0,1]$. For problem (S^0)–(BC_3^0), instead of functions h and k from the proof of Theorem 1.6.2, the solutions of problems (1.75) and (1.84) are the functions h and k_1, respectively, which satisfy $h(t) > 0$ for all $t \in (0,1]$ and $k_1(t) > 0$ for all $t \in [0,1)$.

Therefore, we also obtain the following results:

Theorem 1.6.4. *Assume that assumptions (J1)–(J5) hold. Then problem* (S^0)–(BC_1^0) *has at least one positive solution* ($u(t) > 0$ *and* $v(t) > 0$ *for all* $t \in [0,1)$) *for* $a_0 > 0$ *and* $b_0 > 0$ *sufficiently small.*

Theorem 1.6.5. *Assume that assumptions (J1)–(J4) and (J6) hold. Then problem* (S^0)–(BC_1^0) *has no positive solution* ($u(t) > 0$ *and* $v(t) > 0$ *for all* $t \in [0,1)$) *for* a_0 *and* b_0 *sufficiently large.*

Theorem 1.6.6. *Assume that assumptions (J1)–(J5) hold. Then problem* (S^0)–(BC_2^0) *has at least one positive solution* ($u(t) > 0$ *for all* $t \in [0,1)$, *and* $v(t) > 0$ *for all* $t \in (0,1]$) *for* $a_0 > 0$ *and* $b_0 > 0$ *sufficiently small.*

Theorem 1.6.7. *Assume that assumptions (J1)–(J4) and (J6) hold. Then problem* (S^0)–(BC_2^0) *has no positive solution* ($u(t) > 0$ *for all* $t \in [0,1)$, *and* $v(t) > 0$ *for all* $t \in (0,1]$) *for* a_0 *and* b_0 *sufficiently large.*

Theorem 1.6.8. *Assume that assumptions (J1)–(J5) hold. Then problem* (S^0)–(BC_3^0) *has at least one positive solution* ($u(t) > 0$ *for all* $t \in (0,1]$, *and* $v(t) > 0$ *for all* $t \in [0,1)$) *for* $a_0 > 0$ *and* $b_0 > 0$ *sufficiently small.*

Theorem 1.6.9. *Assume that assumptions (J1)–(J4) and (J6) hold. Then problem* (S^0)–(BC_3^0) *has no positive solution* ($u(t) > 0$ *for all* $t \in (0,1]$, *and* $v(t) > 0$ *for all* $t \in [0,1)$) *for* a_0 *and* b_0 *sufficiently large.*

1.6.4 An example

Example 1.6.1. We consider $p(t) = at$, $q(t) = bt$ for all $t \in [0,1]$ with $a, b > 0$, $\alpha = 3$, $\beta = 2$, $\gamma = 2$, $\delta = 1$, $\tilde{\alpha} = 5$, $\tilde{\beta} = 2$, $\tilde{\gamma} = 2$, $\tilde{\delta} = 1/2$, $H_1(t) = t$, $H_2(t) = t^2$, $K_1(t) = t^3$, and $K_2(t) = \sqrt{t}$ for all $t \in [0,1]$. We also consider the functions $f, g : [0,\infty) \to [0,\infty)$, $f(x) = \frac{cx^3}{x+1}$, $g(x) = \frac{dx^4}{2x+3}$ for all $x \in [0,\infty)$, with $c, d > 0$. We have $\lim_{x\to\infty} f(x)/x = \lim_{x\to\infty} g(x)/x = \infty$.

Therefore, we consider the nonlinear second-order differential system

$$\begin{cases} u''(t) + at\dfrac{cv^3(t)}{v(t)+1} = 0, & t \in (0,1), \\[2ex] v''(t) + bt\dfrac{du^4(t)}{2u(t)+3} = 0, & t \in (0,1), \end{cases} \tag{$\widetilde{S^0}$}$$

with the boundary conditions

$$\begin{cases} 3u(0) - 2u'(0) = \displaystyle\int_0^1 u(s)\,ds, & 2u(1) + u'(1) = 2\displaystyle\int_0^1 su(s)\,ds + a_0, \\[2ex] 5v(0) - 2v'(0) = 3\displaystyle\int_0^1 s^2 v(s)\,ds, & 2v(1) + \dfrac{1}{2}v'(1) = \dfrac{1}{2}\displaystyle\int_0^1 \dfrac{1}{\sqrt{s}} v(s)\,ds + b_0. \end{cases} \tag{$\widetilde{BC^0}$}$$

We obtain $\alpha - \int_0^1 dH_1(\tau) = 2 > 0$, $\gamma - \int_0^1 dH_2(\tau) = 1 > 0$, $\tilde{\alpha} - \int_0^1 dK_1(\tau) = 4 > 0$, $\tilde{\gamma} - \int_0^1 dK_2(\tau) = 1 > 0$, $\psi(t) = 3t + 2$, $\phi(t) = -2t + 3$ for all $t \in [0,1]$, $\tau_1 = 13$, $\tilde{\Delta}_1 = \frac{43}{6}$, $\Delta_1 = \frac{559}{6}$, $\tilde{\psi}(t) = 5t + 2$, $\tilde{\phi}(t) = -2t + \frac{5}{2}$ for all $t \in [0,1]$, $\tau_2 = \frac{33}{2}$, $\tilde{\Delta}_2 = \frac{137}{12}$, and $\Delta_2 = \frac{1507}{8}$. So assumptions (J1)–(J4) and (J6) are satisfied.

In addition, we have

$$g_1(t,s) = \frac{1}{13}\begin{cases} (-2t+3)(3s+2), & 0 \le s \le t \le 1, \\ (-2s+3)(3t+2), & 0 \le t \le s \le 1, \end{cases}$$

$$g_2(t,s) = \frac{2}{33}\begin{cases} (-2t+\frac{5}{2})(5s+2), & 0 \le s \le t \le 1, \\ (-2s+\frac{5}{2})(5t+2), & 0 \le t \le s \le 1, \end{cases}$$

and the functions J_1 and J_2 are of the form

$$J_1(s) = g_1(s,s) + \frac{212}{559}\int_0^1 g_1(\tau,s)\,d\tau + \frac{786}{559}\int_0^1 \tau g_1(\tau,s)\,d\tau$$
$$= \frac{424 + 464s - 364s^2 - 131s^3}{559},$$

$$J_2(s) = g_2(s,s) + \frac{98}{137}\int_0^1 \tau^2 g_2(\tau,s)\,d\tau + \frac{983}{3014}\int_0^1 \frac{1}{\sqrt{\tau}} g_2(\tau,s)\,d\tau$$

$$= \frac{4312 + 8588s - 3932s^{3/2} - 5480s^2 - 539s^4}{9042}.$$

Then we deduce $L = \max\{a\int_0^1 sJ_1(s)\,ds, b\int_0^1 sJ_2(s)\,ds\}$, with $\int_0^1 sJ_1(s)\,ds \approx 0.4462731$ and $\int_0^1 sJ_2(s)\,ds \approx 0.2693436$. We choose $c_0 = 1$, and if we select c and d satisfying the conditions $c < \frac{2}{L}$, $d < \frac{5}{L}$, then we obtain $f(x) \le \frac{c}{2} < \frac{1}{L}$, $g(x) \le \frac{d}{5} < \frac{1}{L}$ for all $x \in [0,1]$. For example, if $a = 1$ and $b = 1/2$, then for $c \le 4.48$ and $d \le 11.2$ the above conditions for f and g are satisfied. So, assumption $\widetilde{(J5)}$ is also satisfied. By Theorems 1.6.2 and 1.6.3, we conclude that problem $(\widetilde{S}^0)$–(BC^0) has at least one positive solution for sufficiently small $a_0 > 0$ and $b_0 > 0$, and no positive solution for sufficiently large a_0 and b_0.

Systems of higher-order ordinary differential equations with multipoint boundary conditions

2

2.1 Existence and nonexistence of positive solutions for systems with parameters

In recent decades, nonlocal boundary value problems, including multipoint boundary value problems, for ordinary differential or difference equations have become a rapidly growing area of research. The study of these types of problems is driven not only by a theoretical interest but also by the fact that several phenomena in engineering, physics, and the life sciences can be modeled in this way. For example, problems with feedback controls such as the steady states of a thermostat, where a controller at one of its ends adds or removes heat, depending on the temperature registered at another point, can be interpreted with a second-order ordinary differential equation subject to a three-point boundary condition. Another example is represented by the vibrations of a guy wire of uniform cross section and composed of N parts of different densities, which can be set up as a multipoint boundary value problem (Moshinsky, 1950).

The study of multipoint boundary value problems for second-order differential equations was initiated by Il'in and Moiseev (1987a,b). Since then, such multipoint boundary value problems (continuous or discrete cases) have been studied by many authors by using different methods, such as fixed point theorems in cones, the Leray–Schauder continuation theorem, nonlinear alternatives of Leray–Schauder type, and coincidence degree theory.

2.1.1 Presentation of the problem

In this section, we consider the system of nonlinear higher-order ordinary differential equations

$$\begin{cases} u^{(n)}(t) + \lambda a(t) f(t, u(t), v(t)) = 0, & t \in (0, T), \\ v^{(m)}(t) + \mu b(t) g(t, u(t), v(t)) = 0, & t \in (0, T), \end{cases} \tag{S}$$

with the multipoint boundary conditions

Boundary Value Problems for Systems of Differential, Difference and Fractional Equations.
http://dx.doi.org/10.1016/B978-0-12-803652-5.00002-8

$$\begin{cases} u(0)=\sum_{i=1}^{p} a_i u(\xi_i), \quad u'(0)=\cdots=u^{(n-2)}(0)=0, \quad u(T)=\sum_{i=1}^{q} b_i u(\eta_i), \\ v(0)=\sum_{i=1}^{r} c_i v(\zeta_i), \quad v'(0)=\cdots=v^{(m-2)}(0)=0, \quad v(T)=\sum_{i=1}^{l} d_i v(\rho_i), \end{cases} \tag{BC}$$

where $n, m, p, q, r, l \in \mathbb{N}$, $n, m \geq 2$; $a_i, \xi_i \in \mathbb{R}$ for all $i=1,\ldots,p$; $b_i, \eta_i \in \mathbb{R}$ for all $i=1,\ldots,q$; $c_i, \zeta_i \in \mathbb{R}$ for all $i=1,\ldots,r$; $d_i, \rho_i \in \mathbb{R}$ for all $i=1,\ldots,l$; $0 < \xi_1 < \cdots < \xi_p < T$, $0 < \eta_1 < \cdots < \eta_q < T$, $0 < \zeta_1 < \cdots < \zeta_r < T$, $0 < \rho_1 < \cdots < \rho_l < T$. In the case $n=2$ or $m=2$, the boundary conditions above are of the form $u(0)=\sum_{i=1}^{p} a_i u(\xi_i)$, $u(T)=\sum_{i=1}^{q} b_i u(\eta_i)$, or $v(0)=\sum_{i=1}^{r} c_i v(\zeta_i)$, $v(T)=\sum_{i=1}^{l} d_i v(\rho_i)$, respectively—that is, without conditions on the derivatives of u and v at the point 0.

We shall give sufficient conditions on parameters λ and μ and on functions f and g such that positive solutions of (S)–(BC) exist. The nonexistence of positive solutions of (S)–(BC) will also be investigated. By a positive solution of problem (S)–(BC), we mean a pair of functions $(u, v) \in C^n([0,T]) \times C^m([0,T])$ satisfying (S) and (BC) with $u(t) \geq 0$, $v(t) \geq 0$ for all $t \in [0,T]$ and $(u, v) \neq (0,0)$. This problem is a generalization of the one studied in Henderson and Luca (2011), where $a_i=0$ for $i=1,\ldots,p$, and $c_i=0$ for $i=1,\ldots,r$. Some particular cases of the problem studied in Henderson and Luca (2011) were investigated in Luca (2012b) (where $n=m$, $q=l$, $b_i=d_i$, and $\eta_i=\rho_i$ for all $i=1,\ldots,q$), in Luca (2010) (where $n=m$, $f(t,u,v)=\tilde{f}(v)$, $g(t,u,v)=\tilde{g}(u)$ (denoted by $(\tilde{\mathrm{S}})$), $q=l$, $b_i=d_i$, $\eta_i=\rho_i$ for all $i=1,\ldots,q$), and in Henderson and Ntouyas (2007) (where the authors studied system $(\tilde{\mathrm{S}})$ with $T=1$ and the boundary conditions $u(0)=u'(0)=\cdots=u^{(n-2)}(0)=0$, $u(1)=\alpha u(\eta)$, $v(0)=v'(0)=\cdots=v^{(n-2)}(0)=0$, $v(1)=\alpha v(\eta)$ with $0<\eta<1$, $0<\alpha\eta^{n-1}<1$). We also mention Su et al. (2007), where the authors used the fixed point index theory to prove the existence of positive solutions for system (S) with $\lambda=\mu=1$, $T=1$ and (BC) with $a_i=0$ for $i=1,\ldots,p$, $c_i=0$ for $i=1,\ldots,r$ and $1/2 \leq \eta_1 < \eta_2 < \cdots < \eta_q < 1$, $1/2 \leq \rho_1 < \rho_2 < \cdots < \rho_l < 1$.

2.1.2 Auxiliary results

In this section, we present some auxiliary results related to the following nth-order differential equation with multipoint boundary conditions:

$$u^{(n)}(t) + y(t) = 0, \quad t \in (0,T), \tag{2.1}$$

$$u(0)=\sum_{i=1}^{p} a_i u(\xi_i), \quad u'(0)=\cdots=u^{(n-2)}(0)=0, \quad u(T)=\sum_{i=1}^{q} b_i u(\eta_i), \tag{2.2}$$

where $n, p, q \in \mathbb{N}$, $n \geq 2$, $a_i, \xi_i \in \mathbb{R}$ for all $i=1,\ldots,p$; $b_i, \eta_i \in \mathbb{R}$ for all $i=1,\ldots,q$; $0 < \xi_1 < \cdots < \xi_p < T$, $0 < \eta_1 < \cdots < \eta_q < T$. If $n=2$, condition (2.2) has the form $u(0)=\sum_{i=1}^{p} a_i u(\xi_i)$, $u(T)=\sum_{i=1}^{q} b_i u(\eta_i)$.

Lemma 2.1.1. *If $a_i, \xi_i \in \mathbb{R}$ for all $i=1,\ldots,p$, $b_i, \eta_i \in \mathbb{R}$ for all $i=1,\ldots,q$, $0<\xi_1<\cdots<\xi_p<T$, $0<\eta_1<\cdots<\eta_q<T$, $\Delta_1=(1-\sum_{i=1}^q b_i)\sum_{i=1}^p a_i\xi_i^{n-1}+(1-\sum_{i=1}^p a_i)(T^{n-1}-\sum_{i=1}^q b_i\eta_i^{n-1})\neq 0$, and $y\in C(0,T)\cap L^1(0,T)$, then the unique solution of* (2.1)–(2.2) *is given by*

$$u(t)=-\int_0^t \frac{(t-s)^{n-1}}{(n-1)!}y(s)\,\mathrm{d}s+\frac{t^{n-1}}{\Delta_1}\left\{\left(1-\sum_{i=1}^q b_i\right)\sum_{i=1}^p a_i\int_0^{\xi_i}\frac{(\xi_i-s)^{n-1}}{(n-1)!}y(s)\,\mathrm{d}s\right.$$

$$\left.+\left(1-\sum_{i=1}^p a_i\right)\frac{1}{(n-1)!}\left[\int_0^T (T-s)^{n-1}y(s)\,\mathrm{d}s-\sum_{i=1}^q b_i\int_0^{\eta_i}(\eta_i-s)^{n-1}y(s)\,\mathrm{d}s\right]\right\}$$

$$+\frac{1}{\Delta_1}\left\{\left(\sum_{i=1}^p a_i\xi_i^{n-1}\right)\frac{1}{(n-1)!}\left[\int_0^T (T-s)^{n-1}y(s)\,\mathrm{d}s-\sum_{i=1}^q b_i\int_0^{\eta_i}(\eta_i-s)^{n-1}y(s)\,\mathrm{d}s\right]\right.$$

$$\left.-\left(T^{n-1}-\sum_{i=1}^q b_i\eta_i^{n-1}\right)\sum_{i=1}^p a_i\int_0^{\xi_i}\frac{(\xi_i-s)^{n-1}}{(n-1)!}y(s)\,\mathrm{d}s\right\}. \tag{2.3}$$

Proof. If $n\geq 3$, then the solution of (2.1) is

$$u(t)=-\int_0^t \frac{(t-s)^{n-1}}{(n-1)!}y(s)\,\mathrm{d}s+At^{n-1}+\sum_{i=1}^{n-2}A_it^i+B,$$

with $A, B, A_i\in\mathbb{R}$ for all $i=1,\ldots,n-2$. By using the condition $u'(0)=\cdots=u^{(n-2)}(0)=0$, we obtain $A_i=0$ for all $i=1,\ldots,n-2$. Then we conclude

$$u(t)=-\int_0^t \frac{(t-s)^{n-1}}{(n-1)!}y(s)\,\mathrm{d}s+At^{n-1}+B.$$

If $n=2$, the solution of our problem is given directly by the above expression where n is replaced by 2.

Therefore, for a general $n\geq 2$, by using the conditions $u(0)=\sum_{i=1}^p a_iu(\xi_i)$ and $u(T)=\sum_{i=1}^q b_iu(\eta_i)$, we deduce

$$\begin{cases} B=\displaystyle\sum_{i=1}^p a_i\left[-\int_0^{\xi_i}\frac{(\xi_i-s)^{n-1}}{(n-1)!}y(s)\,\mathrm{d}s+A\xi_i^{n-1}+B\right], \\ -\displaystyle\int_0^T\frac{(T-s)^{n-1}}{(n-1)!}y(s)\,\mathrm{d}s+AT^{n-1} \\ +B=\displaystyle\sum_{i=1}^q b_i\left[-\int_0^{\eta_i}\frac{(\eta_i-s)^{n-1}}{(n-1)!}y(s)\,\mathrm{d}s+A\eta_i^{n-1}+B\right] \end{cases}$$

or

$$\begin{cases} A\sum_{i=1}^{p} a_i\xi_i^{n-1} + \left(\sum_{i=1}^{p} a_i - 1\right)B = \sum_{i=1}^{p} a_i \int_0^{\xi_i} \frac{(\xi_i - s)^{n-1}}{(n-1)!}y(s)\,ds, \\ A\left(T^{n-1} - \sum_{i=1}^{q} b_i\eta_i^{n-1}\right) + B\left(1 - \sum_{i=1}^{q} b_i\right) = \int_0^{T} \frac{(T-s)^{n-1}}{(n-1)!}y(s)\,ds \\ -\sum_{i=1}^{q} b_i \int_0^{\eta_i} \frac{(\eta_i - s)^{n-1}}{(n-1)!}y(s)\,ds. \end{cases} \tag{2.4}$$

The above system with the unknowns A and B has the determinant

$$\Delta_1 = \left(1 - \sum_{i=1}^{q} b_i\right)\sum_{i=1}^{p} a_i\xi_i^{n-1} + \left(1 - \sum_{i=1}^{p} a_i\right)\left(T^{n-1} - \sum_{i=1}^{q} b_i\eta_i^{n-1}\right).$$

By using the assumptions of this lemma, we have $\Delta_1 \neq 0$. Hence, system (2.4) has a unique solution—namely,

$$\begin{aligned} A = \frac{1}{\Delta_1}\Bigg\{&\left(1 - \sum_{i=1}^{q} b_i\right)\sum_{i=1}^{p} a_i \int_0^{\eta_i} \frac{(\xi_i - s)^{n-1}}{(n-1)!}y(s)\,ds \\ &+ \left(1 - \sum_{i=1}^{p} a_i\right)\frac{1}{(n-1)!}\left[\int_0^{T}(T-s)^{n-1}y(s)\,ds - \sum_{i=1}^{q} b_i\int_0^{\eta_i}(\eta_i - s)^{n-1}y(s)\,ds\right]\Bigg\}, \\ B = \frac{1}{\Delta_1}\Bigg\{&\left(\sum_{i=1}^{p} a_i\xi_i^{n-1}\right)\frac{1}{(n-1)!}\left[\int_0^{T}(T-s)^{n-1}y(s)\,ds - \sum_{i=1}^{q} b_i\int_0^{\eta_i}(\eta_i - s)^{n-1}y(s)\,ds\right] \\ &- \left(T^{n-1} - \sum_{i=1}^{q} b_i\eta_i^{n-1}\right)\sum_{i=1}^{p} a_i\int_0^{\xi_i}\frac{(\xi_i - s)^{n-1}}{(n-1)!}y(s)\,ds\Bigg\}. \end{aligned}$$

Therefore, we obtain expression (2.3) for the solution $u(t)$ of problem (2.1)–(2.2). □

Lemma 2.1.2. *Under the assumptions of Lemma 2.1.1, the Green's function for the boundary value problem* (2.1)–(2.2) *is given by*

$$\begin{aligned} G_1(t,s) = g_1(t,s) &+ \frac{1}{\Delta_1}\left[\left(T^{n-1} - t^{n-1}\right)\left(1 - \sum_{i=1}^{q} b_i\right) + \sum_{i=1}^{q} b_i\left(T^{n-1} - \eta_i^{n-1}\right)\right]\sum_{i=1}^{p} a_i g_1(\xi_i, s) \\ &+ \frac{1}{\Delta_1}\left[t^{n-1}\left(1 - \sum_{i=1}^{p} a_i\right) + \sum_{i=1}^{p} a_i\xi_i^{n-1}\right]\sum_{i=1}^{q} b_i g_1(\eta_i, s), \quad (t,s) \in [0,T]\times[0,T], \end{aligned} \tag{2.5}$$

where

$$g_1(t,s) = \frac{1}{(n-1)!T^{n-1}}\begin{cases} t^{n-1}(T-s)^{n-1} - T^{n-1}(t-s)^{n-1}, & 0 \le s \le t \le T, \\ t^{n-1}(T-s)^{n-1}, & 0 \le t \le s \le T. \end{cases} \tag{2.6}$$

Proof. By Lemma 2.1.1 and relation (2.3), we conclude

$$
\begin{aligned}
u(t) = & \frac{1}{(n-1)!T^{n-1}} \left\{ \int_0^t \left[t^{n-1}(T-s)^{n-1} - T^{n-1}(t-s)^{n-1} \right] y(s)\,\mathrm{d}s \right. \\
& + \int_t^T t^{n-1}(T-s)^{n-1} y(s)\,\mathrm{d}s - \int_0^T t^{n-1}(T-s)^{n-1} y(s)\,\mathrm{d}s \\
& + \frac{T^{n-1}t^{n-1}}{\Delta_1} \left\{ \left(1 - \sum_{i=1}^q b_i\right) \sum_{i=1}^p a_i \int_0^{\xi_i} (\xi_i - s)^{n-1} y(s)\,\mathrm{d}s \right. \\
& \left. + \left(1 - \sum_{i=1}^p a_i\right) \left[\int_0^T (T-s)^{n-1} y(s)\,\mathrm{d}s - \sum_{i=1}^q b_i \int_0^{\eta_i} (\eta_i - s)^{n-1} y(s)\,\mathrm{d}s \right] \right\} \\
& + \frac{T^{n-1}}{\Delta_1} \left\{ \left(\sum_{i=1}^p a_i \xi_i^{n-1} \right) \left[\int_0^T (T-s)^{n-1} y(s)\,\mathrm{d}s - \sum_{i=1}^q b_i \int_0^{\eta_i} (\eta_i - s)^{n-1} y(s)\,\mathrm{d}s \right] \right. \\
& \left. \left. - \left(T^{n-1} - \sum_{i=1}^q b_i \eta_i^{n-1} \right) \sum_{i=1}^p a_i \int_0^{\xi_i} (\xi_i - s)^{n-1} y(s)\,\mathrm{d}s \right\} \right\} \\
= & \frac{1}{(n-1)!T^{n-1}} \left\{ \int_0^t \left[t^{n-1}(T-s)^{n-1} - T^{n-1}(t-s)^{n-1} \right] y(s)\,\mathrm{d}s \right. \\
& + \int_t^T t^{n-1}(T-s)^{n-1} y(s)\,\mathrm{d}s - \frac{1}{\Delta_1} \left[\left(1 - \sum_{i=1}^q b_i\right) \sum_{i=1}^p a_i \xi_i^{n-1} \right. \\
& \left. + \left(1 - \sum_{i=1}^p a_i\right) \left(T^{n-1} - \sum_{i=1}^q b_i \eta_i^{n-1} \right) \right] \int_0^T t^{n-1}(T-s)^{n-1} y(s)\,\mathrm{d}s \\
& + \frac{T^{n-1}t^{n-1}}{\Delta_1} \left\{ \left(1 - \sum_{i=1}^q b_i\right) \sum_{i=1}^p a_i \int_0^{\xi_i} (\xi_i - s)^{n-1} y(s)\,\mathrm{d}s \right. \\
& \left. + \left(1 - \sum_{i=1}^p a_i\right) \left[\int_0^T (T-s)^{n-1} y(s)\,\mathrm{d}s - \sum_{i=1}^q b_i \int_0^{\eta_i} (\eta_i - s)^{n-1} y(s)\,\mathrm{d}s \right] \right\} \\
& + \frac{T^{n-1}}{\Delta_1} \left\{ \left(\sum_{i=1}^p a_i \xi_i^{n-1} \right) \left[\int_0^T (T-s)^{n-1} y(s)\,\mathrm{d}s - \sum_{i=1}^q b_i \int_0^{\eta_i} (\eta_i - s)^{n-1} y(s)\,\mathrm{d}s \right] \right. \\
& \left. \left. - \left(T^{n-1} - \sum_{i=1}^q b_i \eta_i^{n-1} \right) \sum_{i=1}^p a_i \int_0^{\xi_i} (\xi_i - s)^{n-1} y(s)\,\mathrm{d}s \right\} \right\}.
\end{aligned}
$$

Therefore, we deduce

$$
\begin{aligned}
u(t) = & \frac{1}{(n-1)!T^{n-1}} \left\{ \int_0^t \left[t^{n-1}(T-s)^{n-1} - T^{n-1}(t-s)^{n-1} \right] y(s)\,ds \right. \\
& + \int_t^T t^{n-1}(T-s)^{n-1} y(s)\,ds + \frac{t^{n-1}}{\Delta_1} \left[-\left(1-\sum_{i=1}^q b_i\right) \sum_{i=1}^p a_i \xi_i^{n-1} \int_0^T (T-s)^{n-1} y(s)\,ds \right. \\
& - \left(1-\sum_{i=1}^p a_i\right) \int_0^T T^{n-1}(T-s)^{n-1} y(s)\,ds + \left(1-\sum_{i=1}^p a_i\right) \sum_{i=1}^q b_i \eta_i^{n-1} \int_0^T (T-s)^{n-1} y(s)\,ds \\
& + T^{n-1}\left(1-\sum_{i=1}^q b_i\right) \sum_{i=1}^p a_i \int_0^{\xi_i} (\xi_i-s)^{n-1} y(s)\,ds + T^{n-1}\left(1-\sum_{i=1}^p a_i\right) \int_0^T (T-s)^{n-1} y(s)\,ds \\
& \left. - T^{n-1}\left(1-\sum_{i=1}^p a_i\right) \sum_{i=1}^q b_i \int_0^{\eta_i} (\eta_i-s)^{n-1} y(s)\,ds \right] \\
& + \frac{T^{n-1}}{\Delta_1} \left[\left(\sum_{i=1}^p a_i \xi_i^{n-1} \right) \int_0^T (T-s)^{n-1} y(s)\,ds - \left(\sum_{i=1}^p a_i \xi_i^{n-1} \right) \sum_{i=1}^q b_i \int_0^{\eta_i} (\eta_i-s)^{n-1} y(s)\,ds \right. \\
& \left.\left. - T^{n-1} \sum_{i=1}^p a_i \int_0^{\xi_i} (\xi_i-s)^{n-1} y(s)\,ds + \left(\sum_{i=1}^q b_i \eta_i^{n-1} \right) \sum_{i=1}^p a_i \int_0^{\xi_i} (\xi_i-s)^{n-1} y(s)\,ds \right] \right\} \\
= & \frac{1}{(n-1)!T^{n-1}} \left\{ \int_0^t \left[t^{n-1}(T-s)^{n-1} - T^{n-1}(t-s)^{n-1} \right] y(s)\,ds \right. \\
& + \int_t^T t^{n-1}(T-s)^{n-1} y(s)\,ds + \frac{t^{n-1}}{\Delta_1} \left[\left(1-\sum_{j=1}^q b_j\right) \sum_{i=1}^p a_i \left(\int_0^{\xi_i} T^{n-1}(\xi_i-s)^{n-1} y(s)\,ds \right. \right. \\
& \left. - \int_0^{\xi_i} \xi_i^{n-1}(T-s)^{n-1} y(s)\,ds \right) - \left(1-\sum_{j=1}^q b_j\right) \sum_{i=1}^p a_i \int_{\xi_i}^T \xi_i^{n-1}(T-s)^{n-1} y(s)\,ds \\
& - \left(1-\sum_{j=1}^p a_j\right) \sum_{i=1}^q b_i \left(\int_0^{\eta_i} T^{n-1}(\eta_i-s)^{n-1} y(s)\,ds - \int_0^{\eta_i} \eta_i^{n-1}(T-s)^{n-1} y(s)\,ds \right) \\
& \left. + \left(1-\sum_{j=1}^p a_j\right) \sum_{i=1}^q b_i \int_{\eta_i}^T \eta_i^{n-1}(T-s)^{n-1} y(s)\,ds \right] \\
& + \frac{T^{n-1}}{\Delta_1} \left[\sum_{i=1}^p a_i \int_0^{\xi_i} \xi_i^{n-1}(T-s)^{n-1} y(s)\,ds - \sum_{i=1}^p a_i \int_0^{\xi_i} T^{n-1}(\xi_i-s)^{n-1} y(s)\,ds \right. \\
& + \sum_{i=1}^p a_i \int_{\xi_i}^T \xi_i^{n-1}(T-s)^{n-1} y(s)\,ds - \left(\sum_{i=1}^p a_i \xi_i^{n-1} \right) \sum_{i=1}^q b_i \int_0^{\eta_i} (\eta_i-s)^{n-1} y(s)\,ds \\
& \left.\left. + \left(\sum_{i=1}^q b_i \eta_i^{n-1} \right) \sum_{i=1}^p a_i \int_0^{\xi_i} (\xi_i-s)^{n-1} y(s)\,ds \right] \right\}.
\end{aligned}
$$

Hence, we obtain

$$
\begin{aligned}
u(t) = & \frac{1}{(n-1)!T^{n-1}} \left\{ \int_0^t \left[t^{n-1}(T-s)^{n-1} - T^{n-1}(t-s)^{n-1} \right] y(s)\, \mathrm{d}s \right. \\
& + \int_t^T t^{n-1}(T-s)^{n-1} y(s)\, \mathrm{d}s + \frac{t^{n-1}}{\Delta_1} \left[-\left(1 - \sum_{j=1}^q b_j\right) \sum_{i=1}^p a_i \left(\int_0^{\xi_i} \left[\xi_i^{n-1}(T-s)^{n-1} \right.\right.\right. \\
& \left.\left. - T^{n-1}(\xi_i - s)^{n-1} \right] y(s)\, \mathrm{d}s \right) - \left(1 - \sum_{j=1}^q b_j\right) \sum_{i=1}^p a_i \int_{\xi_i}^T \xi_i^{n-1}(T-s)^{n-1} y(s)\, \mathrm{d}s \\
& + \left(1 - \sum_{j=1}^p a_j\right) \sum_{i=1}^q b_i \left(\int_0^{\eta_i} \left[\eta_i^{n-1}(T-s)^{n-1} - T^{n-1}(\eta_i - s)^{n-1} \right] y(s)\, \mathrm{d}s \right) \\
& \left. + \left(1 - \sum_{j=1}^p a_j\right) \sum_{i=1}^q b_i \int_{\eta_i}^T \eta_i^{n-1}(T-s)^{n-1} y(s)\, \mathrm{d}s \right] \\
& + \frac{T^{n-1}}{\Delta_1} \left[\sum_{i=1}^p a_i \int_0^{\xi_i} \left[\xi_i^{n-1}(T-s)^{n-1} - T^{n-1}(\xi_i - s)^{n-1} \right] y(s)\, \mathrm{d}s \right. \\
& \left. + \sum_{i=1}^p a_i \int_{\xi_i}^T \xi_i^{n-1}(T-s)^{n-1} y(s)\, \mathrm{d}s \right] - \frac{1}{\Delta_1} \left(\sum_{i=1}^p a_i \xi_i^{n-1} \right) \sum_{i=1}^q b_i \int_0^{\eta_i} T^{n-1}(\eta_i - s)^{n-1} y(s)\, \mathrm{d}s \\
& + \frac{1}{\Delta_1} \left(\sum_{i=1}^q b_i \eta_i^{n-1} \right) \sum_{i=1}^p a_i \int_0^{\xi_i} T^{n-1}(\xi_i - s)^{n-1} y(s)\, \mathrm{d}s \\
& + \frac{1}{\Delta_1} \left(\sum_{i=1}^p a_i \xi_i^{n-1} \right) \sum_{i=1}^q b_i \int_0^T \eta_i^{n-1}(T-s)^{n-1} y(s)\, \mathrm{d}s \\
& \left. - \frac{1}{\Delta_1} \left(\sum_{i=1}^q b_i \eta_i^{n-1} \right) \sum_{i=1}^p a_i \int_0^T \xi_i^{n-1}(T-s)^{n-1} y(s)\, \mathrm{d}s \right\} \\
= & \frac{1}{(n-1)!T^{n-1}} \left\{ \int_0^t \left[t^{n-1}(T-s)^{n-1} - T^{n-1}(t-s)^{n-1} \right] y(s)\, \mathrm{d}s \right. \\
& + \int_t^T t^{n-1}(T-s)^{n-1} y(s)\, \mathrm{d}s + \frac{t^{n-1}}{\Delta_1} \left[-\left(1 - \sum_{j=1}^q b_j\right) \sum_{i=1}^p a_i \left(\int_0^{\xi_i} \left[\xi_i^{n-1}(T-s)^{n-1} \right.\right.\right. \\
& \left.\left. - T^{n-1}(\xi_i - s)^{n-1} \right] y(s)\, \mathrm{d}s \right) - \left(1 - \sum_{j=1}^q b_j\right) \sum_{i=1}^p a_i \int_{\xi_i}^T \xi_i^{n-1}(T-s)^{n-1} y(s)\, \mathrm{d}s \\
& + \left(1 - \sum_{j=1}^p a_j\right) \sum_{i=1}^q b_i \left(\int_0^{\eta_i} \left[\eta_i^{n-1}(T-s)^{n-1} - T^{n-1}(\eta_i - s)^{n-1} \right] y(s)\, \mathrm{d}s \right) \\
& \left. + \left(1 - \sum_{j=1}^p a_j\right) \sum_{i=1}^q b_i \int_{\eta_i}^T \eta_i^{n-1}(T-s)^{n-1} y(s)\, \mathrm{d}s \right]
\end{aligned}
$$

$$
\begin{aligned}
&+\frac{T^{n-1}}{\Delta_1}\left[\sum_{i=1}^{p} a_i \int_0^{\xi_i}\left[\xi_i^{n-1}(T-s)^{n-1}-T^{n-1}(\xi_i-s)^{n-1}\right]y(s)\,\mathrm{d}s\right.\\
&\left.+\sum_{i=1}^{p} a_i \int_{\xi_i}^{T} \xi_i^{n-1}(T-s)^{n-1}y(s)\,\mathrm{d}s\right]+\frac{1}{\Delta_1}\left(\sum_{i=1}^{p} a_i\xi_i^{n-1}\right)\sum_{i=1}^{q} b_i\left[\int_0^{\eta_i}\left[\eta_i^{n-1}(T-s)^{n-1}\right.\right.\\
&\left.\left.-T^{n-1}(\eta_i-s)^{n-1}\right]y(s)\,\mathrm{d}s+\int_{\eta_i}^{T}\eta_i^{n-1}(T-s)^{n-1}y(s)\,\mathrm{d}s\right]\\
&-\frac{1}{\Delta_1}\left(\sum_{i=1}^{q} b_i\eta_i^{n-1}\right)\sum_{i=1}^{p} a_i\left[\int_0^{\xi_i}\left[\xi_i^{n-1}(T-s)^{n-1}-T^{n-1}(\xi_i-s)^{n-1}\right]y(s)\,\mathrm{d}s\right.\\
&\left.\left.+\int_{\xi_i}^{T}\xi_i^{n-1}(T-s)^{n-1}y(s)\,\mathrm{d}s\right]\right\}.
\end{aligned}
$$

Then the solution of problem (2.1)–(2.2) is

$$
\begin{aligned}
u(t)=&\int_0^T\left\{g_1(t,s)+\frac{t^{n-1}}{\Delta_1}\left[-\left(1-\sum_{j=1}^{q} b_j\right)\sum_{i=1}^{p} a_i g_1(\xi_i,s)+\left(1-\sum_{j=1}^{p} a_j\right)\right.\right.\\
&\left.\times\sum_{i=1}^{q} b_i g_1(\eta_i,s)\right]+\frac{T^{n-1}}{\Delta_1}\sum_{i=1}^{p} a_i g_1(\xi_i,s)+\frac{1}{\Delta_1}\left(\sum_{i=1}^{p} a_i\xi_i^{n-1}\right)\sum_{i=1}^{q} b_i g_1(\eta_i,s)\\
&\left.-\frac{1}{\Delta_1}\left(\sum_{i=1}^{q} b_i\eta_i^{n-1}\right)\sum_{i=1}^{p} a_i g_1(\xi_i,s)\right\}y(s)\,\mathrm{d}s\\
=&\int_0^T\left\{g_1(t,s)+\frac{1}{\Delta_1}\left[-t^{n-1}\left(1-\sum_{j=1}^{q} b_j\right)+T^{n-1}-\sum_{i=1}^{q} b_i\eta_i^{n-1}\right]\sum_{i=1}^{p} a_i g_1(\xi_i,s)\right.\\
&\left.+\frac{1}{\Delta_1}\left[t^{n-1}\left(1-\sum_{j=1}^{p} a_j\right)+\left(\sum_{i=1}^{p} a_i\xi_i^{n-1}\right)\right]\sum_{i=1}^{q} b_i g_1(\eta_i,s)\right\}y(s)\,\mathrm{d}s\\
=&\int_0^T\left\{g_1(t,s)+\frac{1}{\Delta_1}\left[\left(T^{n-1}-t^{n-1}\right)\left(1-\sum_{j=1}^{q} b_j\right)+\sum_{i=1}^{q} b_i\left(T^{n-1}-\eta_i^{n-1}\right)\right]\sum_{i=1}^{p} a_i g_1(\xi_i,s)\right.\\
&\left.+\frac{1}{\Delta_1}\left[t^{n-1}\left(1-\sum_{j=1}^{p} a_j\right)+\sum_{i=1}^{p} a_i\xi_i^{n-1}\right]\sum_{i=1}^{q} b_i g_1(\eta_i,s)\right\}y(s)\,\mathrm{d}s,
\end{aligned}
$$

where g_1 is given by (2.6).

Hence, $u(t)=\int_0^T G_1(t,s)y(s)\,\mathrm{d}s$, where G_1 is given in (2.5). □

Lemma 2.1.3 (Ji and Guo, 2009; see also Henderson and Luca, 2013b)**.** *The function g_1 given by* (2.6) *has the following properties:*

(a) $g_1\colon[0,T]\times[0,T]\to\mathbb{R}_+$ *is a continuous function, and* $g_1(t,s)>0$ *for all* $(t,s)\in(0,T)\times(0,T)$.

(b) $g_1(t,s)\le g_1(\theta_1(s),s)$ *for all* $(t,s)\in[0,T]\times[0,T]$.

(c) *For any* $c \in (0, \frac{T}{2})$,

$\min_{t\in[c,T-c]} g_1(t,s) \geq \frac{c^{n-1}}{T^{n-1}} g_1(\theta_1(s), s)$ *for all* $s \in [0,T]$,

where $\theta_1(s) = s$ *if* $n=2$ *and* $\theta_1(s) = \begin{cases} \dfrac{s}{1-(1-\frac{s}{T})^{\frac{n-1}{n-2}}}, & s \in (0,T], \\ \dfrac{T(n-2)}{n-1}, & s=0, \end{cases}$ *if* $n \geq 3$.

Lemma 2.1.4. *If* $a_i \geq 0$ *for all* $i = 1, \ldots, p$, $\sum_{i=1}^{p} a_i < 1$, $b_i \geq 0$ *for all* $i = 1, \ldots, q$, $\sum_{i=1}^{q} b_i < 1$, $0 < \xi_1 < \cdots < \xi_p < T$, $0 < \eta_1 < \cdots < \eta_q < T$, *then the Green's function* G_1 *of problem* (2.1)–(2.2) *is continuous on* $[0,T] \times [0,T]$ *and satisfies* $G_1(t,s) \geq 0$ *for all* $(t,s) \in [0,T] \times [0,T]$. *Moreover, if* $y \in C(0,T) \cap L^1(0,T)$ *satisfies* $y(t) \geq 0$ *for all* $t \in (0,T)$, *then the solution* u *of problem* (2.1)–(2.2) *satisfies* $u(t) \geq 0$ *for all* $t \in [0,T]$.

Proof. By using the assumptions of this lemma, we have $\Delta_1 > 0$ and $G_1(t,s) \geq 0$ for all $(t,s) \in [0,T] \times [0,T]$, and so $u(t) \geq 0$ for all $t \in [0,T]$. □

Lemma 2.1.5. *Assume that* $a_i \geq 0$ *for all* $i = 1, \ldots, p$, $\sum_{i=1}^{p} a_i < 1$, $b_i \geq 0$ *for all* $i = 1, \ldots, q$, $\sum_{i=1}^{q} b_i < 1$, $0 < \xi_1 < \cdots < \xi_p < T$, $0 < \eta_1 < \cdots < \eta_q < T$. *Then the Green's function* G_1 *of problem* (2.1)–(2.2) *satisfies the following inequalities:*

(a) $G_1(t,s) \leq J_1(s), \forall (t,s) \in [0,T] \times [0,T]$, *where*

$$J_1(s) = g_1(\theta_1(s), s) + \frac{1}{\Delta_1}\left[T^{n-1}\left(1 - \sum_{j=1}^{q} b_j\right) + \sum_{i=1}^{q} b_i \left(T^{n-1} - \eta_i^{n-1}\right)\right] \times \sum_{i=1}^{p} a_i g_1(\xi_i, s) + \frac{1}{\Delta_1}\left[T^{n-1}\left(1 - \sum_{j=1}^{p} a_j\right) + \sum_{i=1}^{p} a_i \xi_i^{n-1}\right] \sum_{i=1}^{q} b_i g_1(\eta_i, s).$$

(b) *For every* $c \in (0, T/2)$, *we have*

$$\min_{t\in[c,T-c]} G_1(t,s) \geq \gamma_1 J_1(s) \geq \gamma_1 G_1(t', s), \quad \forall t', \quad s \in [0,T],$$

where $\gamma_1 = c^{n-1}/T^{n-1}$.

Proof. The first inequality, (a), is evident. For the second inequality, (b), for $c \in (0, T/2)$, $t \in [c, T-c]$, and $s \in [0,T]$, we deduce

$$G_1(t,s) \geq \frac{c^{n-1}}{T^{n-1}} g_1(\theta_1(s), s) + \frac{1}{\Delta_1}\left[\left(T^{n-1} - (T-c)^{n-1}\right)\left(1 - \sum_{j=1}^{q} b_j\right) + \sum_{i=1}^{q} b_i \left(T^{n-1} - \eta_i^{n-1}\right)\right] \sum_{i=1}^{p} a_i g_1(\xi_i, s) + \frac{1}{\Delta_1}\left[c^{n-1}\left(1 - \sum_{j=1}^{p} a_j\right) + \sum_{i=1}^{p} a_i \xi_i^{n-1}\right] \times \sum_{i=1}^{q} b_i g_1(\eta_i, s) = \frac{c^{n-1}}{T^{n-1}} g_1(\theta_1(s), s) + \frac{1}{\Delta_1}\left[T^{n-1}\left(1 - \sum_{j=1}^{q} b_j\right) + \sum_{i=1}^{q} b_i \left(T^{n-1} - \eta_i^{n-1}\right)\right]$$

$$\times \frac{\left[\left(T^{n-1}-(T-c)^{n-1}\right)\left(1-\sum_{j=1}^{q} b_j\right)+\sum_{i=1}^{q} b_i\left(T^{n-1}-\eta_i^{n-1}\right)\right]}{\left[T^{n-1}\left(1-\sum_{j=1}^{q} b_j\right)+\sum_{i=1}^{q} b_i\left(T^{n-1}-\eta_i^{n-1}\right)\right]} \sum_{i=1}^{p} a_i g_1(\xi_i, s)$$

$$+\frac{1}{\Delta_1}\left[T^{n-1}\left(1-\sum_{j=1}^{p} a_j\right)+\sum_{i=1}^{p} a_i \xi_i^{n-1}\right] \frac{c^{n-1}\left(1-\sum_{j=1}^{p} a_j\right)+\sum_{i=1}^{p} a_i \xi_i^{n-1}}{T^{n-1}\left(1-\sum_{j=1}^{p} a_j\right)+\sum_{i=1}^{p} a_i \xi_i^{n-1}} \sum_{i=1}^{q} b_i g_1(\eta_i, s)$$

$$\geq \frac{c^{n-1}}{T^{n-1}} J_1(s) = \gamma_1 J_1(s).$$

□

Lemma 2.1.6. *Assume that $a_i \geq 0$ for all $i=1,\ldots,p$, $\sum_{i=1}^{p} a_i < 1$, $b_i \geq 0$ for all $i=1,\ldots,q$, $\sum_{i=1}^{q} b_i < 1$, $0 < \xi_1 < \cdots < \xi_p < T$, $0 < \eta_1 < \cdots < \eta_q < T$, $c \in (0, T/2)$, and $y \in C(0,T) \cap L^1(0,T)$, $y(t) \geq 0$ for all $t \in (0,T)$. Then the solution $u(t)$, $t \in [0,T]$ of problem (2.1)–(2.2) satisfies the inequality $\inf_{t\in[c,T-c]} u(t) \geq \gamma_1 \sup_{t'\in[0,T]} u(t')$.*

Proof. For $c \in (0, T/2)$, $t \in [c, T-c]$, and $t' \in [0,T]$, we have

$$u(t) = \int_0^T G_1(t,s)y(s)\,ds \geq \gamma_1 \int_0^T J_1(s)y(s)\,ds \geq \gamma_1 \int_0^T G_1(t',s)y(s)\,ds = \gamma_1 u(t').$$

Then $\inf_{t\in[c,T-c]} u(t) \geq \gamma_1 \sup_{t'\in[0,T]} u(t')$. □

We can also formulate results similar to those of Lemmas 2.1.1–2.1.6 for the boundary value problem

$$v^{(m)}(t) + h(t) = 0, \quad t \in (0,T), \tag{2.7}$$

$$v(0) = \sum_{i=1}^{r} c_i v(\zeta_i), \quad v'(0) = \cdots = v^{(m-2)}(0) = 0, \quad v(T) = \sum_{i=1}^{l} d_i v(\rho_i), \tag{2.8}$$

where $m, r, l \in \mathbb{N}$, $m \geq 2$, $c_i, \zeta_i \in \mathbb{R}$ for all $i=1,\ldots,r$, $d_i, \rho_i \in \mathbb{R}$ for all $i=1,\ldots,l$, $0 < \zeta_1 < \cdots < \zeta_r < T$, $0 < \rho_1 < \cdots < \rho_l < T$, and $h \in C(0,T) \cap L^1(0,T)$. We denote by Δ_2, γ_2, g_2, θ_2, G_2, and J_2 the corresponding constants and functions for problem (2.7)–(2.8) defined in a similar manner as Δ_1, γ_1, g_1, θ_1, G_1, and J_1, respectively.

2.1.3 Main results

In this section, we give sufficient conditions on λ, μ, f, and g such that positive solutions with respect to a cone for our problem (S)–(BC) exist. We also investigate the nonexistence of positive solutions of (S)–(BC).

We present the assumptions that we shall use in the sequel:

(H1) $0 < \xi_1 < \cdots < \xi_p < T$, $a_i \geq 0$ for all $i=1,\ldots,p$, $\sum_{i=1}^{p} a_i < 1$, $0 < \eta_1 < \cdots < \eta_q < T$, $b_i \geq 0$ for all $i=1,\ldots,q$, $\sum_{i=1}^{q} b_i < 1$, $0 < \zeta_1 < \cdots < \zeta_r < T$, $c_i \geq 0$ for all $i=1,\ldots,r$, $\sum_{i=1}^{r} c_i < 1$, $0 < \rho_1 < \cdots < \rho_l < T$, $d_i \geq 0$ for all $i=1,\ldots,l$, $\sum_{i=1}^{l} d_i < 1$.

(H2) The functions a, b: $[0,T] \to [0,\infty)$ are continuous and there exist $t_1, t_2 \in (0,T)$ such that $a(t_1) > 0$, $b(t_2) > 0$.

(H3) The functions f, g: $[0,T]\times[0,\infty)\times[0,\infty)\to[0,\infty)$ are continuous.

From assumption (H2), there exists $c\in(0,T/2)$ such that $t_1,t_2\in(c,T-c)$. We shall work in this section with this number c. This implies that $\int_c^{T-c} g_1(t,s)a(s)\,ds>0$, $\int_c^{T-c} g_2(t,s)b(s)\,ds>0$, $\int_c^{T-c} J_1(s)a(s)\,ds>0$, and $\int_c^{T-c} J_2(s)b(s)\,ds>0$, for all $t\in[0,T]$, where g_1, g_2, J_1, and J_2 are defined in Section 2.1.2.

For c defined above, we introduce the following extreme limits:

$$f_0^{\rm s}=\limsup_{u+v\to 0^+}\max_{t\in[0,T]}\frac{f(t,u,v)}{u+v},\quad g_0^{\rm s}=\limsup_{u+v\to 0^+}\max_{t\in[0,T]}\frac{g(t,u,v)}{u+v},$$

$$f_0^{\rm i}=\liminf_{u+v\to 0^+}\min_{t\in[c,T-c]}\frac{f(t,u,v)}{u+v},\quad g_0^{\rm i}=\liminf_{u+v\to 0^+}\min_{t\in[c,T-c]}\frac{g(t,u,v)}{u+v},$$

$$f_\infty^{\rm s}=\limsup_{u+v\to\infty}\max_{t\in[0,T]}\frac{f(t,u,v)}{u+v},\quad g_\infty^{\rm s}=\limsup_{u+v\to\infty}\max_{t\in[0,T]}\frac{g(t,u,v)}{u+v},$$

$$f_\infty^{\rm i}=\liminf_{u+v\to\infty}\min_{t\in[c,T-c]}\frac{f(t,u,v)}{u+v},\quad g_\infty^{\rm i}=\liminf_{u+v\to\infty}\min_{t\in[c,T-c]}\frac{g(t,u,v)}{u+v}.$$

In the definitions of the extreme limits above, the variables u and v are nonnegative.

By using the Green's functions G_1 and G_2 from Section 2.1.2, we can write our problem (S)–(BC) equivalently as the following nonlinear system of integral equations:

$$\begin{cases} u(t)=\lambda\displaystyle\int_0^T G_1(t,s)a(s)f(s,u(s),v(s))\,ds, & 0\le t\le T,\\ v(t)=\mu\displaystyle\int_0^T G_2(t,s)b(s)g(s,u(s),v(s))\,ds, & 0\le t\le T.\end{cases}$$

We consider the Banach space $X=C([0,T])$ with the supremum norm $\|u\|=\sup_{t\in[0,T]}|u(t)|$ and the Banach space $Y=X\times X$ with the norm $\|(u,v)\|_Y=\|u\|+\|v\|$. We define the cone $P\subset Y$ by

$$P=\Big\{(u,v)\in Y;\quad u(t)\ge 0,\ v(t)\ge 0,\ \ \forall t\in[0,T]\quad\text{and}$$
$$\inf_{t\in[c,T-c]}(u(t)+v(t))\ge\gamma\|(u,v)\|_Y\Big\},$$

where $\gamma=\min\{\gamma_1,\gamma_2\}$, and γ_1 and γ_2 are the constants defined in Section 2.1.2 with respect to the above c.

For $\lambda,\mu>0$, we introduce the operators Q_1,Q_2: $Y\to X$ and Q: $Y\to Y$ defined by

$$Q_1(u,v)(t)=\lambda\int_0^T G_1(t,s)a(s)f(s,u(s),v(s))\,ds,\quad 0\le t\le T,$$

$$Q_2(u,v)(t)=\mu\int_0^T G_2(t,s)b(s)g(s,u(s),v(s))\,ds,\quad 0\le t\le T,$$

and $\mathcal{Q}(u,v) = (\mathcal{Q}_1(u,v), \mathcal{Q}_2(u,v)), (u,v) \in Y$. The solutions of our problem (S)–(BC) coincide with the fixed points of the operator $\mathcal{Q}$. We can easily prove that under assumptions (H1)–(H3), the operator $\mathcal{Q}$: $P \to P$ is completely continuous.

We denote $A = \int_c^{T-c} J_1(s)a(s)\,ds$, $B = \int_0^T J_1(s)a(s)\,ds$, $C = \int_c^{T-c} J_2(s)b(s)\,ds$, and $D = \int_0^T J_2(s)b(s)\,ds$.

First, for $f_0^s, g_0^s, f_\infty^i, g_\infty^i \in (0,\infty)$ and numbers $\alpha_1, \alpha_2 \geq 0$, and $\tilde{\alpha}_1, \tilde{\alpha}_2 > 0$ such that $\alpha_1 + \alpha_2 = 1$ and $\tilde{\alpha}_1 + \tilde{\alpha}_2 = 1$, we define the numbers L_1, L_2, L_3, L_4, L_2', and L_4' by

$$L_1 = \frac{\alpha_1}{\gamma\gamma_1 f_\infty^i A}, \quad L_2 = \frac{\tilde{\alpha}_1}{f_0^s B}, \quad L_3 = \frac{\alpha_2}{\gamma\gamma_2 g_\infty^i C}, \quad L_4 = \frac{\tilde{\alpha}_2}{g_0^s D},$$

$$L_2' = \frac{1}{f_0^s B}, \quad L_4' = \frac{1}{g_0^s D}.$$

Theorem 2.1.1. *Assume that (H1)–(H3) hold, and $\alpha_1, \alpha_2 \geq 0$ and $\tilde{\alpha}_1, \tilde{\alpha}_2 > 0$ such that $\alpha_1 + \alpha_2 = 1$ and $\tilde{\alpha}_1 + \tilde{\alpha}_2 = 1$.*

(1) If $f_0^s, g_0^s, f_\infty^i, g_\infty^i \in (0,\infty)$, $L_1 < L_2$, and $L_3 < L_4$, then for each $\lambda \in (L_1, L_2)$ and $\mu \in (L_3, L_4)$ there exists a positive solution $(u(t), v(t)), t \in [0,T]$ for (S)–(BC).

(2) If $f_0^s = 0$, $g_0^s, f_\infty^i, g_\infty^i \in (0,\infty)$, and $L_3 < L_4'$, then for each $\lambda \in (L_1, \infty)$ and $\mu \in (L_3, L_4')$ there exists a positive solution $(u(t), v(t)), t \in [0,T]$ for (S)–(BC).

(3) If $g_0^s = 0$, $f_0^s, f_\infty^i, g_\infty^i \in (0,\infty)$, and $L_1 < L_2'$, then for each $\lambda \in (L_1, L_2')$ and $\mu \in (L_3, \infty)$ there exists a positive solution $(u(t), v(t)), t \in [0,T]$ for (S)–(BC).

(4) If $f_0^s = g_0^s = 0$ and $f_\infty^i, g_\infty^i \in (0,\infty)$, then for each $\lambda \in (L_1, \infty)$ and $\mu \in (L_3, \infty)$ there exists a positive solution $(u(t), v(t)), t \in [0,T]$ for (S)–(BC).

(5) If $\{f_0^s, g_0^s, f_\infty^i \in (0,\infty), g_\infty^i = \infty\}$ or $\{f_0^s, g_0^s, g_\infty^i \in (0,\infty), f_\infty^i = \infty\}$ or $\{f_0^s, g_0^s \in (0,\infty), f_\infty^i = g_\infty^i = \infty\}$, then for each $\lambda \in (0, L_2)$ and $\mu \in (0, L_4)$ there exists a positive solution $(u(t), v(t)), t \in [0,T]$ for (S)–(BC).

(6) If $\{f_0^s = 0, g_0^s, f_\infty^i \in (0,\infty), g_\infty^i = \infty\}$ or $\{f_0^s = 0, f_\infty^i = \infty, g_0^s, g_\infty^i \in (0,\infty)\}$ or $\{f_0^s = 0, g_0^s \in (0,\infty), f_\infty^i = g_\infty^i = \infty\}$, then for each $\lambda \in (0,\infty)$ and $\mu \in (0, L_4')$ there exists a positive solution $(u(t), v(t)), t \in [0,T]$ for (S)–(BC).

(7) If $\{f_0^s, f_\infty^i \in (0,\infty), g_0^s = 0, g_\infty^i = \infty\}$ or $\{f_0^s, g_\infty^i \in (0,\infty), g_0^s = 0, f_\infty^i = \infty\}$ or $\{f_0^s \in (0,\infty), g_0^s = 0, f_\infty^i = g_\infty^i = \infty\}$, then for each $\lambda \in (0, L_2')$ and $\mu \in (0,\infty)$ there exists a positive solution $(u(t), v(t)), t \in [0,T]$ for (S)–(BC).

(8) If $\{f_0^s = g_0^s = 0, f_\infty^i \in (0,\infty), g_\infty^i = \infty\}$ or $\{f_0^s = g_0^s = 0, f_\infty^i = \infty, g_\infty^i \in (0,\infty)\}$ or $\{f_0^s = g_0^s = 0, f_\infty^i = g_\infty^i = \infty\}$, then for each $\lambda \in (0,\infty)$ and $\mu \in (0,\infty)$ there exists a positive solution $(u(t), v(t)), t \in [0,T]$ for (S)–(BC).

Proof. We consider the above cone P and the operators $\mathcal{Q}_1, \mathcal{Q}_2$, and $\mathcal{Q}$. Because the proofs of the above 16 cases are similar, we shall only prove one of them—namely, the first case of (7). So, we suppose $f_0^s, f_\infty^i \in (0,\infty)$, $g_0^s = 0$, and $g_\infty^i = \infty$. Let $\lambda \in (0, L_2')$—that is, $\lambda \in (0, \frac{1}{f_0^s B})$ and $\mu \in (0,\infty)$. We choose $\alpha_1' > 0$, $\alpha_1' < \min\{\lambda\gamma\gamma_1 f_\infty^i A, 1\}$ and $\tilde{\alpha}_1' \in (\lambda f_0^s B, 1)$. Let $\alpha_2' = 1 - \alpha_1'$ and $\tilde{\alpha}_2' = 1 - \tilde{\alpha}_1'$, and let $\varepsilon > 0$ be a positive number such that $\varepsilon < f_\infty^i$ and

$$\frac{\alpha_1'}{\gamma\gamma_1(f_\infty^i - \varepsilon)A} \leq \lambda, \quad \frac{\varepsilon\alpha_2'}{\gamma\gamma_2 C} \leq \mu, \quad \frac{\tilde{\alpha}_1'}{(f_0^s + \varepsilon)B} \geq \lambda, \quad \frac{\tilde{\alpha}_2'}{\varepsilon D} \geq \mu.$$

By the definitions of $f_0^{\rm s}$ and $g_0^{\rm s}$, we deduce that there exists $R_1 > 0$ such that $f(t,u,v) \le (f_0^{\rm s}+\varepsilon)(u+v)$ and $g(t,u,v) \le \varepsilon(u+v)$ for all $t \in [0,T]$ and $u, v \ge 0$, $u+v \in [0,R_1]$. We define the set $\Omega_1 = \{(u,v) \in Y, \|(u,v)\|_Y < R_1\}$. Let $(u,v) \in P \cap \partial\Omega_1$—that is, $\|(u,v)\|_Y = R_1$ or $\|u\| + \|v\| = R_1$. Then $u(t)+v(t) \le R_1$ for all $t \in [0,T]$, and by Lemma 2.1.5, we obtain

$$Q_1(u,v)(t) \le \lambda \int_0^T J_1(s)a(s)f(s,u(s),v(s))\,{\rm d}s \le \lambda \int_0^T J_1(s)a(s)(f_0^{\rm s}+\varepsilon)(u(s)+v(s))\,{\rm d}s$$

$$\le \lambda(f_0^{\rm s}+\varepsilon)\int_0^T J_1(s)a(s)(\|u\|+\|v\|)\,{\rm d}s = \lambda(f_0^{\rm s}+\varepsilon)B\|(u,v)\|_Y \le \tilde{\alpha}_1'\|(u,v)\|_Y,$$

$$Q_2(u,v)(t) \le \mu \int_0^T J_2(s)b(s)g(s,u(s),v(s))\,{\rm d}s \le \mu \int_0^T J_2(s)b(s)\varepsilon(u(s)+v(s))\,{\rm d}s$$

$$\le \mu\varepsilon \int_0^T J_2(s)b(s)(\|u\|+\|v\|)\,{\rm d}s = \mu\varepsilon D\|(u,v)\|_Y \le \tilde{\alpha}_2'\|(u,v)\|_Y,$$

for all $t \in [0,T]$. Therefore, $\|Q_1(u,v)\| \le \tilde{\alpha}_1'\|(u,v)\|_Y$, $\|Q_2(u,v)\| \le \tilde{\alpha}_2'\|(u,v)\|_Y$, and then

$$\|\mathcal{Q}(u,v)\|_Y \le \tilde{\alpha}_1'\|(u,v)\|_Y + \tilde{\alpha}_2'\|(u,v)\|_Y = \|(u,v)\|_Y.$$

Hence, for all $(u,v) \in P \cap \partial\Omega_1$, we deduce $\|\mathcal{Q}(u,v)\|_Y \le \|(u,v)\|_Y$.

By the definitions of $f_\infty^{\rm i}$ and $g_\infty^{\rm i}$, there exists $\bar{R}_2 > 0$ such that $f(t,u,v) \ge (f_\infty^{\rm i} - \varepsilon)(u+v)$ and $g(t,u,v) \ge \frac{1}{\varepsilon}(u+v)$ for all $u, v \ge 0$, $u+v \ge \bar{R}_2$, and $t \in [c, T-c]$. We consider $R_2 = \max\{2R_1, \bar{R}_2/\gamma\}$, and we define $\Omega_2 = \{(u,v) \in Y, \|(u,v)\|_Y < R_2\}$. Let $(u,v) \in P \cap \partial\Omega_2$—that is, $\|(u,v)\|_Y = R_2$. Then we obtain $u(t)+v(t) \ge \gamma\|(u,v)\|_Y = \gamma R_2 \ge \bar{R}_2$ for all $t \in [c, T-c]$. By Lemma 2.1.5, we have

$$Q_1(u,v)(c) \ge \lambda\gamma_1 \int_0^T J_1(s)a(s)f(s,u(s),v(s))\,{\rm d}s \ge \lambda\gamma_1 \int_c^{T-c} J_1(s)a(s)f(s,u(s),v(s))\,{\rm d}s$$

$$\ge \lambda\gamma_1 \int_c^{T-c} J_1(s)a(s)(f_\infty^{\rm i} - \varepsilon)(u(s)+v(s))\,{\rm d}s \ge \lambda\gamma_1(f_\infty^{\rm i}-\varepsilon)A\gamma\|(u,v)\|_Y \ge \alpha_1'\|(u,v)\|_Y,$$

$$Q_2(u,v)(c) \ge \mu\gamma_2 \int_0^T J_2(s)b(s)g(s,u(s),v(s))\,{\rm d}s \ge \mu\gamma_2 \int_c^{T-c} J_2(s)b(s)g(s,u(s),v(s))\,{\rm d}s$$

$$\ge \mu\gamma_2 \int_c^{T-c} J_2(s)b(s)\frac{1}{\varepsilon}(u(s)+v(s))\,{\rm d}s \ge \frac{\mu\gamma\gamma_2 C}{\varepsilon}\|(u,v)\|_Y \ge \alpha_2'\|(u,v)\|_Y.$$

Therefore, $\|Q_1(u,v)\| \ge \alpha_1'\|(u,v)\|_Y$, $\|Q_2(u,v)\| \ge \alpha_2'\|(u,v)\|_Y$, and then

$$\|\mathcal{Q}(u,v)\|_Y \ge \alpha_1'\|(u,v)\|_Y + \alpha_2'\|(u,v)\|_Y = \|(u,v)\|_Y.$$

Hence, for all $(u,v) \in P \cap \partial\Omega_2$, we obtain $\|\mathcal{Q}(u,v)\|_Y \ge \|(u,v)\|_Y$.

By using Theorem 1.1.1, we deduce that the operator $\mathcal{Q}$ has a fixed point $(u,v) \in P \cap (\bar{\Omega}_2 \setminus \Omega_1)$. □

Remark 2.1.1. We mention that in Theorem 2.1.1 we have the possibility to choose $\alpha_1 = 0$ or $\alpha_2 = 0$. Therefore, each of the first four cases contains three subcases. For example, in the third case, $g_0^{\rm s} = 0$ and $f_0^{\rm s}, f_\infty^{\rm i}, g_\infty^{\rm i} \in (0,\infty)$, we have the following situations:

(a) If $\alpha_1, \alpha_2 \in (0,1)$, $\alpha_1 + \alpha_2 = 1$, and $L_1 < L_2'$, then $\lambda \in (L_1, L_2')$ and $\mu \in (L_3, \infty)$.
(b) If $\alpha_1 = 1$, $\alpha_2 = 0$, and $L_1' < L_2'$, then $\lambda \in (L_1', L_2')$ and $\mu \in (0,\infty)$, where $L_1' = \frac{1}{\gamma\gamma_1 f_\infty^{\rm i} A}$.
(c) If $\alpha_1 = 0$ and $\alpha_2 = 1$, then $\lambda \in (0, L_2')$ and $\mu \in (L_3', \infty)$, where $L_3' = \frac{1}{\gamma\gamma_2 g_\infty^{\rm i} C}$.

In what follows, for $f_0^{\rm i}, g_0^{\rm i}, f_\infty^{\rm s}, g_\infty^{\rm s} \in (0,\infty)$ and numbers $\alpha_1, \alpha_2 \geq 0$ and $\tilde{\alpha}_1, \tilde{\alpha}_2 > 0$ such that $\alpha_1 + \alpha_2 = 1$ and $\tilde{\alpha}_1 + \tilde{\alpha}_2 = 1$, we define the numbers $\tilde{L}_1, \tilde{L}_2, \tilde{L}_3$, $\tilde{L}_4$, $\tilde{L}_2'$, and $\tilde{L}_4'$ by

$$\tilde{L}_1 = \frac{\alpha_1}{\gamma\gamma_1 f_0^{\rm i} A}, \quad \tilde{L}_2 = \frac{\tilde{\alpha}_1}{f_\infty^{\rm s} B}, \quad \tilde{L}_3 = \frac{\alpha_2}{\gamma\gamma_2 g_0^{\rm i} C}, \quad \tilde{L}_4 = \frac{\tilde{\alpha}_2}{g_\infty^{\rm s} D},$$
$$\tilde{L}_2' = \frac{1}{f_\infty^{\rm s} B}, \quad \tilde{L}_4' = \frac{1}{g_\infty^{\rm s} D}.$$

Theorem 2.1.2. *Assume that (H1)–(H3) hold, $\alpha_1, \alpha_2 \geq 0$, and $\tilde{\alpha}_1, \tilde{\alpha}_2 > 0$ such that $\alpha_1 + \alpha_2 = 1$, $\tilde{\alpha}_1 + \tilde{\alpha}_2 = 1$.*

(1) If $f_0^{\rm i}, g_0^{\rm i}, f_\infty^{\rm s}, g_\infty^{\rm s} \in (0,\infty)$, $\tilde{L}_1 < \tilde{L}_2$, and $\tilde{L}_3 < \tilde{L}_4$, then for each $\lambda \in (\tilde{L}_1, \tilde{L}_2)$ and $\mu \in (\tilde{L}_3, \tilde{L}_4)$ there exists a positive solution $(u(t), v(t)), t \in [0,T]$ for (S)–(BC).
(2) If $f_0^{\rm i}, g_0^{\rm i}, f_\infty^{\rm s} \in (0,\infty)$, $g_\infty^{\rm s} = 0$, and $\tilde{L}_1 < \tilde{L}_2'$, then for each $\lambda \in (\tilde{L}_1, \tilde{L}_2')$ and $\mu \in (\tilde{L}_3, \infty)$ there exists a positive solution $(u(t), v(t)), t \in [0,T]$ for (S)–(BC).
(3) If $f_0^{\rm i}, g_0^{\rm i}, g_\infty^{\rm s} \in (0,\infty)$, $f_\infty^{\rm s} = 0$, and $\tilde{L}_3 < \tilde{L}_4'$, then for each $\lambda \in (\tilde{L}_1, \infty)$ and $\mu \in (\tilde{L}_3, \tilde{L}_4')$ there exists a positive solution $(u(t), v(t)), t \in [0,T]$ for (S)–(BC).
(4) If $f_0^{\rm i}, g_0^{\rm i} \in (0,\infty)$ and $f_\infty^{\rm s} = g_\infty^{\rm s} = 0$, then for each $\lambda \in (\tilde{L}_1, \infty)$ and $\mu \in (\tilde{L}_3, \infty)$ there exists a positive solution $(u(t), v(t)), t \in [0,T]$ for (S)–(BC).
(5) If $\{f_0^{\rm i} = \infty,\ g_0^{\rm i}, f_\infty^{\rm s}, g_\infty^{\rm s} \in (0,\infty)\}$ or $\{f_0^{\rm i}, f_\infty^{\rm s}, g_\infty^{\rm s} \in (0,\infty),\ g_0^{\rm i} = \infty\}$ or $\{f_0^{\rm i} = g_0^{\rm i} = \infty,\ f_\infty^{\rm s}, g_\infty^{\rm s} \in (0,\infty)\}$, then for each $\lambda \in (0, \tilde{L}_2)$ and $\mu \in (0, \tilde{L}_4)$ there exists a positive solution $(u(t), v(t)), t \in [0,T]$ for (S)–(BC).
(6) If $\{f_0^{\rm i} = \infty,\ g_0^{\rm i}, f_\infty^{\rm s} \in (0,\infty),\ g_\infty^{\rm s} = 0\}$ or $\{f_0^{\rm i}, f_\infty^{\rm s} \in (0,\infty),\ g_0^{\rm i} = \infty,\ g_\infty^{\rm s} = 0\}$ or $\{f_0^{\rm i} = g_0^{\rm i} = \infty,\ f_\infty^{\rm s} \in (0,\infty),\ g_\infty^{\rm s} = 0\}$, then for each $\lambda \in (0, \tilde{L}_2')$ and $\mu \in (0,\infty)$ there exists a positive solution $(u(t), v(t)), t \in [0,T]$ for (S)–(BC).
(7) If $\{f_0^{\rm i} = \infty,\ g_0^{\rm i}, g_\infty^{\rm s} \in (0,\infty),\ f_\infty^{\rm s} = 0\}$ or $\{f_0^{\rm i}, g_\infty^{\rm s} \in (0,\infty),\ g_0^{\rm i} = \infty,\ f_\infty^{\rm s} = 0\}$ or $\{f_0^{\rm i} = g_0^{\rm i} = \infty,\ f_\infty^{\rm s} = 0,\ g_\infty^{\rm s} \in (0,\infty)\}$, then for each $\lambda \in (0,\infty)$ and $\mu \in (0, \tilde{L}_4')$ there exists a positive solution $(u(t), v(t)), t \in [0,T]$ for (S)–(BC).
(8) If $\{f_0^{\rm i} = \infty,\ g_0^{\rm i} \in (0,\infty),\ f_\infty^{\rm s} = g_\infty^{\rm s} = 0\}$ or $\{f_0^{\rm i} \in (0,\infty),\ g_0^{\rm i} = \infty,\ f_\infty^{\rm s} = g_\infty^{\rm s} = 0\}$ or $\{f_0^{\rm i} = g_0^{\rm i} = \infty,\ f_\infty^{\rm s} = g_\infty^{\rm s} = 0\}$, then for each $\lambda \in (0,\infty)$ and $\mu \in (0,\infty)$ there exists a positive solution $(u(t), v(t)), t \in [0,T]$ for (S)–(BC).

Proof. We consider the above cone P and the operators Q_1, Q_2, and $\mathcal{Q}$. Because the proofs of the above 16 cases are similar, we shall only prove one of them—namely, the first case of (5). So, we suppose $f_0^{\rm i} = \infty$ and $g_0^{\rm i}, f_\infty^{\rm s}, g_\infty^{\rm s} \in (0,\infty)$. Let $\lambda \in (0, \tilde{L}_2)$—that is, $\lambda \in \left(0, \frac{\tilde{\alpha}_1}{f_\infty^{\rm s} B}\right)$—and let $\mu \in (0, \tilde{L}_4)$—that is, $\mu \in \left(0, \frac{\tilde{\alpha}_2}{g_\infty^{\rm s} D}\right)$. Let $\varepsilon > 0$ be a positive number such that $\varepsilon < g_0^{\rm i}$ and

$$\frac{\varepsilon\alpha_1}{\gamma\gamma_1 A} \leq \lambda, \quad \frac{\alpha_2}{\gamma\gamma_2(g_0^{\rm i} - \varepsilon)C} \leq \mu, \quad \frac{\tilde{\alpha}_1}{(f_\infty^{\rm s} + \varepsilon)B} \geq \lambda, \quad \frac{\tilde{\alpha}_2}{(g_\infty^{\rm s} + \varepsilon)D} \geq \mu.$$

By the definitions of $f_0^{\rm i}$ and $g_0^{\rm i}$, we deduce that there exists $R_3 > 0$ such that $f(t,u,v) \geq \frac{1}{\varepsilon}(u+v)$, $g(t,u,v) \geq (g_0^{\rm i} - \varepsilon)(u+v)$ for all $u, v \geq 0$ with $u+v \in [0, R_3]$ and

$t \in [c, T-c]$. We denote $\Omega_3 = \{(u,v) \in Y, \|(u,v)\|_Y < R_3\}$. Let $(u,v) \in P \cap \partial\Omega_3$—that is, $\|(u,v)\|_Y = R_3$. Because $u(t) + v(t) \le \|u\| + \|v\| = R_3$ for all $t \in [0,T]$, then by using Lemma 2.1.5, we obtain

$$\begin{aligned}
Q_1(u,v)(c) &\ge \lambda\gamma_1 \int_c^{T-c} J_1(s)a(s)f(s,u(s),v(s))\,\mathrm{d}s \\
&\ge \lambda\gamma_1 \int_c^{T-c} J_1(s)a(s)\frac{1}{\varepsilon}(u(s)+v(s))\,\mathrm{d}s \\
&\ge \frac{\lambda\gamma\gamma_1}{\varepsilon} \int_c^{T-c} J_1(s)a(s)(\|u\|+\|v\|)\,\mathrm{d}s \\
&= \frac{\lambda\gamma\gamma_1 A}{\varepsilon}\|(u,v)\|_Y \ge \alpha_1\|(u,v)\|_Y,
\end{aligned}$$

$$\begin{aligned}
Q_2(u,v)(c) &\ge \mu\gamma_2 \int_c^{T-c} J_2(s)b(s)g(s,u(s),v(s))\,\mathrm{d}s \\
&\ge \mu\gamma_2 \int_c^{T-c} J_2(s)b(s)(g_0^{\mathrm{i}}-\varepsilon)(u(s)+v(s))\,\mathrm{d}s \\
&\ge \mu\gamma\gamma_2(g_0^{\mathrm{i}}-\varepsilon) \int_c^{T-c} J_2(s)b(s)(\|u\|+\|v\|)\,\mathrm{d}s \\
&= \mu\gamma\gamma_2(g_0^{\mathrm{i}}-\varepsilon)C\|(u,v)\|_Y \ge \alpha_2\|(u,v)\|_Y.
\end{aligned}$$

Therefore, $\|Q_1(u,v)\| \ge \alpha_1\|(u,v)\|_Y$, $\|Q_2(u,v)\| \ge \alpha_2\|(u,v)\|_Y$, and then

$$\|\mathcal{Q}(u,v)\|_Y \ge \alpha_1\|(u,v)\|_Y + \alpha_2\|(u,v)\|_Y = \|(u,v)\|_Y.$$

Hence, for all $(u,v) \in P \cap \partial\Omega_2$, we have $\|\mathcal{Q}(u,v)\|_Y \ge \|(u,v)\|_Y$.

We define now the functions $f^*, g^*\colon [0,T] \times \mathbb{R}_+ \to \mathbb{R}_+$, $f^*(t,x) = \max_{0\le u+v\le x} f(t,u,v)$, $g^*(t,x) = \max_{0\le u+v\le x} g(t,u,v)$ for all $t \in [0,T]$ and $x \in [0,\infty)$. For the above ε, there exists $\bar{R}_4 > 0$ such that $f^*(t,x) \le (f_\infty^{\mathrm{s}} + \varepsilon)x$ and $g^*(t,x) \le (g_\infty^{\mathrm{s}} + \varepsilon)x$ for all $x \ge \bar{R}_4$ and $t \in [0,T]$. We consider $R_4 = \max\{2R_3, \bar{R}_4\}$, and we denote $\Omega_4 = \{(u,v) \in Y, \|(u,v)\|_Y < R_4\}$. Let $(u,v) \in P \cap \partial\Omega_4$. Then for all $t \in [0,T]$, we obtain

$$\begin{aligned}
Q_1(u,v)(t) &\le \lambda \int_0^T J_1(s)a(s)f(s,u(s),v(s))\,\mathrm{d}s \\
&\le \lambda \int_0^T J_1(s)a(s)f^*(s,\|(u,v)\|_Y)\,\mathrm{d}s \\
&\le \lambda(f_\infty^{\mathrm{s}}+\varepsilon) \int_0^T J_1(s)a(s)\|(u,v)\|_Y\,\mathrm{d}s = \lambda(f_\infty^{\mathrm{s}}+\varepsilon)B\|(u,v)\|_Y \\
&\le \tilde{\alpha}_1\|(u,v)\|_Y,
\end{aligned}$$

$$
\begin{aligned}
Q_2(u,v)(t) &\le \mu\int_0^T J_2(s)b(s)g(s,u(s),v(s))\,ds \\
&\le \mu\int_0^T J_2(s)b(s)g^*(s,\|(u,v)\|_Y)\,ds \\
&\le \mu(g_\infty^s+\varepsilon)\int_0^T J_2(s)b(s)\|(u,v)\|_Y\,ds = \mu(g_\infty^s+\varepsilon)D\|(u,v)\|_Y \\
&\le \tilde{\alpha}_2\|(u,v)\|_Y.
\end{aligned}
$$

Therefore, $\|Q_1(u,v)\| \le \tilde{\alpha}_1\|(u,v)\|_Y$, $\|Q_2(u,v)\| \le \tilde{\alpha}_2\|(u,v)\|_Y$, and then

$$\|\mathcal{Q}(u,v)\|_Y \le \tilde{\alpha}_1\|(u,v)\|_Y + \tilde{\alpha}_2\|(u,v)\|_Y = \|(u,v)\|_Y.$$

Hence, for all $(u,v)\in P\cap\partial\Omega_4$ we have $\|\mathcal{Q}(u,v)\|_Y \le \|(u,v)\|_Y$.

By using Theorem 1.1.1, we deduce that the operator $\mathcal{Q}$ has a fixed point $(u,v)\in P\cap(\bar{\Omega}_4\setminus\Omega_3)$. □

Now, we present intervals for λ and μ for which there exists no positive solution of problem (S)–(BC).

Theorem 2.1.3. *Assume that (H1)–(H3) hold. If $f_0^s, f_\infty^s, g_0^s, g_\infty^s < \infty$, then there exist positive constants λ_0 and μ_0 such that for every $\lambda\in(0,\lambda_0)$ and $\mu\in(0,\mu_0)$ the boundary value problem (S)–(BC) has no positive solution.*

Proof. From the definitions of f_0^s, f_∞^s, g_0^s, and g_∞^s, we deduce that there exist $M_1, M_2>0$ such that $f(t,u,v)\le M_1(u+v)$, $g(t,u,v)\le M_2(u+v)$ for all $t\in[0,T]$ and $u,v\ge 0$. As in the proof of Theorem 1.2.1, one can prove that $\lambda_0=\frac{1}{2M_1B}$ and $\mu_0=\frac{1}{2M_2D}$ satisfy our theorem. □

Theorem 2.1.4. *Assume that (H1)–(H3) hold.*

(a) *If $f_0^i, f_\infty^i>0$ and $f(t,u,v)>0$ for all $t\in[c,T-c]$, $u\ge 0$, $v\ge 0$, and $u+v>0$, then there exists a positive constant $\tilde{\lambda}_0$ such that for every $\lambda>\tilde{\lambda}_0$ and $\mu>0$ the boundary value problem (S)–(BC) has no positive solution.*

(b) *If $g_0^i, g_\infty^i>0$ and $g(t,u,v)>0$ for all $t\in[c,T-c]$, $u\ge 0$, $v\ge 0$, and $u+v>0$, then there exists a positive constant $\tilde{\mu}_0$ such that for every $\mu>\tilde{\mu}_0$ and $\lambda>0$ the boundary value problem (S)–(BC) has no positive solution.*

(c) *If $f_0^i, f_\infty^i, g_0^i, g_\infty^i>0$ and $f(t,u,v)>0$, $g(t,u,v)>0$ for all $t\in[c,T-c]$, $u\ge 0$, $v\ge 0$, and $u+v>0$, then there exist positive constants $\hat{\lambda}_0$ and $\hat{\mu}_0$ such that for every $\lambda>\hat{\lambda}_0$ and $\mu>\hat{\mu}_0$ the boundary value problem (S)–(BC) has no positive solution.*

Proof.

(a) From the assumptions of the theorem, we deduce that there exists $m_1>0$ such that $f(t,u,v)\ge m_1(u+v)$ for all $t\in[c,T-c]$ and $u,v\ge 0$. By using arguments similar to those used in the proof of Theorem 1.2.2, we can prove that $\tilde{\lambda}_0=\frac{1}{\gamma\gamma_1 m_1 A}$ satisfies our theorem (a).

(b) From the assumptions of the theorem, we deduce that there exists $m_2>0$ such that $g(t,u,v)\ge m_2(u+v)$ for all $t\in[c,T-c]$ and $u,v\ge 0$. As in the proof of Theorem 1.2.3, we show that $\tilde{\mu}_0=\frac{1}{\gamma\gamma_2 m_2 C}$ satisfies our theorem (b).

(c) We define $\hat{\lambda}_0=\frac{\tilde{\lambda}_0}{2}$ and $\hat{\mu}_0=\frac{\tilde{\mu}_0}{2}$, and by using arguments similar to those used in the proof of Theorem 1.2.4, we find that they satisfy our theorem (c).

□

2.1.4 Examples

Example 2.1.1. Let $T=1, n=3, m=2, p=2, q=1, r=1, l=3, \xi_1=1/3, \xi_2=2/3, a_1=1/3, a_2=1/2, \eta_1=2/5, b_1=1/2, \zeta_1=1/2, c_1=3/4, \rho_1=1/4, \rho_2=1/2, \rho_3=3/4, d_1=1/3, d_2=1/4, d_3=1/5$, and $a(t)=a_0t$, $b(t)=b_0t$, $t\in[0,1]$, with $a_0>0$, $b_0>0$. We have $\sum_{i=1}^p a_i<1$, $\sum_{i=1}^q b_i<1$, $\sum_{i=1}^r c_i<1$, and $\sum_{i=1}^l d_i<1$. We consider the higher-order differential system

$$\begin{cases} u'''(t)+\lambda a_0 t f(t,u(t),v(t))=0, & t\in(0,1),\\ v''(t)+\mu b_0 t g(t,u(t),v(t))=0, & t\in(0,1), \end{cases} \tag{S_1}$$

with the boundary conditions

$$\begin{cases} u(0)=\frac{1}{3}u\left(\frac{1}{3}\right)+\frac{1}{2}u\left(\frac{2}{3}\right), \quad u'(0)=0, \quad u(1)=\frac{1}{2}u\left(\frac{2}{5}\right),\\ v(0)=\frac{3}{4}v\left(\frac{1}{2}\right), \quad v(1)=\frac{1}{3}v\left(\frac{1}{4}\right)+\frac{1}{4}v\left(\frac{1}{2}\right)+\frac{1}{5}v\left(\frac{3}{4}\right), \end{cases} \tag{BC_1}$$

where the functions f and g are given by

$$f(t,u,v)=\frac{(u+v)[p_1(u+v)+1](q_1+\sin v)}{u+v+1},$$

$$g(t,u,v)=\frac{(u+v)[p_2(u+v)+1](q_2+\cos u)}{u+v+1},$$

for all $t\in[0,1]$, $u,v\in[0,\infty)$, and $p_1,p_2>0$, $q_1,q_2>1$. For $c=1/4$, it follows that $f_0^{\rm s}=f_0^{\rm i}=q_1$, $g_0^{\rm s}=g_0^{\rm i}=q_2+1$, $f_\infty^{\rm s}=p_1(q_1+1)$, $f_\infty^{\rm i}=p_1(q_1-1)$, $g_\infty^{\rm s}=p_2(q_2+1)$, and $g_\infty^{\rm i}=p_2(q_2-1)$.

We also obtain

$$g_1(t,s)=\frac{1}{2}\begin{cases} t^2(1-s)^2-(t-s)^2, & 0\le s\le t\le 1,\\ t^2(1-s)^2, & 0\le t\le s\le 1, \end{cases}$$

$$g_2(t,s)=\begin{cases} s(1-t), & 0\le s\le t\le 1,\\ t(1-s), & 0\le t\le s\le 1, \end{cases}$$

$\theta_1(s)=\frac{1}{2-s}$, $\theta_2(s)=s$ for $s\in[0,1]$, $\Delta_1=\frac{191}{675}$, and $\Delta_2=\frac{29}{120}$. For the functions J_1 and J_2, we deduce

$$J_1(s)=\begin{cases} \dfrac{3156s-6542s^2+2673s^3}{1528(2-s)}, & 0\le s<\frac{1}{3},\\[2mm] \dfrac{184+1960s-4334s^2+1845s^3}{1528(2-s)}, & \frac{1}{3}\le s<\frac{2}{5},\\[2mm] \dfrac{184+474s-1362s^2+635s^3}{764(2-s)}, & \frac{2}{5}\le s<\frac{2}{3},\\[2mm] \dfrac{(s-1)^2(7s+368)}{382(2-s)}, & \frac{2}{3}\le s\le 1, \end{cases}$$

and

$$J_2(s)=\begin{cases}\dfrac{359s-116s^2}{116}, & 0\le s<\frac{1}{4},\\ \dfrac{25+259s-116s^2}{116}, & \frac{1}{4}\le s<\frac{1}{2},\\ \dfrac{178-47s-116s^2}{116}, & \frac{1}{2}\le s<\frac{3}{4},\\ \dfrac{(1-s)(223+116s)}{116}, & \frac{3}{4}\le s\le 1.\end{cases}$$

We also obtain $\gamma_1=\frac{1}{16}$, $\gamma_2=\frac{1}{4}$, and $\gamma=\frac{1}{16}$, and for $\alpha_1,\alpha_2>0$ and $\alpha_1+\alpha_2=1$, we have

$$L_1=\frac{256\alpha_1}{p_1(q_1-1)a_0}\left(\int_{1/4}^{3/4} sJ_1(s)\,ds\right)^{-1},\quad L_2=\frac{\alpha_1}{q_1a_0}\left(\int_0^1 sJ_1(s)\,ds\right)^{-1},$$

$$L_3=\frac{64\alpha_2}{p_2(q_2-1)b_0}\left(\int_{1/4}^{3/4} sJ_2(s)\,ds\right)^{-1},\quad L_4=\frac{\alpha_2}{(q_2+1)b_0}\left(\int_0^1 sJ_2(s)\,ds\right)^{-1}.$$

After some computations, we obtain $\int_{1/4}^{3/4} sJ_1(s)\,ds\approx 0.029386$, $\int_0^1 sJ_1(s)\,ds\approx 0.036694$, $\int_{1/4}^{3/4} sJ_2(s)\,ds\approx 0.22315$, and $\int_0^1 sJ_2(s)\,ds\approx 0.310165$. The conditions $L_1<L_2$ and $L_3<L_4$ become

$$\frac{p_1(q_1-1)}{256q_1}>\frac{\int_0^1 sJ_1(s)\,ds}{\int_{1/4}^{3/4} sJ_1(s)\,ds},\quad \frac{p_2(q_2-1)}{64(q_2+1)}>\frac{\int_0^1 sJ_2(s)\,ds}{\int_{1/4}^{3/4} sJ_2(s)\,ds}.$$

For example, if $\frac{p_1(q_1-1)}{q_1}\ge 320$ and $\frac{p_2(q_2-1)}{q_2+1}\ge 89$, then the above conditions are satisfied. Therefore, by Theorem 2.1.1 (1), for each $\lambda\in(L_1,L_2)$ and $\mu\in(L_3,L_4)$, there exists a positive solution $(u(t),v(t))$, $t\in[0,1]$ for problem (S_1)–(BC_1).

Example 2.1.2. Let $T=1$, $n=4$, $m=3$, $p=1$, $q=2$, $r=2$, $l=1$, $\xi_1=1/2$, $a_1=3/4$, $\eta_1=1/4$, $\eta_2=3/4$, $b_1=1/2$, $b_2=1/3$, $\zeta_1=1/3$, $\zeta_2=2/3$, $c_1=4/5$, $c_2=1/6$, $\rho_1=3/5$, $d_1=6/7$, $a(t)=1$, $b(t)=1$ for all $t\in[0,1]$, and $f(t,u,v)=e^t(u+v)(2+\sin v)$, $g(t,u,v)=\sqrt[3]{1-t}\,(u^2+v^2)$ for all $t\in[0,1]$, $u,v\in[0,\infty)$. We have $\sum_{i=1}^p a_i<1$, $\sum_{i=1}^q b_i<1$, $\sum_{i=1}^r c_i<1$, and $\sum_{i=1}^l d_i<1$. We consider the higher-order differential system

$$\begin{cases}u^{(4)}(t)+\lambda e^t(u(t)+v(t))(2+\sin v(t))=0, & t\in(0,1),\\ v^{(3)}(t)+\mu\sqrt[3]{1-t}\,(u^2(t)+v^2(t))=0, & t\in(0,1),\end{cases}\tag{S_2}$$

with the boundary conditions

$$\begin{cases}u(0)=\frac{3}{4}u\left(\frac{1}{2}\right),\quad u'(0)=u''(0)=0,\quad u(1)=\frac{1}{2}u\left(\frac{1}{4}\right)+\frac{1}{3}u\left(\frac{3}{4}\right),\\ v(0)=\frac{4}{5}v\left(\frac{1}{3}\right)+\frac{1}{6}v\left(\frac{2}{3}\right),\quad v'(0)=0,\quad v(1)=\frac{6}{7}v\left(\frac{3}{5}\right).\end{cases}\tag{BC_2}$$

For $c = 1/4$, we deduce that $f_0^s = 2e$, $f_0^i = 2e^{1/4}$, $g_0^s = 0$, $g_0^i = 0$, $f_\infty^s = 3e$, $f_\infty^i = e^{1/4}$, $g_\infty^s = \infty$, and $g_\infty^i = \infty$.

We also obtain

$$g_1(t,s) = \frac{1}{6}\begin{cases} t^3(1-s)^3 - (t-s)^3, & 0 \le s \le t \le 1, \\ t^3(1-s)^3, & 0 \le t \le s \le 1, \end{cases}$$

$$g_2(t,s) = \frac{1}{2}\begin{cases} t^2(1-s)^2 - (t-s)^2, & 0 \le s \le t \le 1, \\ t^2(1-s)^2, & 0 \le t \le s \le 1, \end{cases}$$

$$\theta_1(s) = \begin{cases} \frac{s}{1-(1-s)^{3/2}}, & s \in (0,1], \\ 2/3, & s = 0, \end{cases}, \ \theta_2(s) = \frac{1}{2-s}, \ s \in [0,1], \ \Delta_1 = \frac{117}{512}, \ \Delta_2 = \frac{2189}{47{,}250},$$

$\gamma = \gamma_1 = \frac{1}{64}$, $\gamma_2 = \frac{1}{16}$. For the functions J_1 and J_2, we deduce

$$J_1(s) = \begin{cases} 0, & s = 0, \\ \dfrac{s^3(1-s)^3}{6[1-(1-s)^{3/2}]^2} + \dfrac{1917s - 5850s^2 + 4880s^3}{8424}, & 0 < s < \frac{1}{4}, \\ \dfrac{s^3(1-s)^3}{6[1-(1-s)^{3/2}]^2} + \dfrac{33 + 3438s - 10{,}116s^2 + 7648s^3}{16{,}848}, & \frac{1}{4} \le s < \frac{1}{2}, \\ \dfrac{s^3(1-s)^3}{6[1-(1-s)^{3/2}]^2} + \dfrac{507 - 1224s + 828s^2 - 100s^3}{8424}, & \frac{1}{2} \le s < \frac{3}{4}, \\ \dfrac{s^3(1-s)^3}{6[1-(1-s)^{3/2}]^2} + \dfrac{67(1-s)^3}{702}, & \frac{3}{4} \le s \le 1, \end{cases}$$

and

$$J_2(s) = \begin{cases} \dfrac{37{,}893s - 84{,}920s^2 + 33{,}534s^3}{4378(2-s)}, & 0 \le s < \frac{1}{3}, \\ \dfrac{5808 + 141s - 15{,}224s^2 + 7398s^3}{4378(2-s)}, & \frac{1}{3} \le s < \frac{3}{5}, \\ \dfrac{11{,}532 - 21{,}801s + 10{,}216s^2 - 552s^3}{4378(2-s)}, & \frac{3}{5} \le s < \frac{2}{3}, \\ \dfrac{(1-s)^2(16{,}372 - 5997s)}{4378(2-s)}, & \frac{2}{3} \le s \le 1. \end{cases}$$

After some computations, we obtain $B = \int_0^1 J_1(s)\,ds \approx 0.01437846$ and $L_2' = 1/(2eB) \approx 12.79272803$. Therefore, by Theorem 2.1.1 (7), for each $\lambda \in (0, L_2')$ and $\mu \in (0, \infty)$ there exists a positive solution $(u(t), v(t))$, $t \in [0,1]$ for problem (S_2)–(BC_2). By using Theorem 2.1.4 (a) ($m_1 = e^{1/4}$), we deduce that there exists $\tilde{\lambda}_0 \ge L_2'$ such that for any $\lambda > \tilde{\lambda}_0$ and $\mu > 0$ problem (S_2)–(BC_2) has no positive solution.

Remark 2.1.2. The results presented in this section were published in Henderson and Luca (2012b, 2014d).

2.2 Existence and multiplicity of positive solutions for systems without parameters

In this section, we investigate the existence and multiplicity of positive solutions for problem (S)–(BC) from Section 2.1 with $\lambda = \mu = 1$, $a(t) = 1$, $b(t) = 1$ for all $t \in [0, T]$, and where f and g are dependent only on t and v, and t and u, respectively. The nonlinearities f and g are nonsingular functions, or singular functions at $t = 0$ and/or $t = T$.

2.2.1 *Nonsingular nonlinearities*

We consider the system of nonlinear higher-order ordinary differential equations

$$\begin{cases} u^{(n)}(t) + f(t, v(t)) = 0, & t \in (0, T), \\ v^{(m)}(t) + g(t, u(t)) = 0, & t \in (0, T), \end{cases} \tag{S$'$}$$

with the multipoint boundary conditions

$$\begin{cases} u(0) = \sum_{i=1}^{p} a_i u(\xi_i), \quad u'(0) = \cdots = u^{(n-2)}(0) = 0, \quad u(T) = \sum_{i=1}^{q} b_i u(\eta_i), \\ v(0) = \sum_{i=1}^{r} c_i v(\zeta_i), \quad v'(0) = \cdots = v^{(m-2)}(0) = 0, \quad v(T) = \sum_{i=1}^{l} d_i v(\rho_i), \end{cases} \tag{BC}$$

where $n, m, p, q, r, l \in \mathbb{N}$, $n, m \geq 2$; $a_i, \xi_i \in \mathbb{R}$ for all $i = 1, \ldots, p$; $b_i, \eta_i \in \mathbb{R}$ for all $i = 1, \ldots, q$; $c_i, \zeta_i \in \mathbb{R}$ for all $i = 1, \ldots, r$; $d_i, \rho_i \in \mathbb{R}$ for all $i = 1, \ldots, l$; $0 < \xi_1 < \cdots < \xi_p < T$, $0 < \eta_1 < \cdots < \eta_q < T$, $0 < \zeta_1 < \cdots < \zeta_r < T$, $0 < \rho_1 < \cdots < \rho_l < T$. In the case $n = 2$ or $m = 2$, the boundary conditions above are of the form $u(0) = \sum_{i=1}^{p} a_i u(\xi_i)$, $u(T) = \sum_{i=1}^{q} b_i u(\eta_i)$, or $v(0) = \sum_{i=1}^{r} c_i v(\zeta_i)$, $v(T) = \sum_{i=1}^{l} d_i v(\rho_i)$, respectively—that is, without conditions on the derivatives of u and v at the point 0.

Under sufficient conditions on f and g, we prove the existence and multiplicity of positive solutions of the above problem, by applying the fixed point index theory. By a positive solution of (S$'$)–(BC) we mean a pair of functions $(u, v) \in C^n([0, T]) \times C^m([0, T])$ satisfying (S$'$) and (BC) with $u(t) \geq 0$, $v(t) \geq 0$ for all $t \in [0, T]$, and $\sup_{t \in [0,T]} u(t) > 0$, $\sup_{t \in [0,T]} v(t) > 0$. This problem is a generalization of the problem studied in Henderson and Luca (2013b), where in the boundary conditions we have $a_i = 0$ for all $i = 1, \ldots, p$, and $c_i = 0$ for all $i = 1, \ldots, r$.

We present the basic assumptions that we shall use in the sequel:

(I1) $0 < \xi_1 < \cdots < \xi_p < T, a_i \geq 0$ for all $i = 1, \ldots, p$, $\sum_{i=1}^{p} a_i < 1, 0 < \eta_1 < \cdots < \eta_q < T$, $b_i \geq 0$ for all $i = 1, \ldots, q$, $\sum_{i=1}^{q} b_i < 1$, $0 < \zeta_1 < \cdots < \zeta_r < T$, $c_i \geq 0$ for all $i = 1, \ldots, r$, $\sum_{i=1}^{r} c_i < 1$, $0 < \rho_1 < \cdots < \rho_l < T$, and $d_i \geq 0$ for all $i = 1, \ldots, l$, $\sum_{i=1}^{l} d_i < 1$.

(I2) The functions $f, g \in C([0,T]\times[0,\infty),[0,\infty))$ and $f(t,0)=0, g(t,0)=0$ for all $t \in [0,T]$.

Under assumption (I1), we have all auxiliary results in Lemmas 2.1.1–2.1.6 from Section 2.1.2.

The pair of functions $(u,v) \in C^n([0,T]) \times C^m([0,T])$ is a solution for our problem (S′)–(BC) if and only if $(u,v) \in C([0,T]) \times C([0,T])$ is a solution for the nonlinear integral system

$$\begin{cases} u(t) = \displaystyle\int_0^T G_1(t,s) f\left(s, \int_0^T G_2(s,\tau) g(\tau, u(\tau))\, \mathrm{d}\tau\right) \mathrm{d}s, & t \in [0,T], \\ v(t) = \displaystyle\int_0^T G_2(t,s) g(s,u(s))\, \mathrm{d}s, & t \in [0,T]. \end{cases}$$

We consider the Banach space $X = C([0,T])$ with the supremum norm $\|\cdot\|$ and define the cone $P' \subset X$ by $P' = \{u \in X, u(t) \geq 0, \forall\, t \in [0,T]\}$.

We also define the operator $\mathcal{A}$: $P' \to X$ by

$$(\mathcal{A}u)(t) = \int_0^T G_1(t,s) f\left(s, \int_0^T G_2(s,\tau) g(\tau, u(\tau))\, \mathrm{d}\tau\right) \mathrm{d}s, \quad t \in [0,T],$$

and the operators $\mathcal{B}$: $P' \to X$ and $\mathcal{C}$: $P' \to X$ by

$$(\mathcal{B}u)(t) = \int_0^T G_1(t,s) u(s)\, \mathrm{d}s, \quad (\mathcal{C}u)(t) = \int_0^T G_2(t,s) u(s)\, \mathrm{d}s, \quad t \in [0,T].$$

Under assumptions (I1) and (I2), using also Lemma 2.1.4, we can easily see that $\mathcal{A}$, $\mathcal{B}$, and $\mathcal{C}$ are completely continuous from P' to P'. Thus, the existence and multiplicity of positive solutions of system (S′)–(BC) are equivalent to the existence and multiplicity of fixed points of the operator $\mathcal{A}$.

Theorem 2.2.1. *Assume that (I1) and (I2) hold. If the functions f and g also satisfy the following conditions (I3) and (I4), then problem* (S′)–(BC) *has at least one positive solution* $(u(t), v(t)), t \in [0,T]$*:*

(I3) *There exist* $c \in (0, T/2)$ *and a positive constant* $p_1 \in (0,1]$ *such that*

$$(1)\ \tilde{f}^{\mathrm{i}}_{\infty} = \liminf_{u\to\infty} \inf_{t\in[c,T-c]} \frac{f(t,u)}{u^{p_1}} \in (0,\infty] \quad \textit{and} \quad (2)\ \tilde{g}^{\mathrm{i}}_{\infty} = \liminf_{u\to\infty} \inf_{t\in[c,T-c]} \frac{g(t,u)}{u^{1/p_1}} = \infty.$$

(I4) *There exists a positive constant* $q_1 \in (0,\infty)$ *such that*

$$(1)\ \tilde{f}^{\mathrm{s}}_{0} = \limsup_{u\to 0+} \sup_{t\in[0,T]} \frac{f(t,u)}{u^{q_1}} \in [0,\infty) \quad \textit{and} \quad (2)\ \tilde{g}^{\mathrm{s}}_{0} = \limsup_{u\to 0+} \sup_{t\in[0,T]} \frac{g(t,u)}{u^{1/q_1}} = 0.$$

Proof. From (1) of assumption (I3), we conclude that there exist $C_1, C_2 > 0$ such that

$$f(t,u) \geq C_1 u^{p_1} - C_2, \quad \forall\, (t,u) \in [c, T-c] \times [0,\infty). \tag{2.9}$$

Then for $u \in P'$, by using (2.9), the reverse form of Hölder's inequality, and Lemma 2.1.5, we have for $p_1 \in (0,1)$

$$
\begin{aligned}
(\mathcal{A}u)(t) &= \int_0^T G_1(t,s) f\left(s, \int_0^T G_2(s,\tau) g(\tau, u(\tau))\,\mathrm{d}\tau\right) \mathrm{d}s \\
&\geq \int_c^{T-c} G_1(t,s) \left[C_1 \left(\int_0^T G_2(s,\tau) g(\tau,u(\tau))\,\mathrm{d}\tau \right)^{p_1} - C_2 \right] \mathrm{d}s \\
&\geq \int_c^{T-c} G_1(t,s) \left[C_1 \int_0^T (G_2(s,\tau) g(\tau,u(\tau)))^{p_1}\,\mathrm{d}\tau \left(\int_0^T \mathrm{d}\tau \right)^{p_1/q_0} \right] \mathrm{d}s \\
&\quad - C_2 \int_0^T J_1(s)\,\mathrm{d}s \geq C_1 T^{p_1/q_0} \int_c^{T-c} G_1(t,s) \\
&\quad \times \left(\int_c^{T-c} (G_2(s,\tau))^{p_1} (g(\tau,u(\tau)))^{p_1}\,\mathrm{d}\tau \right) \mathrm{d}s - C_3, \forall t \in [0,T],
\end{aligned}
$$

where $q_0 = p_1/(p_1-1)$ and $C_3 = C_2 \int_0^T J_1(s)\,\mathrm{d}s$.

Therefore, for $u \in P'$ and $p_1 \in (0,1]$, we have

$$
(\mathcal{A}u)(t) \geq \tilde{C}_1 \int_c^{T-c} G_1(t,s) \left(\int_c^{T-c} (G_2(s,\tau))^{p_1} (g(\tau,u(\tau)))^{p_1}\,\mathrm{d}\tau \right) \mathrm{d}s - C_3, \quad \forall t \in [0,T], \tag{2.10}
$$

where $\tilde{C}_1 = C_1 T^{p_1/q_0}$ for $p_1 \in (0,1)$ and $\tilde{C}_1 = C_1$ for $p_1 = 1$.

For c from (I3), we define the cone $P_0 = \{u \in P';\ \inf_{t\in[c,T-c]} u(t) \geq \gamma \|u\|\}$, where $\gamma = \min\{\gamma_1, \gamma_2\}$ and γ_1 and γ_2 are defined in Section 2.1.2.

From our assumptions and Lemma 2.1.6, it can be shown that for any $y \in P'$ the functions $u(t) = (\mathcal{B}y)(t)$ and $v(t) = (\mathcal{C}y)(t)$ for $t \in [0,T]$ satisfy the inequalities

$$
\inf_{t\in[c,T-c]} u(t) \geq \gamma_1 \|u\| \geq \gamma \|u\|, \qquad \inf_{t\in[c,T-c]} v(t) \geq \gamma_2 \|v\| \geq \gamma \|v\|.
$$

So, $u = \mathcal{B}y \in P_0$ and $v = \mathcal{C}y \in P_0$. Therefore, we deduce that $\mathcal{B}(P') \subset P_0$, $\mathcal{C}(P') \subset P_0$.

Now we consider the function $u_0(t), t \in [0,T]$, the solution of problem (2.1)–(2.2) with $y = y_0$, where $y_0(t) = 1$ for all $t \in [0,T]$. Then $u_0(t) = \int_0^T G_1(t,s)\,\mathrm{d}s = (\mathcal{B}y_0)(t)$, $t \in [0,T]$. Obviously, we have $u_0(t) \geq 0$ for all $t \in [0,T]$. We also consider the set

$$
M = \left\{u \in P';\quad \text{there exists} \quad \lambda \geq 0 \quad \text{such that} \quad u = \mathcal{A}u + \lambda u_0 \right\}.
$$

We shall show that $M \subset P_0$ and M is a bounded subset of X. If $u \in M$, then there exists $\lambda \geq 0$ such that $u(t) = (\mathcal{A}u)(t) + \lambda u_0(t), t \in [0,T]$. From the definition of u_0, we have

$$
u(t) = (\mathcal{A}u)(t) + \lambda(\mathcal{B}y_0)(t) = \mathcal{B}(Fu(t)) + \lambda(\mathcal{B}y_0)(t) = \mathcal{B}(Fu(t) + \lambda y_0(t)) \in P_0,
$$

where $F\colon P' \to P'$ is defined by $(Fu)(t) = f\left(t, \int_0^T G_2(t,s) g(s,u(s))\,\mathrm{d}s\right)$. Therefore, $M \subset P_0$, and from the definition of P_0, we have

$$\|u\| \le \frac{1}{\gamma} \inf_{t\in[c,T-c]} u(t), \quad \forall\, u \in M. \tag{2.11}$$

From (2) of assumption (I3), we conclude that for $\varepsilon_0 = (2/(\tilde{C}_1 m_1 m_2 \gamma_1 \gamma_2^{p_1}))^{1/p_1} > 0$ there exists $C_4 > 0$ such that

$$(g(t,u))^{p_1} \ge \varepsilon_0^{p_1} u - C_4, \quad \forall\,(t,u) \in [c, T-c] \times [0,\infty), \tag{2.12}$$

where $m_1 = \int_c^{T-c} J_1(\tau)\,d\tau > 0$, $m_2 = \int_c^{T-c} (J_2(\tau))^{p_1}\,d\tau > 0$.

For $u \in M$ and $t \in [c, T-c]$, by using Lemma 2.1.5 and relations (2.10) and (2.12), we obtain

$$\begin{aligned}
u(t) &= (\mathcal{A}u)(t) + \lambda u_0(t) \ge (\mathcal{A}u)(t)\\
&\ge \tilde{C}_1 \int_c^{T-c} G_1(t,s)\left[\int_c^{T-c} (G_2(s,\tau))^{p_1} (g(\tau,u(\tau)))^{p_1}\,d\tau\right] ds - C_3\\
&\ge \tilde{C}_1 \gamma_1 \gamma_2^{p_1} \left(\int_c^{T-c} J_1(s)\,ds\right)\left(\int_c^{T-c} (J_2(\tau))^{p_1}\left(\varepsilon_0^{p_1} u(\tau) - C_4\right) d\tau\right) - C_3\\
&\ge \tilde{C}_1 \varepsilon_0^{p_1} \gamma_1 \gamma_2^{p_1} \left(\int_c^{T-c} J_1(s)\,ds\right)\left(\int_c^{T-c} (J_2(\tau))^{p_1}\,d\tau\right) \inf_{\tau\in[c,T-c]} u(\tau) - C_5\\
&= 2 \inf_{\tau\in[c,T-c]} u(\tau) - C_5,
\end{aligned}$$

where $C_5 = C_3 + C_4 \tilde{C}_1 m_1 m_2 \gamma_1 \gamma_2^{p_1} > 0$.

Hence, $\inf_{t\in[c,T-c]} u(t) \ge 2 \inf_{t\in[c,T-c]} u(t) - C_5$, and so

$$\inf_{t\in[c,T-c]} u(t) \le C_5, \quad \forall\, u \in M. \tag{2.13}$$

Now from relations (2.11) and (2.13), it can be shown that $\|u\| \le \frac{1}{\gamma} \inf_{t\in[c,T-c]} u(t) \le C_5/\gamma$ for all $u \in M$—that is, M is a bounded subset of X.

Besides, there exists a sufficiently large $L > 0$ such that

$$u \ne \mathcal{A}u + \lambda u_0, \quad \forall\, u \in \partial B_L \cap P', \quad \forall\, \lambda \ge 0.$$

From Theorem 1.3.2, we deduce that the fixed point index of the operator $\mathcal{A}$ is

$$i(\mathcal{A}, B_L \cap P', P') = 0. \tag{2.14}$$

Next, from assumption (I4), we conclude that there exists $M_0 > 0$ and $\delta_1 \in (0,1)$ such that

$$\begin{aligned}
&f(t,u) \le M_0 u^{q_1}, \quad \forall\,(t,u) \in [0,T] \times [0,1]; \quad g(t,u) \le \varepsilon_1 u^{1/q_1},\\
&\forall\,(t,u) \in [0,T] \times [0,\delta_1],
\end{aligned} \tag{2.15}$$

where $\varepsilon_1 = \min\{1/M_2, (1/(2M_0 M_1 M_2^{q_1}))^{1/q_1}\} > 0$, $M_1 = \int_0^T J_1(s)\,ds > 0$, $M_2 = \int_0^T J_2(s)\,ds > 0$. Hence, for any $u \in \bar{B}_{\delta_1} \cap P'$ and $t \in [0,T]$ we obtain

$$\int_0^T G_2(t,s)g(s,u(s))\,\mathrm{d}s \le \varepsilon_1 \int_0^T J_2(s)(u(s))^{1/q_1}\,\mathrm{d}s \le \varepsilon_1 M_2\|u\|^{1/q_1} \le 1. \quad (2.16)$$

Therefore, by (2.15) and (2.16), we deduce that for any $u \in \bar{B}_{\delta_1} \cap P'$ and $t \in [0,T]$

$$(\mathcal{A}u)(t) \le M_0 \int_0^T G_1(t,s)\left(\int_0^T G_2(s,\tau)g(\tau,u(\tau))\,\mathrm{d}\tau\right)^{q_1} \mathrm{d}s$$

$$\le M_0\varepsilon_1^{q_1}M_2^{q_1}\|u\| \int_0^T J_1(s)\,\mathrm{d}s = M_0\varepsilon_1^{q_1}M_1M_2^{q_1}\|u\| \le \frac{1}{2}\|u\|.$$

This implies that $\|\mathcal{A}u\| \le \|u\|/2, \forall\, u \in \partial B_{\delta_1} \cap P'$. From Theorem 1.3.1, we conclude that the fixed point index of the operator $\mathcal{A}$ is

$$i(\mathcal{A}, B_{\delta_1} \cap P', P') = 1. \quad (2.17)$$

Combining (2.14) and (2.17), we obtain

$$i(\mathcal{A}, (B_L \setminus \bar{B}_{\delta_1}) \cap P', P') = i(\mathcal{A}, B_L \cap P', P') - i(\mathcal{A}, B_{\delta_1} \cap P', P') = -1.$$

We conclude that $\mathcal{A}$ has at least one fixed point $u_1 \in (B_L \setminus \bar{B}_{\delta_1}) \cap P'$—that is, $\delta_1 < \|u_1\| < L$.

Let $v_1(t) = \int_0^T G_2(t,s)g(s,u_1(s))\,\mathrm{d}s$. Then $(u_1,v_1) \in P' \times P'$ is a solution of (S′)–(BC). In addition, $\|v_1\| > 0$. If we suppose that $v_1(t) = 0$, for all $t \in [0,T]$, then by using (I2), we have $f(s,v_1(s)) = f(s,0) = 0$ for all $s \in [0,T]$. This implies $u_1(t) = \int_0^T G_1(t,s)f(s,v_1(s))\,\mathrm{d}s = 0$ for all $t \in [0,T]$, which contradicts $\|u_1\| > 0$. By using Theorem 1.1 from Ji et al. (2009) (see also Eloe and Henderson, 1997), we obtain $u_1(t) > 0$ and $v_1(t) > 0$ for all $t \in (0, T-c]$. The proof of Theorem 2.2.1 is completed. □

Theorem 2.2.2. *Assume that (I1) and (I2). If the functions f and g also satisfy the following conditions (I5) and (I6), then problem* (S′)–(BC) *has at least one positive solution* $(u(t), v(t))$, $t \in [0,T]$*:*

(I5) *There exists a positive constant* $r_1 \in (0,\infty)$ *such that*

$$(1)\ \tilde{f}_\infty^{\rm s} = \limsup_{u\to\infty}\ \sup_{t\in[0,T]} \frac{f(t,u)}{u^{r_1}} \in [0,\infty) \quad \textit{and} \quad (2)\ \tilde{g}_\infty^{\rm s} = \limsup_{u\to\infty}\ \sup_{t\in[0,T]} \frac{g(t,u)}{u^{1/r_1}} = 0.$$

(I6) *There exists* $c \in (0, T/2)$ *such that*

$$(1)\ \tilde{f}_0^{\rm i} = \liminf_{u\to 0+}\ \inf_{t\in[c,T-c]} \frac{f(t,u)}{u} \in (0,\infty] \quad \textit{and} \quad (2)\ \tilde{g}_0^{\rm i} = \liminf_{u\to 0+}\ \inf_{t\in[c,T-c]} \frac{g(t,u)}{u} = \infty.$$

Proof. From assumption (I5), we deduce that there exist $C_6, C_7, C_8 > 0$ such that

$$f(t,u) \le C_6u^{r_1} + C_7, \quad g(t,u) \le \varepsilon_2 u^{1/r_1} + C_8, \quad \forall\,(t,u) \in [0,T]\times[0,\infty), \quad (2.18)$$

where $\varepsilon_2 = (1/(2C_6M_1M_2^{r_1}))^{1/r_1}$.

Hence, for $u \in P'$, by using (2.18), we obtain

$$
\begin{aligned}
(\mathcal{A}u)(t) &\le \int_0^T G_1(t,s)\left[C_6\left(\int_0^T G_2(s,\tau)g(\tau,u(\tau))\,\mathrm{d}\tau\right)^{r_1} + C_7\right]\mathrm{d}s \\
&\le C_6\int_0^T G_1(t,s)\left[\int_0^T G_2(s,\tau)\left(\varepsilon_2\|u\|^{1/r_1} + C_8\right)\mathrm{d}\tau\right]^{r_1}\mathrm{d}s + M_1C_7 \\
&= C_6\left(\varepsilon_2\|u\|^{1/r_1} + C_8\right)^{r_1}\left(\int_0^T J_1(s)\,\mathrm{d}s\right)\left(\int_0^T J_2(\tau)\,\mathrm{d}\tau\right)^{r_1} + M_1C_7, \quad \forall t \in [0,T].
\end{aligned}
$$

Therefore, we have

$$
(\mathcal{A}u)(t) \le C_6M_1M_2^{r_1}\left(\varepsilon_2\|u\|^{1/r_1} + C_8\right)^{r_1} + M_1C_7, \quad \forall t \in [0,T]. \tag{2.19}
$$

After some computations, we can show that there exists a sufficiently large $R > 0$ such that

$$
C_6M_1M_2^{r_1}\left(\varepsilon_2\|u\|^{1/r_1} + C_8\right)^{r_1} + M_1C_7 \le \frac{3}{4}\|u\|, \quad \forall u \in P', \quad \|u\| \ge R. \tag{2.20}
$$

Hence, from (2.19) and (2.20), we obtain $\|\mathcal{A}u\| \le \frac{3}{4}\|u\| < \|u\|$ for all $u \in \partial B_R \cap P'$. Therefore, from Theorem 1.3.1, we have

$$
i(\mathcal{A}, B_R \cap P', P') = 1. \tag{2.21}
$$

On the other hand, from assumption (I6) we deduce that there exist positive constants $C_9 > 0$ and $u_1 > 0$ such that

$$
f(t,u) \ge C_9u, \quad g(t,u) \ge \frac{C_0}{C_9}u, \quad \forall (t,u) \in [c, T-c] \times [0,u_1], \tag{2.22}
$$

where $C_0 = (\gamma_1\gamma_2m_1m_3)^{-1}$ and $m_3 = \int_c^{T-c} J_2(\tau)\,\mathrm{d}\tau > 0$.

Because $g(t,0) = 0$ for all $t \in [0,T]$, and g is continuous, it can be shown that there exists a sufficiently small $\delta_2 \in (0,u_1)$ such that $g(t,u) \le \frac{u_1}{M_2}$ for all $(t,u) \in [c, T-c] \times [0,\delta_2]$. Hence,

$$
\int_c^{T-c} G_2(s,\tau)g(\tau,u(\tau))\,\mathrm{d}\tau \le \int_c^{T-c} J_2(\tau)g(\tau,u(\tau))\,\mathrm{d}\tau \le u_1, \tag{2.23}
$$

$\forall u \in \bar{B}_{\delta_2} \cap P', \quad s \in [c, T-c]$.

From (2.22), (2.23), and Lemma 2.1.5, we deduce that for any $u \in \bar{B}_{\delta_2} \cap P'$ we have

$$
\begin{aligned}
(\mathcal{A}u)(t) &\ge C_9\int_c^{T-c} G_1(t,s)\left(\int_c^{T-c} G_2(s,\tau)g(\tau,u(\tau))\,\mathrm{d}\tau\right)\mathrm{d}s \\
&\ge C_0\gamma_2\int_c^{T-c} G_1(t,s)\left(\int_c^{T-c} J_2(\tau)u(\tau)\,\mathrm{d}\tau\right)\mathrm{d}s =: (\mathcal{L}u)(t), \quad \forall t \in [0,T].
\end{aligned}
$$

Hence, by using the above linear operator $\mathcal{L}$: $P' \to P'$, we obtain

$$\mathcal{A}u \geq \mathcal{L}u, \quad \forall\, u \in \partial B_{\delta_2} \cap P'. \tag{2.24}$$

For $w_0(t) = \int_c^{T-c} G_1(t,s)\,ds, t \in [0,T]$, we have $w_0 \in P'$, $w_0 \neq 0$ and

$$\begin{aligned}(\mathcal{L}w_0)(t) &\geq C_0\gamma_1\gamma_2\left(\int_c^{T-c} J_2(\tau)\,d\tau\right)\left(\int_c^{T-c} J_1(\tau)\,d\tau\right)\left(\int_c^{T-c} G_1(t,s)\,ds\right)\\ &= C_0\gamma_1\gamma_2 m_1 m_3 \int_c^{T-c} G_1(t,s)\,ds = \int_c^{T-c} G_1(t,s)\,ds\\ &= w_0(t), \quad t \in [0,T].\end{aligned}$$

Therefore,

$$\mathcal{L}w_0 \geq w_0. \tag{2.25}$$

We may suppose that $\mathcal{A}$ has no fixed point on $\partial B_{\delta_2} \cap P'$ (otherwise the proof is finished). From (2.24), (2.25), and Theorem 1.3.3, we conclude that

$$i(\mathcal{A}, B_{\delta_2} \cap P', P') = 0. \tag{2.26}$$

Therefore, from (2.21) and (2.26), we have

$$i(\mathcal{A}, (B_R \setminus \bar{B}_{\delta_2}) \cap P', P') = i(\mathcal{A}, B_R \cap P', P') - i(\mathcal{A}, B_{\delta_2} \cap P', P') = 1.$$

Then $\mathcal{A}$ has at least one fixed point in $(B_R \setminus \bar{B}_{\delta_2}) \cap P'$. Thus, problem (S′)–(BC) has at least one positive solution $(u,v) \in P' \times P'$ with $u(t) \geq 0, v(t) \geq 0$ for all $t \in [0,T]$ and $\|u\| > 0$, $\|v\| > 0$. This completes the proof of Theorem 2.2.2. □

Using arguments similar to those used in the proof of Theorem 1.3.6 from Section 1.3, we obtain the following theorem:

Theorem 2.2.3. *Assume that (I1) and (I2) hold. If the functions f and g also satisfy conditions (I3) and (I6) and the following condition (I7), then problem* (S′)–(BC) *has at least two positive solutions $(u_1(t), v_1(t))$, $(u_2(t), v_2(t))$, $t \in [0,T]$:*

(I7) For each $t \in [0,T]$, $f(t,u)$ and $g(t,u)$ are nondecreasing with respect to u, and there exists a constant $N > 0$ such that

$$f\left(t, m_0 \int_0^T g(s,N)\,ds\right) < \frac{N}{m_0}, \quad \forall\, t \in [0,T],$$

where $m_0 = \max\{K_1T, K_2\}$, $K_1 = \max_{s\in[0,T]} J_1(s)$, $K_2 = \max_{s\in[0,T]} J_2(s)$.

We present now some examples which illustrate our results above.

Example 2.2.1. Let $f(t,u) = (at^\alpha + b)u^{1/2}(2 + \sin u)$ and $g(t,u) = (\sigma t^\beta + \delta)u^3(3 + \cos u)$ for all $t \in [0,T]$, $u \in [0,\infty)$, with $a, \sigma \geq 0$, $b, \delta > 0$, $\alpha, \beta > 0$, and $p_1 = 1/2$ and $q_1 = 1/2$. Then assumptions (I3) and (I4) are satisfied; for $c \in (0, T/2)$ we have $\tilde{f}_\infty^{\rm i} = (ac^\alpha + b)$, $\tilde{g}_\infty^{\rm i} = \infty$, and $\tilde{f}_0^{\rm s} = 2(aT^\alpha + b)$, $\tilde{g}_0^{\rm s} = 0$. Under assumptions (I1) and (I2), by Theorem 2.2.1, we deduce that problem (S′)–(BC) has at least one positive solution.

Example 2.2.2. Let $f(t,u)=u^{1/2}$ and $g(t,u)=u^{1/3}$ for all $t\in[0,T]$, $u\in[0,\infty)$, and $r_1=1/2$. Then assumptions (I5) and (I6) are satisfied; for $c\in(0,T/2)$ we have $\tilde{f}^{\rm s}_\infty=1$, $\tilde{g}^{\rm s}_\infty=0$, $\tilde{f}^{\rm i}_0=\infty$, and $\tilde{g}^{\rm i}_0=\infty$. Under assumptions (I1) and (I2), by Theorem 2.2.2, we conclude that problem (S′)–(BC) has at least one positive solution.

Example 2.2.3. We consider the following problem

$$\begin{cases} u^{(4)}(t)+a(v^\alpha(t)+v^\beta(t))=0, & t\in(0,1),\\ v^{(3)}(t)+b(u^\gamma(t)+u^\delta(t))=0, & t\in(0,1), \end{cases} \tag{S_3}$$

with the multipoint boundary conditions

$$\begin{cases} u(0)=\dfrac{3}{4}u\left(\dfrac{1}{2}\right), \quad u'(0)=u''(0)=0, \quad u(1)=\dfrac{1}{2}u\left(\dfrac{1}{4}\right)+\dfrac{1}{3}u\left(\dfrac{3}{4}\right),\\ v(0)=\dfrac{4}{5}v\left(\dfrac{1}{3}\right)+\dfrac{1}{6}v\left(\dfrac{2}{3}\right), \quad v'(0)=0, \quad v(1)=\dfrac{6}{7}v\left(\dfrac{3}{5}\right), \end{cases} \tag{BC_3}$$

where $\alpha>1$, $\beta<1$, $\gamma>2$, $\delta<1$, and $a,b>0$. Here $T=1$, and $f(t,x)=a(x^\alpha+x^\beta)$ and $g(t,x)=b(x^\gamma+x^\delta)$ for all $t\in[0,1]$, $x\in[0,\infty)$, and the boundary conditions (BC_3) are the same as the boundary conditions (BC_2) from Example 2.1.2 in Section 2.1.4.

By using the expressions for J_1 and J_2 given in Example 2.1.2 from Section 2.1.4, we obtain $K_1=\max_{s\in[0,1]}J_1(s)\approx 0.03150703$ and $K_2=\max_{s\in[0,1]}J_2(s)\approx 0.62315711$. Then $m_0=\max\{K_1T,K_2\}=K_2$. The functions $f(t,u)$ and $g(t,u)$ are nondecreasing with respect to u for any $t\in[0,1]$, and for $p_1=1/2$ assumptions (I3) and (I6) are satisfied; for $c\in(0,1/2)$ we obtain $\tilde{f}^{\rm i}_\infty=\infty$, $\tilde{g}^{\rm i}_\infty=\infty$, $\tilde{f}^{\rm i}_0=\infty$, and $\tilde{g}^{\rm i}_0=\infty$. We take $N=1$ and then $\int_0^1 g(s,1)\,{\rm d}s=2b$ and $f(t,2bm_0)=a[(2bm_0)^\alpha+(2bm_0)^\beta]$. If $a[(2bm_0)^\alpha+(2bm_0)^\beta]<\frac{1}{m_0}$, which is equivalent to $a[m_0^{\alpha+1}(2b)^\alpha+m_0^{\beta+1}(2b)^\beta]<1$, then assumption (I7) is satisfied. For example, if $\alpha=2$, $\beta=1/2$, $b=1/2$, and $a<\frac{1}{m_0^3+m_0^{3/2}}$ (e.g., $a\le 1.36$), then the above inequality is satisfied. By Theorem 2.2.3, we deduce that problem (S_3)–(BC_3) has at least two positive solutions.

Remark 2.2.1. The results presented in this section were published in Henderson and Luca (2013c).

2.2.2 Singular nonlinearities

We consider the system of nonlinear higher-order singular ordinary differential equations

$$\begin{cases} u^{(n)}(t)+f(t,v(t))=0, & t\in(0,T),\\ v^{(m)}(t)+g(t,u(t))=0, & t\in(0,T), \end{cases} \tag{S′}$$

with the multipoint boundary conditions

$$\begin{cases} u(0)=\sum_{i=1}^{p} a_i u(\xi_i), \quad u'(0)=\cdots=u^{(n-2)}(0)=0, \quad u(T)=\sum_{i=1}^{q} b_i u(\eta_i), \\ v(0)=\sum_{i=1}^{r} c_i v(\zeta_i), \quad v'(0)=\cdots=v^{(m-2)}(0)=0, \quad v(T)=\sum_{i=1}^{l} d_i v(\rho_i), \end{cases} \tag{BC}$$

where $n, m, p, q, r, l \in \mathbb{N}$, $n, m \geq 2$; $a_i, \xi_i \in \mathbb{R}$ for all $i=1,\ldots,p$; $b_i, \eta_i \in \mathbb{R}$ for all $i=1,\ldots,q$; $c_i, \zeta_i \in \mathbb{R}$ for all $i=1,\ldots,r$; $d_i, \rho_i \in \mathbb{R}$ for all $i=1,\ldots,l$; $0 < \xi_1 < \cdots < \xi_p < T$, $0 < \eta_1 < \cdots < \eta_q < T$, $0 < \zeta_1 < \cdots < \zeta_r < T$, $0 < \rho_1 < \cdots < \rho_l < T$. In the case $n=2$ or $m=2$, the boundary conditions above are of the form $u(0)=\sum_{i=1}^{p} a_i u(\xi_i)$, $u(T)=\sum_{i=1}^{q} b_i u(\eta_i)$, or $v(0)=\sum_{i=1}^{r} c_i v(\zeta_i)$, $v(T)=\sum_{i=1}^{l} d_i v(\rho_i)$, respectively—that is, without conditions on the derivatives of u and v at the point 0.

We present some weaker assumptions on f and g, which do not possess any sublinear or superlinear growth conditions and may be singular at $t=0$ and/or $t=T$, such that positive solutions for problem (S′)–(BC) exist. By a positive solution of (S′)–(BC), we mean a pair of functions $(u, v) \in (C([0,T];\mathbb{R}_+) \cap C^n(0,T)) \times (C([0,T];\mathbb{R}_+) \cap C^m(0,T))$ satisfying (S′) and (BC) with $\sup_{t\in[0,T]} u(t) > 0$, $\sup_{t\in[0,T]} v(t) > 0$. This problem is a generalization of the problem studied in Henderson and Luca (2013h), where in (BC) we have $a_i=0$ for all $i=1,\ldots,p$ and $c_i=0$ for all $i=1,\ldots,r$.

We present the basic assumptions that we shall use in the sequel:

(L1) $0 < \xi_1 < \cdots < \xi_p < T$, $a_i \geq 0$ for all $i=1,\ldots,p$, $\sum_{i=1}^{p} a_i < 1$, $0 < \eta_1 < \cdots < \eta_q < T$, $b_i \geq 0$ for all $i=1,\ldots,q$, $\sum_{i=1}^{q} b_i < 1$, $0 < \zeta_1 < \cdots < \zeta_r < T$, $c_i \geq 0$ for all $i=1,\ldots,r$, $\sum_{i=1}^{r} c_i < 1$, $0 < \rho_1 < \cdots < \rho_l < T$, and $d_i \geq 0$ for all $i=1,\ldots,l$, $\sum_{i=1}^{l} d_i < 1$.

(L2) The functions $f, g \in C((0,T) \times \mathbb{R}_+, \mathbb{R}_+)$, and there exists $p_i \in C((0,T), \mathbb{R}_+)$ and $q_i \in C(\mathbb{R}_+, \mathbb{R}_+)$, $i=1,2$, with $0 < \int_0^T p_i(t)\,dt < \infty$, $i=1,2$, $q_1(0)=0$, $q_2(0)=0$ such that

$$f(t,x) \leq p_1(t)q_1(x), \quad g(t,x) \leq p_2(t)q_2(x), \quad \forall t \in (0,T), \ x \in \mathbb{R}_+.$$

Under assumption (L1), we have all auxiliary results in Lemmas 2.1.1–2.1.6 from Section 2.1.2.

The pair of functions $(u, v) \in (C([0,T]) \cap C^n(0,T)) \times (C([0,T]) \cap C^m(0,T))$ is a solution for our problem (S′)–(BC) if and only if $(u, v) \in C([0,T]) \times C([0,T])$ is a solution for the nonlinear integral equations

$$\begin{cases} u(t) = \int_0^T G_1(t,s) f(s, v(s))\,ds, \quad t \in [0,T], \\ v(t) = \int_0^T G_2(t,s) g(s, u(s))\,ds, \quad t \in [0,T]. \end{cases} \tag{2.27}$$

System (2.27) can be written as the nonlinear integral system

$$\begin{cases} u(t) = \int_0^T G_1(t,s)f\left(s, \int_0^T G_2(s,\tau)g(\tau,u(\tau))\,d\tau\right) ds, & t \in [0,T], \\ v(t) = \int_0^T G_2(t,s)g(s,u(s))\,ds, & t \in [0,T]. \end{cases}$$

We consider the Banach space $X = C([0,T])$ with the supremum norm $\|\cdot\|$, and define the cone $P' \subset X$ by $P' = \{u \in X, u(t) \geq 0, \forall t \in [0,T]\}$.

We also define the operator $\mathcal{D}$: $P' \to X$ by

$$(\mathcal{D}u)(t) = \int_0^T G_1(t,s)f\left(s, \int_0^T G_2(s,\tau)g(\tau,u(\tau))\,d\tau\right) ds, \quad u \in P', t \in [0,T].$$

Lemma 2.2.1. *Assume that (L1) and (L2) hold. Then $\mathcal{D}$: $P' \to P'$ is completely continuous.*

Proof. We denote $\alpha = \int_0^T J_1(s)p_1(s)\,ds$ and $\beta = \int_0^T J_2(s)p_2(s)\,ds$. Using (L2), we deduce that $0 < \alpha < \infty$ and $0 < \beta < \infty$. By Lemma 2.1.4 and the corresponding lemma for G_2, we get that $\mathcal{D}$ maps P' into P'.

We shall prove that $\mathcal{D}$ maps bounded sets into relatively compact sets. Suppose $E \subset P'$ is an arbitrary bounded set. First, we prove that $\mathcal{D}(E)$ is a bounded set. Because E is bounded, there exists $M_1 > 0$ such that $\|u\| \leq M_1$ for all $u \in E$. By the continuity of q_2, there exists $M_2 > 0$ such that $M_2 = \sup_{x \in [0,M_1]} q_2(x)$. By using Lemma 2.1.5 for G_2, for any $u \in E$ and $s \in [0,T]$, we obtain

$$\int_0^T G_2(s,\tau)g(\tau,u(\tau))\,d\tau \leq \int_0^T G_2(s,\tau)p_2(\tau)q_2(u(\tau))\,d\tau \leq \beta M_2. \tag{2.28}$$

Because q_1 is continuous, there exists $M_3 > 0$ such that $M_3 = \sup_{x \in [0,\beta M_2]} q_1(x)$. Therefore, from (2.28), (L2), and Lemma 2.1.5, we deduce

$$\begin{aligned} (\mathcal{D}u)(t) &\leq \int_0^T G_1(t,s)p_1(s)q_1\left(\int_0^T G_2(s,\tau)g(\tau,u(\tau))\,d\tau\right) ds \\ &\leq M_3 \int_0^T J_1(s)p_1(s)\,ds = \alpha M_3, \quad \forall t \in [0,T], \ u \in E. \end{aligned} \tag{2.29}$$

So, $\|\mathcal{D}u\| \leq \alpha M_3$ for all $u \in E$. Therefore, $\mathcal{D}(E)$ is bounded.

In what follows, we shall prove that $\mathcal{D}(E)$ is equicontinuous. By using Lemma 2.1.2, for all $t \in [0,T]$, we have

$$\begin{aligned} (\mathcal{D}u)(t) = \int_0^T &\left\{ g_1(t,s) + \frac{1}{\Delta_1}\left[\left(T^{n-1} - t^{n-1}\right)\left(1 - \sum_{i=1}^q b_i\right) + \sum_{i=1}^q b_i\left(T^{n-1} - \eta_i^{n-1}\right)\right]\right. \\ &\times \sum_{i=1}^p a_i g_1(\xi_i,s) + \left.\frac{1}{\Delta_1}\left[t^{n-1}\left(1 - \sum_{i=1}^p a_i\right) + \sum_{i=1}^p a_i \xi_i^{n-1}\right] \sum_{i=1}^q b_i g_1(\eta_i,s)\right\} \\ &\times f\left(s, \int_0^T G_2(s,\tau)g(\tau,u(\tau))\,d\tau\right) ds \end{aligned}$$

$$
\begin{aligned}
&= \frac{1}{(n-1)!T^{n-1}} \int_0^T \left[t^{n-1}(T-s)^{n-1} - T^{n-1}(t-s)^{n-1}\right] \\
&\times f\left(s, \int_0^T G_2(s,\tau)g(\tau,u(\tau))\,d\tau\right) ds + \frac{1}{(n-1)!T^{n-1}} \int_t^T t^{n-1}(T-s)^{n-1} \\
&\times f\left(s, \int_0^T G_2(s,\tau)g(\tau,u(\tau))\,d\tau\right) ds + \frac{1}{\Delta_1}\left[(T^{n-1}-t^{n-1})\left(1-\sum_{i=1}^q b_i\right)\right. \\
&\left.+ \sum_{i=1}^q b_i(T^{n-1}-\eta_i^{n-1})\right] \sum_{i=1}^p a_i \int_0^T g_1(\xi_i,s) f\left(s, \int_0^T G_2(s,\tau)g(\tau,u(\tau))\,d\tau\right) ds \\
&+ \frac{1}{\Delta_1}\left[t^{n-1}\left(1-\sum_{i=1}^p a_i\right) + \sum_{i=1}^p a_i\xi_i^{n-1}\right] \sum_{i=1}^q b_i \int_0^T g_1(\eta_i,s) \\
&\times f\left(s, \int_0^T G_2(s,\tau)g(\tau,u(\tau))\,d\tau\right) ds.
\end{aligned}
$$

Therefore, we obtain

$$
\begin{aligned}
(\mathcal{D}u)'(t) = &\int_0^t \frac{t^{n-2}(T-s)^{n-1} - T^{n-1}(t-s)^{n-2}}{(n-2)!T^{n-1}} f\left(s, \int_0^T G_2(s,\tau)g(\tau,u(\tau))\,d\tau\right) ds \\
&+ \int_t^T \frac{t^{n-2}(T-s)^{n-1}}{(n-2)!T^{n-1}} f\left(s, \int_0^T G_2(s,\tau)g(\tau,u(\tau))\,d\tau\right) ds \\
&+ \frac{1}{\Delta_1}\left[-(n-1)t^{n-2}\left(1-\sum_{i=1}^q b_i\right)\right] \\
&\times \sum_{i=1}^p a_i \int_0^T g_1(\xi_i,s) f\left(s, \int_0^T G_2(s,\tau)g(\tau,u(\tau))\,d\tau\right) ds \\
&+ \frac{1}{\Delta_1}\left[(n-1)t^{n-2}\left(1-\sum_{i=1}^p a_i\right)\right] \\
&\times \sum_{i=1}^q b_i \int_0^T g_1(\eta_i,s) f\left(s, \int_0^T G_2(s,\tau)g(\tau,u(\tau))\,d\tau\right) ds, \quad \forall t \in (0,T).
\end{aligned}
$$

So, for any $t \in (0,T)$ we deduce

$$
\begin{aligned}
|(\mathcal{D}u)'(t)| \le &\int_0^t \frac{t^{n-2}(T-s)^{n-1} + T^{n-1}(t-s)^{n-2}}{(n-2)!T^{n-1}} \\
&\times p_1(s)q_1\left(\int_0^T G_2(s,\tau)g(\tau,u(\tau))\,d\tau\right) ds \\
&+ \int_t^T \frac{t^{n-2}(T-s)^{n-1}}{(n-2)!T^{n-1}} p_1(s)q_1\left(\int_0^T G_2(s,\tau)g(\tau,u(\tau))\,d\tau\right) ds
\end{aligned}
$$

$$
\begin{aligned}
&+\frac{(n-1)t^{n-2}}{\Delta_1}\left(1-\sum_{i=1}^{q} b_i\right)\sum_{i=1}^{p} a_i \int_0^T g_1(\xi_i,s)p_1(s)\\
&\times q_1\left(\int_0^T G_2(s,\tau)g(\tau,u(\tau))\,\mathrm{d}\tau\right)\mathrm{d}s+\frac{(n-1)t^{n-2}}{\Delta_1}\left(1-\sum_{i=1}^{p} a_i\right)\\
&\times\sum_{i=1}^{q} b_i\int_0^T g_1(\eta_i,s)p_1(s)q_1\left(\int_0^T G_2(s,\tau)g(\tau,u(\tau))\,\mathrm{d}\tau\right)\mathrm{d}s\\
&\le M_3\left(\int_0^t \frac{t^{n-2}(T-s)^{n-1}+T^{n-1}(t-s)^{n-2}}{(n-2)!T^{n-1}}p_1(s)\,\mathrm{d}s\right.\\
&+\int_t^T \frac{t^{n-2}(T-s)^{n-1}}{(n-2)!T^{n-1}}p_1(s)\,\mathrm{d}s+\frac{(n-1)t^{n-2}}{\Delta_1}\left(1-\sum_{i=1}^{q} b_i\right)\\
&\times\sum_{i=1}^{p} a_i\int_0^T g_1(\xi_i,s)p_1(s)\,\mathrm{d}s+\frac{(n-1)t^{n-2}}{\Delta_1}\left(1-\sum_{i=1}^{p} a_i\right)\\
&\left.\times\sum_{i=1}^{q} b_i\int_0^T g_1(\eta_i,s)p_1(s)\,\mathrm{d}s\right). \qquad (2.30)
\end{aligned}
$$

We denote

$$
\begin{aligned}
h(t)&=\int_0^t \frac{t^{n-2}(T-s)^{n-1}+T^{n-1}(t-s)^{n-2}}{(n-2)!T^{n-1}}p_1(s)\,\mathrm{d}s+\int_t^T \frac{t^{n-2}(T-s)^{n-1}}{(n-2)!T^{n-1}}p_1(s)\,\mathrm{d}s,\\
\mu(t)&=h(t)+\frac{(n-1)t^{n-2}}{\Delta_1}\left(1-\sum_{i=1}^{q} b_i\right)\sum_{i=1}^{p} a_i\int_0^T g_1(\xi_i,s)p_1(s)\,\mathrm{d}s\\
&+\frac{(n-1)t^{n-2}}{\Delta_1}\left(1-\sum_{i=1}^{p} a_i\right)\sum_{i=1}^{q} b_i\int_0^T g_1(\eta_i,s)p_1(s)\,\mathrm{d}s,\quad t\in(0,T).
\end{aligned}
$$

For the integral of the function h, by exchanging the order of integration, we obtain after some computations

$$
\begin{aligned}
\int_0^T h(t)\,\mathrm{d}t&=\int_0^T \frac{(T-s)^{n-1}}{(n-2)!T^{n-1}}\left(\frac{T^{n-1}-s^{n-1}}{n-1}\right)p_1(s)\,\mathrm{d}s\\
&+\int_0^T \frac{p_1(s)(T-s)^{n-1}}{(n-1)!}\,\mathrm{d}s+\int_0^T \frac{(T-s)^{n-1}s^{n-1}}{(n-1)!T^{n-1}}p_1(s)\,\mathrm{d}s\\
&=\frac{2}{(n-1)!}\int_0^T (T-s)^{n-1}p_1(s)\,\mathrm{d}s<\infty.
\end{aligned}
$$

For the integral of the function μ, we have

$$\begin{aligned}
\int_0^T \mu(t)\,\mathrm{d}t \le{} & \frac{2}{(n-1)!}\int_0^T (T-s)^{n-1}p_1(s)\,\mathrm{d}s + \frac{T^{n-1}}{\Delta_1}\left(1-\sum_{i=1}^q b_i\right) \\
& \times \sum_{i=1}^p a_i \int_0^T g_1(\theta_1(s),s)p_1(s)\,\mathrm{d}s + \frac{T^{n-1}}{\Delta_1}\left(1-\sum_{i=1}^p a_i\right) \\
& \times \sum_{i=1}^q b_i \int_0^T g_1(\theta_1(s),s)p_1(s)\,\mathrm{d}s \\
\le{} & \frac{1}{(n-1)!}\left[2+\frac{T^{n-1}}{\Delta_1}\left(1-\sum_{i=1}^q b_i\right)\left(\sum_{i=1}^p a_i\right)\right. \\
& \left. + \frac{T^{n-1}}{\Delta_1}\left(1-\sum_{i=1}^p a_i\right)\left(\sum_{i=1}^q b_i\right)\right]\int_0^T (T-s)^{n-1}p_1(s)\,\mathrm{d}s < \infty.
\end{aligned} \tag{2.31}$$

We deduce that $\mu \in L^1(0,T)$. Thus, for any given $t_1, t_2 \in [0,T]$ with $t_1 \le t_2$ and $u \in E$, by (2.30), we obtain

$$|(\mathcal{D}u)(t_1) - (\mathcal{D}u)(t_2)| = \left|\int_{t_1}^{t_2} (\mathcal{D}u)'(t)\,\mathrm{d}t\right| \le M_3 \int_{t_1}^{t_2} \mu(t)\,\mathrm{d}t. \tag{2.32}$$

From (2.31), (2.32), and absolute continuity of the integral function, we conclude that $\mathcal{D}(E)$ is equicontinuous. This conclusion together with (2.29) and the Ascoli–Arzelà theorem yields that $\mathcal{D}(E)$ is relatively compact. Therefore, $\mathcal{D}$ is a compact operator.

By using arguments similar to those used in the proof of Lemma 1.4.1, we can show that $\mathcal{D}$ is continuous on P'. Therefore, $\mathcal{D}$: $\mathcal{P}' \to \mathcal{P}'$ is completely continuous. □

For $c \in (0, \frac{T}{2})$, we define the cone

$$P_0 = \left\{ u \in X, \quad u(t) \ge 0, \quad \forall t \in [0,T], \quad \inf_{t\in[c,T-c]} u(t) \ge \gamma\|u\| \right\},$$

where $\gamma = \min\{\gamma_1, \gamma_2\}$, and γ_1 and γ_2 are given in Section 2.1.2. Under assumptions (L1) and (L2), we have $\mathcal{D}(P') \subset P_0$. For $u \in P'$, let $v = \mathcal{D}(u)$. By Lemma 2.1.6, we have $\inf_{t\in[c,T-c]} v(t) \ge \gamma_1\|v\| \ge \gamma\|v\|$—that is, $v \in P_0$.

Using Lemma 2.2.1 and arguments similar to those used in the proofs of Theorems 1.4.1 and 1.4.2, we obtain for problem (S′)–(BC) the following results:

Theorem 2.2.4. *Assume that (L1) and (L2) hold. If the functions f and g also satisfy the following conditions (L3) and (L4), then problem* (S′)–(BC) *has at least one positive solution* $(u(t), v(t)), t \in [0, T]$*:*

(L3) *There exist* $r_1, r_2 \in (0, \infty)$ *with* $r_1 r_2 \geq 1$ *such that*

$$(1)\ q_{10}^s = \limsup_{x\to 0^+} \frac{q_1(x)}{x^{r_1}} \in [0, \infty) \quad \textit{and} \quad (2)\ q_{20}^s = \limsup_{x\to 0^+} \frac{q_2(x)}{x^{r_2}} = 0.$$

(L4) *There exist* $l_1, l_2 \in (0, \infty)$ *with* $l_1 l_2 \geq 1$ *and* $c \in (0, T/2)$ *such that*

$$(1)\ \hat{f}_\infty^{\rm i} = \liminf_{x\to\infty} \inf_{t\in[c,T-c]} \frac{f(t,x)}{x^{l_1}} \in (0, \infty] \quad \textit{and} \quad (2)\ \hat{g}_\infty^{\rm i} = \liminf_{x\to\infty} \inf_{t\in[c,T-c]} \frac{g(t,x)}{x^{l_2}} = \infty.$$

Theorem 2.2.5. *Assume that (L1) and (L2) hold. If the functions f and g also satisfy the following conditions (L5) and (L6), then problem* (S′)–(BC) *has at least one positive solution* $(u(t), v(t)), t \in [0, T]$*:*

(L5) *There exist* $\alpha_1, \alpha_2 \in (0, \infty)$ *with* $\alpha_1\alpha_2 \leq 1$ *such that*

$$(1)\ q_{1\infty}^s = \limsup_{x\to\infty} \frac{q_1(x)}{x^{\alpha_1}} \in [0, \infty) \quad \textit{and} \quad (2)\ q_{2\infty}^s = \limsup_{x\to\infty} \frac{q_2(x)}{x^{\alpha_2}} = 0.$$

(L6) *There exist* $\beta_1, \beta_2 \in (0, \infty)$ *with* $\beta_1\beta_2 \leq 1$ *and* $c \in (0, T/2)$ *such that*

$$(1)\ \hat{f}_0^{\rm i} = \liminf_{x\to 0^+} \inf_{t\in[c,T-c]} \frac{f(t,x)}{x^{\beta_1}} \in (0, \infty] \quad \textit{and} \quad (2)\ \hat{g}_0^{\rm i} = \liminf_{x\to 0^+} \inf_{t\in[c,T-c]} \frac{g(t,x)}{x^{\beta_2}} = \infty.$$

By using Theorem 1.1 from Ji et al. (2009), in the above theorems, we obtain $u(t) > 0$ and $v(t) > 0$ for all $t \in (0, T - c]$.

Remark 2.2.2. The results presented in this section were published in Henderson and Luca (2014f).

2.3 Remarks on a particular case

In this section, we present briefly some existence results for the positive solutions for a particular case of problem (S)–(BC) from Section 2.1.

2.3.1 Auxiliary results

We investigate in this section system (S)–(BC) from Section 2.1 with $a_i = 0$ for all $i = 1, \ldots, p$ and $c_i = 0$ for all $i = 1, \ldots, r$.

Namely, we consider here the system of nonlinear higher-order ordinary differential equations

$$\begin{cases} u^{(n)}(t) + \lambda c(t) f(u(t), v(t)) = 0, & t \in (0, T), \\ v^{(m)}(t) + \mu d(t) g(u(t), v(t)) = 0, & t \in (0, T), \end{cases} \tag{S$_0$}$$

with the multipoint boundary conditions

$$\begin{cases} u(0)=u'(0)=\cdots=u^{(n-2)}(0)=0, \quad u(T)=\sum_{i=1}^{p-2} a_i u(\xi_i), \\ v(0)=v'(0)=\cdots=v^{(m-2)}(0)=0, \quad v(T)=\sum_{i=1}^{q-2} b_i v(\eta_i), \end{cases} \tag{BC$_0$}$$

where $n,m,p,q \in \mathbb{N}$, $n,m \geq 2$, $p,q \geq 3$; $a_i, \xi_i \in \mathbb{R}$ for all $i=1,\ldots,p-2$; $b_i, \eta_i \in \mathbb{R}$ for all $i=1,\ldots,q-2$; $0 < \xi_1 < \cdots < \xi_{p-2} < T$, $0 < \eta_1 < \cdots < \eta_{q-2} < T$. In the case $n=2$ or $m=2$, the boundary conditions above are of the form $u(0)=0, u(T)=\sum_{i=1}^{p-2} a_i u(\xi_i)$, or $v(0)=0, v(T)=\sum_{i=1}^{q-2} b_i v(\eta_i)$, respectively—that is, without conditions on the derivatives of u and v at the point 0.

By a positive solution of problem (S$_0$)–(BC$_0$) we mean a pair of functions $(u,v) \in C^n([0,T]) \times C^m([0,T])$ satisfying (S$_0$) and (BC$_0$) with $u(t) \geq 0, v(t) \geq 0$ for all $t \in [0,T]$ and $(u,v) \neq (0,0)$.

First, we consider the following nth-order differential equation with p-point boundary conditions:

$$u^{(n)}(t) + y(t) = 0, \quad t \in (0,T), \tag{2.33}$$

$$u(0)=u'(0)=\cdots=u^{(n-2)}(0)=0, \quad u(T)=\sum_{i=1}^{p-2} a_i u(\xi_i), \tag{2.34}$$

where $n,p \in \mathbb{N}$, $n \geq 2$, $p \geq 3$; $a_i, \xi_i \in \mathbb{R}$ for all $i=1,\ldots,p-2$; $0 < \xi_1 < \cdots < \xi_{p-2} < T$. If $n=2$, the boundary conditions above are of the form $u(0)=0$, $u(T)=\sum_{i=1}^{p-2} a_i u(\xi_i)$.

Lemma 2.3.1 (Ji et al., 2009; Luca, 2010). *If $a_i \in \mathbb{R}$ for all $i=1,\ldots,p-2$, $0<\xi_1<\cdots<\xi_{p-2}<T$, $d=T^{n-1}-\sum_{i=1}^{p-2} a_i \xi_i^{n-1} \neq 0$, and $y \in C(0,T) \cap L^1(0,T)$, then the unique solution of problem* (2.33)–(2.34) *is given by*

$$u(t) = \frac{t^{n-1}}{d(n-1)!}\int_0^T (T-s)^{n-1} y(s)\,ds - \frac{t^{n-1}}{d(n-1)!}\sum_{i=1}^{p-2} a_i \int_0^{\xi_i} (\xi_i - s)^{n-1} y(s)\,ds$$
$$- \frac{1}{(n-1)!}\int_0^t (t-s)^{n-1} y(s)\,ds, \quad 0 \leq t \leq T.$$

Lemma 2.3.2 (Ji et al., 2009; Luca, 2010). *Under the assumptions of Lemma 2.3.1, the Green's function for the boundary value problem* (2.33)–(2.34) *is given by*

$$\tilde{G}_1(t,s) = \begin{cases} \dfrac{t^{n-1}}{d(n-1)!}\left[(T-s)^{n-1} - \displaystyle\sum_{i=j+1}^{p-2} a_i(\xi_i - s)^{n-1}\right] - \dfrac{1}{(n-1)!}(t-s)^{n-1}, \\ \text{if } \xi_j \le s < \xi_{j+1}, \quad s \le t, \\ \dfrac{t^{n-1}}{d(n-1)!}\left[(T-s)^{n-1} - \displaystyle\sum_{i=j+1}^{p-2} a_i(\xi_i - s)^{n-1}\right], \\ \text{if } \xi_j \le s < \xi_{j+1}, \ s \ge t, \ j=0,\dots p-3, \\ \dfrac{t^{n-1}}{d(n-1)!}(T-s)^{n-1} - \dfrac{1}{(n-1)!}(t-s)^{n-1}, \quad \text{if } \xi_{p-2} \le s \le T, \ s \le t, \\ \dfrac{t^{n-1}}{d(n-1)!}(T-s)^{n-1}, \quad \text{if } \xi_{p-2} \le s \le T, \ s \ge t, \ (\xi_0 = 0). \end{cases}$$

Using the above Green's function, we can express the solution of problem (2.33)–(2.34) as $u(t) = \int_0^T \tilde{G}_1(t,s)y(s)\,\mathrm{d}s$.

Lemma 2.3.3 (Ji et al., 2009; Luca, 2010). *If $a_i \ge 0$ for all $i=1,\dots,p-2$, $0 < \xi_1 < \cdots < \xi_{p-2} < T$, $d > 0$ and $y \in C(0,T) \cap L^1(0,T)$, $y(t) \ge 0$ for all $t \in (0,T)$, then the solution u of problem* (2.33)–(2.34) *satisfies $u(t) \ge 0$ for all $t \in [0,T]$.*

Lemma 2.3.4 (Luca, 2010). *If $a_i \ge 0$ for all $i=1,\dots,p-2$, $0 < \xi_1 < \cdots < \xi_{p-2} < T$, $d > 0$ and $y \in C(0,T) \cap L^1(0,T)$, $y(t) \ge 0$ for all $t \in (0,T)$, then the solution u of problem* (2.33)–(2.34) *satisfies*

$$\begin{cases} u(t) \le \dfrac{T^{n-1}}{d(n-1)!}\displaystyle\int_0^T (T-s)^{n-1}y(s)\,\mathrm{d}s, & \forall t \in [0,T], \\ u(\xi_j) \ge \dfrac{\xi_j^{n-1}}{d(n-1)!}\displaystyle\int_{\xi_{p-2}}^T (T-s)^{n-1}y(s)\,\mathrm{d}s, & \forall j = \overline{1,p-2}. \end{cases}$$

Lemma 2.3.5 (Ji et al., 2009). *Assume that $a_i > 0$ for all $i=1,\dots,p-2$, $0 < \xi_1 < \cdots < \xi_{p-2} < T$, $d > 0$, and $y \in C(0,T) \cap L^1(0,T)$, $y(t) \ge 0$ for all $t \in (0,T)$. Then the solution u of problem* (2.33)–(2.34) *satisfies* $\inf_{t\in[\xi_{p-2},T]} u(t) \ge \tilde{\gamma}_1 \|u\|$, *where*

$$\tilde{\gamma}_1 = \begin{cases} \min\left\{\dfrac{a_{p-2}(T-\xi_{p-2})}{T-a_{p-2}\xi_{p-2}}, \dfrac{a_{p-2}\xi_{p-2}^{n-1}}{T^{n-1}}\right\}, & \text{if } \sum_{i=1}^{p-2} a_i < 1, \\ \min\left\{\dfrac{a_1\xi_1^{n-1}}{T^{n-1}}, \dfrac{\xi_{p-2}^{n-1}}{T^{n-1}}\right\}, & \text{if } \sum_{i=1}^{p-2} a_i \ge 1. \end{cases}$$

We can also formulate results similar to those of Lemmas 2.3.1–2.3.5 for the boundary value problem:

$$v^{(m)}(t) + h(t) = 0, \quad t \in (0,T), \tag{2.35}$$

$$v(0)=v'(0)=\cdots=v^{(m-2)}(0)=0, \quad v(T)=\sum_{i=1}^{q-2} b_i v(\eta_i), \tag{2.36}$$

where $m, q \in \mathbb{N}$, $m \geq 2$, $q \geq 3$; $b_i, \eta_i \in \mathbb{R}$ for all $i=1,\ldots,q-2$; $0<\eta_1<\cdots<\eta_{q-2}<T$. If $m=2$, the boundary conditions above are of the form $v(0)=0$, $v(T)=\sum_{i=1}^{q-2} b_i v(\eta_i)$.

If $e=T^{m-1}-\sum_{i=1}^{q-2} b_i \eta_i^{m-1} \neq 0$, and $h \in C(0,T) \cap L^1(0,T)$, we denote by $\tilde{G}_2$ the Green's function corresponding to problem (2.35)–(2.36)—that is,

$$\tilde{G}_2(t,s)=\begin{cases} \dfrac{t^{m-1}}{e(m-1)!}\left[(T-s)^{m-1}-\displaystyle\sum_{i=j+1}^{q-2} b_i(\eta_i-s)^{m-1}\right]-\dfrac{1}{(m-1)!}(t-s)^{m-1}, \\ \quad \text{if } \eta_j \leq s<\eta_{j+1},\ s \leq t, \\ \dfrac{t^{m-1}}{e(m-1)!}\left[(T-s)^{m-1}-\displaystyle\sum_{i=j+1}^{q-2} b_i(\eta_i-s)^{m-1}\right], \\ \quad \text{if } \eta_j \leq s<\eta_{j+1},\ s \geq t,\ j=0,\ldots q-3, \\ \dfrac{t^{m-1}}{e(m-1)!}(T-s)^{m-1}-\dfrac{1}{(m-1)!}(t-s)^{m-1}, \quad \text{if } \eta_{q-2} \leq s \leq T,\ s \leq t, \\ \dfrac{t^{m-1}}{e(m-1)!}(T-s)^{m-1}, \quad \text{if } \eta_{q-2} \leq s \leq T,\ s \geq t,\ (\eta_0=0). \end{cases}$$

Under assumptions similar to those from Lemma 2.3.5, we have the inequality $\inf_{t\in[\eta_{q-2},T]} v(t) \geq \tilde{\gamma}_2\|v\|$, where v is the solution of problem (2.35)–(2.36), and $\tilde{\gamma}_2$ is given by

$$\tilde{\gamma}_2=\begin{cases} \min\left\{\dfrac{b_{q-2}(T-\eta_{q-2})}{T-b_{q-2}\eta_{q-2}}, \dfrac{b_{q-2}\eta_{q-2}^{m-1}}{T^{m-1}}\right\}, & \text{if } \sum_{i=1}^{q-2} b_i<1, \\ \min\left\{\dfrac{b_1\eta_1^{m-1}}{T^{m-1}}, \dfrac{\eta_{q-2}^{m-1}}{T^{m-1}}\right\}, & \text{if } \sum_{i=1}^{q-2} b_i \geq 1. \end{cases}$$

2.3.2 Main results

In this section, we give sufficient conditions on λ, μ, f, and g such that positive solutions with respect to a cone for problem (S_0)–(BC_0) exist.

We present the assumptions that we shall use in the sequel:

($\widetilde{H1}$) $0<\xi_1<\cdots<\xi_{p-2}<T$, $a_i>0$ for all $i=1,\ldots,p-2$, $d=T^{n-1}-\sum_{i=1}^{p-2} a_i\xi_i^{n-1}>0$, $0<\eta_1<\cdots<\eta_{q-2}<T$, $b_i>0$ for all $i=1,\ldots,q-2$, $e=T^{m-1}-\sum_{i=1}^{q-2} b_i\eta_i^{m-1}>0$.

($\widetilde{H2}$) The functions c, d: $[0,T] \to [0,\infty)$ are continuous, and there exist $t_1, t_2 \in [\theta_0, T]$ such that $c(t_1)>0$ and $d(t_2)>0$, where $\theta_0=\max\{\xi_{p-2}, \eta_{q-2}\}$.

($\widetilde{H3}$) The functions f, g: $[0,\infty) \times [0,\infty) \to [0,\infty)$ are continuous.

We introduce the following extreme limits:

$$f_0^s = \limsup_{u+v\to 0^+} \frac{f(u,v)}{u+v}, \quad g_0^s = \limsup_{u+v\to 0^+} \frac{g(u,v)}{u+v}, \quad f_0^i = \liminf_{u+v\to 0^+} \frac{f(u,v)}{u+v},$$

$$g_0^i = \liminf_{u+v\to 0^+} \frac{g(u,v)}{u+v}, \quad f_\infty^s = \limsup_{u+v\to \infty} \frac{f(u,v)}{u+v}, \quad g_\infty^s = \limsup_{u+v\to \infty} \frac{g(u,v)}{u+v},$$

$$f_\infty^i = \liminf_{u+v\to \infty} \frac{f(u,v)}{u+v}, \quad g_\infty^i = \liminf_{u+v\to \infty} \frac{g(u,v)}{u+v}.$$

In the definitions of the extreme limits above, the variables u and v are nonnegative.

We consider the Banach space $X = C([0,T])$ with the supremum norm $\|\cdot\|$ and the Banach space $Y = X \times X$ with the norm $\|(u,v)\|_Y = \|u\| + \|v\|$.

We define the cone $C \subset Y$ by

$$C = \Big\{(u,v) \in Y;\quad u(t) \geq 0, v(t) \geq 0, \forall t \in [0,T] \quad \text{and} \quad \inf_{t\in[\theta_0,T]} (u(t)+v(t)) \geq \tilde{\gamma}\|(u,v)\|_Y\Big\},$$

where $\tilde{\gamma} = \min\{\tilde{\gamma}_1, \tilde{\gamma}_2\}$, and $\tilde{\gamma}_1$ and $\tilde{\gamma}_2$ are defined in Section 2.3.1, and $\theta_0 = \max\{\xi_{p-2}, \eta_{q-2}\}$.

Let P_1, P_2: $Y \to X$ and $\mathcal{P}$: $Y \to Y$ be the operators defined by

$$P_1(u,v)(t) = \lambda \int_0^T \tilde{G}_1(t,s)c(s)f(u(s),v(s))\,ds, \quad t \in [0,T],$$

$$P_2(u,v)(t) = \mu \int_0^T \tilde{G}_2(t,s)d(s)g(u(s),v(s))\,ds, \quad t \in [0,T],$$

and $\mathcal{P}(u,v) = (P_1(u,v), P_2(u,v))$, $(u,v) \in Y$, where $\tilde{G}_1$ and $\tilde{G}_2$ are the Green's functions defined in Section 2.3.1.

The solutions of problem (S_0)–(BC_0) coincide with the fixed points of the operator $\mathcal{P}$. In addition, the operator $\mathcal{P}$: $C \to C$ is a completely continuous operator.

First, for $f_0^s, g_0^s, f_\infty^i, g_\infty^i \in (0,\infty)$ and positive numbers $\alpha_1, \alpha_2 > 0$ such that $\alpha_1 + \alpha_2 = 1$, we define the positive numbers L_1, L_2, L_3, and L_4 by

$$L_1 = \alpha_1 \left(\frac{\tilde{\gamma}\xi_{p-2}^{n-1}}{d(n-1)!} \int_{\theta_0}^T (T-s)^{n-1} c(s) f_\infty^i\,ds \right)^{-1},$$

$$L_2 = \alpha_1 \left(\frac{T^{n-1}}{d(n-1)!} \int_0^T (T-s)^{n-1} c(s) f_0^s\,ds \right)^{-1},$$

$$L_3 = \alpha_2 \left(\frac{\tilde{\gamma}\eta_{q-2}^{m-1}}{e(m-1)!} \int_{\theta_0}^{T} (T-s)^{m-1} d(s) g_\infty^{\rm i} \, {\rm d}s \right)^{-1},$$

$$L_4 = \alpha_2 \left(\frac{T^{m-1}}{e(m-1)!} \int_0^T (T-s)^{m-1} d(s) g_0^{\rm s} \, {\rm d}s \right)^{-1}.$$

By using Lemmas 2.3.1–2.3.5, the operator $\mathcal{P}$, Theorem 1.1.1, and arguments similar to those used in the proofs of Theorems 2.1.1 and 2.1.2, we obtain the following results:

Theorem 2.3.1. *Assume that* $(\widetilde{\rm H1})$–$(\widetilde{\rm H3})$ *hold and that* $\alpha_1, \alpha_2 > 0$ *are positive numbers such that* $\alpha_1 + \alpha_2 = 1$.

(a) *If* $f_0^{\rm s}, g_0^{\rm s}, f_\infty^{\rm i}, g_\infty^{\rm i} \in (0,\infty)$, $L_1 < L_2$, *and* $L_3 < L_4$, *then for each* $\lambda \in (L_1, L_2)$ *and* $\mu \in (L_3, L_4)$ *there exists a positive solution* $(u(t), v(t)), t \in [0,T]$ *for* $({\rm S}_0)$–$({\rm BC}_0)$.
(b) *If* $f_0^{\rm s} = g_0^{\rm s} = 0$ *and* $f_\infty^{\rm i}, g_\infty^{\rm i} \in (0,\infty)$, *then for each* $\lambda \in (L_1, \infty)$ *and* $\mu \in (L_3, \infty)$ *there exists a positive solution* $(u(t), v(t)), t \in [0,T]$ *for* $({\rm S}_0)$–$({\rm BC}_0)$.
(c) *If* $f_0^{\rm s}, g_0^{\rm s} \in (0,\infty)$ *and* $f_\infty^{\rm i} = g_\infty^{\rm i} = \infty$, *then for each* $\lambda \in (0, L_2)$ *and* $\mu \in (0, L_4)$ *there exists a positive solution* $(u(t), v(t)), t \in [0,T]$ *for* $({\rm S}_0)$–$({\rm BC}_0)$.
(d) *If* $f_0^{\rm s} = g_0^{\rm s} = 0$ *and* $f_\infty^{\rm i} = g_\infty^{\rm i} = \infty$, *then for each* $\lambda \in (0,\infty)$ *and* $\mu \in (0,\infty)$ *there exists a positive solution* $(u(t), v(t)), t \in [0,T]$ *for* $({\rm S}_0)$–$({\rm BC}_0)$.

In what follows, for $f_0^{\rm i}, g_0^{\rm i}, f_\infty^{\rm s}, g_\infty^{\rm s} \in (0,\infty)$ and positive numbers $\alpha_1, \alpha_2 > 0$ such that $\alpha_1 + \alpha_2 = 1$, we define the positive numbers $\tilde{L}_1, \tilde{L}_2, \tilde{L}_3$, and $\tilde{L}_4$ by

$$\tilde{L}_1 = \alpha_1 \left(\frac{\tilde{\gamma}\xi_{p-2}^{n-1}}{d(n-1)!} \int_{\theta_0}^{T} (T-s)^{n-1} c(s) f_0^{\rm i} \, {\rm d}s \right)^{-1},$$

$$\tilde{L}_2 = \alpha_1 \left(\frac{T^{n-1}}{d(n-1)!} \int_0^T (T-s)^{n-1} c(s) f_\infty^{\rm s} \, {\rm d}s \right)^{-1},$$

$$\tilde{L}_3 = \alpha_2 \left(\frac{\tilde{\gamma}\eta_{q-2}^{m-1}}{e(m-1)!} \int_{\theta_0}^{T} (T-s)^{m-1} d(s) g_0^{\rm i} \, {\rm d}s \right)^{-1},$$

$$\tilde{L}_4 = \alpha_2 \left(\frac{T^{m-1}}{e(m-1)!} \int_0^T (T-s)^{m-1} d(s) g_\infty^{\rm s} \, {\rm d}s \right)^{-1}.$$

Theorem 2.3.2. *Assume that* $(\widetilde{\rm H1})$–$(\widetilde{\rm H3})$ *hold and that* $\alpha_1, \alpha_2 > 0$ *are positive numbers such that* $\alpha_1 + \alpha_2 = 1$.

(a) *If* $f_0^{\rm i}, g_0^{\rm i}, f_\infty^{\rm s}, g_\infty^{\rm s} \in (0,\infty)$, $\tilde{L}_1 < \tilde{L}_2$, *and* $\tilde{L}_3 < \tilde{L}_4$, *then for each* $\lambda \in (\tilde{L}_1, \tilde{L}_2)$ *and* $\mu \in (\tilde{L}_3, \tilde{L}_4)$ *there exists a positive solution* $(u(t), v(t)), t \in [0,T]$ *for* $({\rm S}_0)$–$({\rm BC}_0)$.
(b) *If* $f_\infty^{\rm s} = g_\infty^{\rm s} = 0$ *and* $f_0^{\rm i}, g_0^{\rm i} \in (0,\infty)$, *then for each* $\lambda \in (\tilde{L}_1, \infty)$ *and* $\mu \in (\tilde{L}_3, \infty)$ *there exists a positive solution* $(u(t), v(t)), t \in [0,T]$ *for* $({\rm S}_0)$–$({\rm BC}_0)$.
(c) *If* $f_\infty^{\rm s}, g_\infty^{\rm s} \in (0,\infty)$ *and* $f_0^{\rm i} = g_0^{\rm i} = \infty$, *then for each* $\lambda \in (0, \tilde{L}_2)$ *and* $\mu \in (0, \tilde{L}_4)$ *there exists a positive solution* $(u(t), v(t)), t \in [0,T]$ *for* $({\rm S}_0)$–$({\rm BC}_0)$.
(d) *If* $f_\infty^{\rm s} = g_\infty^{\rm s} = 0$ *and* $f_0^{\rm i} = g_0^{\rm i} = \infty$, *then for each* $\lambda \in (0,\infty)$ *and* $\mu \in (0,\infty)$ *there exists a positive solution* $(u(t), v(t)), t \in [0,T]$ *for* $({\rm S}_0)$–$({\rm BC}_0)$.

We present now an example which illustrates our results above.

Example 2.3.1. Let $T=1$, $n=3$, $m=4$, $p=5$, $q=4$, and $c(t)=c_0t$ and $d(t)=d_0t$ for all $t\in[0,1]$, with $c_0,d_0>0$, $\xi_1=\frac{1}{4}$, $\xi_2=\frac{1}{2}$, $\xi_3=\frac{3}{4}$, $\eta_1=\frac{1}{3}$, $\eta_2=\frac{2}{3}$, $a_1=1$, $a_2=\frac{1}{2}$, $a_3=\frac{1}{3}$, $b_1=1$, and $b_2=2$.

We consider the higher-order differential system

$$\begin{cases} u^{(3)}(t)+\lambda c_0tf(u(t),v(t))=0, & t\in(0,1),\\ v^{(4)}(t)+\mu d_0tg(u(t),v(t))=0, & t\in(0,1),\end{cases} \tag{S$_4$}$$

with the boundary conditions

$$\begin{cases} u(0)=u'(0)=0, & u(1)=u\left(\frac{1}{4}\right)+\frac{1}{2}u\left(\frac{1}{2}\right)+\frac{1}{3}u\left(\frac{3}{4}\right),\\ v(0)=v'(0)=v''(0)=0, & v(1)=v\left(\frac{1}{3}\right)+2v\left(\frac{2}{3}\right).\end{cases} \tag{BC$_4$}$$

The functions f and g are given by

$$f(u,v)=p_1|\sin\beta_1(u+v)|+320p_2(u+v)e^{-1/(u+v)^{\gamma_1}},\quad \forall u,v\in[0,\infty),$$

$$g(u,v)=q_1|\sin\beta_2(u+v)|+400q_2(u+v)e^{-1/(u+v)^{\gamma_2}},\quad \forall u,v\in[0,\infty),$$

with $p_1,p_2,q_1,q_2,\beta_1,\beta_2,\gamma_1,\gamma_2>0$. It follows that $f_0^{\rm s}=\beta_1p_1$, $f_\infty^{\rm i}=320p_2$, $g_0^{\rm s}=\beta_2q_1$, and $g_\infty^{\rm i}=400q_2$.

We also obtain $d=\frac{5}{8}$, $e=\frac{10}{27}$, $\theta_0=\frac{3}{4}$, $\tilde{\gamma}_1=\frac{1}{16}$, $\tilde{\gamma}_2=\frac{1}{27}$, and $\tilde{\gamma}=\frac{1}{27}$. The constants $L_i, i=1,\dots,4$ above are $L_1=\frac{576\alpha_1}{13c_0p_2}$, $L_2=\frac{15\alpha_1}{c_0\beta_1p_1}$, $L_3=\frac{648\alpha_2}{d_0q_2}$, and $L_4=\frac{400\alpha_2}{9d_0\beta_2q_1}$, and the conditions $L_1<L_2$ and $L_3<L_4$ are equivalent to $\beta_1p_1/p_2<65/192$ and $\beta_2q_1/q_2<50/729$, respectively.

We apply Theorem 2.3.1 (a) for $\alpha_1,\alpha_2>0$ with $\alpha_1+\alpha_2=1$. If the above conditions are satisfied, then for each $\lambda\in(L_1,L_2)$ and $\mu\in(L_3,L_4)$ there exists a positive solution $(u(t),v(t)),t\in[0,1]$ for problem (S$_4$)–(BC$_4$).

Remark 2.3.1. The results presented in this section were published in Henderson and Luca (2011).

2.4 Boundary conditions with additional positive constants

In this section, we investigate the existence and nonexistence of positive solutions for a system of higher-order ordinary differential equations with multipoint boundary conditions in which some positive constants appear.

2.4.1 Presentation of the problem

We consider the system of nonlinear higher-order ordinary differential equations

$$\begin{cases} u^{(n)}(t)+a(t)f(v(t))=0, & t\in(0,T),\\ v^{(m)}(t)+b(t)g(u(t))=0, & t\in(0,T),\end{cases} \tag{S^0}$$

with the multipoint boundary conditions

$$\begin{cases} u(0) = \sum_{i=1}^{p} a_i u(\xi_i) + a_0, \quad u'(0) = \cdots = u^{(n-2)}(0) = 0, \quad u(T) = \sum_{i=1}^{q} b_i u(\eta_i), \\ v(0) = \sum_{i=1}^{r} c_i v(\zeta_i) + b_0, \quad v'(0) = \cdots = v^{(m-2)}(0) = 0, \quad v(T) = \sum_{i=1}^{l} d_i v(\rho_i), \end{cases} \tag{BC^0}$$

where $n, m, p, q, r, l \in \mathbb{N}$, $n, m \geq 2$, $a_i, \xi_i \in \mathbb{R}$ for all $i = 1, \ldots, p$; $b_i, \eta_i \in \mathbb{R}$ for all $i = 1, \ldots, q$; $c_i, \zeta_i \in \mathbb{R}$ for all $i = 1, \ldots, r$; $d_i, \rho_i \in \mathbb{R}$ for all $i = 1, \ldots, l$; $0 < \xi_1 < \cdots < \xi_p < T$, $0 < \eta_1 < \cdots < \eta_q < T$, $0 < \zeta_1 < \cdots < \zeta_r < T$, $0 < \rho_1 < \cdots < \rho_l < T$, and a_0 and b_0 are positive constants. In the case $n = 2$ or $m = 2$, the boundary conditions above are of the form $u(0) = \sum_{i=1}^{p} a_i u(\xi_i) + a_0$, $u(T) = \sum_{i=1}^{q} b_i u(\eta_i)$, or $v(0) = \sum_{i=1}^{r} c_i v(\zeta_i) + b_0$, $v(T) = \sum_{i=1}^{l} d_i v(\rho_i)$, respectively—that is, without conditions on the derivatives of u and v at the point 0.

By using the Schauder fixed point theorem (Theorem 1.6.1), we shall prove the existence of positive solutions of problem (S^0)–(BC^0). By a positive solution of (S^0)–(BC^0) we mean a pair of functions $(u, v) \in C^n([0,T]; \mathbb{R}_+) \times C^m([0,T]; \mathbb{R}_+)$ satisfying (S^0) and (BC^0) with $u(t) > 0$, $v(t) > 0$ for all $t \in [0, T)$. We shall also give sufficient conditions for the nonexistence of positive solutions for this problem. System (S^0) with the multipoint boundary conditions $u(0) = u'(0) = \cdots = u^{(n-2)}(0) = 0$, $u(T) = \sum_{i=1}^{p-2} a_i u(\xi_i) + a_0$, $v(0) = v'(0) = \cdots = v^{(m-2)}(0) = 0$, $v(T) = \sum_{i=1}^{q-2} b_i v(\eta_i)$ $+ b_0$, $(a_0, b_0 > 0)$ was investigated in Henderson and Luca (2012c).

We present the assumptions that we shall use in the sequel:

(J1) $0 < \xi_1 < \cdots < \xi_p < T$, $a_i \geq 0$ for all $i = 1, \ldots, p$, $\sum_{i=1}^{p} a_i < 1$, $0 < \eta_1 < \cdots < \eta_q < T$, $b_i \geq 0$ for all $i = 1, \ldots, q$, $\sum_{i=1}^{q} b_i < 1$, $0 < \zeta_1 < \cdots < \zeta_r < T$, $c_i \geq 0$ for all $i = 1, \ldots, r$, $\sum_{i=1}^{r} c_i < 1$, $0 < \rho_1 < \cdots < \rho_l < T$, $d_i \geq 0$ for all $i = 1, \ldots, l$, $\sum_{i=1}^{l} d_i < 1$.

(J2) The functions $a, b\colon [0,T] \to [0, \infty)$ are continuous, and there exist $t_1, t_2 \in (0, T)$ such that $a(t_1) > 0$, $b(t_2) > 0$.

(J3) $f, g\colon [0, \infty) \to [0, \infty)$ are continuous functions, and there exists $c_0 > 0$ such that $f(u) < \frac{c_0}{L}$, $g(u) < \frac{c_0}{L}$ for all $u \in [0, c_0]$, where $L = \max\{\int_0^T a(s) J_1(s)\, ds, \int_0^T b(s) J_2(s)\, ds\}$, and J_1 and J_2 are defined in Section 2.1.2.

(J4) $f, g\colon [0, \infty) \to [0, \infty)$ are continuous functions and satisfy the conditions

$$\lim_{u\to\infty} \frac{f(u)}{u} = \infty, \ \lim_{u\to\infty} \frac{g(u)}{u} = \infty.$$

Under assumption (J1) we have all auxiliary results in Lemmas 2.1.1–2.1.6 from Section 2.1.2. Besides, by (J2) we deduce that $\int_0^T a(s) J_1(s)\, ds > 0$ and $\int_0^T b(s) J_2(s)$ $ds > 0$—that is, the constant L from J3 is positive.

2.4.2 Main results

Our first theorem is the following existence result for problem (S^0)–(BC^0).

Theorem 2.4.1. *Assume that assumptions (J1)–(J3) hold. Then problem* (S^0)–(BC^0) *has at least one positive solution for* $a_0 > 0$ *and* $b_0 > 0$ *sufficiently small.*

Proof. We consider the problems

$$\begin{cases} h^{(n)}(t) = 0, \quad t \in (0,T), \\ h(0) = \sum_{i=1}^{p} a_i h(\xi_i) + 1, \quad h'(0) = \cdots = h^{(n-2)}(0) = 0, \quad h(T) = \sum_{i=1}^{q} b_i h(\eta_i), \end{cases} \tag{2.37}$$

$$\begin{cases} k^{(m)}(t) = 0, \quad t \in (0,T), \\ k(0) = \sum_{i=1}^{r} c_i k(\zeta_i) + 1, \quad k'(0) = \cdots = k^{(m-2)}(0) = 0, \quad k(T) = \sum_{i=1}^{l} d_i k(\rho_i). \end{cases} \tag{2.38}$$

Problems (2.37) and (2.38) have the solutions

$$h(t) = \frac{1}{\Delta_1}\left[-t^{n-1}\left(1 - \sum_{i=1}^{q} b_i\right) + T^{n-1} - \sum_{i=1}^{q} b_i \eta_i^{n-1}\right], \quad t \in [0,T],$$

$$k(t) = \frac{1}{\Delta_2}\left[-t^{m-1}\left(1 - \sum_{i=1}^{l} d_i\right) + T^{m-1} - \sum_{i=1}^{l} d_i \rho_i^{m-1}\right], \quad t \in [0,T], \tag{2.39}$$

respectively, where Δ_1 and Δ_2 are defined in Section 2.1.2. By assumption (J1) we obtain $h(t) > 0$ and $k(t) > 0$ for all $t \in [0,T)$.

We define the functions $x(t)$ and $y(t)$, $t \in [0,T]$ by

$$x(t) = u(t) - a_0 h(t), \quad y(t) = v(t) - b_0 k(t), \quad t \in [0,T],$$

where (u,v) is a solution of (S^0)–(BC^0). Then (S^0)–(BC^0) can be equivalently written as

$$\begin{cases} x^{(n)}(t) + a(t) f(y(t) + b_0 k(t)) = 0, \quad t \in (0,T), \\ y^{(m)}(t) + b(t) g(x(t) + a_0 h(t)) = 0, \quad t \in (0,T), \end{cases} \tag{2.40}$$

with the boundary conditions

$$\begin{cases} x(0) = \sum_{i=1}^{p} a_i x(\xi_i), \quad x'(0) = \cdots = x^{(n-2)}(0) = 0, \quad x(T) = \sum_{i=1}^{q} b_i x(\eta_i), \\ y(0) = \sum_{i=1}^{r} c_i y(\zeta_i), \quad y'(0) = \cdots = y^{(m-2)}(0) = 0, \quad y(T) = \sum_{i=1}^{l} d_i y(\rho_i). \end{cases} \tag{2.41}$$

Using the Green's functions G_1 and G_2 from Section 2.1.2, we find a pair (x,y) is a solution of problem (2.40)–(2.41) if and only if (x,y) is a solution for the nonlinear integral equations

$$\begin{cases} x(t) = \displaystyle\int_0^T G_1(t,s)a(s)f\left(\int_0^T G_2(s,\tau)b(\tau)g(x(\tau)+a_0h(\tau))\,\mathrm{d}\tau + b_0k(s)\right)\mathrm{d}s, \\ y(t) = \displaystyle\int_0^T G_2(t,s)b(s)g(x(s)+a_0h(s))\,\mathrm{d}s, \qquad 0 \le t \le T, \end{cases} \tag{2.42}$$

where $h(t), k(t), t \in [0,T]$ are given by (2.39).

We consider the Banach space $X = C([0,T])$ with the supremum norm $\|\cdot\|$, and define the set

$$E = \{x \in C([0,T]), \quad 0 \le x(t) \le c_0, \quad \forall t \in [0,T]\} \subset X.$$

We also define the operator $\mathcal{S}: E \to X$ by

$$(\mathcal{S}x)(t) = \int_0^T G_1(t,s)a(s)f\left(\int_0^T G_2(s,\tau)b(\tau)g(x(\tau)+a_0h(\tau))\,\mathrm{d}\tau + b_0k(s)\right)\mathrm{d}s,$$

$$0 \le t \le T, \quad x \in E.$$

For sufficiently small $a_0 > 0$ and $b_0 > 0$, by (J3), we deduce

$$f(y(t)+b_0k(t)) \le \frac{c_0}{L}, \quad g(x(t)+a_0h(t)) \le \frac{c_0}{L}, \quad \forall t \in [0,T], \quad x, y \in E.$$

Then, by using Lemma 2.1.4, we obtain $(\mathcal{S}x)(t) \ge 0$ for all $t \in [0,T]$ and $x \in E$. By Lemma 2.1.5, for all $x \in E$, we have

$$\int_0^T G_2(s,\tau)b(\tau)g(x(\tau)+a_0h(\tau))\,\mathrm{d}\tau \le \int_0^T J_2(\tau)b(\tau)g(x(\tau)+a_0h(\tau))\,\mathrm{d}\tau$$

$$\le \frac{c_0}{L}\int_0^T J_2(\tau)b(\tau)\,\mathrm{d}\tau \le c_0, \quad \forall s \in [0,T],$$

and

$$(\mathcal{S}x)(t) \le \int_0^T J_1(s)a(s)f\left(\int_0^T G_2(s,\tau)b(\tau)g(x(\tau)+a_0h(\tau))\,\mathrm{d}\tau + b_0k(s)\right)\mathrm{d}s$$

$$\le \frac{c_0}{L}\int_0^T J_1(s)a(s)\,\mathrm{d}s \le c_0, \quad \forall t \in [0,T].$$

Therefore, $\mathcal{S}(E) \subset E$.

Using standard arguments, we deduce that $\mathcal{S}$ is completely continuous. By Theorem 1.6.1, we conclude that $\mathcal{S}$ has a fixed point $x \in E$. This element together with y given by (2.42) represents a solution for (2.40)–(2.41). This shows that our problem (S^0)–(BC^0) has a positive solution (u, v) with $u = x + a_0h$, $v = y + b_0k$ for sufficiently small $a_0 > 0$ and $b_0 > 0$. □

In what follows, we present sufficient conditions for the nonexistence of positive solutions of (S^0)–(BC^0).

Theorem 2.4.2. *Assume that assumptions (J1), (J2), and (J4) hold. Then problem* (S^0)–(BC^0) *has no positive solution for* a_0 *and* b_0 *sufficiently large.*

Proof. We suppose that (u, v) is a positive solution of (S^0)–(BC^0). Then (x, y) with $x = u - a_0 h$, $y = v - b_0 k$ is a solution for (2.40)–(2.41), where h and k are the solutions of problems (2.37) and (2.38), respectively, (given by (2.39)). By (J2) there exists $c \in (0, T/2)$ such that $t_1, t_2 \in (c, T - c)$, and then $\int_c^{T-c} a(s)J_1(s)\,ds > 0$, $\int_c^{T-c} b(s)J_2(s)\,ds > 0$. Now by using Lemma 2.1.4, we have $x(t) \geq 0$, $y(t) \geq 0$ for all $t \in [0, T]$, and by Lemma 2.1.6 we obtain $\inf_{t\in[c,T-c]} x(t) \geq \gamma_1\|x\|$ and $\inf_{t\in[c,T-c]} y(t) \geq \gamma_2\|y\|$.

Using now (2.39), we deduce that

$$\inf_{t\in[c,T-c]} h(t) = h(T-c) = \frac{h(T-c)}{h(0)}\|h\|, \qquad \inf_{t\in[c,T-c]} k(t) = k(T-c) = \frac{k(T-c)}{k(0)}\|k\|.$$

Therefore, we obtain

$$\inf_{t\in[c,T-c]}(x(t) + a_0 h(t)) \geq \gamma_1\|x\| + a_0\frac{h(T-c)}{h(0)}\|h\| \geq r_1(\|x\| + a_0\|h\|) \geq r_1\|x + a_0 h\|,$$

$$\inf_{t\in[c,T-c]}(y(t) + b_0 k(t)) \geq \gamma_2\|y\| + b_0\frac{k(T-c)}{k(0)}\|k\| \geq r_2(\|y\| + b_0\|k\|) \geq r_2\|y + b_0 k\|,$$

where $r_1 = \min\{\gamma_1, \frac{h(T-c)}{h(0)}\}$, $r_2 = \min\{\gamma_2, \frac{k(T-c)}{k(0)}\}$.

We now consider $R = (\min\{\gamma_2 r_1 \int_c^{T-c} b(s)J_2(s)\,ds, \gamma_1 r_2 \int_c^{T-c} a(s)J_1(s)\,ds\})^{-1} > 0$.

By using (J4), for R defined above, we conclude that there exists $M > 0$ such that $f(u) > 2Ru$, $g(u) > 2Ru$ for all $u \geq M$. We consider $a_0 > 0$ and $b_0 > 0$ sufficiently large such that $\inf_{t\in[c,T-c]}(x(t) + a_0 h(t)) \geq M$ and $\inf_{t\in[c,T-c]}(y(t) + b_0 k(t)) \geq M$. By (J2), (2.40), (2.41), and the above inequalities, we deduce that $\|x\| > 0$ and $\|y\| > 0$.

Now by using Lemma 2.1.5 and the above considerations, we have

$$\begin{aligned}
y(c) &= \int_0^T G_2(c, s)b(s)g(x(s) + a_0 h(s))\,ds \geq \gamma_2 \int_0^T J_2(s)b(s)g(x(s) + a_0 h(s))\,ds \\
&\geq \gamma_2 \int_c^{T-c} J_2(s)b(s)g(x(s) + a_0 h(s))\,ds \geq 2R\gamma_2 \int_c^{T-c} J_2(s)b(s)(x(s) + a_0 h(s))\,ds \\
&\geq 2R\gamma_2 \int_c^{T-c} J_2(s)b(s) \inf_{\tau\in[c,T-c]}(x(\tau) + a_0 h(\tau))\,ds \\
&\geq 2R\gamma_2 r_1 \int_c^{T-c} J_2(s)b(s)\|x + a_0 h\|\,ds \geq 2\|x + a_0 h\| \geq 2\|x\|.
\end{aligned}$$

Therefore, we obtain

$$\|x\| \leq y(c)/2 \leq \|y\|/2. \tag{2.43}$$

In a similar manner, we deduce

$$
\begin{aligned}
x(c) &= \int_0^T G_1(c,s)a(s)f(y(s)+b_0k(s))\,ds \ge \gamma_1 \int_0^T J_1(s)a(s)f(y(s)+b_0k(s))\,ds \\
&\ge \gamma_1 \int_c^{T-c} J_1(s)a(s)f(y(s)+b_0k(s))\,ds \ge 2R\gamma_1 \int_c^{T-c} J_1(s)a(s)(y(s)+b_0k(s))\,ds \\
&\ge 2R\gamma_1 \int_c^{T-c} J_1(s)a(s) \inf_{\tau\in[c,T-c]} (y(\tau)+b_0k(\tau))\,ds \\
&\ge 2R\gamma_1 r_2 \int_c^{T-c} J_1(s)a(s)\|y+b_0k\|\,ds \ge 2\|y+b_0k\| \ge 2\|y\|.
\end{aligned}
$$

So, we obtain

$$\|y\| \le x(c)/2 \le \|x\|/2. \tag{2.44}$$

By (2.43) and (2.44), we conclude that $\|x\| \le \|y\|/2 \le \|x\|/4$, which is a contradiction, because $\|x\| > 0$. Then, for a_0 and b_0 sufficiently large, our problem (S^0)–(BC^0) has no positive solution. □

Results similar to Theorems 2.4.1 and 2.4.2 can be obtained if instead of boundary conditions (BC^0) we have

$$
\begin{cases}
u(0) = \sum_{i=1}^{p} a_i u(\xi_i), \quad u'(0) = \cdots = u^{(n-2)}(0) = 0, \quad u(T) = \sum_{i=1}^{q} b_i u(\eta_i) + a_0, \\
v(0) = \sum_{i=1}^{r} c_i v(\zeta_i), \quad v'(0) = \cdots = v^{(m-2)}(0) = 0, \quad v(T) = \sum_{i=1}^{l} d_i v(\rho_i) + b_0,
\end{cases} \tag{BC_1^0}
$$

or

$$
\begin{cases}
u(0) = \sum_{i=1}^{p} a_i u(\xi_i) + a_0, \quad u'(0) = \cdots = u^{(n-2)}(0) = 0, \quad u(T) = \sum_{i=1}^{q} b_i u(\eta_i), \\
v(0) = \sum_{i=1}^{r} c_i v(\zeta_i), \quad v'(0) = \cdots = v^{(m-2)}(0) = 0, \quad v(T) = \sum_{i=1}^{l} d_i v(\rho_i) + b_0,
\end{cases} \tag{BC_2^0}
$$

or

$$
\begin{cases}
u(0) = \sum_{i=1}^{p} a_i u(\xi_i), \quad u'(0) = \cdots = u^{(n-2)}(0) = 0, \quad u(T) = \sum_{i=1}^{q} b_i u(\eta_i) + a_0, \\
v(0) = \sum_{i=1}^{r} c_i v(\zeta_i) + b_0, \quad v'(0) = \cdots = v^{(m-2)}(0) = 0, \quad v(T) = \sum_{i=1}^{l} d_i v(\rho_i),
\end{cases} \tag{BC_3^0}
$$

where a_0 and b_0 are positive constants.

For problem (S^0)–(BC_1^0), instead of functions h and k from the proof of Theorem 2.4.1, the solutions of problems

$$\begin{cases} h_1^{(n)}(t)=0, \quad t\in(0,T), \\ h_1(0)=\sum_{i=1}^{p} a_i h_1(\xi_i), \quad h_1'(0)=\cdots=h_1^{(n-2)}(0)=0, \quad h_1(T)=\sum_{i=1}^{q} b_i h_1(\eta_i)+1, \end{cases} \tag{2.45}$$

$$\begin{cases} k_1^{(m)}(t)=0, \quad t\in(0,T), \\ k_1(0)=\sum_{i=1}^{r} c_i k_1(\zeta_i), \quad k_1'(0)=\cdots=k_1^{(m-2)}(0)=0, \quad k_1(T)=\sum_{i=1}^{l} d_i k_1(\rho_i)+1 \end{cases} \tag{2.46}$$

are

$$h_1(t)=\frac{1}{\Delta_1}\left[t^{n-1}\left(1-\sum_{i=1}^{p} a_i\right)+\sum_{i=1}^{p} a_i\xi_i^{n-1}\right], \quad t\in[0,T],$$

$$k_1(t)=\frac{1}{\Delta_2}\left[t^{m-1}\left(1-\sum_{i=1}^{r} c_i\right)+\sum_{i=1}^{r} c_i\zeta_i^{m-1}\right], \quad t\in[0,T],$$

respectively. By assumption (J1) we obtain $h_1(t)>0$ and $k_1(t)>0$ for all $t\in(0,T]$.

For problem (S^0)–(BC_2^0), instead of functions h and k from the proof of Theorem 2.4.1, the solutions of problems (2.37) and (2.46) are the functions h and k_1, respectively, which satisfy $h(t)>0$ for all $t\in[0,T)$ and $k_1(t)>0$ for all $t\in(0,T]$. For problem (S^0)–(BC_3^0), instead of functions h and k from the proof of Theorem 2.4.1, the solutions of problems (2.45) and (2.38) are the functions h_1 and k, respectively, which satisfy $h_1(t)>0$ for all $t\in(0,T]$ and $k(t)>0$ for all $t\in[0,T)$.

Therefore, we also obtain the following results:

Theorem 2.4.3. *Assume that assumptions (J1)–(J3) hold. Then problem* (S^0)–(BC_1^0) *has at least one positive solution ($u(t)>0$ and $v(t)>0$ for all $t\in(0,T]$) for $a_0>0$ and $b_0>0$ sufficiently small.*

Theorem 2.4.4. *Assume that assumptions (J1), (J2), and (J4) hold. Then problem* (S^0)–(BC_1^0) *has no positive solution ($u(t)>0$ and $v(t)>0$ for all $t\in(0,T]$) for a_0 and b_0 sufficiently large.*

Theorem 2.4.5. *Assume that assumptions (J1)–(J3) hold. Then problem* (S^0)–(BC_2^0) *has at least one positive solution ($u(t)>0$ for all $t\in[0,T)$, and $v(t)>0$ for all $t\in(0,T]$) for $a_0>0$ and $b_0>0$ sufficiently small.*

Theorem 2.4.6. *Assume that assumptions (J1), (J2), and (J4) hold. Then problem* (S^0)–(BC_2^0) *has no positive solution ($u(t)>0$ for all $t\in[0,T)$, and $v(t)>0$ for all $t\in(0,T]$) for a_0 and b_0 sufficiently large.*

Theorem 2.4.7. *Assume that assumptions (J1)–(J3) hold. Then problem* (S^0)*–*(BC_3^0) *has at least one positive solution (*$u(t) > 0$ *for all* $t \in (0, T]$*, and* $v(t) > 0$ *for all* $t \in [0, T)$*) for* $a_0 > 0$ *and* $b_0 > 0$ *sufficiently small.*

Theorem 2.4.8. *Assume that assumptions (J1), (J2), and (J4) hold. Then problem* (S^0)*–*(BC_3^0) *has no positive solution (*$u(t) > 0$ *for all* $t \in (0, T]$*, and* $v(t) > 0$ *for all* $t \in [0, T)$*) for* a_0 *and* b_0 *sufficiently large.*

Remark 2.4.1. All the results from Sections 2.1, 2.2, and 2.4 can be easily generalized for the case in which the multipoint boundary conditions (BC) or (BC^0) are replaced by Riemann–Stieltjes integral boundary conditions:

$$\begin{cases} u(0) = \int_0^T u(s)\, dH_1(s), \quad u'(0) = \cdots = u^{(n-2)}(0) = 0, \quad u(T) = \int_0^T u(s)\, dH_2(s), \\ v(0) = \int_0^T v(s)\, dK_1(s), \quad v'(0) = \cdots = v^{(m-2)}(0) = 0, \quad v(T) = \int_0^T v(s)\, dK_2(s), \end{cases} \tag{$\widehat{BC}$}$$

or

$$\begin{cases} u(0) = \int_0^T u(s)\, dH_1(s) + a_0, \quad u'(0) = \cdots = u^{(n-2)}(0) = 0, \quad u(T) = \int_0^T u(s)\, dH_2(s), \\ v(0) = \int_0^T v(s)\, dK_1(s) + b_0, \quad v'(0) = \cdots = v^{(n-2)}(0) = 0, \quad v(T) = \int_0^T v(s)\, dK_2(s), \end{cases} \tag{$\widehat{BC^0}$}$$

respectively.

2.4.3 An example

Example 2.4.1. We consider $T = 1$, $a(t) = at^2$, $b(t) = bt^3$ for all $t \in [0, 1]$ with $a, b > 0$, $n = 4$, $m = 3$, $p = 1$, $q = 2$, $r = 2$, $l = 1$, $\xi_1 = 1/2$, $a_1 = 3/4$, $\eta_1 = 1/4$, $\eta_2 = 3/4$, $b_1 = 1/2$, $b_2 = 1/3$, $\zeta_1 = 1/3$, $\zeta_2 = 2/3$, $c_1 = 4/5$, $c_2 = 1/6$, $\rho_1 = 3/5$, and $d_1 = 6/7$. We also consider the functions $f, g \colon [0, \infty) \to [0, \infty)$, $f(x) = \frac{\tilde{a}x^\alpha}{x^\beta + \tilde{c}}$, $g(x) = \frac{\tilde{b}x^\gamma}{x^\delta + \tilde{d}}$ for all $x \in [0, \infty)$, with $\tilde{a}, \tilde{b}, \tilde{c}, \tilde{d} > 0$, $\alpha, \beta, \gamma, \delta > 0$, $\alpha > \beta + 1$, and $\gamma > \delta + 1$. We have $\lim_{x\to\infty} f(x)/x = \lim_{x\to\infty} g(x)/x = \infty$.

Therefore, we consider the nonlinear higher-order differential system

$$\begin{cases} u^{(4)}(t) + at^2 \dfrac{\tilde{a}v^\alpha(t)}{v^\beta(t) + \tilde{c}} = 0, \quad t \in (0, 1), \\ v^{(3)}(t) + bt^3 \dfrac{\tilde{b}u^\gamma(t)}{u^\delta(t) + \tilde{d}} = 0, \quad t \in (0, 1), \end{cases} \tag{$\widetilde{S^0}$}$$

with the boundary conditions

$$\begin{cases} u(0) = \frac{3}{4}u\left(\frac{1}{2}\right) + a_0, & u'(0) = u''(0) = 0, \quad u(1) = \frac{1}{2}u\left(\frac{1}{4}\right) + \frac{1}{3}u\left(\frac{3}{4}\right), \\ v(0) = \frac{4}{5}v\left(\frac{1}{3}\right) + \frac{1}{6}v\left(\frac{2}{3}\right) + b_0, & v'(0) = 0, \quad v(1) = \frac{6}{7}v\left(\frac{3}{5}\right). \end{cases} \tag{$\widetilde{BC^0}$}$$

By using also Example 2.1.2 from Section 2.1.4, we deduce that assumptions (J1), (J2), and (J4) are satisfied. In addition, by using functions J_1 and J_2 from Example 2.1.2 from Section 2.1.4, we obtain $\tilde{A} = \int_0^1 s^2 J_1(s)\,\mathrm{d}s \approx 0.00194573$, $\tilde{B} = \int_0^1 s^3 J_2(s)\,\mathrm{d}s \approx 0.031305845$, and then $L = \max\{a\tilde{A}, b\tilde{B}\}$. We choose $c_0 = 1$, and if we select $\tilde{a}, \tilde{b}, \tilde{c}, \tilde{d}$ satisfying the conditions $\tilde{a} < \frac{1+\tilde{c}}{L} = (1+\tilde{c})\min\left\{\frac{1}{a\tilde{A}}, \frac{1}{b\tilde{B}}\right\}$ and $\tilde{b} < \frac{1+\tilde{d}}{L} = (1+\tilde{d})\min\left\{\frac{1}{a\tilde{A}}, \frac{1}{b\tilde{B}}\right\}$, then we conclude that $f(x) \le \frac{\tilde{a}}{1+\tilde{c}} < \frac{1}{L}, g(x) \le \frac{\tilde{b}}{1+\tilde{d}} < \frac{1}{L}$ for all $x \in [0,1]$. For example, if $a=2$, $b=3$, and $\tilde{c}=\tilde{d}=1$, then for $\tilde{a} \le 21.29$ and $\tilde{b} \le 21.29$ the above conditions for f and g are satisfied. So, assumption $\widetilde{(\mathrm{J3})}$ is also satisfied. By Theorems 2.4.1 and 2.4.2 we deduce that problem (S^0)–(BC^0) has at least one positive solution (here $u(t) > 0$ and $v(t) > 0$ for all $t \in [0,1]$) for sufficiently small $a_0 > 0$ and $b_0 > 0$, and no positive solution for sufficiently large a_0 and b_0.

2.5 A system of semipositone integral boundary value problems

In this section, we investigate the existence of positive solutions for a system of nonlinear higher-order ordinary differential equations with sign-changing nonlinearities, subject to integral boundary conditions.

2.5.1 Presentation of the problem

We consider the system of nonlinear higher-order ordinary differential equations

$$\begin{cases} u^{(n)}(t) + \lambda f(t, u(t), v(t)) = 0, & t \in (0,T), \\ v^{(m)}(t) + \mu g(t, u(t), v(t)) = 0, & t \in (0,T), \end{cases} \tag{$\tilde{S}$}$$

with the integral boundary conditions

$$\begin{cases} u(0) = \displaystyle\int_0^T u(s)\,\mathrm{d}H_1(s), \quad u'(0) = \cdots = u^{(n-2)}(0) = 0, \quad u(T) = \displaystyle\int_0^T u(s)\,\mathrm{d}H_2(s), \\ v(0) = \displaystyle\int_0^T v(s)\,\mathrm{d}K_1(s), \quad v'(0) = \cdots = v^{(m-2)}(0) = 0, \quad v(T) = \displaystyle\int_0^T v(s)\,\mathrm{d}K_2(s), \end{cases} \tag{$\widetilde{\mathrm{BC}}$}$$

where $n, m \in \mathbb{N}$, $n, m \ge 2$. In the case $n=2$ or $m=2$, the boundary conditions above are of the form $u(0) = \int_0^T u(s)\ \mathrm{d}H_1(s)$, $u(T) = \int_0^T u(s)\ \mathrm{d}H_2(s)$, or $v(0) = \int_0^T v(s)\ \mathrm{d}K_1(s)$, $v(T) = \int_0^T v(s)\ \mathrm{d}K_2(s)$, respectively—that is, without conditions on the derivatives of u and v at the point 0. The nonlinearities f and g are sign-changing continuous functions (i.e., we have a so-called system of semipositone boundary value problems), and the integrals from $(\widetilde{\mathrm{BC}})$ are Riemann–Stieltjes integrals.

By using a nonlinear alternative of Leray–Schauder type, we present intervals for parameters λ and μ such that the above problem $(\tilde{S})$–$(\widetilde{BC})$ has at least one positive solution. By a positive solution of problem $(\tilde{S})$–$(\widetilde{BC})$ we mean a pair of functions $(u, v) \in C^n([0,T]) \times C^m([0,T])$ satisfying $(\tilde{S})$ and $(\widetilde{BC})$ with $u(t) \geq 0$, $v(t) \geq 0$ for all $t \in [0,T]$ and $u(t) > 0$, $v(t) > 0$ for all $t \in (0,T)$. In the case when f and g are nonnegative functions and, in the boundary conditions $(\widetilde{BC})$, H_1, H_2, K_1, and K_2 are step functions (denoted by $(\widetilde{BC}_1)$), the existence of positive solutions of the above problem ($u(t) \geq 0$, $v(t) \geq 0$ for all $t \in [0,T]$, $(u,v) \neq (0,0)$) was studied in Section 2.1 using the Guo–Krasnosel'skii fixed point theorem. The positive solutions ($u(t) \geq 0$, $v(t) \geq 0$ for all $t \in [0,T]$, $\sup_{t\in[0,T]} u(t) > 0$, $\sup_{t\in[0,T]} v(t) > 0$) of system $(\tilde{S})$ with $\lambda = \mu = 1$ and with $f(t,u,v)$ and $g(t,u,v)$ replaced by $\tilde{f}(t,v)$ and $\tilde{g}(t,u)$, respectively ($\tilde{f}$ and $\tilde{g}$ are nonnegative functions), with the boundary conditions $(\widetilde{BC}_1)$ were investigated in Section 2.2 (the nonsingular and singular cases). In Section 2.2.1 (nonsingular nonlinearities), we obtained the existence and multiplicity of positive solutions by applying some theorems from the fixed point index theory, and in Section 2.2.2 (singular nonlinearities), we studied the existence of positive solutions by using the Guo–Krasnosel'skii fixed point theorem. We also mention (Eloe and Ahmad, 2005) where the authors investigated the existence of positive solutions for the nonlinear nth-order differential equation $u^{(n)}(t) + a(t)f(u(t)) = 0$, $t \in (0,1)$, subject to the boundary conditions $u(0) = u'(0) = \cdots = u^{(n-2)}(0) = 0$, $\alpha u(\eta) = u(1)$, with $0 < \eta < 1$ and $0 < \alpha\eta^{n-1} < 1$.

In the proof of the main result, we shall use the following nonlinear alternative of Leray–Schauder type (see Agarwal et al., 2001):

Theorem 2.5.1. *Let X be a Banach space with $\Omega \subset X$ closed and convex. Assume U is a relatively open subset of Ω with $0 \in U$, and let S: $\bar{U} \to \Omega$ be a completely continuous operator. Then either*

(1) S has a fixed point in $\bar{U}$ or
(2) there exist $u \in \partial U$ and $\nu \in (0,1)$ such that $u = \nu S u$.

2.5.2 Auxiliary results

In this section, we present some auxiliary results related to the nth-order differential equation

$$u^{(n)}(t) + z(t) = 0, \quad t \in (0,T), \tag{2.47}$$

with the integral boundary conditions

$$u(0) = \int_0^T u(s)\, dH_1(s), \quad u'(0) = \cdots = u^{(n-2)}(0) = 0,$$
$$u(T) = \int_0^T u(s)\, dH_2(s), \tag{2.48}$$

where $n \in \mathbb{N}$, $n \geq 2$, and H_1, H_2: $[0,T] \to \mathbb{R}$ are functions of bounded variation. If $n = 2$, condition (2.48) has the form $u(0) = \int_0^T u(s)\, dH_1(s)$, $u(T) = \int_0^T u(s)\, dH_2(s)$.

Lemma 2.5.1. *If H_1 and H_2 are functions of bounded variation, $\Delta_1 = (1-\int_0^T \mathrm{d}H_2(s))\int_0^T s^{n-1}\mathrm{d}H_1(s) + (1-\int_0^T \mathrm{d}H_1(s))(T^{n-1}-\int_0^T s^{n-1}\mathrm{d}H_2(s)) \neq 0$, and $z \in C(0,T)\cap L^1(0,T)$, then unique the solution of problem* (2.47)–(2.48) *is given by*

$$\begin{aligned} u(t) = &-\int_0^t \frac{(t-s)^{n-1}}{(n-1)!} z(s)\,\mathrm{d}s + \frac{t^{n-1}}{\Delta_1}\left\{\left(1-\int_0^T \mathrm{d}H_2(s)\right)\frac{1}{(n-1)!}\right.\\ &\times \int_0^T \left(\int_0^s (s-\tau)^{n-1} z(\tau)\,\mathrm{d}\tau\right)\mathrm{d}H_1(s) + \left(1-\int_0^T \mathrm{d}H_1(s)\right)\frac{1}{(n-1)!}\\ &\left.\times \left[\int_0^T (T-s)^{n-1} z(s)\,\mathrm{d}s - \int_0^T \left(\int_0^s (s-\tau)^{n-1} z(\tau)\,\mathrm{d}\tau\right)\mathrm{d}H_2(s)\right]\right\}\\ &+ \frac{1}{\Delta_1}\left\{\left(\int_0^T s^{n-1}\mathrm{d}H_1(s)\right)\frac{1}{(n-1)!}\left[\int_0^T (T-s)^{n-1} z(s)\,\mathrm{d}s\right.\right.\\ &\left.-\int_0^T \left(\int_0^s (s-\tau)^{n-1} z(\tau)\,\mathrm{d}\tau\right)\mathrm{d}H_2(s)\right] - \left(T^{n-1} - \int_0^T s^{n-1}\mathrm{d}H_2(s)\right)\\ &\left.\times \frac{1}{(n-1)!}\int_0^T \left(\int_0^s (s-\tau)^{n-1} z(\tau)\,\mathrm{d}\tau\right)\mathrm{d}H_1(s)\right\}. \end{aligned} \tag{2.49}$$

Proof. If $n \geq 3$, then the solution of (2.47) is

$$u(t) = -\int_0^t \frac{(t-s)^{n-1}}{(n-1)!} z(s)\,\mathrm{d}s + At^{n-1} + \sum_{i=1}^{n-2} A_i t^i + B,$$

with $A, B, A_i \in \mathbb{R}$ for all $i = 1, \ldots, n-2$. By using the conditions $u'(0) = \cdots = u^{(n-2)}(0) = 0$, we obtain $A_i = 0$ for all $i = 1, \ldots, n-2$. Then we conclude

$$u(t) = -\int_0^t \frac{(t-s)^{n-1}}{(n-1)!} z(s)\,\mathrm{d}s + At^{n-1} + B.$$

If $n=2$, the solution of our problem is given directly by the above expression, where n is replaced by 2.

Therefore, for a general $n \geq 2$, by using the conditions $u(0) = \int_0^T u(s)\,\mathrm{d}H_1(s)$ and $u(T) = \int_0^T u(s)\,\mathrm{d}H_2(s)$, we deduce

$$\begin{cases} B = \displaystyle\int_0^T \left[-\int_0^s \frac{(s-\tau)^{n-1}}{(n-1)!} z(\tau)\,\mathrm{d}\tau + As^{n-1} + B\right]\mathrm{d}H_1(s), \\ -\displaystyle\int_0^T \frac{(T-s)^{n-1}}{(n-1)!} z(s)\,\mathrm{d}s + AT^{n-1} + B \\ \quad = \displaystyle\int_0^T \left[-\int_0^s \frac{(s-\tau)^{n-1}}{(n-1)!} z(\tau)\,\mathrm{d}\tau + As^{n-1} + B\right]\mathrm{d}H_2(s), \end{cases}$$

or

$$\begin{cases} A\int_0^T s^{n-1}\mathrm{d}H_1(s) + B\left(\int_0^T \mathrm{d}H_1(s) - 1\right) = \int_0^T \left(\int_0^s \frac{(s-\tau)^{n-1}}{(n-1)!} z(\tau)\,\mathrm{d}\tau\right)\mathrm{d}H_1(s), \\ A\left(T^{n-1} - \int_0^T s^{n-1}\mathrm{d}H_2(s)\right) + B\left(1 - \int_0^T \mathrm{d}H_2(s)\right) \\ = \int_0^T \frac{(T-s)^{n-1}}{(n-1)!} z(s)\,\mathrm{d}s - \int_0^T \left(\int_0^s \frac{(s-\tau)^{n-1}}{(n-1)!} z(\tau)\,\mathrm{d}\tau\right)\mathrm{d}H_2(s). \end{cases} \tag{2.50}$$

The above system with the unknowns A and B has the determinant

$$\Delta_1 = \left(1 - \int_0^T \mathrm{d}H_2(s)\right)\int_0^T s^{n-1}\mathrm{d}H_1(s) + \left(1 - \int_0^T \mathrm{d}H_1(s)\right)\left(T^{n-1} - \int_0^T s^{n-1}\mathrm{d}H_2(s)\right).$$

By using the assumptions of this lemma, we have $\Delta_1 \neq 0$. Hence, system (2.50) has a unique solution—namely,

$$\begin{aligned} A = \frac{1}{\Delta_1}\Bigg\{&\left(1 - \int_0^T \mathrm{d}H_2(s)\right)\frac{1}{(n-1)!}\int_0^T \left(\int_0^s (s-\tau)^{n-1} z(\tau)\,\mathrm{d}\tau\right)\mathrm{d}H_1(s) \\ &+ \left(1 - \int_0^T \mathrm{d}H_1(s)\right)\frac{1}{(n-1)!}\Bigg[\int_0^T (T-s)^{n-1} z(s)\,\mathrm{d}s \\ &- \int_0^T \left(\int_0^s (s-\tau)^{n-1} z(\tau)\,\mathrm{d}\tau\right)\mathrm{d}H_2(s)\Bigg]\Bigg\}, \\ B = \frac{1}{\Delta_1}\Bigg\{&\left(\int_0^T s^{n-1}\mathrm{d}H_1(s)\right)\frac{1}{(n-1)!}\Bigg[\int_0^T (T-s)^{n-1} z(s)\,\mathrm{d}s \\ &- \int_0^T \left(\int_0^s (s-\tau)^{n-1} z(\tau)\,\mathrm{d}\tau\right)\mathrm{d}H_2(s)\Bigg] - \left(T^{n-1} - \int_0^T s^{n-1}\mathrm{d}H_2(s)\right) \\ &\times \frac{1}{(n-1)!}\int_0^T \left(\int_0^s (s-\tau)^{n-1} z(\tau)\,\mathrm{d}\tau\right)\mathrm{d}H_1(s)\Bigg\}. \end{aligned}$$

Therefore, we obtain expression (2.49) for the solution $u(t)$ of problem (2.47)–(2.48). □

Lemma 2.5.2. *Under the assumptions of Lemma 2.5.1, the solution of problem (2.47)–(2.48) can be expressed as* $u(t) = \int_0^T G_1(t,s)z(s)\,\mathrm{d}s$, *where the Green's function* G_1 *is defined by*

$$\begin{aligned} G_1(t,s) = g_1(t,s) + \frac{1}{\Delta_1}\Bigg[&\left(T^{n-1} - t^{n-1}\right)\left(1 - \int_0^T \mathrm{d}H_2(\tau)\right) \\ &+ \int_0^T \left(T^{n-1} - \tau^{n-1}\right)\mathrm{d}H_2(\tau)\Bigg]\int_0^T g_1(\tau,s)\,\mathrm{d}H_1(\tau) \end{aligned}$$

$$+\frac{1}{\Delta_1}\left[t^{n-1}\left(1-\int_0^T \mathrm{d}H_1(\tau)\right)\right.$$
$$\left.+\int_0^T \tau^{n-1}\mathrm{d}H_1(\tau)\right]\int_0^T g_1(\tau,s)\,\mathrm{d}H_2(\tau), \tag{2.51}$$

for all $(t,s)\in[0,T]\times[0,T]$, *and*

$$g_1(t,s)=\frac{1}{(n-1)!T^{n-1}}\begin{cases} t^{n-1}(T-s)^{n-1}-T^{n-1}(t-s)^{n-1}, & 0\le s\le t\le T,\\ t^{n-1}(T-s)^{n-1}, & 0\le t\le s\le T.\end{cases} \tag{2.52}$$

Proof. By Lemma 2.5.1 and relation (2.49), we conclude

$$u(t)=\frac{1}{(n-1)!T^{n-1}}\left\{\int_0^t\left[t^{n-1}(T-s)^{n-1}-T^{n-1}(t-s)^{n-1}\right]z(s)\,\mathrm{d}s\right.$$
$$+\int_t^T t^{n-1}(T-s)^{n-1}z(s)\,\mathrm{d}s-\int_0^T t^{n-1}(T-s)^{n-1}z(s)\,\mathrm{d}s$$
$$+\frac{T^{n-1}t^{n-1}}{\Delta_1}\left\{\left(1-\int_0^T \mathrm{d}H_2(s)\right)\int_0^T\left(\int_0^s (s-\tau)^{n-1}z(\tau)\,\mathrm{d}\tau\right)\mathrm{d}H_1(s)\right.$$
$$+\left(1-\int_0^T \mathrm{d}H_1(s)\right)\left[\int_0^T (T-s)^{n-1}z(s)\,\mathrm{d}s\right.$$
$$\left.\left.-\int_0^T\left(\int_0^s (s-\tau)^{n-1}z(\tau)\,\mathrm{d}\tau\right)\mathrm{d}H_2(s)\right]\right\}$$
$$+\frac{T^{n-1}}{\Delta_1}\left\{\left(\int_0^T s^{n-1}\mathrm{d}H_1(s)\right)\left[\int_0^T (T-s)^{n-1}z(s)\,\mathrm{d}s\right.\right.$$
$$\left.-\int_0^T\left(\int_0^s (s-\tau)^{n-1}z(\tau)\,\mathrm{d}\tau\right)\mathrm{d}H_2(s)\right]-\left(T^{n-1}-\int_0^T s^{n-1}\mathrm{d}H_2(s)\right)$$
$$\left.\left.\times\int_0^T\left(\int_0^s (s-\tau)^{n-1}z(\tau)\,\mathrm{d}\tau\right)\mathrm{d}H_1(s)\right\}\right\}$$
$$=\frac{1}{(n-1)!T^{n-1}}\left\{\int_0^t\left[t^{n-1}(T-s)^{n-1}-T^{n-1}(t-s)^{n-1}\right]z(s)\,\mathrm{d}s\right.$$
$$+\int_t^T t^{n-1}(T-s)^{n-1}z(s)\,\mathrm{d}s-\frac{1}{\Delta_1}\left[\left(1-\int_0^T \mathrm{d}H_2(s)\right)\left(\int_0^T s^{n-1}\mathrm{d}H_1(s)\right.\right.$$
$$\left.\left.+\left(1-\int_0^T \mathrm{d}H_1(s)\right)\left(T^{n-1}-\int_0^T s^{n-1}\mathrm{d}H_2(s)\right)\right]\int_0^T t^{n-1}(T-s)^{n-1}z(s)\,\mathrm{d}s$$
$$+\frac{T^{n-1}t^{n-1}}{\Delta_1}\left\{\left(1-\int_0^T \mathrm{d}H_2(s)\right)\int_0^T\left(\int_0^s (s-\tau)^{n-1}z(\tau)\,\mathrm{d}\tau\right)\mathrm{d}H_1(s)\right.$$
$$+\left(1-\int_0^T \mathrm{d}H_1(s)\right)\left[\int_0^T (T-s)^{n-1}z(s)\,\mathrm{d}s\right.$$

$$
\begin{aligned}
&- \int_0^T \left(\int_0^s (s-\tau)^{n-1} z(\tau)\, \mathrm{d}\tau \right) \mathrm{d}H_2(s) \Big] \Big\} \\
&+ \frac{T^{n-1}}{\Delta_1} \Big\{ \left(\int_0^T s^{n-1} \mathrm{d}H_1(s) \right) \Big[\int_0^T (T-s)^{n-1} z(s)\, \mathrm{d}s \\
&- \int_0^T \left(\int_0^s (s-\tau)^{n-1} z(\tau)\, \mathrm{d}\tau \right) \mathrm{d}H_2(s) \Big] - \left(T^{n-1} - \int_0^T s^{n-1} \mathrm{d}H_2(s) \right) \\
&\times \int_0^T \left(\int_0^s (s-\tau)^{n-1} z(\tau)\, \mathrm{d}\tau \right) \mathrm{d}H_1(s) \Big\} \Big\}.
\end{aligned}
$$

Therefore, we deduce

$$
\begin{aligned}
u(t) = {}& \frac{1}{(n-1)!T^{n-1}} \Big\{ \int_0^t [t^{n-1}(T-s)^{n-1} - T^{n-1}(t-s)^{n-1}] z(s)\, \mathrm{d}s \\
&+ \int_t^T t^{n-1}(T-s)^{n-1} z(s)\, \mathrm{d}s + \frac{t^{n-1}}{\Delta_1} \Big[- \left(1 - \int_0^T \mathrm{d}H_2(\tau) \right) \left(\int_0^T s^{n-1} \mathrm{d}H_1(s) \right) \\
&\times \left(\int_0^T (T-\tau)^{n-1} z(\tau)\, \mathrm{d}\tau \right) - \left(1 - \int_0^T \mathrm{d}H_1(\tau) \right) \left(\int_0^T T^{n-1}(T-s)^{n-1} z(s)\, \mathrm{d}s \right) \\
&+ \left(1 - \int_0^T \mathrm{d}H_1(\tau) \right) \left(\int_0^T s^{n-1} \mathrm{d}H_2(s) \right) \left(\int_0^T (T-\tau)^{n-1} z(\tau)\, \mathrm{d}\tau \right) \\
&+ T^{n-1} \left(1 - \int_0^T \mathrm{d}H_2(\tau) \right) \int_0^T \left(\int_0^s (s-\tau)^{n-1} z(\tau)\, \mathrm{d}\tau \right) \mathrm{d}H_1(s) \\
&+ T^{n-1} \left(1 - \int_0^T \mathrm{d}H_1(\tau) \right) \int_0^T (T-s)^{n-1} z(s)\, \mathrm{d}s \\
&- T^{n-1} \left(1 - \int_0^T \mathrm{d}H_1(\tau) \right) \int_0^T \left(\int_0^s (s-\tau)^{n-1} z(\tau)\, \mathrm{d}\tau \right) \mathrm{d}H_2(s) \Big] \\
&+ \frac{T^{n-1}}{\Delta_1} \Big[\left(\int_0^T s^{n-1} \mathrm{d}H_1(s) \right) \int_0^T (T-\tau)^{n-1} z(\tau)\, \mathrm{d}\tau - \left(\int_0^T s^{n-1} \mathrm{d}H_1(s) \right) \\
&\times \int_0^T \left(\int_0^s (s-\tau)^{n-1} z(\tau)\, \mathrm{d}\tau \right) \mathrm{d}H_2(s) - T^{n-1} \int_0^T \left(\int_0^s (s-\tau)^{n-1} z(\tau)\, \mathrm{d}\tau \right) \mathrm{d}H_1(s) \\
&+ \left(\int_0^T s^{n-1} \mathrm{d}H_2(s) \right) \left(\int_0^T \left(\int_0^s (s-\tau)^{n-1} z(\tau)\, \mathrm{d}\tau \right) \mathrm{d}H_1(s) \right) \Big] \Big\} \\
= {}& \frac{1}{(n-1)!T^{n-1}} \Big\{ \int_0^t [t^{n-1}(T-s)^{n-1} - T^{n-1}(t-s)^{n-1}] z(s)\, \mathrm{d}s \\
&+ \int_t^T t^{n-1}(T-s)^{n-1} z(s)\, \mathrm{d}s + \frac{t^{n-1}}{\Delta_1} \Big\{ \left(1 - \int_0^T \mathrm{d}H_2(\tau) \right) \\
&\times \int_0^T \left(\int_0^s T^{n-1}(s-\tau)^{n-1} z(\tau)\, \mathrm{d}\tau - \int_0^T s^{n-1}(T-\tau)^{n-1} z(\tau)\, \mathrm{d}\tau \right) \mathrm{d}H_1(s)
\end{aligned}
$$

$$
\begin{aligned}
&+\left(1-\int_0^T \mathrm{d}H_1(\tau)\right)\left[\int_0^T\left(\int_0^T s^{n-1}(T-\tau)^{n-1}z(\tau)\,\mathrm{d}\tau\right)\,\mathrm{d}H_2(s)\right.\\
&\left.\left.-\ T^{n-1}\int_0^T\left(\int_0^s (s-\tau)^{n-1}z(\tau)\,\mathrm{d}\tau\right)\,\mathrm{d}H_2(s)\right]\right\}\\
&+\frac{T^{n-1}}{\Delta_1}\left[\int_0^T s^{n-1}\left(\int_0^T (T-\tau)^{n-1}z(\tau)\,\mathrm{d}\tau\right)\,\mathrm{d}H_1(s)-\left(\int_0^T \tau^{n-1}\mathrm{d}H_1(\tau)\right)\right.\\
&\times\left(\int_0^T\left(\int_0^s (s-\tau)^{n-1}z(\tau)\,\mathrm{d}\tau\right)\,\mathrm{d}H_2(s)\right)\\
&-\int_0^T\left(\int_0^s T^{n-1}(s-\tau)^{n-1}z(\tau)\,\mathrm{d}\tau\right)\,\mathrm{d}H_1(s)\\
&\left.\left.+\left(\int_0^T \tau^{n-1}\mathrm{d}H_2(\tau)\right)\left(\int_0^T\left(\int_0^s (s-\tau)^{n-1}z(\tau)\,\mathrm{d}\tau\right)\,\mathrm{d}H_1(s)\right)\right]\right\}.
\end{aligned}
$$

Hence, we obtain

$$
\begin{aligned}
u(t)=&\frac{1}{(n-1)!T^{n-1}}\left\{\int_0^t [t^{n-1}(T-s)^{n-1}-T^{n-1}(t-s)^{n-1}]z(s)\,\mathrm{d}s\right.\\
&+\int_t^T t^{n-1}(T-s)^{n-1}z(s)\,\mathrm{d}s+\frac{t^{n-1}}{\Delta_1}\left\{\left(1-\int_0^T \mathrm{d}H_2(\tau)\right)\int_0^T\left[\int_0^s\left(T^{n-1}(s-\tau)^{n-1}\right.\right.\right.\\
&\left.\left.-s^{n-1}(T-\tau)^{n-1}\right)z(\tau)\,\mathrm{d}\tau-\int_s^T s^{n-1}(T-\tau)^{n-1}z(\tau)\,\mathrm{d}\tau\right]\mathrm{d}H_1(s)\\
&+\left(1-\int_0^T \mathrm{d}H_1(\tau)\right)\int_0^T\left[\int_0^s (s^{n-1}(T-\tau)^{n-1}-T^{n-1}(s-\tau)^{n-1})z(\tau)\,\mathrm{d}\tau\right.\\
&\left.\left.+\int_s^T s^{n-1}(T-\tau)^{n-1}z(\tau)\,\mathrm{d}\tau\right]\mathrm{d}H_2(s)\right\}+\frac{T^{n-1}}{\Delta_1}\left[\int_0^T\left(\int_0^s\left(s^{n-1}(T-\tau)^{n-1}\right.\right.\right.\\
&\left.\left.\left.-T^{n-1}(s-\tau)^{n-1}\right)z(\tau)\,\mathrm{d}\tau+\int_s^T s^{n-1}(T-\tau)^{n-1}z(\tau)\,\mathrm{d}\tau\right)\,\mathrm{d}H_1(s)\right]\\
&-\frac{1}{\Delta_1}\left(\int_0^T \tau^{n-1}\mathrm{d}H_1(\tau)\right)\left(\int_0^T\left(\int_0^s T^{n-1}(s-\tau)^{n-1}z(\tau)\,\mathrm{d}\tau\right)\,\mathrm{d}H_2(s)\right)\\
&+\frac{1}{\Delta_1}\left(\int_0^T \tau^{n-1}\mathrm{d}H_2(\tau)\right)\left(\int_0^T\left(\int_0^s T^{n-1}(s-\tau)^{n-1}z(\tau)\,\mathrm{d}\tau\right)\,\mathrm{d}H_1(s)\right)\\
&+\frac{1}{\Delta_1}\left(\int_0^T \tau^{n-1}\mathrm{d}H_1(\tau)\right)\left(\int_0^T\left(\int_0^T s^{n-1}(T-\tau)^{n-1}z(\tau)\,\mathrm{d}\tau\right)\,\mathrm{d}H_2(s)\right)\\
&\left.-\frac{1}{\Delta_1}\left(\int_0^T \tau^{n-1}\mathrm{d}H_2(\tau)\right)\left(\int_0^T\left(\int_0^T s^{n-1}(T-\tau)^{n-1}z(\tau)\,\mathrm{d}\tau\right)\,\mathrm{d}H_1(s)\right)\right\}\\
=&\frac{1}{(n-1)!T^{n-1}}\left\{\int_0^t [t^{n-1}(T-s)^{n-1}-T^{n-1}(t-s)^{n-1}]z(s)\,\mathrm{d}s\right.
\end{aligned}
$$

$$+\int_t^T t^{n-1}(T-s)^{n-1}z(s)\,\mathrm{d}s+\frac{t^{n-1}}{\Delta_1}\left\{-\left(1-\int_0^T \mathrm{d}H_2(\tau)\right)\int_0^T\left[\int_0^s\left(s^{n-1}(T-\tau)^{n-1}\right.\right.\right.$$
$$\left.-T^{n-1}(s-\tau)^{n-1}\right)z(\tau)\,\mathrm{d}\tau+\left.\int_s^T s^{n-1}(T-\tau)^{n-1}z(\tau)\,\mathrm{d}\tau\right]\mathrm{d}H_1(s)$$
$$+\left(1-\int_0^T \mathrm{d}H_1(\tau)\right)\int_0^T\left[\int_0^s(s^{n-1}(T-\tau)^{n-1}-T^{n-1}(s-\tau)^{n-1})z(\tau)\,\mathrm{d}\tau\right.$$
$$\left.\left.+\int_s^T s^{n-1}(T-\tau)^{n-1}z(\tau)\,\mathrm{d}\tau\right]\mathrm{d}H_2(s)\right\}+\frac{T^{n-1}}{\Delta_1}\left[\int_0^T\left(\int_0^s\left(s^{n-1}(T-\tau)^{n-1}\right.\right.\right.$$
$$\left.\left.\left.-T^{n-1}(s-\tau)^{n-1}\right)z(\tau)\,\mathrm{d}\tau+\int_s^T s^{n-1}(T-\tau)^{n-1}z(\tau)\,\mathrm{d}\tau\right)\mathrm{d}H_1(s)\right]$$
$$+\frac{1}{\Delta_1}\left(\int_0^T \tau^{n-1}\mathrm{d}H_1(\tau)\right)\left[\int_0^T\left(\int_0^s\left[s^{n-1}(T-\tau)^{n-1}\right.\right.\right.$$
$$\left.\left.\left.-\ T^{n-1}(s-\tau)^{n-1}\right]z(\tau)\,\mathrm{d}\tau+\int_s^T s^{n-1}(T-\tau)^{n-1}z(\tau)\,\mathrm{d}\tau\right)\mathrm{d}H_2(s)\right]$$
$$-\frac{1}{\Delta_1}\left(\int_0^T \tau^{n-1}\mathrm{d}H_2(\tau)\right)\left[\int_0^T\left(\int_0^s\left[s^{n-1}(T-\tau)^{n-1}\right.\right.\right.$$
$$\left.\left.\left.-\ T^{n-1}(s-\tau)^{n-1}\right]z(\tau)\,\mathrm{d}\tau+\int_s^T s^{n-1}(T-\tau)^{n-1}z(\tau)\,\mathrm{d}\tau\right)\mathrm{d}H_1(s)\right]\Bigg\}.$$

Then the solution u of problem (2.47)–(2.48) is

$$u(t)=\int_0^T g_1(t,s)z(s)\,\mathrm{d}s+\frac{t^{n-1}}{\Delta_1}\left[-\left(1-\int_0^T \mathrm{d}H_2(\tau)\right)\right.$$
$$\times\int_0^T\left(\int_0^T g_1(s,\tau)z(\tau)\right)\mathrm{d}H_1(s)+\left(1-\int_0^T \mathrm{d}H_1(\tau)\right)$$
$$\left.\times\int_0^T\left(\int_0^T g_1(s,\tau)z(\tau)\,\mathrm{d}\tau\right)\mathrm{d}H_2(s)\right]$$
$$+\frac{T^{n-1}}{\Delta_1}\int_0^T\left(\int_0^T g_1(s,\tau)z(\tau)\,\mathrm{d}\tau\right)\mathrm{d}H_1(s)$$
$$+\frac{1}{\Delta_1}\left(\int_0^T \tau^{n-1}\mathrm{d}H_1(\tau)\right)\int_0^T\left(\int_0^T g_1(s,\tau)z(\tau)\,\mathrm{d}\tau\right)\mathrm{d}H_2(s)$$
$$-\frac{1}{\Delta_1}\left(\int_0^T \tau^{n-1}\mathrm{d}H_2(\tau)\right)\int_0^T\left(\int_0^T g_1(s,\tau)z(\tau)\,\mathrm{d}\tau\right)\mathrm{d}H_1(s)$$
$$=\int_0^T g_1(t,s)z(s)\,\mathrm{d}s+\frac{t^{n-1}}{\Delta_1}\left[-\left(1-\int_0^T \mathrm{d}H_2(\tau)\right)\right.$$
$$\times\int_0^T\left(\int_0^T g_1(\tau,s)\,\mathrm{d}H_1(\tau)\right)z(s)\,\mathrm{d}s+\left(1-\int_0^T \mathrm{d}H_1(\tau)\right)$$

$$
\begin{aligned}
&\times \int_0^T \left(\int_0^T g_1(\tau,s)\, \mathrm{d}H_2(\tau) \right) z(s)\, \mathrm{d}s \Bigg] \\
&+ \frac{T^{n-1}}{\Delta_1} \int_0^T \left(\int_0^T g_1(\tau,s)\, \mathrm{d}H_1(\tau) \right) z(s)\, \mathrm{d}s \\
&+ \frac{1}{\Delta_1} \left(\int_0^T \tau^{n-1} \mathrm{d}H_1(\tau) \right) \int_0^T \left(\int_0^T g_1(\tau,s)\, \mathrm{d}H_2(\tau) \right) z(s)\, \mathrm{d}s \\
&- \frac{1}{\Delta_1} \left(\int_0^T \tau^{n-1} \mathrm{d}H_2(\tau) \right) \int_0^T \left(\int_0^T g_1(\tau,s)\, \mathrm{d}H_1(\tau) \right) z(s)\, \mathrm{d}s,
\end{aligned}
$$

where g_1 is given in (2.52).

Therefore, we conclude

$$
\begin{aligned}
u(t) = {} & \int_0^T g_1(t,s) z(s)\, \mathrm{d}s + \frac{1}{\Delta_1} \left[-t^{n-1} \left(1 - \int_0^T \mathrm{d}H_2(\tau) \right) \right. \\
& \left. + T^{n-1} - \int_0^T \tau^{n-1} \mathrm{d}H_2(\tau) \right] \int_0^T \left(\int_0^T g_1(\tau,s)\, \mathrm{d}H_1(\tau) \right) z(s)\, \mathrm{d}s \\
& + \frac{1}{\Delta_1} \left[t^{n-1} \left(1 - \int_0^T \mathrm{d}H_1(\tau) \right) + \int_0^T \tau^{n-1} \mathrm{d}H_1(\tau) \right] \\
& \times \int_0^T \left(\int_0^T g_1(\tau,s)\, \mathrm{d}H_2(\tau) \right) z(s)\, \mathrm{d}s \\
= {} & \int_0^T \left\{ g_1(t,s) + \frac{1}{\Delta_1} \left[\left(T^{n-1} - t^{n-1} \right) \left(1 - \int_0^T \mathrm{d}H_2(\tau) \right) \right. \right. \\
& \left. + \int_0^T \left(T^{n-1} - \tau^{n-1} \right) \mathrm{d}H_2(\tau) \right] \int_0^T g_1(\tau,s)\, \mathrm{d}H_1(\tau) \\
& + \frac{1}{\Delta_1} \left[t^{n-1} \left(1 - \int_0^T \mathrm{d}H_1(\tau) \right) + \int_0^T \tau^{n-1} \mathrm{d}H_1(\tau) \right] \\
& \left. \times \int_0^T g_1(\tau,s)\, \mathrm{d}H_2(\tau) \right\} z(s)\, \mathrm{d}s = \int_0^T G_1(t,s) z(s)\, \mathrm{d}s
\end{aligned}
$$

where G_1 is given in (2.51). □

Using arguments similar to those used in the proof of Lemma 2.2 from Xie et al. (2008), we obtain the following lemma:

Lemma 2.5.3. *For any $n \geq 2$, the function g_1 given by (2.52) has the following properties:*

(a) *$g_1\colon [0,T] \times [0,T] \to \mathbb{R}_+$ is a continuous function, and $g_1(t,s) > 0$ for all $(t,s) \in (0,T) \times (0,T)$.*

(b) *$g_1(t,s) \leq h_1(s)$ for all $(t,s) \in [0,T] \times [0,T]$, where $h_1(s) = \frac{s(T-s)^{n-1}}{(n-2)!T}$.*

(c) *$g_1(t,s) \geq k_1(t) h_1(s)$ for all $(t,s) \in [0,T] \times [0,T]$, where*

$$
k_1(t) = \min \left\{ \frac{(T-t)t^{n-2}}{(n-1)T^{n-1}}, \frac{t^{n-1}}{(n-1)T^{n-1}} \right\} = \begin{cases} \dfrac{t^{n-1}}{(n-1)T^{n-1}}, & 0 \leq t \leq T/2, \\ \dfrac{(T-t)t^{n-2}}{(n-1)T^{n-1}}, & T/2 \leq t \leq T. \end{cases}
$$

Lemma 2.5.4. *Assume that H_1, H_2: $[0,T] \to \mathbb{R}$ are nondecreasing functions, $H_1(T) - H_1(0) < 1$, and $H_2(T) - H_2(0) < 1$. Then the Green's function G_1 of problem (2.47)–(2.48) given by (2.51) is continuous on $[0,T] \times [0,T]$ and satisfies $G_1(t,s) \geq 0$ for all $(t,s) \in [0,T] \times [0,T]$, $G_1(t,s) > 0$ for all $(t,s) \in (0,T) \times (0,T)$. Moreover, if $z \in C(0,T) \cap L^1(0,T)$ satisfies $z(t) \geq 0$ for all $t \in (0,T)$, then the solution u of problem (2.47)–(2.48) satisfies $u(t) \geq 0$ for all $t \in [0,T]$.*

Proof. By using the assumptions of this lemma and Lemmas 2.5.2 and 2.5.3, we obtain $\Delta_1 > 0$, $G_1(t,s) \geq 0$ for all $(t,s) \in [0,T] \times [0,T]$, $G_1(t,s) > 0$ for all $(t,s) \in (0,T) \times (0,T)$, and so $u(t) \geq 0$ for all $t \in [0,T]$. □

Lemma 2.5.5. *Assume that H_1, H_2: $[0,T] \to \mathbb{R}$ are nondecreasing functions, $H_1(T) - H_1(0) < 1$, and $H_2(T) - H_2(0) < 1$. Then the Green's function G_1 of problem (2.47)–(2.48) satisfies the following inequalities:*

(a) *$G_1(t,s) \leq J_1(s), \forall (t,s) \in [0,T] \times [0,T]$, where $J_1(s) = \tau_1 h_1(s), \forall s \in [0,T]$, and*

$$\tau_1 = 1 + \frac{1}{\Delta_1}\left[T^{n-1}(1 - H_2(T) + H_2(0)) + \int_0^T (T^{n-1} - \tau^{n-1})\, dH_2(\tau)\right] \times (H_1(T) - H_1(0)) + \frac{1}{\Delta_1}\left[T^{n-1}(1 - H_1(T) + H_1(0)) + \int_0^T \tau^{n-1} dH_1(\tau)\right] \times (H_2(T) - H_2(0)). \tag{2.53}$$

(b) *$G_1(t,s) \geq \gamma_1(t) J_1(s), \forall (t,s) \in [0,T] \times [0,T]$, where*

$$\gamma_1(t) = \frac{1}{\tau_1}\left\{k_1(t) + \frac{1}{\Delta_1}\left[(T^{n-1} - t^{n-1})(1 - H_2(T) + H_2(0)) + \int_0^T (T^{n-1} - \tau^{n-1})\, dH_2(\tau)\right]\int_0^T k_1(\tau)\, dH_1(\tau) + \frac{1}{\Delta_1}\left[t^{n-1}(1 - H_1(T) + H_1(0)) + \int_0^T \tau^{n-1} dH_1(\tau)\right] \times \int_0^T k_1(\tau)\, dH_2(\tau)\right\}, \quad \forall t \in [0,T]. \tag{2.54}$$

Proof. **(a)** We have

$$G_1(t,s) \leq h_1(s) + \frac{1}{\Delta_1}\left[T^{n-1}\left(1 - \int_0^T dH_2(\tau)\right) + \int_0^T \left(T^{n-1} - \tau^{n-1}\right) dH_2(\tau)\right] \times \int_0^T h_1(s)\, dH_1(\tau) + \frac{1}{\Delta_1}\left[T^{n-1}\left(1 - \int_0^T dH_1(\tau)\right) + \int_0^T \tau^{n-1} dH_1(\tau)\right] \times \int_0^T h_1(s)\, dH_2(\tau) = \tau_1 h_1(s) = J_1(s), \quad \forall (t,s) \in [0,T] \times [0,T],$$

where τ_1 is given in (2.53).

(b) For the second inequality, we obtain

$$G_1(t,s) \geq k_1(t) h_1(s) + \frac{1}{\Delta_1}\left[\left(T^{n-1} - t^{n-1}\right)\left(1 - \int_0^T dH_2(\tau)\right) + \int_0^T \left(T^{n-1} - \tau^{n-1}\right) dH_2(\tau)\right]\int_0^T k_1(\tau) h_1(s)\, dH_1(\tau)$$

$$+\frac{1}{\Delta_1}\left[t^{n-1}\left(1-\int_0^T dH_1(\tau)\right)+\int_0^T \tau^{n-1}dH_1(\tau)\right]\int_0^T k_1(\tau)h_1(s)\,dH_2(\tau)$$

$$=\frac{1}{\tau_1}(\tau_1 h_1(s))\left\{k_1(t)+\frac{1}{\Delta_1}\left[\left(T^{n-1}-t^{n-1}\right)\left(1-\int_0^T dH_2(\tau)\right)\right.\right.$$

$$+\int_0^T\left(T^{n-1}-\tau^{n-1}\right)\,dH_2(\tau)\Bigg]\int_0^T k_1(\tau)\,dH_1(\tau)+\frac{1}{\Delta_1}\left[t^{n-1}\left(1-\int_0^T dH_1(\tau)\right)\right.$$

$$\left.\left.+\int_0^T \tau^{n-1}dH_1(\tau)\right]\int_0^T k_1(\tau)\,dH_2(\tau)\right\}=\gamma_1(t)J_1(s),\quad \forall\,(t,s)\in[0,T]\times[0,T],$$

where $\gamma_1(t)$ is defined in (2.54). □

Lemma 2.5.6. *Assume that H_1, H_2: $[0,T]\to\mathbb{R}$ are nondecreasing functions, $H_1(T)-H_1(0)<1$, $H_2(T)-H_2(0)<1$, and $z\in C(0,T)\cap L^1(0,T)$, $z(t)\geq 0$ for all $t\in(0,T)$. Then the solution $u(t), t\in[0,T]$ of problem* (2.47)–(2.48) *satisfies the inequality $u(t)\geq\gamma_1(t)\sup_{t'\in[0,T]}u(t')$ for all $t\in[0,T]$.*

Proof. For $t\in[0,T]$, we deduce

$$u(t)=\int_0^T G_1(t,s)z(s)\,ds\geq\int_0^T \gamma_1(t)J_1(s)z(s)\,ds=\gamma_1(t)\int_0^T J_1(s)z(s)\,ds$$

$$\geq\gamma_1(t)\int_0^T G_1(t',s)z(s)\,ds=\gamma_1(t)u(t'),\quad \forall t'\in[0,T].$$

Therefore, we conclude that $u(t)\geq\gamma_1(t)\sup_{t'\in[0,T]}u(t')$ for all $t\in[0,T]$. □

We can also formulate results similar to those of Lemmas 2.5.1–2.5.6 for the ordinary differential equation

$$v^{(m)}(t)+\tilde{z}(t)=0,\quad 0<t<T, \tag{2.55}$$

with the integral boundary conditions

$$v(0)=\int_0^T v(s)\,dK_1(s),\quad v'(0)=\cdots=v^{(m-2)}(0)=0,\quad v(T)=\int_0^T v(s)\,dK_2(s), \tag{2.56}$$

where $m\in\mathbb{N}$, $m\geq 2$, K_1, K_2: $[0,T]\to\mathbb{R}$ are nondecreasing functions and $\tilde{z}\in C(0,T)\cap L^1(0,T)$. In the case $m=2$, the boundary conditions have the form $v(0)=\int_0^T v(s)\,dK_1(s)$, $v(T)=\int_0^T v(s)\,dK_2(s)$. We denote by $\Delta_2, g_2, G_2, h_2, k_2, \tau_2, J_2$, and γ_2 the corresponding constants and functions for problem (2.55)–(2.56) defined in a similar manner as $\Delta_1, g_1, G_1, h_1, k_1$, τ_1, J_1, and γ_1, respectively.

2.5.3 Main result

In this section, we investigate the existence of positive solutions for our problem $(\tilde{\mathrm{S}})$–$(\widetilde{\mathrm{BC}})$. We present now the assumptions that we shall use in the sequel:

$(\widehat{\mathrm{H1}})$ H_1, H_2, K_1, K_2: $[0,T]\to\mathbb{R}$ are nondecreasing functions, $H_1(T)-H_1(0)<1$, $H_2(T)-H_2(0)<1$, $K_1(T)-K_1(0)<1$, and $K_2(T)-K_2(0)<1$.

($\widehat{\rm H2}$) The functions $f, g \in C([0,T] \times [0,\infty) \times [0,\infty), (-\infty,+\infty))$, and there exist functions $p_1, p_2 \in C([0,T], (0,\infty))$ such that $f(t,u,v) \geq -p_1(t)$ and $g(t,u,v) \geq -p_2(t)$ for any $t \in [0,T]$ and $u, v \in [0,\infty)$.

($\widehat{\rm H3}$) $f(t,0,0) > 0$, $g(t,0,0) > 0$ for all $t \in [0,T]$.

We consider the system of nonlinear ordinary differential equations

$$\begin{cases} x^{(n)}(t) + \lambda(f(t,[x(t)-q_1(t)]^*,[y(t)-q_2(t)]^*) + p_1(t)) = 0, & 0 < t < T, \\ y^{(m)}(t) + \mu(g(t,[x(t)-q_1(t)]^*,[y(t)-q_2(t)]^*) + p_2(t)) = 0, & 0 < t < T, \end{cases} \tag{2.57}$$

with the integral boundary conditions

$$\begin{cases} x(0) = \int_0^T x(s)\,dH_1(s), \quad x'(0) = \cdots = x^{(n-2)}(0) = 0, \quad x(T) = \int_0^T x(s)\,dH_2(s), \\ y(0) = \int_0^T y(s)\,dK_1(s), \quad y'(0) = \cdots = y^{(m-2)}(0) = 0, \quad y(T) = \int_0^T y(s)\,dK_2(s), \end{cases} \tag{2.58}$$

where

$$z(t)^* = \begin{cases} z(t), & z(t) \geq 0, \\ 0, & z(t) < 0. \end{cases}$$

Here q_1 and q_2 are given by $q_1(t) = \lambda \int_0^T G_1(t,s)p_1(s)\,ds$ and $q_2(t) = \mu \int_0^T G_2(t,s)$ $p_2(s)\,ds$—that is, they are the solutions of the problems

$$\begin{cases} q_1^{(n)}(t) + \lambda p_1(t) = 0, \quad t \in (0,T), \\ q_1(0) = \int_0^T q_1(s)\,dH_1(s), \quad q_1'(0) = \cdots = q_1^{(n-2)}(0) = 0, \quad q_1(T) = \int_0^T q_1(s)\,dH_2(s), \end{cases} \tag{2.59}$$

and

$$\begin{cases} q_2^{(m)}(t) + \mu p_2(t) = 0, \quad t \in (0,T), \\ q_2(0) = \int_0^T q_2(s)\,dK_1(s), \quad q_2'(0) = \cdots = q_2^{(m-2)}(0) = 0, \quad q_2(T) = \int_0^T q_2(s)\,dK_2(s), \end{cases} \tag{2.60}$$

respectively. If $n = 2$ or $m = 2$, then the above conditions do not contain the conditions on the derivatives at the point 0. By ($\widehat{\rm H1}$), ($\widehat{\rm H2}$), and Lemma 2.5.4, we have $q_1(t) \geq 0$, $q_2(t) \geq 0$ for all $t \in [0,T]$, and $q_1(t) > 0$, $q_2(t) > 0$ for all $t \in (0,T)$.

We shall prove that there exists a solution (x,y) for the boundary value problem (2.57)–(2.58) with $x(t) \geq q_1(t)$ and $y(t) \geq q_2(t)$ for all $t \in [0,T]$. In this case, the pair of functions (u, v) with $u(t) = x(t) - q_1(t)$ and $v(t) = y(t) - q_2(t)$, $t \in [0,T]$ represents a positive solution of the boundary value problem ($\tilde{\rm S}$)–($\widetilde{\rm BC}$). By (2.57)–(2.58) and (2.59)–(2.60), we have

$$u^{(n)}(t)=x^{(n)}(t)-q_1^{(n)}(t)=-\lambda f(t,[x(t)-q_1(t)]^*,[y(t)-q_2(t)]^*)$$
$$-\lambda p_1(t)+\lambda p_1(t)=-\lambda f(t,u(t),v(t)),\quad \forall t\in(0,T),$$
$$v^{(m)}(t)=y^{(m)}(t)-q_2^{(m)}(t)=-\mu g(t,[x(t)-q_1(t)]^*,[y(t)-q_2(t)]^*)$$
$$-\mu p_2(t)+\mu p_2(t)=-\mu g(t,u(t),v(t)),\quad \forall t\in(0,T),$$

and

$$u(0)=x(0)-q_1(0)=\int_0^T u(s)\,\mathrm{d}H_1(s),$$
$$u'(0)=x'(0)-q_1'(0)=0,\ldots,u^{(n-2)}(0)=x^{(n-2)}(0)-q_1^{(n-2)}(0)=0,$$
$$u(T)=x(T)-q_1(T)=\int_0^T u(s)\,\mathrm{d}H_2(s),$$
$$v(0)=y(0)-q_2(0)=\int_0^T v(s)\,\mathrm{d}K_1(s),$$
$$v'(0)=y'(0)-q_2'(0)=0,\ldots,v^{(m-2)}(0)=y^{(m-2)}(0)-q_2^{(m-2)}(0)=0,$$
$$v(T)=y(T)-q_2(T)=\int_0^T v(s)\,\mathrm{d}K_2(s).$$

Therefore, in what follows, we shall investigate the boundary value problem (2.57)–(2.58).

By using Lemma 2.5.2, we find problem (2.57)–(2.58) is equivalent to the system

$$\begin{cases} x(t)=\lambda\displaystyle\int_0^T G_1(t,s)\big(f(s,[x(s)-q_1(s)]^*,[y(s)-q_2(s)]^*)+p_1(s)\big)\,\mathrm{d}s, & t\in[0,T],\\ y(t)=\mu\displaystyle\int_0^T G_2(t,s)\big(g(s,[x(s)-q_1(s)]^*,[y(s)-q_2(s)]^*)+p_2(s)\big)\,\mathrm{d}s, & t\in[0,T]. \end{cases}$$

We consider the Banach space $X=C([0,T])$ with the supremum norm $\|\cdot\|$ and the Banach space $Y=X\times X$ with the norm $\|(x,y)\|_Y=\|x\|+\|y\|$. We also define the cones

$$\tilde{P}_1=\{x\in X,\quad x(t)\geq\gamma_1(t)\|x\|,\quad \forall t\in[0,T]\}\subset X,$$
$$\tilde{P}_2=\{y\in X,\quad y(t)\geq\gamma_2(t)\|y\|,\quad \forall t\in[0,T]\}\subset X,$$

and $\tilde{P}=\tilde{P}_1\times\tilde{P}_2\subset Y$.

For $\lambda,\mu>0$, we define the operator $\tilde{Q}\colon\tilde{P}\to Y$ by $\tilde{Q}(x,y)=(\tilde{Q}_1(x,y),\tilde{Q}_2(x,y))$ with

$$\tilde{Q}_1(x,y)(t)=\lambda\int_0^T G_1(t,s)\big(f(s,[x(s)-q_1(s)]^*,[y(s)-q_2(s)]^*)+p_1(s)\big)\,\mathrm{d}s,\ 0\leq t\leq T,$$
$$\tilde{Q}_2(x,y)(t)=\mu\int_0^T G_2(t,s)\big(g(s,[x(s)-q_1(s)]^*,[y(s)-q_2(s)]^*)+p_2(s)\big)\,\mathrm{d}s,\ 0\leq t\leq T.$$

Lemma 2.5.7. *If* $(\widehat{H1})$ *and* $(\widehat{H2})$ *hold, then the operator* $\tilde{Q}$: $\tilde{P} \to \tilde{P}$ *is a completely continuous operator.*

Proof. The operators $\tilde{Q}_1$ and $\tilde{Q}_2$ are well defined. For every $(x,y) \in \tilde{P}$, by Lemma 2.5.5, we have $\tilde{Q}_1(x,y)(t) < \infty$ and $\tilde{Q}_2(x,y)(t) < \infty$ for all $t \in [0,T]$. Then, by Lemma 2.5.6, we obtain

$$\tilde{Q}_1(x,y)(t) \geq \gamma_1(t) \sup_{t' \in [0,T]} \tilde{Q}_1(x,y)(t'), \quad \tilde{Q}_2(x,y)(t) \geq \gamma_2(t) \sup_{t' \in [0,T]} \tilde{Q}_2(x,y)(t'),$$

for all $t \in [0,T]$. Therefore, we conclude

$$\tilde{Q}_1(x,y)(t) \geq \gamma_1(t)\|\tilde{Q}_1(x,y)\|, \quad \tilde{Q}_2(x,y)(t) \geq \gamma_2(t)\|\tilde{Q}_2(x,y)\|, \quad \forall t \in [0,T],$$

and $\tilde{Q}(x,y) = (\tilde{Q}_1(x,y), \tilde{Q}_2(x,y)) \in \tilde{P}$.

By using standard arguments, we deduce that the operator $\tilde{Q}$: $\tilde{P} \to \tilde{P}$ is a completely continuous operator. □

Then $(x,y) \in \tilde{P}$ is a solution of problem (2.57)–(2.58) if and only if (x,y) is a fixed point of the operator $\tilde{Q}$.

Theorem 2.5.2. *Assume that* $(\widehat{H1})$–$(\widehat{H3})$ *hold. Then there exist constants* $\lambda_0 > 0$ *and* $\mu_0 > 0$ *such that for any* $\lambda \in (0, \lambda_0]$ *and* $\mu \in (0, \mu_0]$ *the boundary value problem* $(\tilde{\mathrm{S}})$–$(\widetilde{\mathrm{BC}})$ *has at least one positive solution.*

Proof. Let $\delta \in (0,1)$ be fixed. From $(\widehat{\mathrm{H3}})$, there exists $R_0 > 0$ such that

$$f(t,u,v) \geq \delta f(t,0,0) > 0, \quad g(t,u,v) \geq \delta g(t,0,0) > 0, \quad \forall t \in [0,T], \quad u,v \in [0,R_0]. \tag{2.61}$$

We define

$$\bar{f}(R_0) = \max_{0 \leq t \leq T, 0 \leq u,v \leq R_0} \{f(t,u,v) + p_1(t)\} \geq \max_{0 \leq t \leq T} \{\delta f(t,0,0) + p_1(t)\} > 0,$$
$$\bar{g}(R_0) = \max_{0 \leq t \leq T, 0 \leq u,v \leq R_0} \{g(t,u,v) + p_2(t)\} \geq \max_{0 \leq t \leq T} \{\delta g(t,0,0) + p_2(t)\} > 0,$$

$$c_1 = \int_0^T J_1(s)\,ds > 0, \quad c_2 = \int_0^T J_2(s)\,ds > 0, \quad \lambda_0 = \frac{R_0}{4c_1\bar{f}(R_0)} > 0, \quad \mu_0 = \frac{R_0}{4c_2\bar{g}(R_0)} > 0.$$

We shall show that for any $\lambda \in (0, \lambda_0]$ and $\mu \in (0, \mu_0]$ problem (2.57)–(2.58) has at least one positive solution.

So, let $\lambda \in (0, \lambda_0]$ and $\mu \in (0, \mu_0]$ be arbitrary, but fixed for the moment. We define the set $U = \{(x,y) \in \tilde{P}, \|(x,y)\|_Y < R_0\}$. We suppose that there exist $(x_0, y_0) \in \partial U$, $(\|(x_0,y_0)\|_Y = R_0$ or $\|x_0\| + \|y_0\| = R_0)$, and $\nu \in (0,1)$ such that $(x_0, y_0) = \nu\tilde{Q}(x_0, y_0)$ or $x_0 = \nu\tilde{Q}_1(x_0, y_0)$, $y_0 = \nu\tilde{Q}_2(x_0, y_0)$.

We deduce that

$$[x_0(t) - q_1(t)]^* = x_0(t) - q_1(t) \leq x_0(t) \leq R_0, \quad \text{if } x_0(t) - q_1(t) \geq 0,$$
$$[x_0(t) - q_1(t)]^* = 0, \text{ for } x_0(t) - q_1(t) < 0, \quad \forall t \in [0,T],$$
$$[y_0(t) - q_2(t)]^* = y_0(t) - q_2(t) \leq y_0(t) \leq R_0, \quad \text{if } y_0(t) - q_2(t) \geq 0,$$
$$[y_0(t) - q_2(t)]^* = 0, \text{ for } y_0(t) - q_2(t) < 0, \quad \forall t \in [0,T].$$

Then, by Lemma 2.5.5, for all $t \in [0,T]$ we obtain

$$x_0(t) = \nu \tilde{Q}_1(x_0, y_0)(t) \leq \tilde{Q}_1(x_0, y_0)(t) = \lambda \int_0^T G_1(t,s) \big(f(s, [x_0(s) - q_1(s)]^*, [y_0(s) - q_2(s)]^*) + p_1(s)\big)\, ds \leq \lambda \int_0^T G_1(t,s)\bar{f}(R_0)\, ds$$
$$\leq \lambda \int_0^T J_1(s)\bar{f}(R_0)\, ds \leq \lambda_0 c_1 \bar{f}(R_0) = R_0/4.$$

In a similar manner, we conclude $y_0(t) \leq \mu_0 c_2 \bar{g}(R_0) = R_0/4$, for all $t \in [0,T]$.

Hence $\|x_0\| \leq R_0/4$ and $\|y_0\| \leq R_0/4$. Then $R_0 = \|(x_0, y_0)\|_Y = \|x_0\| + \|y_0\| \leq \frac{R_0}{4} + \frac{R_0}{4} = \frac{R_0}{2}$, which is a contradiction.

Therefore, by Lemma 2.5.7 and Theorem 2.5.1, we deduce that $\tilde{Q}$ has a fixed point $(x,y) \in \bar{U} \cap \tilde{P}$. That is, $(x,y) = \tilde{Q}(x,y)$ or equivalently $x = \tilde{Q}_1(x,y), y = \tilde{Q}_2(x,y)$, and $\|x\| + \|y\| \leq R_0$, with $x(t) \geq \gamma_1(t)\|x\| \geq 0$ and $y(t) \geq \gamma_2(t)\|y\| \geq 0$ for all $t \in [0,T]$.

Moreover, by (2.61), we obtain

$$x(t) = \tilde{Q}_1(x,y)(t) = \lambda \int_0^T G_1(t,s) \big(f(s, [x(s) - q_1(s)]^*, [y(s) - q_2(s)]^*) + p_1(s)\big)\, ds$$
$$\geq \lambda \int_0^T G_1(t,s)(\delta f(s,0,0) + p_1(s))\, ds > \lambda \int_0^T G_1(t,s) p_1(s)\, ds$$
$$= q_1(t) > 0, \forall t \in (0,T).$$

In a similar manner, we have $y(t) > q_2(t) > 0$ for all $t \in (0,T)$.

Let $u(t) = x(t) - q_1(t) \geq 0$ and $v(t) = y(t) - q_2(t) \geq 0$ for all $t \in [0,T]$, with $u(t) > 0$, $v(t) > 0$ on $(0,T)$. Then, (u,v) is a positive solution of the boundary value problem $(\tilde{S})$–$(\widetilde{BC})$. □

2.5.4 Examples

Let $T=1$, $n=3$, $m=4$, $H_1(t) = t^4/3$, $K_2(t) = t^3/2$, for $t \in [0,1]$, and

$$H_2(t) = \begin{cases} 0, & t \in [0,1/3), \\ 1/3, & t \in [1/3, 2/3), \\ 5/6, & t \in [2/3, 1], \end{cases} \qquad K_1(t) = \begin{cases} 0, & t \in [0,1/2), \\ 1/2, & t \in [1/2, 1]. \end{cases}$$

Then, we have $\int_0^1 u(s)\, dH_1(s) = \frac{4}{3}\int_0^1 s^3 u(s)\, ds$, $\int_0^1 u(s)\, dH_2(s) = \frac{1}{3}u\left(\frac{1}{3}\right) + \frac{1}{2}u\left(\frac{2}{3}\right)$, $\int_0^1 v(s)\, dK_1(s) = \frac{1}{2}v\left(\frac{1}{2}\right)$, and $\int_0^1 v(s)\, dK_2(s) = \frac{3}{2}\int_0^1 s^2 v(s)\, ds$.

We consider the system of differential equations

$$\begin{cases} u^{(3)}(t) + \lambda f(t, u(t), v(t)) = 0, & t \in (0,1), \\ v^{(4)}(t) + \mu g(t, u(t), v(t)) = 0, & t \in (0,1), \end{cases} \qquad (\tilde{S}_0)$$

with the boundary conditions

$$\begin{cases} u(0)=\frac{4}{3}\int_0^1 s^3u(s)\,\mathrm{d}s, \quad u'(0)=0, \quad u(1)=\frac{1}{3}u\left(\frac{1}{3}\right)+\frac{1}{2}u\left(\frac{2}{3}\right), \\ v(0)=\frac{1}{2}v\left(\frac{1}{2}\right), \quad v'(0)=v''(0)=0, \quad v(1)=\frac{3}{2}\int_0^1 s^2v(s)\,\mathrm{d}s. \end{cases} \qquad (\widetilde{\mathrm{BC}}_0)$$

Then, we obtain $H_1(1)-H_1(0)=\frac{1}{3}<1$, $H_2(1)-H_2(0)=\frac{5}{6}<1$, $K_1(1)-K_1(0)=\frac{1}{2}<1$, and $K_2(1)-K_2(0)=\frac{1}{2}<1$.

We also deduce

$$g_1(t,s)=\frac{1}{2}\begin{cases} t^2(1-s)^2-(t-s)^2, & 0\le s\le t\le 1, \\ t^2(1-s)^2, & 0\le t\le s\le 1, \end{cases}$$

$$g_2(t,s)=\frac{1}{6}\begin{cases} t^3(1-s)^3-(t-s)^3, & 0\le s\le t\le 1, \\ t^3(1-s)^3, & 0\le t\le s\le 1, \end{cases}$$

$\Delta_1=\frac{43}{81}$, $\Delta_2=\frac{13}{32}$, $\tau_1=\frac{123}{43}$, $\tau_2=\frac{34}{13}$, $h_1(s)=s(1-s)^2$, $h_2(s)=\frac{1}{2}s(1-s)^3$, $J_1(s)=\frac{123}{43}s(1-s)^2$, $J_2(s)=\frac{17}{13}s(1-s)^3$, $s\in[0,1]$, $c_1=\int_0^1 J_1(s)\,\mathrm{d}s=\frac{123}{516}$, and $c_2=\int_0^1 J_2(s)\,\mathrm{d}s=\frac{17}{260}$.

Example 2.5.1. We consider the functions

$$f(t,u,v)=(u-1)(u-2)+\cos(3v), \quad g(t,u,v)=(v-2)(v-3)+\sin(2u),$$

for all $t\in[0,1]$, $u,v\in[0,\infty)$. There exists $M_0>0$ ($M_0=\frac{5}{4}$) such that $f(t,u,v)+M_0\ge 0$, $g(t,u,v)+M_0\ge 0$ for all $t\in[0,1]$ and $u,v\in[0,\infty)$ ($p_1(t)=p_2(t)=M_0$, for all $t\in[0,1]$).

Let $\delta=\frac{1}{4}<1$ and $R_0=\frac{1}{2}$. Then

$$f(t,u,v)\ge\delta f(t,0,0)=\frac{3}{4}, \quad g(t,u,v)\ge\delta g(t,0,0)=\frac{3}{2}, \quad \forall t\in[0,1], u,v\in[0,1/2].$$

Besides,

$$\bar{f}(R_0)=\max_{0\le t\le 1, 0\le u,v\le R_0}\{f(t,u,v)+p_1(t)\}=4.25,$$

$$\bar{g}(R_0)=\max_{0\le t\le 1, 0\le u,v\le R_0}\{g(t,u,v)+p_2(t)\}=7.25+\sin 1.$$

Then $\lambda_0=\frac{1}{34c_1}\approx 0.12338594$ and $\mu_0=\frac{1}{8c_2(7.25+\sin 1)}\approx 0.23626911$. By Theorem 2.5.2, for any $\lambda\in(0,\lambda_0]$ and $\mu\in(0,\mu_0]$ we conclude that problem $(\tilde{\mathrm{S}}_0)$–$(\widetilde{\mathrm{BC}}_0)$ has a positive solution (u,v).

Example 2.5.2. We consider the functions

$$f(t,u,v)=v^3+\cos(2u), \quad g(t,u,v)=u^{1/4}+\cos(3v), \quad \forall t\in[0,1], \quad u,v\in[0,\infty).$$

There exists $M_0>0$ ($M_0=1$) such that $f(t,u,v)+M_0\ge 0$, $g(t,u,v)+M_0\ge 0$ for all $t\in[0,1]$ and $u,v\in[0,\infty)$ ($p_1(t)=p_2(t)=M_0$, for all $t\in[0,1]$).

Let $\delta = \frac{1}{2} < 1$ and $R_0 = \frac{\pi}{9}$. Then

$$f(t,u,v) \geq \delta f(t,0,0) = \frac{1}{2}, \quad g(t,u,v) \geq \delta g(t,0,0) = \frac{1}{2}, \ \forall t \in [0,1], \ u,v \in [0,\pi/9].$$

Besides, $\bar{f}(R_0) = \frac{\pi^3}{81} + 2$ and $\bar{g}(R_0) = \left(\frac{\pi}{9}\right)^{1/4} + 2$. Then $\lambda_0 \approx 0.15364044$ and $\mu_0 \approx 0.48206348$.

By Theorem 2.5.2, for any $\lambda \in (0, \lambda_0]$ and $\mu \in (0, \mu_0]$ we deduce that problem $(\tilde{\mathrm{S}}_0)$–$(\widetilde{\mathrm{BC}}_0)$ has a positive solution (u, v).

Remark 2.5.1. The results presented in this section were published in Luca and Tudorache (2015).

Systems of second-order difference equations with multipoint boundary conditions

3

3.1 Existence and nonexistence of positive solutions for systems with parameters

The mathematical modeling of many nonlinear problems from computer science, economics, mechanical engineering, control systems, biological neural networks, and other fields leads to the consideration of nonlinear difference equations (Kelley and Peterson, 2001; Lakshmikantham and Trigiante, 1988). In recent decades, many authors have investigated such problems by using various methods, such as fixed point theorems, the critical point theory, upper and lower solutions, the fixed point index theory, and the topological degree theory.

In this section, we shall investigate the existence and nonexistence of positive solutions for a system of nonlinear second-order difference equations which contains some parameters and is subject to multipoint boundary conditions.

3.1.1 Presentation of the problem

We consider the system of nonlinear second-order difference equations

$$\begin{cases} \Delta^2 u_{n-1} + \lambda s_n f(n, u_n, v_n) = 0, & n = \overline{1, N-1}, \\ \Delta^2 v_{n-1} + \mu t_n g(n, u_n, v_n) = 0, & n = \overline{1, N-1}, \end{cases} \tag{S}$$

with the multipoint boundary conditions

$$u_0 = \sum_{i=1}^{p} a_i u_{\xi_i}, \quad u_N = \sum_{i=1}^{q} b_i u_{\eta_i}, \quad v_0 = \sum_{i=1}^{r} c_i v_{\zeta_i}, \quad v_N = \sum_{i=1}^{l} d_i v_{\rho_i}, \tag{BC}$$

where $N, p, q, r, l \in \mathbb{N}$, $N \geq 2$, Δ is the forward difference operator with step size 1, $\Delta u_n = u_{n+1} - u_n$, $\Delta^2 u_{n-1} = u_{n+1} - 2u_n + u_{n-1}$, $n = \overline{k, m}$ means that $n = k, k+1, \ldots, m$ for $k, m \in \mathbb{N}$, $a_i \in \mathbb{R}$ for all $i = \overline{1, p}$, $b_i \in \mathbb{R}$ for all $i = \overline{1, q}$, $c_i \in \mathbb{R}$ for all $i = \overline{1, r}$, $d_i \in \mathbb{R}$ for all $i = \overline{1, l}$, $\xi_i \in \mathbb{N}$ for all $i = \overline{1, p}$, $\eta_i \in \mathbb{N}$ for all $i = \overline{1, q}$, $\zeta_i \in \mathbb{N}$ for all $i = \overline{1, r}$, $\rho_i \in \mathbb{N}$ for all $i = \overline{1, l}$, $1 \leq \xi_1 < \cdots < \xi_p \leq N-1$, $1 \leq \eta_1 < \cdots < \eta_q \leq N-1$, $1 \leq \zeta_1 < \cdots < \zeta_r \leq N-1$, and $1 \leq \rho_1 < \cdots < \rho_l \leq N-1$.

We shall give sufficient conditions on λ, μ, f, and g such that positive solutions of (S)–(BC) exist. In addition, we present some nonexistence results for the positive solutions of the above problem. By a positive solution of problem (S)–(BC), we mean

Boundary Value Problems for Systems of Differential, Difference and Fractional Equations.
http://dx.doi.org/10.1016/B978-0-12-803652-5.00003-X

a pair of sequences $((u_n)_{n=\overline{0,N}}, (v_n)_{n=\overline{0,N}})$ satisfying (S) and (BC) with $u_n \geq 0$, $v_n \geq 0$ for all $n = \overline{0,N}$ and $(u,v) \neq (0,0)$. This problem is a generalization of the one studied in Henderson and Luca (2012f), where in (S) we had $f(n,u,v) = \tilde{f}(u,v)$, $g(n,u,v) = \tilde{g}(u,v)$ (denoted by $(\tilde{S})$) and in (BC) we had $p = r = 1$, $\xi_1 = 1$, $\zeta_1 = 1$, $a_1 = \beta/(\alpha+\beta)$, $c_1 = \delta/(\gamma+\delta)$ (denoted by $(\widetilde{BC})$). Other particular cases of $(\tilde{S}) - (\widetilde{BC})$ were investigated in Luca (2012a), where $q = l$, $b_i = d_i$, $\eta_i = \rho_i$ for all $i = \overline{1,q}$, $\alpha = \gamma$, and $\beta = \delta$, and in Luca (2009), where in $(\tilde{S})$ we had $\tilde{f}(u,v) = \tilde{\tilde{f}}(v)$, $\tilde{g}(u,v) = \tilde{\tilde{g}}(u)$ (denoted by $(\tilde{\tilde{S}})$), and $q = l$, $b_i = d_i$, $\eta_i = \rho_i$ for all $i = \overline{1,q}$, $\alpha = \gamma$, and $\beta = \delta$. System $(\tilde{S})$ with the three-point boundary conditions $u_0 = \beta u_\eta$, $u_N = \alpha u_\eta$, $v_0 = \beta v_\eta$, $v_N = \alpha v_\eta$, $\eta \in \{1,\ldots,N-1\}$, $\alpha, \beta > 0$ was investigated in Henderson et al. (2009). System $(\tilde{\tilde{S}})$ with the boundary conditions $u_0 = v_0 = 0$, $u_N = \alpha u_\eta$, $v_N = \alpha v_\eta$, $\eta \in \{1,\ldots,N-1\}$ with $0 < \alpha < N/\eta$ or $u_0 - \beta\Delta u_0 = 0$, $v_0 - \beta\Delta v_0 = 0$, $u_N = \alpha u_\eta$, $v_N = \alpha v_\eta$ was studied in Henderson et al. (2008c).

3.1.2 Auxiliary results

In this section, we present some auxiliary results related to the following second-order difference equation with multipoint boundary conditions:

$$\Delta^2 u_{n-1} + y_n = 0, \quad n = \overline{1,N-1}, \tag{3.1}$$

$$u_0 = \sum_{i=1}^{p} a_i u_{\xi_i}, \quad u_N = \sum_{i=1}^{q} b_i u_{\eta_i}, \tag{3.2}$$

where $N, p, q \in \mathbb{N}$, $N \geq 2$, $a_i \in \mathbb{R}$ for all $i = \overline{1,p}$, $b_i \in \mathbb{R}$ for all $i = \overline{1,q}$, $\xi_i \in \mathbb{N}$ for all $i = \overline{1,p}$, $\eta_i \in \mathbb{N}$ for all $i = \overline{1,q}$, $1 \leq \xi_1 < \cdots < \xi_p \leq N-1$, and $1 \leq \eta_1 < \cdots < \eta_q \leq N-1$.

Lemma 3.1.1. *If $a_i \in \mathbb{R}$ for all $i = \overline{1,p}$, $b_i \in \mathbb{R}$ for all $i = \overline{1,q}$, $\xi_i \in \mathbb{N}$ for all $i = \overline{1,p}$, $\eta_i \in \mathbb{N}$ for all $i = \overline{1,q}$, $1 \leq \xi_1 < \cdots < \xi_p \leq N-1$, $1 \leq \eta_1 < \cdots < \eta_q \leq N-1$, $\Delta_1 = \left(1 - \sum_{i=1}^q b_i\right)\sum_{i=1}^p a_i\xi_i + \left(1 - \sum_{i=1}^p a_i\right)\left(N - \sum_{i=1}^q b_i\eta_i\right) \neq 0$, and $y_n \in \mathbb{R}$ for all $n = \overline{1,N-1}$, then the unique solution of problem* (3.1)–(3.2) *is given by*

$$\begin{aligned} u_n = &-\sum_{j=1}^{n-1}(n-j)y_j + \frac{n}{\Delta_1}\left\{\left(1 - \sum_{i=1}^q b_i\right)\sum_{i=1}^p a_i\left(\sum_{j=1}^{\xi_i-1}(\xi_i - j)y_j\right)\right. \\ &\left.+\left(1 - \sum_{i=1}^p a_i\right)\left[\sum_{j=1}^{N-1}(N-j)y_j - \sum_{i=1}^q b_i\left(\sum_{j=1}^{\eta_i-1}(\eta_i - j)y_j\right)\right]\right\} \\ &+\frac{1}{\Delta_1}\left\{\left(\sum_{i=1}^p a_i\xi_i\right)\left[\sum_{j=1}^{N-1}(N-j)y_j - \sum_{i=1}^q b_i\left(\sum_{j=1}^{\eta_i-1}(\eta_i - j)y_j\right)\right]\right. \\ &\left.-\left(N - \sum_{i=1}^q b_i\eta_i\right)\sum_{i=1}^p a_i\left(\sum_{j=1}^{\xi_i-1}(\xi_i - j)y_j\right)\right\}, \quad n = \overline{0,N}. \end{aligned} \tag{3.3}$$

Proof. By equation (3.1), for u_1 fixed, we obtain

$$u_n = nu_1 - (n-1)u_0 - \sum_{j=1}^{n-1}(n-j)y_j, \quad n = \overline{2,N}.$$

By (3.2), we have $u_0 = \sum_{i=1}^{p} a_i u_{\xi_i}$, $u_N = \sum_{i=1}^{q} b_i u_{\eta_i}$. From the above relations, we deduce

$$\begin{cases} u_0 = \sum_{i=1}^{p}\left[\xi_i u_1 - (\xi_i - 1)u_0 - \sum_{j=1}^{\xi_i - 1}(\xi_i - j)y_j\right], \\ Nu_1 - (N-1)u_0 - \sum_{j=1}^{N-1}(N-j)y_j = \sum_{i=1}^{q} b_i\left[\eta_i u_1 - (\eta_i - 1)u_0 - \sum_{j=1}^{\eta_i - 1}(\eta_i - j)y_j\right]. \end{cases}$$

The above system can be written as

$$\begin{cases} \left(\sum_{i=1}^{p} a_i\xi_i\right)u_1 + \left[-1 - \sum_{i=1}^{p} a_i(\xi_i - 1)\right]u_0 = \sum_{i=1}^{p} a_i\left(\sum_{j=1}^{\xi_i - 1}(\xi_i - j)y_j\right), \\ \left(N - \sum_{i=1}^{q} b_i\eta_i\right)u_1 + \left[-N + 1 + \sum_{i=1}^{q} b_i(\eta_i - 1)\right]u_0 = \sum_{j=1}^{N-1}(N-j)y_j \\ -\sum_{i=1}^{q} b_i\left(\sum_{j=1}^{\eta_i - 1}(\eta_i - j)y_j\right). \end{cases} \tag{3.4}$$

The determinant of the above system in the unknowns u_1 and u_0 is

$$\begin{aligned} \Delta_1 &= \begin{vmatrix} \sum_{i=1}^{p} a_i\xi_i & -1 - \sum_{i=1}^{p} a_i(\xi_i - 1) \\ N - \sum_{i=1}^{q} b_i\eta_i & -N + 1 + \sum_{i=1}^{q} b_i(\eta_i - 1) \end{vmatrix} \\ &= \left(\sum_{i=1}^{p} a_i\xi_i\right)\left(1 - N + \sum_{i=1}^{q} b_i(\eta_i - 1)\right) + \left(N - \sum_{i=1}^{q} b_i\eta_i\right)\left(1 + \sum_{i=1}^{p} a_i(\xi_i - 1)\right) \\ &= \left(\sum_{i=1}^{p} a_i\xi_i\right)\left(1 - \sum_{i=1}^{q} b_i\right) + \left(\sum_{i=1}^{p} a_i\xi_i\right)\left(\sum_{i=1}^{q} b_i\eta_i\right) - N\sum_{i=1}^{p} a_i\xi_i \\ &\quad + \left(N - \sum_{i=1}^{q} b_i\eta_i\right)\left(1 - \sum_{i=1}^{p} a_i\right) + N\sum_{i=1}^{p} a_i\xi_i - \left(\sum_{i=1}^{q} b_i\eta_i\right)\left(\sum_{i=1}^{p} a_i\xi_i\right) \\ &= \left(1 - \sum_{i=1}^{q} b_i\right)\sum_{i=1}^{p} a_i\xi_i + \left(1 - \sum_{i=1}^{p} a_i\right)\left(N - \sum_{i=1}^{q} b_i\eta_i\right). \end{aligned}$$

Because $\Delta_1 \neq 0$ (by the assumption of this lemma), system (3.4) has a unique solution—namely,

$$u_1 = \frac{1}{\Delta_1}\left\{\left(1 - N + \sum_{i=1}^{q} b_i(\eta_i - 1)\right)\sum_{i=1}^{p} a_i\left(\sum_{j=1}^{\xi_i-1}(\xi_i - j)y_j\right)\right.$$
$$\left.+\left(1 + \sum_{i=1}^{p} a_i(\xi_i - 1)\right)\left[\sum_{j=1}^{N-1}(N-j)y_j - \sum_{i=1}^{q} b_i\left(\sum_{j=1}^{\eta_i-1}(\eta_i - j)y_j\right)\right]\right\},$$

$$u_0 = \frac{1}{\Delta_1}\left\{\left(\sum_{i=1}^{p} a_i\xi_i\right)\left[\sum_{j=1}^{N-1}(N-j)y_j - \sum_{i=1}^{q} b_i\left(\sum_{j=1}^{\eta_i-1}(\eta_i - j)y_j\right)\right]\right.$$
$$\left.-\left(N - \sum_{i=1}^{q} b_i\eta_i\right)\sum_{i=1}^{p} a_i\left(\sum_{j=1}^{\xi_i-1}(\xi_i - j)y_j\right)\right\}.$$

Then the solution of problem (3.1)–(3.2) is

$$u_n = -\sum_{j=1}^{n-1}(n-j)y_j + \frac{n}{\Delta_1}\left\{\left(1 - N + \sum_{i=1}^{q} b_i(\eta_i - 1)\right)\sum_{i=1}^{p} a_i\left(\sum_{j=1}^{\xi_i-1}(\xi_i - j)y_j\right)\right.$$
$$\left.+\left(1 + \sum_{i=1}^{p} a_i(\xi_i - 1)\right)\left[\sum_{j=1}^{N-1}(N-j)y_j - \sum_{i=1}^{q} b_i\left(\sum_{j=1}^{\eta_i-1}(\eta_i - j)y_j\right)\right]\right\}$$
$$-\frac{n-1}{\Delta_1}\left\{\sum_{i=1}^{p} a_i\xi_i\left[\sum_{j=1}^{N-1}(N-j)y_j - \sum_{i=1}^{q} b_i\left(\sum_{j=1}^{\eta_i-1}(\eta_i - j)y_j\right)\right]\right.$$
$$\left.-\left(N - \sum_{i=1}^{q} b_i\eta_i\right)\sum_{i=1}^{p} a_i\left(\sum_{j=1}^{\xi_i-1}(\xi_i - j)y_j\right)\right\}$$
$$= -\sum_{j=1}^{n-1}(n-j)y_j + \frac{n}{\Delta_1}\left\{\left(1 - N + \sum_{i=1}^{q} b_i(\eta_i - 1)\right)\sum_{i=1}^{p} a_i\left(\sum_{j=1}^{\xi_i-1}(\xi_i - j)y_j\right)\right.$$
$$+\left(1 + \sum_{i=1}^{p} a_i(\xi_i - 1)\right)\left[\sum_{j=1}^{N-1}(N-j)y_j - \sum_{i=1}^{q} b_i\left(\sum_{j=1}^{\eta_i-1}(\eta_i - j)y_j\right)\right]$$
$$-\left(\sum_{i=1}^{p} a_i\xi_i\right)\left[\sum_{j=1}^{N-1}(N-j)y_j - \sum_{i=1}^{q} b_i\left(\sum_{j=1}^{\eta_i-1}(\eta_i - j)y_j\right)\right]$$

$$+\left(N-\sum_{i=1}^{q} b_i\eta_i\right)\sum_{i=1}^{p} a_i\left(\sum_{j=1}^{\xi_i-1}(\xi_i-j)y_j\right)\Bigg\}$$

$$+\frac{1}{\Delta_1}\left\{\left(\sum_{i=1}^{p} a_i\xi_i\right)\left[\sum_{j=1}^{N-1}(N-j)y_j-\sum_{i=1}^{q} b_i\left(\sum_{j=1}^{\eta_i-1}(\eta_i-j)y_j\right)\right]\right.$$

$$\left.-\left(N-\sum_{i=1}^{q} b_i\eta_i\right)\sum_{i=1}^{p} a_i\left(\sum_{j=1}^{\xi_i-1}(\xi_i-j)y_j\right)\right\}.$$

Therefore, we obtain

$$u_n=-\sum_{j=1}^{n-1}(n-j)y_j+\frac{n}{\Delta_1}\left\{\left(1-\sum_{i=1}^{q} b_i\right)\sum_{i=1}^{p} a_i\left(\sum_{j=1}^{\xi_i-1}(\xi_i-j)y_j\right)\right.$$

$$+\left(1-\sum_{i=1}^{p} a_i\right)\left[\sum_{j=1}^{N-1}(N-j)y_j-\sum_{i=1}^{q} b_i\left(\sum_{j=1}^{\eta_i-1}(\eta_i-j)y_j\right)\right]$$

$$+\left(\sum_{i=1}^{q} b_i\eta_i-N\right)\sum_{i=1}^{p} a_i\left(\sum_{j=1}^{\xi_i-1}(\xi_i-j)y_j\right)$$

$$+\left(\sum_{i=1}^{p} a_i\xi_i\right)\left[\sum_{j=1}^{N-1}(N-j)y_j-\sum_{i=1}^{q} b_i\left(\sum_{j=1}^{\eta_i-1}(\eta_i-j)y_j\right)\right]$$

$$-\left(\sum_{i=1}^{p} a_i\xi_i\right)\left[\sum_{j=1}^{N-1}(N-j)y_j-\sum_{i=1}^{q} b_i\left(\sum_{j=1}^{\eta_i-1}(\eta_i-j)y_j\right)\right]$$

$$\left.+\left(N-\sum_{i=1}^{q} b_i\eta_i\right)\sum_{i=1}^{p} a_i\left(\sum_{j=1}^{\xi_i-1}(\xi_i-j)y_j\right)\right\}$$

$$+\frac{1}{\Delta_1}\left\{\left(\sum_{i=1}^{p} a_i\xi_i\right)\left[\sum_{j=1}^{N-1}(N-j)y_j-\sum_{i=1}^{q} b_i\left(\sum_{j=1}^{\eta_i-1}(\eta_i-j)y_j\right)\right]\right.$$

$$\left.-\left(N-\sum_{i=1}^{q} b_i\eta_i\right)\sum_{i=1}^{p} a_i\left(\sum_{j=1}^{\xi_i-1}(\xi_i-j)y_j\right)\right\},\quad n=\overline{0,N}.$$

So, we deduce for u_n, $n=\overline{0,N}$, the expression given by (3.3). Here we consider that $\sum_{i=1}^{0} z_i=0$ and $\sum_{i=1}^{-1} z_i=0$. □

Lemma 3.1.2. *Under the assumptions of Lemma 3.1.1, the Green's function for the boundary value problem* (3.1)–(3.2) *is given by*

$$G_1(n,j) = g_0(n,j) + \frac{1}{\Delta_1}\left[(N-n)\left(1-\sum_{k=1}^{q} b_k\right) + \sum_{i=1}^{q} b_i(N-\eta_i)\right]$$
$$\times \sum_{i=1}^{p} a_i g_0(\xi_i,j) + \frac{1}{\Delta_1}\left[n\left(1-\sum_{k=1}^{p} a_k\right) + \sum_{i=1}^{p} a_i\xi_i\right]$$
$$\times \sum_{i=1}^{q} b_i g_0(\eta_i,j), \quad n=\overline{0,N}, \quad j=\overline{1,N-1}, \tag{3.5}$$

where

$$g_0(n,j) = \frac{1}{N}\begin{cases} j(N-n), & 1 \le j \le n \le N, \\ n(N-j), & 0 \le n \le j \le N-1. \end{cases} \tag{3.6}$$

Proof. By Lemma 3.1.1 and relation (3.3), we have

$$u_n = \frac{1}{N}\left\{\sum_{j=1}^{n-1} j(N-n)y_j + \sum_{j=n}^{N-1} n(N-j)y_j - \sum_{j=1}^{N-1} n(N-j)y_j\right.$$
$$+ \frac{nN}{\Delta_1}\left\{\left(1-\sum_{i=1}^{q} b_i\right)\sum_{i=1}^{p} a_i\left[\sum_{j=1}^{\xi_i-1}(\xi_i-j)y_j\right]\right.$$
$$\left.+ \left(1-\sum_{i=1}^{p} a_i\right)\left[\sum_{j=1}^{N-1}(N-j)y_j - \sum_{i=1}^{q} b_i\left(\sum_{j=1}^{\eta_i-1}(\eta_i-j)y_j\right)\right]\right\}$$
$$+ \frac{N}{\Delta_1}\left\{\left(\sum_{i=1}^{p} a_i\xi_i\right)\left[\sum_{j=1}^{N-1}(N-j)y_j - \sum_{i=1}^{q} b_i\left(\sum_{j=1}^{\eta_i-1}(\eta_i-j)y_j\right)\right]\right.$$
$$\left.\left.- \left(N-\sum_{i=1}^{q} b_i\eta_i\right)\sum_{i=1}^{p} a_i\left(\sum_{j=1}^{\xi_i-1}(\xi_i-j)y_j\right)\right\}\right\}$$
$$= \frac{1}{N}\left\{\sum_{j=1}^{n-1} j(N-n)y_j + \sum_{j=n}^{N-1} n(N-j)y_j - \frac{1}{\Delta_1}\left[\left(1-\sum_{i=1}^{q} b_i\right)\sum_{i=1}^{p} a_i\xi_i\right.\right.$$
$$\left.+ \left(1-\sum_{i=1}^{p} a_i\right)\left(N-\sum_{i=1}^{q} b_i\eta_i\right)\right]\sum_{j=1}^{N-1} n(N-j)y_j$$
$$+ \frac{nN}{\Delta_1}\left\{\left(1-\sum_{i=1}^{q} b_i\right)\sum_{i=1}^{p} a_i\left(\sum_{j=1}^{\xi_i-1}(\xi_i-j)y_j\right)\right.$$

$$+\left(1-\sum_{i=1}^{p}a_i\right)\left[\sum_{j=1}^{N-1}(N-j)y_j-\sum_{i=1}^{q}b_i\left(\sum_{j=1}^{\eta_i-1}(\eta_i-j)y_j\right)\right]\Bigg\}$$
$$+\frac{N}{\Delta_1}\left\{\left(\sum_{i=1}^{p}a_i\xi_i\right)\left[\sum_{j=1}^{N-1}(N-j)y_j-\sum_{i=1}^{q}b_i\left(\sum_{j=1}^{\eta_i-1}(\eta_i-j)y_j\right)\right]\right.$$
$$\left.-\left(N-\sum_{i=1}^{q}b_i\eta_i\right)\sum_{i=1}^{p}a_i\left(\sum_{j=1}^{\xi_i-1}(\xi_i-j)y_j\right)\right\}\Bigg\}.$$

Therefore, we obtain

$$u_n=\frac{1}{N}\left\{\sum_{j=1}^{n-1}j(N-n)y_j+\sum_{j=n}^{N-1}n(N-j)y_j\right.$$
$$+\frac{n}{\Delta_1}\left[-\left(1-\sum_{i=1}^{q}b_i\right)\left(\sum_{i=1}^{p}a_i\xi_i\right)\left(\sum_{j=1}^{N-1}(N-j)y_j\right)\right.$$
$$-\left(1-\sum_{i=1}^{p}a_i\right)\left(\sum_{j=1}^{N-1}N(N-j)y_j\right)+\left(1-\sum_{i=1}^{p}a_i\right)\left(\sum_{i=1}^{q}b_i\eta_i\right)$$
$$\times\left(\sum_{j=1}^{N-1}(N-j)y_j\right)+N\left(1-\sum_{i=1}^{q}b_i\right)\sum_{i=1}^{p}a_i\left(\sum_{j=1}^{\xi_i-1}(\xi_i-j)y_j\right)$$
$$\left.+N\left(1-\sum_{i=1}^{p}a_i\right)\left(\sum_{j=1}^{N-1}(N-j)y_j\right)-N\left(1-\sum_{i=1}^{p}a_i\right)\sum_{i=1}^{q}b_i\left(\sum_{j=1}^{\eta_i-1}(\eta_i-j)y_j\right)\right]$$
$$+\frac{N}{\Delta_1}\left[\left(\sum_{i=1}^{p}a_i\xi_i\right)\left(\sum_{j=1}^{N-1}(N-j)y_j\right)-\left(\sum_{i=1}^{p}a_i\xi_i\right)\sum_{i=1}^{q}b_i\left(\sum_{j=1}^{\eta_i-1}(\eta_i-j)y_j\right)\right.$$
$$\left.\left.-N\sum_{i=1}^{p}a_i\left(\sum_{j=1}^{\xi_i-1}(\xi_i-j)y_j\right)+\left(\sum_{i=1}^{q}b_i\eta_i\right)\sum_{i=1}^{p}a_i\left(\sum_{j=1}^{\xi_i-1}(\xi_i-j)y_j\right)\right]\right\}$$
$$=\frac{1}{N}\left\{\sum_{j=1}^{n-1}j(N-n)y_j+\sum_{j=n}^{N-1}n(N-j)y_j\right.$$
$$+\frac{n}{\Delta_1}\left[\left(1-\sum_{k=1}^{q}b_k\right)\sum_{i=1}^{p}a_i\left(\sum_{j=1}^{\xi_i-1}N(\xi_i-j)y_j-\sum_{j=1}^{\xi_i-1}\xi_i(N-j)y_j\right)\right.$$
$$-\left(1-\sum_{k=1}^{q}b_k\right)\sum_{i=1}^{p}a_i\left(\sum_{j=\xi_i}^{N-1}\xi_j(N-j)y_j\right)$$

$$
\begin{aligned}
&- \left(1 - \sum_{k=1}^{p} a_k\right) \sum_{i=1}^{q} b_i \left(\sum_{j=1}^{\eta_i - 1} N(\eta_i - j) y_j - \sum_{j=1}^{\eta_i - 1} \eta_i (N - j) y_j \right) \\
&+ \left(1 - \sum_{k=1}^{p} a_k\right) \sum_{i=1}^{q} b_i \left(\sum_{j=\eta_i}^{N-1} \eta_i (N - j) y_j \right) \Bigg] \\
&+ \frac{N}{\Delta_1} \left[\sum_{i=1}^{p} a_i \left(\sum_{j=1}^{\xi_i - 1} \xi_i (N - j) y_j \right) - \sum_{i=1}^{p} a_i \left(\sum_{j=1}^{\xi_i - 1} N(\xi_i - j) y_j \right) \right. \\
&+ \sum_{i=1}^{p} a_i \left(\sum_{j=\xi_i}^{N-1} \xi_i (N - j) y_j \right) - \left(\sum_{i=1}^{p} a_i \xi_i \right) \sum_{i=1}^{q} b_i \left(\sum_{j=1}^{\eta_i - 1} (\eta_i - j) y_j \right) \\
&\left. + \left(\sum_{i=1}^{q} b_i \eta_i \right) \sum_{i=1}^{p} a_i \left(\sum_{j=1}^{\xi_i - 1} (\xi_i - j) y_j \right) \right] \Bigg\}.
\end{aligned}
$$

Hence, we deduce

$$
\begin{aligned}
u_n = \frac{1}{N} \Bigg\{ & \sum_{j=1}^{n-1} j(N - n) y_j + \sum_{j=n}^{N-1} n(N - j) y_j \\
&+ \frac{n}{\Delta_1} \left[- \left(1 - \sum_{k=1}^{q} b_k\right) \sum_{i=1}^{p} a_i \left(\sum_{j=1}^{\xi_i - 1} j(N - \xi_i) y_j \right) \right. \\
&- \left(1 - \sum_{k=1}^{q} b_k\right) \sum_{i=1}^{p} a_i \left(\sum_{j=\xi_i}^{N-1} \xi_i (N - j) y_j \right) \\
&+ \left(1 - \sum_{k=1}^{p} a_k\right) \sum_{i=1}^{q} b_i \left(\sum_{j=1}^{\eta_i - 1} j(N - \eta_i) y_j \right) \\
&\left. + \left(1 - \sum_{k=1}^{p} a_k\right) \sum_{i=1}^{q} b_i \left(\sum_{j=\eta_i}^{N-1} \eta_i (N - j) y_j \right) \right] \\
&+ \frac{N}{\Delta_1} \left[\sum_{i=1}^{p} a_i \left(\sum_{j=1}^{\xi_i - 1} j(N - \xi_i) y_j \right) + \sum_{i=1}^{p} a_i \left(\sum_{j=\xi_i}^{N-1} \xi_i (N - j) y_j \right) \right] \\
&- \frac{1}{\Delta_1} \left(\sum_{i=1}^{p} a_i \xi_i \right) \sum_{i=1}^{q} b_i \left(\sum_{j=1}^{\eta_i - 1} N(\eta_i - j) y_j \right) \\
&+ \frac{1}{\Delta_1} \left(\sum_{i=1}^{q} b_i \eta_i \right) \sum_{i=1}^{p} a_i \left(\sum_{j=1}^{\xi_i - 1} N(\xi_i - j) y_j \right)
\end{aligned}
$$

$$
\begin{aligned}
&+\frac{1}{\Delta_1}\left(\sum_{i=1}^{p} a_i\xi_i\right)\sum_{i=1}^{q} b_i\left(\sum_{j=1}^{N-1}\eta_i(N-j)y_j\right)\\
&-\frac{1}{\Delta_1}\left(\sum_{i=1}^{q} b_i\eta_i\right)\sum_{i=1}^{p} a_i\left(\sum_{j=1}^{N-1}\xi_i(N-j)y_j\right)\Bigg\}\\
=&\ \frac{1}{N}\left\{\sum_{j=1}^{n-1} j(N-n)y_j+\sum_{j=n}^{N-1} n(N-j)y_j\right.\\
&+\frac{n}{\Delta_1}\left[-\left(1-\sum_{k=1}^{q} b_k\right)\sum_{i=1}^{p} a_i\left(\sum_{j=1}^{\xi_i-1} j(N-\xi_i)y_j\right)\right.\\
&-\left(1-\sum_{k=1}^{q} b_k\right)\sum_{i=1}^{p} a_i\left(\sum_{j=\xi_i}^{N-1}\xi_i(N-j)y_j\right)\\
&+\left(1-\sum_{k=1}^{p} a_k\right)\sum_{i=1}^{q} b_i\left(\sum_{j=1}^{\eta_i-1} j(N-\eta_i)y_j\right)\\
&+\left.\left(1-\sum_{k=1}^{p} a_k\right)\sum_{i=1}^{q} b_i\left(\sum_{j=\eta_i}^{N-1}\eta_i(N-j)y_j\right)\right]\\
&+\frac{N}{\Delta_1}\left[\sum_{i=1}^{p} a_i\left(\sum_{j=1}^{\xi_i-1} j(N-\xi_i)y_j\right)+\sum_{i=1}^{p} a_i\left(\sum_{j=\xi_i}^{N-1}\xi_i(N-j)y_j\right)\right]\\
&+\frac{1}{\Delta_1}\left(\sum_{i=1}^{p} a_i\xi_i\right)\sum_{i=1}^{q} b_i\left[\left(\sum_{j=1}^{\eta_i-1} j(N-\eta_i)y_j\right)+\left(\sum_{j=\eta_i}^{N-1}\eta_i(N-j)y_j\right)\right]\\
&\left.-\frac{1}{\Delta_1}\left(\sum_{i=1}^{q} b_i\eta_i\right)\sum_{i=1}^{p} a_i\left[\left(\sum_{j=1}^{\xi_i-1} j(N-\xi_i)y_j\right)+\left(\sum_{j=\xi_i}^{N-1}\xi_i(N-j)y_j\right)\right]\right\}.
\end{aligned}
$$

Then the solution of problem (3.1)–(3.2) is

$$
\begin{aligned}
u_n=&\sum_{j=1}^{N-1}\left\{g_0(n,j)+\frac{n}{\Delta_1}\left[-\left(1-\sum_{k=1}^{q} b_k\right)\left(\sum_{i=1}^{p} a_i g_0(\xi_i,j)\right)\right.\right.\\
&+\left.\left(1-\sum_{k=1}^{p} a_k\right)\left(\sum_{i=1}^{q} b_i g_0(\eta_i,j)\right)\right]+\frac{N}{\Delta_1}\sum_{i=1}^{p} a_i g_0(\xi_i,j)\\
&\left.+\frac{1}{\Delta_1}\left(\sum_{i=1}^{p} a_i\xi_i\right)\left(\sum_{i=1}^{q} b_i g_0(\eta_i,j)\right)-\frac{1}{\Delta_1}\left(\sum_{i=1}^{q} b_i\eta_i\right)\left(\sum_{i=1}^{p} a_i g_0(\xi_i,j)\right)\right\}y_j
\end{aligned}
$$

$$= \sum_{j=1}^{N-1} \left\{ g_0(n,j) + \frac{1}{\Delta_1} \left[(N-n) \left(1 - \sum_{k=1}^{q} b_k\right) + \sum_{i=1}^{q} b_i(N-\eta_i) \right] \right.$$
$$\left. \times \sum_{i=1}^{p} a_i g_0(\xi_i, j) + \frac{1}{\Delta_1} \left[n \left(1 - \sum_{k=1}^{p} a_k\right) + \left(\sum_{i=1}^{p} a_i \xi_i\right) \right] \sum_{i=1}^{q} b_i g_0(\eta_i, j) \right\} y_j,$$

where g_0 is given by (3.6). Hence, $u_n = \sum_{j=1}^{N-1} G_1(n,j) y_j$ for all $n = \overline{0,N}$, where G_1 is given in (3.5). □

From the definition of the function g_0, we can easily prove the following properties:

Lemma 3.1.3. *The function g_0 given in* (3.6) *has the following properties:*

(a) $g_0(n,j) \geq 0, \quad \forall n = \overline{0,N}, \quad j = \overline{1,N-1}.$

(b) $g_0(n,j) \leq g_0(j,j) = j(N-j)/N, \quad \forall n = \overline{0,N}, \quad j = \overline{1,N-1}.$

(c) *For every $c \in \{1,\ldots,[\![N/2]\!]\}$, we have $\min_{n=\overline{c,N-c}} g_0(n,j) \geq \frac{c}{N-1} g_0(j,j)$ for all $j = \overline{1,N-1}$, where $[\![N/2]\!]$ is the largest integer not greater than $N/2$.*

Lemma 3.1.4. *If $a_i \geq 0$ for all $i = \overline{1,p}$, $\sum_{i=1}^{p} a_i < 1$, $b_i \geq 0$ for all $i = \overline{1,q}$, $\sum_{i=1}^{q} b_i < 1$, $\xi_i \in \mathbb{N}$ for all $i = \overline{1,p}$, $1 \leq \xi_1 < \xi_2 < \cdots < \xi_p \leq N-1$, $\eta_i \in \mathbb{N}$ for all $i = \overline{1,q}$, and $1 \leq \eta_1 < \eta_2 < \cdots < \eta_q \leq N-1$, then the Green's function G_1 of problem* (3.1)–(3.2) *satisfies $G_1(n,j) \geq 0$ for all $n = \overline{0,N}$, $j = \overline{1,N-1}$. Moreover, if $y_n \geq 0$ for all $n = \overline{1,N-1}$, then the solution u_n, $n = \overline{0,N}$, of problem* (3.1)–(3.2) *satisfies $u_n \geq 0$ for all $n = \overline{0,N}$.*

Proof. By using the assumptions of this lemma, we have $\Delta_1 > 0$ and $G_1(n,j) \geq 0$ for all $n = \overline{0,N}$, $j = \overline{1,N-1}$, and so $u_n \geq 0$ for all $n = \overline{0,N}$. □

Lemma 3.1.5. *Assume that $a_i \geq 0$ for all $i = \overline{1,p}$, $\sum_{i=1}^{p} a_i < 1$, $b_i \geq 0$ for all $i = \overline{1,q}$, $\sum_{i=1}^{q} b_i < 1$, $\xi_i \in \mathbb{N}$ for all $i = \overline{1,p}$, $1 \leq \xi_1 < \xi_2 < \cdots < \xi_p \leq N-1$, $\eta_i \in \mathbb{N}$ for all $i = \overline{1,q}$, and $1 \leq \eta_1 < \eta_2 < \cdots < \eta_q \leq N-1$. Then the Green's function G_1 of problem* (3.1)–(3.2) *satisfies the following inequalities:*

(a) $G_1(n,j) \leq I_1(j), \quad \forall n = \overline{0,N}, \quad j = \overline{1,N-1}$, *where*

$$I_1(j) = g_0(j,j) + \frac{1}{\Delta_1} \left(N - \sum_{i=1}^{q} b_i \eta_i\right) \sum_{i=1}^{p} a_i g_0(\xi_i, j)$$
$$+ \frac{1}{\Delta_1} \left(N - \sum_{i=1}^{p} a_i (N - \xi_i)\right) \sum_{i=1}^{q} b_i g_0(\eta_i, j).$$

(b) *For every $c \in \{1,\ldots,[\![N/2]\!]\}$, we have*

$$\min_{n=\overline{c,N-c}} G_1(n,j) \geq \gamma_1 I_1(j) \geq \gamma_1 G_1(n',j), \quad \forall n' = \overline{0,N}, \quad j = \overline{1,N-1},$$

where

$$\gamma_1 = \min \left\{ \frac{c}{N-1}, \frac{c\left(1 - \sum_{k=1}^{q} b_k\right) + \sum_{i=1}^{q} b_i (N - \eta_i)}{N - \sum_{i=1}^{q} b_i \eta_i}, \frac{c\left(1 - \sum_{k=1}^{p} a_k\right) + \sum_{i=1}^{p} a_i \xi_i}{N - \sum_{i=1}^{p} a_i (N - \xi_i)} \right\} > 0.$$

Proof. The first inequality, (a), is evident. For the second inequality, (b), for $c \in \{1,\dots,[\![N/2]\!]\}$ and $n = \overline{c, N-c}$, $j = \overline{1, N-1}$, $n' = \overline{0, N}$, we deduce

$$\begin{aligned}
G_1(n,j) &\geq \frac{c}{N-1}g_0(j,j) + \frac{1}{\Delta_1}\left[c\left(1-\sum_{k=1}^{q}b_k\right)+\sum_{i=1}^{q}b_i(N-\eta_i)\right] \\
&\quad\times\sum_{i=1}^{p}a_i g_0(\xi_i,j) + \frac{1}{\Delta_1}\left[c\left(1-\sum_{k=1}^{p}a_k\right)+\sum_{i=1}^{p}a_i\xi_i\right]\sum_{i=1}^{q}b_i g_0(\eta_i,j) \\
&= \frac{c}{N-1}g_0(j,j) + \frac{1}{\Delta_1}\left[N\left(1-\sum_{k=1}^{q}b_k\right)+\sum_{i=1}^{q}b_i(N-\eta_i)\right] \\
&\quad\times\frac{c\left(1-\sum_{k=1}^{q}b_k\right)+\sum_{i=1}^{q}b_i(N-\eta_i)}{N\left(1-\sum_{k=1}^{q}b_k\right)+\sum_{i=1}^{q}b_i(N-\eta_i)}\sum_{i=1}^{p}a_i g_0(\xi_i,j) \\
&\quad+\frac{1}{\Delta_1}\left[N\left(1-\sum_{k=1}^{p}a_k\right)+\sum_{i=1}^{p}a_i\xi_i\right] \\
&\quad\times\frac{c\left(1-\sum_{k=1}^{p}a_k\right)+\sum_{i=1}^{p}a_i\xi_i}{N\left(1-\sum_{k=1}^{p}a_k\right)+\sum_{i=1}^{p}a_i\xi_i}\sum_{i=1}^{q}b_i g_0(\eta_i,j) \geq \gamma_1 I_1(j) \geq \gamma_1 G_1(n',j).
\end{aligned}$$

□

Lemma 3.1.6. *Assume that $a_i \geq 0$ for all $i = \overline{1,p}$, $\sum_{i=1}^{p} a_i < 1$, $b_i \geq 0$ for all $i = \overline{1,q}$, $\sum_{i=1}^{q} b_i < 1$, $\xi_i \in \mathbb{N}$ for all $i = \overline{1,p}$, $1 \leq \xi_1 < \xi_2 < \cdots < \xi_p \leq N-1$, $\eta_i \in \mathbb{N}$ for all $i = \overline{1,q}$, $1 \leq \eta_1 < \eta_2 < \cdots < \eta_q \leq N-1$, $c \in \{1,\dots,[\![N/2]\!]\}$, and $y_n \geq 0$ for all $n = \overline{1,N-1}$. Then the solution u_n, $n = \overline{0,N}$, of problem* (3.1)–(3.2) *satisfies the inequality* $\min_{n=\overline{c,N-c}} u_n \geq \gamma_1 \max_{m=\overline{0,N}} u_m$.

Proof. For $c \in \{1,\dots,[\![N/2]\!]\}$, $n = \overline{c,N-c}$, $m = \overline{0,N}$, we have

$$u_n = \sum_{j=1}^{N-1} G_1(n,j)y_j \geq \gamma_1 \sum_{j=1}^{N-1} I_1(j)y_j \geq \gamma_1 \sum_{j=1}^{N-1} G_1(m,j)y_j = \gamma_1 u_m.$$

Then $\min_{n=\overline{c,N-c}} u_n \geq \gamma_1 \max_{m=\overline{0,N}} u_m$. □

We can also formulate results similar to those in Lemmas 3.1.1–3.1.6 for the discrete boundary value problem:

$$\Delta^2 v_{n-1} + h_n = 0, \quad n = \overline{1,N-1}, \tag{3.7}$$

$$v_0 = \sum_{i=1}^{r} c_i v_{\zeta_i}, \quad v_N = \sum_{i=1}^{l} d_i v_{\rho_i}, \tag{3.8}$$

where N, r, $l \in \mathbb{N}$, $N \geq 2$, $c_i \in \mathbb{R}$ for all $i = \overline{1,r}$, $d_i \in \mathbb{R}$ for all $i = \overline{1,l}$, $\zeta_i \in \mathbb{N}$ for all $i = \overline{1,r}$, $\rho_i \in \mathbb{N}$ for all $i = \overline{1,l}$, $1 \leq \zeta_1 < \cdots < \zeta_r \leq N-1$, $1 \leq \rho_1 < \cdots < \rho_l \leq N-1$, and $h_n \in \mathbb{R}$ for all $n = \overline{1,N-1}$. We denote by Δ_2, γ_2, G_2, and I_2

the corresponding constants and functions for problem (3.7)–(3.8) defined in a similar manner as Δ_1, γ_1, G_1, and I_1, respectively.

3.1.3 Main results

In this section, we determine intervals for λ and μ for which there exist positive solutions with respect to a cone for our problem (S)–(BC). We also investigate the nonexistence of positive solutions of (S)–(BC).

We present the assumptions that we shall use in the sequel:

(A1) $\xi_i \in \mathbb{N}$ for all $i = \overline{1,p}$, $1 \le \xi_1 < \cdots < \xi_p \le N-1$, $a_i \ge 0$ for all $i = \overline{1,p}$, $\sum_{i=1}^{p} a_i < 1$, $\eta_i \in \mathbb{N}$ for all $i = \overline{1,q}$, $1 \le \eta_1 < \cdots < \eta_q \le N-1$, $b_i \ge 0$ for all $i = \overline{1,q}$, $\sum_{i=1}^{q} b_i < 1$, $\zeta_i \in \mathbb{N}$ for all $i = \overline{1,r}$, $1 \le \zeta_1 < \cdots < \zeta_r \le N-1$, $c_i \ge 0$ for all $i = \overline{1,r}$, $\sum_{i=1}^{r} c_i < 1$, $\rho_i \in \mathbb{N}$ for all $i = \overline{1,l}$, $1 \le \rho_1 < \cdots < \rho_l \le N-1$, $d_i \ge 0$ for all $i = \overline{1,l}$, $\sum_{i=1}^{l} d_i < 1$.

(A2) The constants s_n, $t_n \ge 0$ for all $n = \overline{1,N-1}$, and there exist $i_0, j_0 \in \{1,\dots,N-1\}$ such that $s_{i_0} > 0$, $t_{j_0} > 0$.

(A3) The functions $f, g : \{1,\dots,N-1\} \times [0,\infty) \times [0,\infty) \to [0,\infty)$ are continuous.

From assumption (A2), there exists $c \in \{1,\dots,[\![N/2]\!]\}$ such that $i_0, j_0 \in \{c,\dots, N-c\}$. We shall work in this section with this number c. This implies that $\sum_{j=c}^{N-c} s_j I_1(j) > 0$ and $\sum_{j=c}^{N-c} t_j I_2(j) > 0$.

For c defined above, we introduce the following extreme limits:

$$f_0^{\rm s} = \limsup_{u+v\to 0^+} \max_{n=\overline{1,N-1}} \frac{f(n,u,v)}{u+v}, \quad g_0^{\rm s} = \limsup_{u+v\to 0^+} \max_{n=\overline{1,N-1}} \frac{g(n,u,v)}{u+v},$$

$$f_0^{\rm i} = \liminf_{u+v\to 0^+} \min_{n=\overline{c,N-c}} \frac{f(n,u,v)}{u+v}, \quad g_0^{\rm i} = \liminf_{u+v\to 0^+} \min_{n=\overline{c,N-c}} \frac{g(n,u,v)}{u+v},$$

$$f_\infty^{\rm s} = \limsup_{u+v\to \infty} \max_{n=\overline{1,N-1}} \frac{f(n,u,v)}{u+v}, \quad g_\infty^{\rm s} = \limsup_{u+v\to \infty} \max_{n=\overline{1,N-1}} \frac{g(n,u,v)}{u+v},$$

$$f_\infty^{\rm i} = \liminf_{u+v\to \infty} \min_{n=\overline{c,N-c}} \frac{f(n,u,v)}{u+v}, \quad g_\infty^{\rm i} = \liminf_{u+v\to \infty} \min_{n=\overline{c,N-c}} \frac{g(n,u,v)}{u+v}.$$

In the definitions of the extreme limits above, the variables u and v are nonnegative.

By using the Green's functions G_1 and G_2 from Section 3.1.2, we can write our problem (S)–(BC) equivalently as the following system:

$$\begin{cases} u_n = \lambda \sum_{i=1}^{N-1} G_1(n,i) s_i f(i,u_i,v_i), & n = \overline{0,N}, \\ v_n = \mu \sum_{i=1}^{N-1} G_2(n,i) t_i g(i,u_i,v_i), & n = \overline{0,N}. \end{cases}$$

We consider the Banach space $X = \mathbb{R}^{N+1} = \{u = (u_0,u_1,\dots,u_N),\ u_i \in \mathbb{R},\ i = \overline{0,N}\}$ with the norm $\|u\| = \max_{n=\overline{0,N}} |u_n|$ and the Banach space $Y = X \times X$ with the norm $\|(u,v)\|_Y = \|u\| + \|v\|$.

For c above, we define the cone $P \subset Y$ by

$$P = \{(u,v) \in Y;\quad u = (u_n)_{n=\overline{0,N}},\quad v = (v_n)_{n=\overline{0,N}},\quad u_n \geq 0,\quad v_n \geq 0,$$
$$\forall n = \overline{0,N} \quad \text{and} \quad \min_{n=\overline{c,N-c}} (u_n + v_n) \geq \gamma \|(u,v)\|_Y\},$$

where $\gamma = \min\{\gamma_1, \gamma_2\}$, and γ_1 and γ_2 are defined in Section 3.1.2.

For λ, $\mu > 0$, we introduce the operators $T_1, T_2 : Y \to X$ and $\mathcal{T} : Y \to Y$ defined by

$$T_1(u,v) = (T_1(u,v))_{n=\overline{0,N}},\quad (T_1(u,v))_n = \lambda \sum_{i=1}^{N-1} G_1(n,i) s_i f(i,u_i,v_i),\quad n = \overline{0,N},$$

$$T_2(u,v) = (T_2(u,v))_{n=\overline{0,N}},\quad (T_2(u,v))_n = \mu \sum_{i=1}^{N-1} G_2(n,i) t_i g(i,u_i,v_i),\quad n = \overline{0,N},$$

and $\mathcal{T}(u,v) = (T_1(u,v), T_2(u,v))$, $(u,v) = ((u_n)_{n=\overline{0,N}}, (v_n)_{n=\overline{0,N}}) \in Y$, where G_1 and G_2 are the Green's functions defined in Section 3.1.2. The solutions of problem (S)–(BC) coincide with the fixed points of the operator $\mathcal{T}$ in the space Y. We can easily prove that under assumptions (A1)–(A3), the operator $\mathcal{T} : P \to P$ is completely continuous.

We denote $A = \sum_{j=c}^{N-c} s_j I_1(j)$, $B = \sum_{j=1}^{N-1} s_j I_1(j)$, $C = \sum_{j=c}^{N-c} t_j I_2(j)$, $D = \sum_{j=1}^{N-1} t_j I_2(j)$.

First, for f_0^s, g_0^s, f_∞^i, $g_\infty^i \in (0,\infty)$ and numbers α_1, $\alpha_2 \geq 0$ and $\tilde{\alpha}_1$, $\tilde{\alpha}_2 > 0$ such that $\alpha_1 + \alpha_2 = 1$ and $\tilde{\alpha}_1 + \tilde{\alpha}_2 = 1$, we define the numbers L_1, L_2, L_3, L_4, L_2', and L_4' by

$$L_1 = \frac{\alpha_1}{\gamma\gamma_1 f_\infty^i A},\quad L_2 = \frac{\tilde{\alpha}_1}{f_0^s B},\quad L_3 = \frac{\alpha_2}{\gamma\gamma_2 g_\infty^i C},\quad L_4 = \frac{\tilde{\alpha}_2}{g_0^s D},$$
$$L_2' = \frac{1}{f_0^s B},\quad L_4' = \frac{1}{g_0^s D}.$$

Theorem 3.1.1. *Assume that (A1)–(A3) hold, and that α_1, $\alpha_2 \geq 0$ and $\tilde{\alpha}_1$, $\tilde{\alpha}_2 > 0$ such that $\alpha_1 + \alpha_2 = 1$ and $\tilde{\alpha}_1 + \tilde{\alpha}_2 = 1$.*

(1) If f_0^s, g_0^s, f_∞^i, $g_\infty^i \in (0,\infty)$, $L_1 < L_2$, and $L_3 < L_4$, then for each $\lambda \in (L_1, L_2)$ and $\mu \in (L_3, L_4)$ there exists a positive solution $((u_n)_{n=\overline{0,N}}, (v_n)_{n=\overline{0,N}})$ for (S)–(BC).

(2) If $f_0^s = 0$, g_0^s, f_∞^i, $g_\infty^i \in (0,\infty)$ and $L_3 < L_4'$, then for each $\lambda \in (L_1, \infty)$ and $\mu \in (L_3, L_4')$ there exists a positive solution $((u_n)_{n=\overline{0,N}}, (v_n)_{n=\overline{0,N}})$ for (S)–(BC).

(3) If $g_0^s = 0$, f_0^s, f_∞^i, $g_\infty^i \in (0,\infty)$ and $L_1 < L_2'$, then for each $\lambda \in (L_1, L_2')$ and $\mu \in (L_3, \infty)$ there exists a positive solution $((u_n)_{n=\overline{0,N}}, (v_n)_{n=\overline{0,N}})$ for (S)–(BC).

(4) If $f_0^s = g_0^s = 0$ and f_∞^i, $g_\infty^i \in (0,\infty)$, then for each $\lambda \in (L_1, \infty)$ and $\mu \in (L_3, \infty)$ there exists a positive solution $((u_n)_{n=\overline{0,N}}, (v_n)_{n=\overline{0,N}})$ for (S)–(BC).

(5) If $\{f_0^s, g_0^s, f_\infty^i \in (0,\infty), g_\infty^i = \infty\}$ or $\{f_0^s, g_0^s, g_\infty^i \in (0,\infty), f_\infty^i = \infty\}$ or $\{f_0^s, g_0^s \in (0,\infty), f_\infty^i = g_\infty^i = \infty\}$, then for each $\lambda \in (0, L_2)$ and $\mu \in (0, L_4)$ there exists a positive solution $((u_n)_{n=\overline{0,N}}, (v_n)_{n=\overline{0,N}})$ for (S)–(BC).

(6) If $\{f_0^s = 0, g_0^s, f_\infty^i \in (0,\infty), g_\infty^i = \infty\}$ or $\{f_0^s = 0, f_\infty^i = \infty, g_0^s, g_\infty^i \in (0,\infty)\}$ or $\{f_0^s = 0, g_0^s \in (0,\infty), f_\infty^i = g_\infty^i = \infty\}$, then for each $\lambda \in (0,\infty)$ and $\mu \in (0, L_4')$ there exists a positive solution $((u_n)_{n=\overline{0,N}}, (v_n)_{n=\overline{0,N}})$ for (S)–(BC).

(7) If $\{f_0^s, f_\infty^i \in (0,\infty),\ g_0^s = 0,\ g_\infty^i = \infty\}$ *or* $\{f_0^s, g_\infty^i \in (0,\infty),\ g_0^s = 0,\ f_\infty^i = \infty\}$ *or* $\{f_0^s \in (0,\infty),\ g_0^s = 0,\ f_\infty^i = g_\infty^i = \infty\}$*, then for each* $\lambda \in (0, L_2')$ *and* $\mu \in (0,\infty)$ *there exists a positive solution* $((u_n)_{n=\overline{0,N}}, (v_n)_{n=\overline{0,N}})$ *for (S)–(BC).*

(8) If $\{f_0^s = g_0^s = 0,\ f_\infty^i \in (0,\infty),\ g_\infty^i = \infty\}$ *or* $\{f_0^s = g_0^s = 0,\ f_\infty^i = \infty,\ g_\infty^i \in (0,\infty)\}$ *or* $\{f_0^s = g_0^s = 0,\ f_\infty^i = g_\infty^i = \infty\}$*, then for each* $\lambda \in (0,\infty)$ *and* $\mu \in (0,\infty)$ *there exists a positive solution* $((u_n)_{n=\overline{0,N}}, (v_n)_{n=\overline{0,N}})$ *for (S)–(BC).*

Proof. We consider the above cone $P \subset Y$ and the operators T_1, T_2, and $\mathcal{T}$. Because the proofs of the above cases are similar, in what follows we shall prove one of them—namely, the second case of (7). So, we suppose f_0^s, $g_\infty^i \in (0,\infty)$, $g_0^s = 0$, and $f_\infty^i = \infty$. Let $\lambda \in (0, L_2')$—that is, $\lambda \in \left(0, \frac{1}{f_0^s B}\right)$—and $\mu \in (0,\infty)$. We choose $\tilde{\alpha}_1' \in (\lambda f_0^s B, 1)$ and $\alpha_2' > 0$, $\alpha_2' < \min\{\mu\gamma\gamma_2 g_\infty^i C, 1\}$. Let $\alpha_1' = 1 - \alpha_2'$, $\tilde{\alpha}_2' = 1 - \tilde{\alpha}_1'$ and $\varepsilon > 0$, $\varepsilon < g_\infty^i$ be positive numbers such that

$$\frac{\varepsilon\alpha_1'}{\gamma\gamma_1 A} \le \lambda, \quad \frac{\alpha_2'}{\gamma\gamma_2(g_\infty^i - \varepsilon)C} \le \mu, \quad \frac{\tilde{\alpha}_1'}{(f_0^s + \varepsilon)B} \ge \lambda, \quad \frac{\tilde{\alpha}_2'}{\varepsilon D} \ge \mu.$$

By using (A3) and the definition of f_0^s and g_0^s, we deduce that there exists $K_1 > 0$ such that $f(n,u,v) \le (f_0^s + \varepsilon)(u+v)$ and $g(n,u,v) \le \varepsilon(u+v)$ for all $n = \overline{1, N-1}$ and $u,\ v \ge 0$ with $0 \le u + v \le K_1$. We define the set $\Omega_1 = \{(u,v) \in Y,\ \|(u,v)\|_Y < K_1\}$. Now let $(u,v) \in P \cap \partial\Omega_1$—that is, $(u,v) \in P$ with $\|(u,v)\|_Y = K_1$ or equivalently $\|u\| + \|v\| = K_1$. Then $u_n + v_n \le K_1$ for all $n = \overline{0,N}$, and by Lemma 3.1.5, we obtain

$$\begin{aligned}(T_1(u,v))_n &\le \lambda \sum_{j=1}^{N-1} s_j I_1(j) f(j,u_j,v_j) \le \lambda \sum_{j=1}^{N-1} s_j I_1(j)(f_0^s + \varepsilon)(u_j + v_j)\\ &\le \lambda(f_0^s + \varepsilon) \sum_{j=1}^{N-1} s_j I_1(j)(\|u\| + \|v\|) \le \tilde{\alpha}_1'(\|u\| + \|v\|) = \tilde{\alpha}_1' \|(u,v)\|_Y,\\ &\forall n = \overline{0,N}.\end{aligned}$$

Therefore, $\|T_1(u,v)\| \le \tilde{\alpha}_1' \|(u,v)\|_Y$. In a similar manner, we conclude

$$\begin{aligned}(T_2(u,v))_n &\le \mu \sum_{j=1}^{N-1} t_j I_2(j) g(j,u_j,v_j) \le \mu \sum_{j=1}^{N-1} t_j I_2(j)\varepsilon(u_j + v_j)\\ &\le \mu\varepsilon \sum_{j=1}^{N-1} t_j I_2(j)(\|u\| + \|v\|) \le \tilde{\alpha}_2'(\|u\| + \|v\|) = \tilde{\alpha}_2' \|(u,v)\|_Y,\\ &\forall n = \overline{0,N}.\end{aligned}$$

Hence, $\|T_2(u,v)\| \le \tilde{\alpha}_2' \|(u,v)\|_Y$.

Then for $(u,v) \in P \cap \partial\Omega_1$, we deduce

$$\|\mathcal{T}(u,v)\|_Y = \|T_1(u,v)\| + \|T_2(u,v)\| \le \tilde{\alpha}_1' \|(u,v)\|_Y + \tilde{\alpha}_2' \|(u,v)\|_Y = \|(u,v)\|_Y. \tag{3.9}$$

By the definitions of $f_\infty^{\rm i}$ and $g_\infty^{\rm i}$, there exists $\bar{K}_2 > 0$ such that $f(n,u,v) \geq \frac{1}{\varepsilon}(u+v)$ and $g(n,u,v) \geq (g_\infty^{\rm i} - \varepsilon)(u+v)$ for all $u,\ v \geq 0$ with $u+v \geq \bar{K}_2$ and $n = \overline{c, N-c}$. We consider $K_2 = \max\{2K_1, \bar{K}_2/\gamma\}$, and we define $\Omega_2 = \{(u,v) \in Y, \quad \|(u,v)\|_Y < K_2\}$. Then for $(u,v) \in P$ with $\|(u,v)\|_Y = K_2$, we obtain

$$u_n + v_n \geq \min_{n=\overline{c,N-c}} (u_n + v_n) \geq \gamma \|(u,v)\|_Y = \gamma K_2 \geq \bar{K}_2, \quad \forall n = \overline{c, N-c}.$$

Then, by Lemma 3.1.5, we conclude

$$(T_1(u,v))_c \geq \lambda\gamma_1 \sum_{j=1}^{N-1} s_j I_1(j) f(j, u_j, v_j) \geq \lambda\gamma_1 \sum_{j=c}^{N-c} s_j I_1(j) f(j, u_j, v_j)$$
$$\geq \lambda\gamma_1 \sum_{j=c}^{N-c} s_j I_1(j) \frac{1}{\varepsilon}(u_j+v_j) \geq \frac{\lambda\gamma_1}{\varepsilon} \sum_{j=c}^{N-c} s_j I_1(j) \gamma \|(u,v)\|_Y \geq \alpha_1' \|(u,v)\|_Y.$$

So, $\|T_1(u,v)\| \geq (T_1(u,v))_c \geq \alpha_1' \|(u,v)\|_Y$.

In a similar manner, we deduce

$$(T_2(u,v))_c \geq \mu\gamma_2 \sum_{j=1}^{N-1} t_j I_2(j) g(j, u_j, v_j) \geq \mu\gamma_2 \sum_{j=c}^{N-c} t_j I_2(j) g(j, u_j, v_j)$$
$$\geq \mu\gamma_2 \sum_{j=c}^{N-c} t_j I_2(j) (g_\infty^{\rm i} - \varepsilon)(u_j + v_j) \geq \mu\gamma_2 (g_\infty^{\rm i} - \varepsilon)$$
$$\times \sum_{j=c}^{N-c} t_j I_2(j) \gamma \|(u,v)\|_Y \geq \alpha_2' \|(u,v)\|_Y.$$

So, $\|T_2(u,v)\| \geq (T_2(u,v))_c \geq \alpha_2' \|(u,v)\|_Y$.

Hence, for $(u,v) \in P \cap \partial\Omega_2$, we obtain

$$\|\mathcal{T}(u,v)\|_Y = \|T_1(u,v)\| + \|T_2(u,v)\| \geq (\alpha_1' + \alpha_2') \|(u,v)\|_Y = \|(u,v)\|_Y. \quad (3.10)$$

By using (3.9), (3.10), and Theorem 1.1.1 (1), we conclude that $\mathcal{T}$ has a fixed point $(u,v) \in P \cap (\overline{\Omega}_2 \setminus \Omega_1)$ such that $K_1 \leq \|u\| + \|v\| \leq K_2$. □

Remark 3.1.1. We mention that in Theorem 3.1.1 we have the possibility to choose $\alpha_1 = 0$ or $\alpha_2 = 0$. Therefore, each of the first four cases contains three subcases. For example, in the first case $f_0^{\rm s},\ g_0^{\rm s},\ f_\infty^{\rm i},\ g_\infty^{\rm i} \in (0,\infty)$, we have the following situations:

(a) If $\alpha_1, \alpha_2 \in (0,1)$, $\alpha_1 + \alpha_2 = 1$, and $L_1 < L_2$, $L_3 < L_4$, then $\lambda \in (L_1, L_2)$ and $\mu \in (L_3, L_4)$.
(b) If $\alpha_1 = 1$, $\alpha_2 = 0$ and $L_1' < L_2$, then $\lambda \in (L_1', L_2)$ and $\mu \in (0, L_4)$, where $L_1' = \frac{1}{\gamma\gamma_1 f_\infty^{\rm i} A}$.
(c) If $\alpha_1 = 0$, $\alpha_2 = 1$ and $L_3' < L_4$, then $\lambda \in (0, L_2)$ and $\mu \in (L_3', L_4)$, where $L_3' = \frac{1}{\gamma\gamma_2 g_\infty^{\rm i} C}$.

In what follows, for $f_0^{\rm i},\ g_0^{\rm i},\ f_\infty^{\rm s},\ g_\infty^{\rm s} \in (0,\infty)$ and numbers $\alpha_1, \alpha_2 \geq 0$ and $\tilde{\alpha}_1$, $\tilde{\alpha}_2 > 0$ such that $\alpha_1 + \alpha_2 = 1$ and $\tilde{\alpha}_1 + \tilde{\alpha}_2 = 1$, we define the numbers $\tilde{L}_1$, $\tilde{L}_2$, $\tilde{L}_3$, $\tilde{L}_4$, $\tilde{L}_2'$, and $\tilde{L}_4'$ by

$$\tilde{L}_1 = \frac{\alpha_1}{\gamma\gamma_1 f_0^{\rm i} A}, \quad \tilde{L}_2 = \frac{\tilde{\alpha}_1}{f_\infty^{\rm s} B}, \quad \tilde{L}_3 = \frac{\alpha_2}{\gamma\gamma_2 g_0^{\rm i} C}, \quad \tilde{L}_4 = \frac{\tilde{\alpha}_2}{g_\infty^{\rm s} D},$$
$$\tilde{L}_2' = \frac{1}{f_\infty^{\rm s} B}, \quad \tilde{L}_4' = \frac{1}{g_\infty^{\rm s} D}.$$

Theorem 3.1.2. *Assume that (A1)–(A3) hold, and that $\alpha_1, \alpha_2 \ge 0$ and $\tilde{\alpha}_1, \tilde{\alpha}_2 > 0$ such that $\alpha_1 + \alpha_2 = 1$ and $\tilde{\alpha}_1 + \tilde{\alpha}_2 = 1$.*

(1) If $f_0^{\rm i}, g_0^{\rm i}, f_\infty^{\rm s}, g_\infty^{\rm s} \in (0,\infty)$, $\tilde{L}_1 < \tilde{L}_2$, and $\tilde{L}_3 < \tilde{L}_4$, then for each $\lambda \in (\tilde{L}_1, \tilde{L}_2)$ and $\mu \in (\tilde{L}_3, \tilde{L}_4)$ there exists a positive solution $((u_n)_{n=\overline{0,N}}, (v_n)_{n=\overline{0,N}})$ for (S)–(BC).

(2) If $f_0^{\rm i}, g_0^{\rm i}, f_\infty^{\rm s} \in (0,\infty)$, $g_\infty^{\rm s} = 0$, and $\tilde{L}_1 < \tilde{L}_2'$, then for each $\lambda \in (\tilde{L}_1, \tilde{L}_2')$ and $\mu \in (\tilde{L}_3, \infty)$ there exists a positive solution $((u_n)_{n=\overline{0,N}}, (v_n)_{n=\overline{0,N}})$ for (S)–(BC).

(3) If $f_0^{\rm i}, g_0^{\rm i}, g_\infty^{\rm s} \in (0,\infty)$, $f_\infty^{\rm s} = 0$, and $\tilde{L}_3 < \tilde{L}_4'$, then for each $\lambda \in (\tilde{L}_1, \infty)$ and $\mu \in (\tilde{L}_3, \tilde{L}_4')$ there exists a positive solution $((u_n)_{n=\overline{0,N}}, (v_n)_{n=\overline{0,N}})$ for (S)–(BC).

(4) If $f_0^{\rm i}, g_0^{\rm i} \in (0,\infty)$, and $f_\infty^{\rm s} = g_\infty^{\rm s} = 0$, then for each $\lambda \in (\tilde{L}_1, \infty)$ and $\mu \in (\tilde{L}_3, \infty)$ there exists a positive solution $((u_n)_{n=\overline{0,N}}, (v_n)_{n=\overline{0,N}})$ for (S)–(BC).

(5) If $\{f_0^{\rm i} = \infty, g_0^{\rm i}, f_\infty^{\rm s}, g_\infty^{\rm s} \in (0,\infty)\}$ or $\{f_0^{\rm i}, f_\infty^{\rm s}, g_\infty^{\rm s} \in (0,\infty), g_0^{\rm i} = \infty\}$ or $\{f_0^{\rm i} = g_0^{\rm i} = \infty, f_\infty^{\rm s}, g_\infty^{\rm s} \in (0,\infty)\}$, then for each $\lambda \in (0, \tilde{L}_2)$ and $\mu \in (0, \tilde{L}_4)$ there exists a positive solution $((u_n)_{n=\overline{0,N}}, (v_n)_{n=\overline{0,N}})$ for (S)–(BC).

(6) If $\{f_0^{\rm i} = \infty, g_0^{\rm i}, f_\infty^{\rm s} \in (0,\infty), g_\infty^{\rm s} = 0\}$ or $\{f_0^{\rm i}, f_\infty^{\rm s} \in (0,\infty), g_0^{\rm i} = \infty, g_\infty^{\rm s} = 0\}$ or $\{f_0^{\rm i} = g_0^{\rm i} = \infty, f_\infty^{\rm s} \in (0,\infty), g_\infty^{\rm s} = 0\}$, then for each $\lambda \in (0, \tilde{L}_2')$ and $\mu \in (0,\infty)$ there exists a positive solution $((u_n)_{n=\overline{0,N}}, (v_n)_{n=\overline{0,N}})$ for (S)–(BC).

(7) If $\{f_0^{\rm i} = \infty, g_0^{\rm i}, g_\infty^{\rm s} \in (0,\infty), f_\infty^{\rm s} = 0\}$ or $\{f_0^{\rm i}, g_\infty^{\rm s} \in (0,\infty), g_0^{\rm i} = \infty, f_\infty^{\rm s} = 0\}$ or $\{f_0^{\rm i} = g_0^{\rm i} = \infty, f_\infty^{\rm s} = 0, g_\infty^{\rm s} \in (0,\infty)\}$, then for each $\lambda \in (0,\infty)$ and $\mu \in (0, \tilde{L}_4')$ there exists a positive solution $((u_n)_{n=\overline{0,N}}, (v_n)_{n=\overline{0,N}})$ for (S)–(BC).

(8) If $\{f_0^{\rm i} = \infty, g_0^{\rm i} \in (0,\infty), f_\infty^{\rm s} = g_\infty^{\rm s} = 0\}$ or $\{f_0^{\rm i} \in (0,\infty), g_0^{\rm i} = \infty, f_\infty^{\rm s} = g_\infty^{\rm s} = 0\}$ or $\{f_0^{\rm i} = g_0^{\rm i} = \infty, f_\infty^{\rm s} = g_\infty^{\rm s} = 0\}$, then for each $\lambda \in (0,\infty)$ and $\mu \in (0,\infty)$ there exists a positive solution $((u_n)_{n=\overline{0,N}}, (v_n)_{n=\overline{0,N}})$ for (S)–(BC).

Proof. We consider the above cone $P \subset Y$ and the operators T_1, T_2, and $\mathcal{T}$. Because the proofs of the above cases are similar, in what follows we shall prove one of them—namely, case (3). So, we suppose $f_0^{\rm i}, g_0^{\rm i}, g_\infty^{\rm s} \in (0,\infty)$, $f_\infty^{\rm s} = 0$, and $\tilde{L}_3 < \tilde{L}_4'$. Let $\lambda \in (\tilde{L}_1, \infty)$ and $\mu \in (\tilde{L}_3, \tilde{L}_4')$—that is, $\lambda \in \left(\frac{\alpha_1}{\gamma\gamma_1 f_0^{\rm i} A}, \infty\right)$ and $\mu \in \left(\frac{\alpha_2}{\gamma\gamma_2 g_0^{\rm i} C}, \frac{1}{g_\infty^{\rm s} D}\right)$. We choose $\tilde{\alpha}_2' \in (\mu g_\infty^{\rm s} D, 1)$. Let $\tilde{\alpha}_1' = 1 - \tilde{\alpha}_2'$ and let $\varepsilon > 0$ be a positive number such that $\varepsilon < \min\{f_0^{\rm i}, g_0^{\rm i}\}$ and

$$\frac{\alpha_1}{\gamma\gamma_1 (f_0^{\rm i} - \varepsilon) A} \le \lambda, \quad \frac{\alpha_2}{\gamma\gamma_2 (g_0^{\rm i} - \varepsilon) C} \le \mu, \quad \frac{\tilde{\alpha}_1'}{\varepsilon B} \ge \lambda, \quad \frac{\tilde{\alpha}_2'}{(g_\infty^{\rm s} + \varepsilon) D} \ge \mu.$$

By the definitions of $f_0^{\rm i}, g_0^{\rm i} \in (0,\infty)$, we deduce that there exists $K_3 > 0$ such that $f(n,u,v) \ge (f_0^{\rm i} - \varepsilon)(u+v)$ and $g(n,u,v) \ge (g_0^{\rm i} - \varepsilon)(u+v)$ for all $u, v \ge 0$ with $0 \le u+v \le K_3$ and $n = \overline{c, N-c}$. We denote $\Omega_3 = \{(u,v) \in Y, \quad \|(u,v)\|_Y < K_3\}$.

Let $(u, v) \in P$ with $\|(u, v)\|_Y = K_3$—that is, $\|u\| + \|v\| = K_3$. Because $u_n + v_n \leq \|u\| + \|v\| = K_3$ for all $n = \overline{0, N}$, then using Lemma 3.1.5, we obtain

$$(T_1(u,v))_c \geq \lambda\gamma_1 \sum_{j=c}^{N-c} s_j I_1(j) f(j, u_j, v_j) \geq \lambda\gamma_1 \sum_{j=c}^{N-c} s_j I_1(j)(f_0^{\mathrm{i}} - \varepsilon)(u_j + v_j)$$

$$\geq \lambda\gamma\gamma_1(f_0^{\mathrm{i}} - \varepsilon) \sum_{j=c}^{N-c} s_j I_1(j)(\|u\| + \|v\|) \geq \alpha_1 \|(u,v)\|_Y.$$

Therefore, $\|T_1(u,v)\| \geq (T_1(u,v))_c \geq \alpha_1\|(u,v)\|_Y$.

In a similar manner, we conclude

$$(T_2(u,v))_c \geq \mu\gamma_2 \sum_{j=c}^{N-c} t_j I_2(j) g(j, u_j, v_j) \geq \mu\gamma_2 \sum_{j=c}^{N-c} t_j I_2(j)(g_0^{\mathrm{i}} - \varepsilon)(u_j + v_j)$$

$$\geq \mu\gamma\gamma_2(g_0^{\mathrm{i}} - \varepsilon) \sum_{j=c}^{N-c} t_j I_2(j)(\|u\| + \|v\|) \geq \alpha_2 \|(u,v)\|_Y.$$

So, $\|T_2(u,v)\| \geq (T_2(u,v))_c \geq \alpha_2\|(u,v)\|_Y$.

Thus, for an arbitrary element $(u, v) \in P \cap \partial\Omega_3$, we deduce that

$$\|\mathcal{T}(u,v)\|_Y \geq (\alpha_1 + \alpha_2)\|(u,v)\|_Y = \|(u,v)\|_Y. \tag{3.11}$$

Now, we define the functions $f^*, g^* : \{1, \ldots, N-1\} \times \mathbb{R}_+ \to \mathbb{R}_+$, $f^*(n,x) = \max_{0 \leq u+v \leq x} f(n,u,v)$, $g^*(n,x) = \max_{0 \leq u+v \leq x} g(n,u,v)$ for all $x \geq 0$ and $n = \overline{1, N-1}$. Then $f(n,u,v) \leq f^*(n,x)$, $g(n,u,v) \leq g^*(n,x)$ for all $n = \overline{1, N-1}$, $u, v \geq 0$, and $u + v \leq x$. The functions $f^*(n, \cdot)$ and $g^*(n, \cdot)$ are nondecreasing for every $n \in \{1, \ldots, N-1\}$, and they satisfy the conditions

$$\lim_{x\to\infty} \max_{n=\overline{1,\,N-1}} \frac{f^*(n,x)}{x} = 0, \quad \limsup_{x\to\infty} \max_{n=\overline{1,\,N-1}} \frac{g^*(n,x)}{x} \leq g_\infty^{\mathrm{s}}.$$

Therefore, for $\varepsilon > 0$ there exists $\bar{K}_4 > 0$ such that

$$\frac{f^*(n,x)}{x} \leq \lim_{x\to\infty} \max_{n=\overline{1,\,N-1}} \frac{f^*(n,x)}{x} + \varepsilon = \varepsilon,$$

$$\frac{g^*(n,x)}{x} \leq \limsup_{x\to\infty} \max_{n=\overline{1,\,N-1}} \frac{g^*(n,x)}{x} + \varepsilon \leq g_\infty^{\mathrm{s}} + \varepsilon,$$

for all $x \geq \bar{K}_4$ and $n = \overline{1, N-1}$, and so $f^*(n,x) \leq \varepsilon x$ and $g^*(n,x) \leq (g_\infty^{\mathrm{s}} + \varepsilon)x$.

We consider $K_4 = \max\{2K_3, \bar{K}_4\}$ and we denote $\Omega_4 = \{(u,v) \in Y, \ \|(u,v)\|_Y < K_4\}$. Let $(u,v) \in P \cap \partial\Omega_4$. By the definitions of f^* and g^*, for any $j = \overline{1, N-1}$ and $n = \overline{0, N}$ we obtain

$$f(j, u_n, v_n) \leq f^*(j, u_n + v_n) \leq f^*(j, \|u\| + \|v\|) = f^*(j, \|(u,v)\|_Y),$$
$$g(j, u_n, v_n) \leq g^*(j, u_n + v_n) \leq g^*(j, \|u\| + \|v\|) = g^*(j, \|(u,v)\|_Y).$$

Then for all $n = \overline{0, N}$, we deduce

$$(T_1(u,v))_n \le \lambda \sum_{j=1}^{N-1} s_j I_1(j) f(j, u_j, v_j) \le \lambda \sum_{j=1}^{N-1} s_j I_1(j) f^*(j, \|(u,v)\|_Y)$$
$$\le \lambda\varepsilon \sum_{j=1}^{N-1} s_j I_1(j) \|(u,v)\|_Y \le \tilde{\alpha}_1' \|(u,v)\|_Y,$$

and so $\|T_1(u,v)\| \le \tilde{\alpha}_1' \|(u,v)\|_Y$.

In a similar manner, we conclude

$$(T_2(u,v))_n \le \mu \sum_{j=1}^{N-1} t_j I_2(j) g(j, u_j, v_j) \le \mu \sum_{j=1}^{N-1} t_j I_2(j) g^*(j, \|(u,v)\|_Y)$$
$$\le \mu(g_\infty^{\mathrm{s}} + \varepsilon) \sum_{j=1}^{N-1} t_j I_2(j) \|(u,v)\|_Y \le \tilde{\alpha}_2' \|(u,v)\|_Y,$$

and so $\|T_2(u,v)\| \le \tilde{\alpha}_2' \|(u,v)\|_Y$.

Therefore, for $(u,v) \in P \cap \partial\Omega_4$, it follows that

$$\|\mathcal{T}(u,v)\|_Y \le (\tilde{\alpha}_1' + \tilde{\alpha}_2') \|(u,v)\|_Y = \|(u,v)\|_Y. \tag{3.12}$$

By using (3.11), (3.12), and Theorem 1.1.1 (2), we deduce that $\mathcal{T}$ has a fixed point $(u,v) \in P \cap (\overline{\Omega_4} \setminus \Omega_3)$ such that $K_3 \le \|(u,v)\|_Y \le K_4$. □

Now, we present intervals for λ and μ for which there exists no positive solution of problem (S)–(BC).

Theorem 3.1.3. *Assume that (A1)–(A3) hold. If* $f_0^{\mathrm{s}}, f_\infty^{\mathrm{s}}, g_0^{\mathrm{s}}, g_\infty^{\mathrm{s}} < \infty$, *then there exist positive constants* λ_0 *and* μ_0 *such that for every* $\lambda \in (0, \lambda_0)$ *and* $\mu \in (0, \mu_0)$ *the boundary value problem (S)–(BC) has no positive solution.*

Proof. From the definitions of $f_0^{\mathrm{s}}, f_\infty^{\mathrm{s}}, g_0^{\mathrm{s}}, g_\infty^{\mathrm{s}} < \infty$, we deduce that there exist $M_1, M_2 > 0$ such that

$$f(j,u,v) \le M_1(u+v), \quad g(j,u,v) \le M_2(u+v), \quad \forall j = \overline{1, N-1}, \quad u, v \ge 0.$$

We define $\lambda_0 = \frac{1}{2M_1 B}$ and $\mu_0 = \frac{1}{2M_2 D}$, where $B = \sum_{j=1}^{N-1} s_j I_1(j)$ and $D = \sum_{j=1}^{N-1} t_j I_2(j)$. We shall show that for every $\lambda \in (0, \lambda_0)$ and $\mu \in (0, \mu_0)$, problem (S)–(BC) has no positive solution.

Let $\lambda \in (0, \lambda_0)$ and $\mu \in (0, \mu_0)$. We suppose that (S)–(BC) has a positive solution $((u_n)_{n=\overline{0,N}}, (v_n)_{n=\overline{0,N}})$. Then we have

$$u_n = (T_1(u,v))_n = \lambda \sum_{j=1}^{N-1} G_1(n,j) s_j f(j, u_j, v_j) \le \lambda \sum_{j=1}^{N-1} s_j I_1(j) f(j, u_j, v_j)$$
$$\le \lambda M_1 \sum_{j=1}^{N-1} s_j I_1(j)(u_j + v_j) \le \lambda M_1 (\|u\| + \|v\|) \sum_{j=1}^{N-1} s_j I_1(j)$$
$$= \lambda M_1 B \|(u,v)\|_Y, \ \forall n = \overline{0, N}.$$

Therefore, we conclude

$$\|u\| \le \lambda M_1 B\|(u,v)\|_Y < \lambda_0 M_1 B\|(u,v)\|_Y = \frac{1}{2}\|(u,v)\|_Y.$$

In a similar manner, we obtain

$$\begin{aligned} v_n &= (T_2(u,v))_n = \mu \sum_{j=1}^{N-1} G_2(n,j) t_j g(j,u_j,v_j) \le \mu \sum_{j=1}^{N-1} t_j I_2(j) g(j,u_j,v_j) \\ &\le \mu M_2 \sum_{j=1}^{N-1} t_j I_2(j)(u_j+v_j) \le \mu M_2(\|u\|+\|v\|) \sum_{j=1}^{N-1} t_j I_2(j) \\ &= \mu M_2 D\|(u,v)\|_Y, \ \forall n = \overline{0,N}. \end{aligned}$$

Therefore, we deduce

$$\|v\| \le \mu M_2 D\|(u,v)\|_Y < \mu_0 M_2 D\|(u,v)\|_Y = \frac{1}{2}\|(u,v)\|_Y.$$

Hence, $\|(u,v)\|_Y = \|u\| + \|v\| < \frac{1}{2}\|(u,v)\|_Y + \frac{1}{2}\|(u,v)\|_Y = \|(u,v)\|_Y$, which is a contradiction. So, the boundary value problem (S)–(BC) has no positive solution. □

Theorem 3.1.4. *Assume that (A1)–(A3) hold. If $f_0^{\rm i}, f_\infty^{\rm i} > 0$ and $f(j,u,v) > 0$ for all $j = \overline{c, N-c}$, $u \ge 0$, $v \ge 0$, $u+v > 0$, then there exists a positive constant $\tilde{\lambda}_0$ such that for every $\lambda > \tilde{\lambda}_0$ and $\mu > 0$ the boundary value problem (S)–(BC) has no positive solution.*

Proof. From the definitions of $f_0^{\rm i}, f_\infty^{\rm i} > 0$, we deduce that there exists $m_1 > 0$ such that $f(j,u,v) \ge m_1(u+v)$, for all $j = \overline{c, N-c}$ and $u, v \ge 0$. We define $\tilde{\lambda}_0 = \frac{1}{\gamma\gamma_1 m_1 A} > 0$, where $A = \sum_{j=c}^{N-c} s_j I_1(j)$. We shall show that for every $\lambda > \tilde{\lambda}_0$ and $\mu > 0$, problem (S)–(BC) has no positive solution. Let $\lambda > \tilde{\lambda}_0$ and $\mu > 0$. We suppose that (S)–(BC) has a positive solution $((u_n)_{n=\overline{0,N}}, (v_n)_{n=\overline{0,N}})$. Then we obtain

$$\begin{aligned} u_c &= (T_1(u,v))_c = \lambda \sum_{j=1}^{N-1} G_1(c,j) s_j f(j,u_j,v_j) \ge \lambda \sum_{j=c}^{N-c} G_1(c,j) s_j f(j,u_j,v_j) \\ &\ge \lambda m_1 \sum_{j=c}^{N-c} G_1(c,j) s_j (u_j+v_j) \ge \lambda m_1 \gamma_1 \sum_{j=c}^{N-c} s_j I_1(j) \gamma(\|u\|+\|v\|) \\ &= \lambda m_1 \gamma\gamma_1 A\|(u,v)\|_Y. \end{aligned}$$

Therefore, we conclude

$$\|u\| \ge u_c \ge \lambda m_1 \gamma\gamma_1 A\|(u,v)\|_Y > \tilde{\lambda}_0 m_1 \gamma\gamma_1 A\|(u,v)\|_Y = \|(u,v)\|_Y,$$

and so $\|(u,v)\|_Y = \|u\|+\|v\| \ge \|u\| > \|(u,v)\|_Y$, which is a contradiction. Therefore, the boundary value problem (S)–(BC) has no positive solution. □

Theorem 3.1.5. *Assume that (A1)–(A3) hold. If $g_0^{\rm i}, g_\infty^{\rm i} > 0$ and $g(j,u,v) > 0$ for all $j = \overline{c, N-c}$, $u \ge 0$, $v \ge 0$, $u+v > 0$, then there exists a positive constant $\tilde{\mu}_0$*

such that for every $\mu > \tilde{\mu}_0$ and $\lambda > 0$ the boundary value problem (S)–(BC) has no positive solution.

Proof. From the definitions of $g_0^{\mathrm{i}}, g_\infty^{\mathrm{i}} > 0$, we deduce that there exists $m_2 > 0$ such that $g(j,u,v) \geq m_2(u+v)$, for all $j = \overline{c, N-c}$ and $u, v \geq 0$. We define $\tilde{\mu}_0 = \frac{1}{\gamma\gamma_2 m_2 C} > 0$, where $C = \sum_{j=c}^{N-c} t_j I_2(j)$. We shall show that for every $\mu > \tilde{\mu}_0$ and $\lambda > 0$, problem (S)–(BC) has no positive solution. Let $\mu > \tilde{\mu}_0$ and $\lambda > 0$. We suppose that (S)–(BC) has a positive solution $((u_n)_{n=\overline{0,N}}, (v_n)_{n=\overline{0,N}})$. Then we obtain

$$\begin{aligned} v_c &= (T_2(u,v))_c = \mu \sum_{j=1}^{N-1} G_2(c,j) t_j g(j,u_j,v_j) \geq \mu \sum_{j=c}^{N-c} G_2(c,j) t_j g(j,u_j,v_j) \\ &\geq \mu m_2 \sum_{j=c}^{N-c} G_2(c,j) t_j (u_j + v_j) \geq \mu m_2 \gamma_2 \sum_{j=c}^{N-c} t_j I_2(j) \gamma (\|u\| + \|v\|) \\ &= \mu m_2 \gamma \gamma_2 C \|(u,v)\|_Y. \end{aligned}$$

Therefore, we conclude

$$\|v\| \geq v_c \geq \mu m_2 \gamma\gamma_2 C \|(u,v)\|_Y > \tilde{\mu}_0 m_2 \gamma\gamma_2 C \|(u,v)\|_Y = \|(u,v)\|_Y,$$

and so $\|(u,v)\|_Y = \|u\| + \|v\| \geq \|v\| > \|(u,v)\|_Y$, which is a contradiction. Therefore, the boundary value problem (S)–(BC) has no positive solution. □

Theorem 3.1.6. *Assume that (A1)–(A3) hold. If $f_0^{\mathrm{i}}, f_\infty^{\mathrm{i}}, g_0^{\mathrm{i}}, g_\infty^{\mathrm{i}} > 0$ and $f(j,u,v) > 0$, $g(j,u,v) > 0$ for all $j = \overline{c, N-c}$, $u \geq 0$, $v \geq 0$, $u+v > 0$, then there exist positive constants $\hat{\lambda}_0$ and $\hat{\mu}_0$ such that for every $\lambda > \hat{\lambda}_0$ and $\mu > \hat{\mu}_0$ the boundary value problem (S)–(BC) has no positive solution.*

Proof. Because $f_0^{\mathrm{i}}, f_\infty^{\mathrm{i}}, g_0^{\mathrm{i}}, g_\infty^{\mathrm{i}} > 0$, we deduce that there exist $m_1, m_2 > 0$ such that

$$f(j,u,v) \geq m_1(u+v), \quad g(j,u,v) \geq m_2(u+v), \quad \forall j = \overline{c, N-c}, \quad u, v \geq 0.$$

We define $\hat{\lambda}_0 = \frac{1}{2\gamma\gamma_1 m_1 A}$ and $\hat{\mu}_0 = \frac{1}{2\gamma\gamma_2 m_2 C}$. Then for every $\lambda > \hat{\lambda}_0$ and $\mu > \hat{\mu}_0$, problem (S)–(BC) has no positive solution. Let $\lambda > \hat{\lambda}_0$ and $\mu > \hat{\mu}_0$. We suppose that (S)–(BC) has a positive solution $((u_n)_{n=\overline{0,N}}, (v_n)_{n=\overline{0,N}})$. In a manner similar to that used in the proofs of Theorems 3.1.4 and 3.1.5, we obtain

$$\|u\| \geq \lambda m_1 \gamma\gamma_1 A \|(u,v)\|_Y, \quad \|v\| \geq \mu m_2 \gamma\gamma_2 C \|(u,v)\|_Y,$$

and so

$$\begin{aligned} \|(u,v)\|_Y &= \|u\| + \|v\| \geq \lambda m_1 \gamma\gamma_1 A \|(u,v)\|_Y + \mu m_2 \gamma\gamma_2 C \|(u,v)\|_Y \\ &> \hat{\lambda}_0 m_1 \gamma\gamma_1 A \|(u,v)\|_Y + \hat{\mu}_0 m_2 \gamma\gamma_2 C \|(u,v)\|_Y = \frac{1}{2}\|(u,v)\|_Y \\ &+ \frac{1}{2}\|(u,v)\|_Y = \|(u,v)\|_Y, \end{aligned}$$

which is a contradiction. Therefore, the boundary value problem (S)–(BC) has no positive solution. □

3.1.4 Examples

Let $N = 20$, $p = 2$, $q = 3$, $r = 1$, $l = 2$, $a_1 = 1/2$, $a_2 = 1/3$, $\xi_1 = 4$, $\xi_2 = 16$, $b_1 = 1/3$, $b_2 = 1/4$, $b_3 = 1/5$, $\eta_1 = 5$, $\eta_2 = 10$, $\eta_3 = 15$, $c_1 = 3/4$, $\zeta_1 = 10$, $d_1 = 1/3$, $d_2 = 1/5$, $\rho_1 = 3$, $\rho_2 = 18$, and $s_n = 1$, $t_n = 1$ for all $n = \overline{1,19}$.

We consider the system of second-order difference equations

$$\begin{cases} \Delta^2 u_{n-1} + \lambda f(n, u_n, v_n) = 0, & n = \overline{1,19}, \\ \Delta^2 v_{n-1} + \mu g(n, u_n, v_n) = 0, & n = \overline{1,19}, \end{cases} \tag{S$_1$}$$

with the multipoint boundary conditions

$$\begin{cases} u_0 = \frac{1}{2} u_4 + \frac{1}{3} u_{16}, \quad u_{20} = \frac{1}{3} u_5 + \frac{1}{4} u_{10} + \frac{1}{5} u_{15}, \\ v_0 = \frac{3}{4} v_{10}, \quad v_{20} = \frac{1}{3} v_3 + \frac{1}{5} v_{18}. \end{cases} \tag{BC$_1$}$$

We have $\sum_{i=1}^{2} a_i = \frac{5}{6} < 1$, $\sum_{i=1}^{3} b_i = \frac{47}{60} < 1$, $\sum_{i=1}^{1} c_i = \frac{3}{4} < 1$, $\sum_{i=1}^{2} d_i = \frac{8}{15} < 1$, $\Delta_1 = \frac{671}{180}$, and $\Delta_2 = \frac{147}{20}$. The functions g_0, I_1, and I_2 are given by

$$g_0(n,j) = \frac{1}{20} \begin{cases} j(20-n), & 1 \le j \le n \le 20, \\ n(20-j), & 0 \le n \le j \le 19, \end{cases}$$

$$I_1(j) = \begin{cases} \frac{2565j}{671} - \frac{j^2}{20}, & 1 \le j \le 4, \\ \frac{7820}{671} + \frac{70j}{61} - \frac{j^2}{20}, & 5 \le j \le 9, \\ \frac{12{,}620}{671} + \frac{290j}{671} - \frac{j^2}{20}, & 10 \le j \le 15, \\ \frac{30{,}700}{671} - \frac{864j}{671} - \frac{j^2}{20}, & 16 \le j \le 19. \end{cases}$$

$$I_2(j) = \begin{cases} \frac{145j}{63} - \frac{j^2}{20}, & 1 \le j \le 2, \\ \frac{250}{147} + \frac{85j}{49} - \frac{j^2}{20}, & 3 \le j \le 9, \\ \frac{2560}{147} + \frac{8j}{49} - \frac{j^2}{20}, & 10 \le j \le 17, \\ \frac{3460}{147} - \frac{26j}{147} - \frac{j^2}{20}, & 18 \le j \le 19. \end{cases}$$

For $c = 1$, we obtain $\gamma_1 = \gamma_2 = \gamma = \frac{1}{19}$. After some computations, we deduce $A = B = \sum_{j=1}^{19} I_1(j) \approx 251.29880775$ and $C = D = \sum_{j=1}^{19} I_2(j) \approx 165.64965986$.

Example 3.1.1. We consider the functions

$$f(j,u,v) = \frac{j(u+v)^2(3+\sin v)}{u+v+1}, \quad g(j,u,v) = \frac{j^2(u+v)^2(2+\cos u)}{u+v+1},$$
$$\forall j = \overline{1,19}, \quad u,\ v \in [0,\infty).$$

We have $f_0^{\mathrm{s}} = g_0^{\mathrm{s}} = 0$, $f_\infty^{\mathrm{i}} = 2$, and $g_\infty^{\mathrm{i}} = 1$. For $\alpha_1 = \alpha_2 = 1/2$, we obtain $L_1 \approx 0.35913421$ and $L_3 \approx 1.08964908$. Then, by Theorem 3.1.1 (4), we deduce that for each $\lambda \in (L_1, \infty)$ and $\mu \in (L_3, \infty)$ there exists a positive solution $((u_n)_{n=\overline{0,20}}, (v_n)_{n=\overline{0,20}})$ for problem (S$_1$)–(BC$_1$).

Because $f_0^s = g_0^s = 0 < \infty$, $f_\infty^s = 76 < \infty$, and $g_\infty^s = 1083 < \infty$, we can also apply Theorem 3.1.3. So, we conclude that there exist $\lambda_0, \mu_0 > 0$ such that for every $\lambda \in (0, \lambda_0)$ and $\mu \in (0, \mu_0)$ problem (S_1)–(BC_1) has no positive solution. By Theorem 3.1.3, the positive constants λ_0 and μ_0 are given by $\lambda_0 = \frac{1}{2M_1B}$ and $\mu_0 = \frac{1}{2M_2D}$. Then we obtain $M_1 = 76$, $M_2 = 1083$, $\lambda_0 \approx 26.18\cdot 10^{-6}$, and $\mu_0 \approx 2.78\cdot 10^{-6}$.

Example 3.1.2. We consider the functions

$$f(j,u,v) = \frac{j}{1+j^2}\left[p_1(u^2+v^2) + \frac{1}{100}(u+v)\right],$$
$$g(j,u,v) = \frac{j^2}{1+j}\left[p_2(u^2+v^2) + \frac{1}{400}(u+v)\right],$$

for all $j = \overline{1,19}$, $u, v \in [0,\infty)$, with $p_1, p_2 > 0$. We have $f_0^s = \frac{1}{200}$, $g_0^s = \frac{361}{8000}$, and $f_\infty^i = g_\infty^i = \infty$. Then for $\tilde{\alpha}_1 = \tilde{\alpha}_2 = 1/2$, we obtain $L_2 \approx 0.39793264$ and $L_4 \approx 0.06689016$. Therefore, by Theorem 3.1.1 (5), we deduce that for each $\lambda \in (0, L_2)$ and $\mu \in (0, L_4)$ there exists a positive solution $((u_n)_{n=\overline{0,20}}, (v_n)_{n=\overline{0,20}})$ for problem (S_1)–(BC_1).

Example 3.1.3. We consider the functions

$$f(j,u,v)=p_1(u^2+v^2), \quad g(j,u,v)=\frac{1}{200}(e^{u+v}-1), \quad \forall j=\overline{1,19}, \quad u,v\in[0,\infty),$$

with $p_1 > 0$. We have $f_0^s = 0$, $g_0^s = \frac{1}{200}$, $f_\infty^i = g_\infty^i = \infty$, and $L_4' \approx 1.2073674$. Then, by Theorem 3.1.1 (6), we conclude that for each $\lambda \in (0,\infty)$ and $\mu \in (0, L_4')$ there exists a positive solution $((u_n)_{n=\overline{0,20}}, (v_n)_{n=\overline{0,20}})$ for problem (S_1)–(BC_1).

Remark 3.1.2. The results presented in this section were published in Henderson and Luca (2013d, 2014b).

3.2 Existence and multiplicity of positive solutions for systems without parameters

In this section, we investigate the existence and multiplicity of positive solutions for problem (S)–(BC) from Section 3.1 with $\lambda = \mu = 1$ and $s_n = t_n = 1$ for all $n = \overline{1, N-1}$, and where f and g are dependent only on n and v, and n and u, respectively.

3.2.1 Presentation of the problem

We consider the system of nonlinear second-order difference equations

$$\begin{cases} \Delta^2 u_{n-1} + f(n, v_n) = 0, & n = \overline{1, N-1}, \\ \Delta^2 v_{n-1} + g(n, u_n) = 0, & n = \overline{1, N-1}, \end{cases} \tag{S'}$$

with the multipoint boundary conditions

$$u_0 = \sum_{i=1}^{p} a_i u_{\xi_i}, \quad u_N = \sum_{i=1}^{q} b_i u_{\eta_i}, \quad v_0 = \sum_{i=1}^{r} c_i v_{\zeta_i}, \quad v_N = \sum_{i=1}^{l} d_i v_{\rho_i}, \tag{BC}$$

where $N, p, q, r, l \in \mathbb{N}$, $N \geq 2$, $a_i \in \mathbb{R}$ for all $i = \overline{1,p}$, $b_i \in \mathbb{R}$ for all $i = \overline{1,q}$, $c_i \in \mathbb{R}$ for all $i = \overline{1,r}$, $d_i \in \mathbb{R}$ for all $i = \overline{1,l}$, $\xi_i \in \mathbb{N}$ for all $i = \overline{1,p}$, $\eta_i \in \mathbb{N}$ for all $i = \overline{1,q}$, $\zeta_i \in \mathbb{N}$ for all $i = \overline{1,r}$, $\rho_i \in \mathbb{N}$ for all $i = \overline{1,l}$, $1 \leq \xi_1 < \cdots < \xi_p \leq N-1$, $1 \leq \eta_1 < \cdots < \eta_q \leq N-1$, $1 \leq \zeta_1 < \cdots < \zeta_r \leq N-1$, and $1 \leq \rho_1 < \cdots < \rho_l \leq N-1$.

Under some assumptions on f and g, we shall study the existence and multiplicity of positive solutions of problem (S′)–(BC) by applying the fixed point index theory. By a positive solution of (S′)–(BC), we mean a pair of sequences $((u_n)_{n=\overline{0,N}}, (v_n)_{n=\overline{0,N}})$ satisfying (S′) and (BC) with $u_n \geq 0$, $v_n \geq 0$ for all $n = \overline{0,N}$, and $\max_{n=\overline{0,N}} u_n > 0$, $\max_{n=\overline{0,N}} v_n > 0$. This problem is a generalization of the one studied in Henderson and Luca (2013a), where $a_i = 0$ for all $i = \overline{1,p}$ and $c_i = 0$ for all $i = \overline{1,r}$.

3.2.2 Main results

In this section, we give sufficient conditions on f and g such that positive solutions with respect to a cone for problem (S′)–(BC) exist.

We present the basic assumptions that we shall use in the sequel:

(J1) $\xi_i \in \mathbb{N}$ for all $i = \overline{1,p}$, $1 \leq \xi_1 < \cdots < \xi_p \leq N-1$, $a_i \geq 0$ for all $i = \overline{1,p}$, $\sum_{i=1}^{p} a_i < 1$, $\eta_i \in \mathbb{N}$ for all $i = \overline{1,q}$, $1 \leq \eta_1 < \cdots < \eta_q \leq N-1$, $b_i \geq 0$ for all $i = \overline{1,q}$, $\sum_{i=1}^{q} b_i < 1$, $\zeta_i \in \mathbb{N}$ for all $i = \overline{1,r}$, $1 \leq \zeta_1 < \cdots < \zeta_r \leq N-1$, $c_i \geq 0$ for all $i = \overline{1,r}$, $\sum_{i=1}^{r} c_i < 1$, $\rho_i \in \mathbb{N}$ for all $i = \overline{1,l}$, $1 \leq \rho_1 < \cdots < \rho_l \leq N-1$, $d_i \geq 0$ for all $i = \overline{1,l}$, $\sum_{i=1}^{l} d_i < 1$.

(J2) The functions $f, g : \{1, \ldots, N-1\} \times [0,\infty) \to [0,\infty)$ are continuous, and $f(n,0) = 0$, $g(n,0) = 0$, for all $n = \overline{1,N-1}$.

Under assumption (J1), we have all auxiliary results in Lemmas 3.1.1–3.1.6 from Section 3.1.2.

The pair of sequences $\left((u_n)_{n=\overline{0,N}}, (v_n)_{n=\overline{0,N}}\right)$ with $u_n \geq 0$, $v_n \geq 0$ for all $n = \overline{0,N}$ is a solution for problem (S′)–(BC) if and only if $\left((u_n)_{n=\overline{0,N}}, (v_n)_{n=\overline{0,N}}\right)$ with $u_n \geq 0$, $v_n \geq 0$ for all $n = \overline{0,N}$ is a solution for the nonlinear system

$$\begin{cases} u_n = \sum_{i=1}^{N-1} G_1(n,i) f(i, v_i), & n = \overline{0,N}, \\ v_n = \sum_{i=1}^{N-1} G_2(n,i) g(i, u_i), & n = \overline{0,N}. \end{cases} \tag{3.13}$$

Besides, system (3.13) can be written as the nonlinear system

$$\begin{cases} u_n = \sum_{i=1}^{N-1} G_1(n,i) f\left(i, \sum_{j=1}^{N-1} G_2(i,j) g(j, u_j)\right), & n = \overline{0,N}, \\ v_n = \sum_{i=1}^{N-1} G_2(n,i) g(i, u_i), & n = \overline{0,N}. \end{cases}$$

We consider the Banach space $X=\mathbb{R}^{N+1}=\{u=(u_0,u_1,\dots,u_N), u_i\in\mathbb{R},\quad i=\overline{0,N}\}$ with the maximum norm $\|u\|=\max_{i=\overline{0,N}}|u_i|$, for $u=(u_n)_{n=\overline{0,N}}$, and define the cone $P'\subset X$ by $P'=\{u\in X,\quad u=(u_n)_{n=\overline{0,N}},\quad u_n\geq 0,\quad n=\overline{0,N}\}$.

We also define the operators $\mathcal{A}:P'\to X$, $\mathcal{B}:P'\to X$, and $\mathcal{C}:P'\to X$ by

$$\mathcal{A}\left((u_n)_{n=\overline{0,N}}\right)=\left(\sum_{i=1}^{N-1}G_1(n,i)f\left(i,\sum_{j=1}^{N-1}G_2(i,j)g(j,u_j)\right)\right)_{n=\overline{0,N}},$$

$$\mathcal{B}\left((u_n)_{n=\overline{0,N}}\right)=\left(\sum_{i=1}^{N-1}G_1(n,i)u_i\right)_{n=\overline{0,N}},$$

$$\mathcal{C}\left((u_n)_{n=\overline{0,N}}\right)=\left(\sum_{i=1}^{N-1}G_2(n,i)u_i\right)_{n=\overline{0,N}}.$$

Under assumptions (J1) and (J2), and using also Lemma 3.1.4, one can easily see that $\mathcal{A}$, $\mathcal{B}$, and $\mathcal{C}$ are completely continuous from P' to P'. Thus, the existence and multiplicity of positive solutions of problem (S′)–(BC) are equivalent to the existence and multiplicity of fixed points of the operator $\mathcal{A}$.

Theorem 3.2.1. *Assume that (J1) and (J2) hold. If the functions f and g also satisfy the following conditions (J3) and (J4), then problem* (S′)–(BC) *has at least one positive solution* $\left((u_n)_{n=\overline{0,N}},(v_n)_{n=\overline{0,N}}\right)$*:*

(J3) There exist $c\in\{1,\dots,[\![N/2]\!]\}$ and $p_1\in(0,1]$ such that

$$(1)\quad \tilde{f}_\infty^{\rm i}=\liminf_{u\to\infty}\min_{n=\overline{c,N-c}}\frac{f(n,u)}{u^{p_1}}\in(0,\infty]\quad\textit{and}\quad(2)\quad \tilde{g}_\infty^{\rm i}=\liminf_{u\to\infty}\min_{n=\overline{c,N-c}}\frac{g(n,u)}{u^{1/p_1}}=\infty.$$

(J4) There exists $q_1\in(0,\infty)$ such that

$$(1)\quad \tilde{f}_0^{\rm s}=\limsup_{u\to0^+}\max_{n=\overline{1,N-1}}\frac{f(n,u)}{u^{q_1}}\in[0,\infty)\quad\textit{and}\quad(2)\quad \tilde{g}_0^{\rm s}=\limsup_{u\to0^+}\max_{n=\overline{1,N-1}}\frac{g(n,u)}{u^{1/q_1}}=0.$$

Proof. From (1) of assumption (J3), we deduce that there exist C_1, $C_2>0$ such that

$$f(n,u)\geq C_1u^{p_1}-C_2,\quad\forall n=\overline{c,N-c},\quad u\in[0,\infty).\tag{3.14}$$

Then for $u\in P'$, by using (3.14) and Lemma 3.1.5, we conclude that there exist $\tilde{C}_1$, $C_3>0$ such that

$$(\mathcal{A}u)_n\geq\tilde{C}_1\sum_{i=c}^{N-c}G_1(n,i)\left(\sum_{j=c}^{N-c}(G_2(i,j))^{p_1}(g(j,u_j))^{p_1}\right)-C_3,\quad\forall n=\overline{0,N}.\tag{3.15}$$

For c given in (J3), we define the cone $P_0=\{u\in P';\quad \min_{n=\overline{c,N-c}}u_n\geq\gamma\|u\|\}$, where $\gamma=\min\{\gamma_1,\gamma_2\}$, and γ_1 and γ_2 are defined in Section 3.1.2. From our assumptions and Lemma 3.1.6, we deduce that for any $y \in P'$, $y = (y_n)_{n=\overline{0,N}}$ the sequences $u = \mathcal{B}(y)$, $u = (u_n)_{n=\overline{0,N}}$ and $v = \mathcal{C}(y)$, $v = (v_n)_{n=\overline{0,N}}$ satisfy the inequalities $\min_{n=\overline{c,N-c}} u_n \geq \gamma_1 \|u\| \geq \gamma \|u\|$ and $\min_{n=\overline{c,N-c}} v_n \geq \gamma_2 \|v\| \geq \gamma \|v\|$. So $u = \mathcal{B}(y)$, $v = \mathcal{C}(y) \in P_0$. Therefore, we conclude that $\mathcal{B}(P') \subset P_0$ and $\mathcal{C}(P') \subset P_0$.

We denote by $u^0 = (u_n^0)_{n=\overline{0,N}}$ the solution of problem (3.1)–(3.2) for $y^0 = (y_n^0)_{n=\overline{1,N-1}}$, $y_n^0 = 1$ for $n = \overline{1, N-1}$. Then by Lemma 3.1.4, we obtain $u_n^0 = \sum_{i=1}^{N-1} G_1(n,i) \geq 0$ for all $n = \overline{0,N}$. So $u^0 = \mathcal{B}(y^0) \in P_0$.

Now let the set $M = \{u \in P';\ \text{there exists}\ \ \lambda \geq 0\ \ \text{such that}\ \ u = \mathcal{A}u + \lambda u^0\}$. We shall show that M is a bounded subset of X. If $u \in M$, then there exists $\lambda \geq 0$ such that $u = \mathcal{A}u + \lambda u^0$, $u = (u_n)_{n=\overline{0,N}}$, with $u_n = (\mathcal{A}u)_n + \lambda u_n^0$ for all $n = \overline{0,N}$. Then we have $u_n = (\mathcal{A}u)_n + \lambda u_n^0 = (\mathcal{B}(F(u) + \lambda y^0))_n$ for all $n = \overline{0,N}$, so $u \in P_0$, where $F : P' \to P'$ is defined by $(Fu)_n = f\left(n, \sum_{i=1}^{N-1} G_2(n,i) g(i,u_i)\right)$, $n = \overline{0,N}$. Therefore $M \subset P_0$, and from the definition of P_0, we deduce

$$\|u\| \leq \frac{1}{\gamma} \min_{n=\overline{c,N-c}} u_n, \quad \forall u = (u_n)_{n=\overline{0,N}} \in M. \tag{3.16}$$

From (2) of assumption (J3), we conclude that for $\varepsilon_0 = 2/(\tilde{C}_1 m_1 m_2 \gamma_1 \gamma_2^{p_1})$ there exists $C_4 > 0$ such that

$$(g(n,u))^{p_1} \geq \varepsilon_0 u - C_4, \quad \forall n = \overline{c, N-c}, \quad u \in [0,\infty), \tag{3.17}$$

where $m_1 = \sum_{i=c}^{N-c} I_1(i) > 0$ and $m_2 = \sum_{i=c}^{N-c} (I_2(i))^{p_1} > 0$.

For $u \in M$ and $n = \overline{c, N-c}$, by using Lemma 3.1.5 and relations (3.15) and (3.17), we obtain

$$\begin{aligned}
u_n &= (\mathcal{A}u)_n + \lambda u_n^0 \geq (\mathcal{A}u)_n \\
&\geq \tilde{C}_1 \sum_{i=c}^{N-c} G_1(n,i) \left(\sum_{j=c}^{N-c} (G_2(i,j))^{p_1} (g(j,u_j))^{p_1} \right) - C_3 \\
&\geq \tilde{C}_1 \sum_{i=c}^{N-c} G_1(n,i) \left(\sum_{j=c}^{N-c} \gamma_2^{p_1} (I_2(j))^{p_1} (g(j,u_j))^{p_1} \right) - C_3 \\
&\geq \tilde{C}_1 \gamma_1 \gamma_2^{p_1} \sum_{i=c}^{N-c} I_1(i) \left(\sum_{j=c}^{N-c} (I_2(j))^{p_1} (\varepsilon_0 u_j - C_4) \right) - C_3 \\
&\geq \tilde{C}_1 \gamma_1 \gamma_2^{p_1} \varepsilon_0 \left(\sum_{i=c}^{N-c} I_1(i) \right) \left(\sum_{j=c}^{N-c} (I_2(j))^{p_1} \right) \min_{j=\overline{c,N-c}} u_j - C_5 \\
&= \tilde{C}_1 \gamma_1 \gamma_2^{p_1} \varepsilon_0 m_1 m_2 \min_{j=\overline{c,N-c}} u_j - C_5 = 2 \min_{j=\overline{c,N-c}} u_j - C_5,
\end{aligned}$$

where $C_5 = C_3 + \tilde{C}_1 C_4 \gamma_1 \gamma_2^{p_1} m_1 m_2 > 0$.

Hence, $\min_{n=\overline{c,N-c}} u_n \geq 2\min_{j=\overline{c,N-c}} u_j - C_5$, and so

$$\min_{n=\overline{c,N-c}} u_n \leq C_5, \quad \forall\, u \in M. \tag{3.18}$$

Now from relations (3.16) and (3.18), we deduce that $\|u\| \leq \min_{n=\overline{c,N-c}} u_n/\gamma \leq C_5/\gamma$ for all $u \in M$—that is, M is a bounded subset of X.

Therefore, there exists a sufficiently large $L > 0$ such that $u \neq \mathcal{A}u + \lambda u_0$ for all $u \in \partial B_L \cap P'$ and $\lambda \geq 0$. From Theorem 1.3.2, we conclude that

$$i(\mathcal{A}, B_L \cap P', P') = 0. \tag{3.19}$$

In what follows, from assumptions (J2) and (J4), we deduce that there exists $M_0 > 0$ and $\delta_1 \in (0,1)$ such that

$$\begin{aligned} &f(n,u) \leq M_0 u^{q_1}, \quad \forall n = \overline{1,N-1}, \quad u \in [0,1], \\ &g(n,u) \leq \varepsilon_1 u^{1/q_1}, \quad \forall n = \overline{1,N-1}, \quad u \in [0,\delta_1], \end{aligned} \tag{3.20}$$

where $\varepsilon_1 = \min\{1/M_2, (1/(2M_0M_1M_2^{q_1}))^{1/q_1}\} > 0$, $M_1 = \sum_{j=1}^{N-1} I_1(j)$, $M_2 = \sum_{j=1}^{N-1} I_2(j)$.

Hence, for all $u \in \overline{B}_{\delta_1} \cap P'$ and $i = \overline{0,N}$, we obtain

$$\sum_{j=1}^{N-1} G_2(i,j)g(j,u_j) \leq \varepsilon_1 \sum_{j=1}^{N-1} G_2(i,j)u_j^{1/q_1} \leq \varepsilon_1 \sum_{j=1}^{N-1} I_2(j)u_j^{1/q_1} \leq \varepsilon_1 M_2 \|u\|^{1/q_1} \leq 1. \tag{3.21}$$

Therefore, by (3.20) and (3.21), we conclude

$$\begin{aligned} (\mathcal{A}u)_n &= \sum_{i=1}^{N-1} G_1(n,i) f\left(i, \sum_{j=1}^{N-1} G_2(i,j)g(j,u_j)\right) \\ &\leq M_0 \sum_{i=1}^{N-1} G_1(n,i)\left(\sum_{j=1}^{N-1} G_2(i,j)g(j,u_j)\right)^{q_1} \leq M_0 \varepsilon_1^{q_1} M_2^{q_1} \|u\| \sum_{i=1}^{N-1} I_1(i) \\ &= M_0 \varepsilon_1^{q_1} M_1 M_2^{q_1} \|u\| \leq \frac{1}{2}\|u\|, \quad \forall\, u \in \overline{B}_{\delta_1} \cap P', \quad n = \overline{0,N}. \end{aligned}$$

This implies that $\|\mathcal{A}u\| \leq \|u\|/2$ for all $u \in \partial B_{\delta_1} \cap P'$. From Theorem 1.3.1, we deduce

$$i(\mathcal{A}, B_{\delta_1} \cap P', P') = 1. \tag{3.22}$$

Combining now (3.19) and (3.22), we obtain

$$i(\mathcal{A}, (B_L \setminus \bar{B}_{\delta_1}) \cap P', P') = i(\mathcal{A}, B_L \cap P', P') - i(\mathcal{A}, B_{\delta_1} \cap P', P') = -1.$$

We conclude that $\mathcal{A}$ has at least one fixed point $u^1 \in (B_L \setminus \bar{B}_{\delta_1}) \cap P'$, $u^1 = (u_n^1)_{n=\overline{0,N}}$—that is, $\delta_1 < \|u^1\| < L$.

Let $v^1 = (v_n^1)_{n=\overline{0,N}}$, $v_n^1 = \sum_{i=1}^{N-1} G_2(n,i)g(i,u_i^1)$, and $n = \overline{0,N}$. Then $(u^1, v^1) \in P' \times P'$ is a solution of (S′)–(BC). Because $u_n^1 = \sum_{i=1}^{N-1} G_1(n,i)f(i,v_i^1)$, $n = \overline{0,N}$ and $\|u^1\| > 0$, we deduce that $\|v^1\| > 0$. If we suppose that $v_n^1 = 0$, for all $n = \overline{0,N}$, then by using (J2), we have $f(i,v_i^1) = f(i,0) = 0$ for all $i = \overline{1,N-1}$. This implies $u_n^1 = 0$ for all $n = \overline{0,N}$, which contradicts $\|u^1\| > 0$. The proof of Theorem 3.2.1 is completed. □

Theorem 3.2.2. *Assume that (J1) and (J2) hold. If the functions f and g also satisfy the following conditions (J5) and (J6), then problem* (S′)–(BC) *has at least one positive solution* $\left((u_n)_{n=\overline{0,N}}, (v_n)_{n=\overline{0,N}}\right)$:

(J5) There exists $r_1 \in (0,\infty)$ such that

$$(1)\quad \tilde{f}_\infty^{s} = \limsup_{u\to\infty} \max_{n=\overline{1,N-1}} \frac{f(n,u)}{u^{r_1}} \in [0,\infty) \quad \text{and} \quad (2)\quad \tilde{g}_\infty^{s} = \limsup_{u\to\infty} \max_{n=\overline{1,N-1}} \frac{g(n,u)}{u^{1/r_1}} = 0.$$

(J6) There exists $c \in \{1,\ldots,[\![N/2]\!]\}$ such that

$$(1)\quad \tilde{f}_0^{i} = \liminf_{u\to 0^+} \min_{n=\overline{c,N-c}} \frac{f(n,u)}{u} \in (0,\infty] \quad \text{and} \quad (2)\quad \tilde{g}_0^{i} = \liminf_{u\to 0^+} \min_{n=\overline{c,N-c}} \frac{g(n,u)}{u} = \infty.$$

Proof. By assumption (J5), we deduce that there exists C_6, C_7, $C_8 > 0$ such that

$$f(n,u) \le C_6 u^{r_1} + C_7, \quad g(n,u) \le \varepsilon_2 u^{1/r_1} + C_8, \quad \forall n = \overline{1,N-1}, \quad u \in [0,\infty), \tag{3.23}$$

where $\varepsilon_2 = (1/(2C_6 M_1 M_2^{r_1}))^{1/r_1}$, and M_1 and M_2 are defined in the proof of Theorem 3.2.1.

Then for $u \in P'$, by using (3.23), we obtain

$$\begin{aligned}
(\mathcal{A}u)_n &= \sum_{i=1}^{N-1} G_1(n,i) f\left(i, \sum_{j=1}^{N-1} G_2(i,j) g(j,u_j)\right) \\
&\le \sum_{i=1}^{N-1} G_1(n,i) \left[C_6 \left(\sum_{j=1}^{N-1} G_2(i,j) g(j,u_j) \right)^{r_1} + C_7 \right] \\
&\le C_6 \sum_{i=1}^{N-1} G_1(n,i) \left[\sum_{j=1}^{N-1} G_2(i,j) \left(\varepsilon_2 u_j^{1/r_1} + C_8 \right) \right]^{r_1} + M_1 C_7 \\
&\le C_6 \sum_{i=1}^{N-1} G_1(n,i) \left[\sum_{j=1}^{N-1} G_2(i,j) \left(\varepsilon_2 \|u\|^{1/r_1} + C_8 \right) \right]^{r_1} + M_1 C_7 \\
&= C_6 \left(\varepsilon_2 \|u\|^{1/r_1} + C_8 \right)^{r_1} \sum_{i=1}^{N-1} G_1(n,i) \left(\sum_{j=1}^{N-1} G_2(i,j) \right)^{r_1} + M_1 C_7, \\
&\quad \forall n = \overline{0,N}.
\end{aligned}$$

Therefore, we have

$$(\mathcal{A}u)_n \leq C_6 M_1 M_2^{r_1} \left(\varepsilon_2 \|u\|^{1/r_1} + C_8\right)^{r_1} + M_1 C_7 =: \mathcal{Q}(u), \quad \forall n = \overline{0,N}. \quad (3.24)$$

Because $\lim_{\|u\|\to\infty} \mathcal{Q}(u)/\|u\| = 1/2$, there exists a sufficiently large $R > 0$ such that

$$\mathcal{Q}(u) \leq \frac{3}{4}\|u\|, \quad \forall u \in P', \quad \|u\| \geq R. \quad (3.25)$$

Hence, from (3.24) and (3.25), we obtain $\|\mathcal{A}u\| \leq \frac{3}{4}\|u\| < \|u\|$ for all $u \in \partial B_R \cap P'$, and from Theorem 1.3.1, we have

$$i(\mathcal{A}, B_R \cap P', P') = 1. \quad (3.26)$$

On the other hand, by (J6), we deduce that there exists $C_9 > 0$ and $\varrho_0 > 0$ such that

$$f(n,u) \geq C_9 u, \quad g(n,u) \geq \frac{C_0}{C_9} u, \quad \forall n = \overline{c, N-c}, \quad u \in [0, \varrho_0], \quad (3.27)$$

where $C_0 = 1/(\gamma_1 \gamma_2 m_1 \tilde{m}_2)$, $m_1 = \sum_{i=c}^{N-c} I_1(i)$, and $\tilde{m}_2 = \sum_{i=c}^{N-c} I_2(i)$.

Because $g(n,0) = 0$ for all $n = \overline{1, N-1}$, and g is continuous, we conclude that there exists a sufficiently small $\delta_2 \in (0, \varrho_0)$ such that $g(n,u) \leq \varrho_0/M_2$ for all $n = \overline{c, N-c}$ and $u \in [0, \delta_2]$. Hence,

$$\sum_{i=c}^{N-c} G_2(n,i) g(i, u_i) \leq \varrho_0, \quad \forall u \in \bar{B}_{\delta_2} \cap P', \quad n = \overline{0,N}. \quad (3.28)$$

From (3.27), (3.28), and Lemma 3.1.5, we deduce that for any $u \in \bar{B}_{\delta_2} \cap P'$ we have

$$\begin{aligned}
(\mathcal{A}u)_n &\geq \sum_{i=c}^{N-c} G_1(n,i) f\left(i, \sum_{j=c}^{N-c} G_2(i,j) g(j,u_j)\right) \\
&\geq C_9 \sum_{i=c}^{N-c} G_1(n,i) \left(\sum_{j=c}^{N-c} G_2(i,j) g(j,u_j)\right) \geq C_0 \sum_{i=c}^{N-c} G_1(n,i) \left(\sum_{j=c}^{N-c} G_2(i,j) u_j\right) \\
&\geq C_0 \sum_{i=c}^{N-c} G_1(n,i) \left(\sum_{j=c}^{N-c} G_2(i,j) u_j\right) \geq C_0 \gamma_2 \sum_{i=c}^{N-c} G_1(n,i) \left(\sum_{j=c}^{N-c} I_2(j) u_j\right) \\
&= C_0 \gamma_2 \left(\sum_{j=c}^{N-c} I_2(j) u_j\right) \left(\sum_{i=c}^{N-c} G_1(n,i)\right) =: (\mathcal{L}u)_n, \quad \forall n = \overline{0,N}.
\end{aligned}$$

Hence, for the linear operator $\mathcal{L} : P' \to P'$ defined as above, we obtain

$$\mathcal{A}u \geq \mathcal{L}u, \quad \forall\, u \in \partial B_{\delta_2} \cap P'. \tag{3.29}$$

For $w^0 = (w_n^0)_n$, $w_n^0 = \sum_{i=c}^{N-c} G_1(n,i)$, $n = \overline{0,N}$, we have $w^0 \in P' \setminus \{0\}$ and

$$\begin{aligned}
(\mathcal{L}w^0)_n &= C_0\gamma_2 \left[\sum_{j=c}^{N-c} I_2(j)\left(\sum_{i=c}^{N-c} G_1(j,i)\right)\right]\left(\sum_{i=c}^{N-c} G_1(n,i)\right) \\
&\geq C_0\gamma_1\gamma_2 \left(\sum_{j=c}^{N-c} I_2(j)\right)\left(\sum_{i=c}^{N-c} I_1(i)\right)\left(\sum_{i=c}^{N-c} G_1(n,i)\right) \\
&= \sum_{i=c}^{N-c} G_1(n,i) = w_n^0, \quad \forall\, n = \overline{0,N}.
\end{aligned}$$

Therefore,

$$\mathcal{L}w^0 \geq w^0. \tag{3.30}$$

We may suppose that $\mathcal{A}$ has no fixed point on $\partial B_{\delta_2} \cap P'$ (otherwise the proof is completed). From (3.29), (3.30), and Theorem 1.3.3, we deduce that

$$i(\mathcal{A}, B_{\delta_2} \cap P', P') = 0. \tag{3.31}$$

Then, from (3.26) and (3.31), we obtain

$$i(\mathcal{A}, (B_R \setminus \bar{B}_{\delta_2}) \cap P', P') = i(\mathcal{A}, B_R \cap P', P') - i(\mathcal{A}, B_{\delta_2} \cap P', P') = 1.$$

We conclude that $\mathcal{A}$ has at least one fixed point u in $(B_R \setminus \bar{B}_{\delta_2}) \cap P'$. Thus, problem (S′)–(BC) has at least one positive solution $(u, v) \in P' \times P'$, $u = (u_n)_{n=\overline{0,N}}$, $v = (v_n)_{n=\overline{0,N}}$, ($\|u\| > 0$, $\|v\| > 0$). This completes the proof of Theorem 3.2.2. □

Theorem 3.2.3. *Assume that (J1) and (J2) hold. If the functions f and g also satisfy conditions (J3) and (J6) and the following condition (J7), then problem* (S′)–(BC) *has at least two positive solutions (u^1, v^1) and (u^2, v^2):*

(J7) For each $n = \overline{1,N-1}$, $f(n,u)$ and $g(n,u)$ are nondecreasing with respect to u, and there exists a constant $R_0 > 0$ such that

$$f\left(n, m_0 \sum_{i=1}^{N-1} g(i, R_0)\right) < \frac{R_0}{m_0}, \quad \forall\, n = \overline{1,N-1},$$

where $m_0 = \max\{K_1, K_2\}$, $K_1 = \sum_{j=1}^{N-1} I_1(j)$, $K_2 = \max_{j=\overline{1,N-1}} I_2(j)$, and I_1 and I_2 are defined in Section 3.1.2.

Proof. From Section 3.1.2, we have $0 \leq G_1(n,i) \leq I_1(i)$, $0 \leq G_2(n,i) \leq I_2(i)$ for all $n = \overline{0,N}$, $i = \overline{1,N-1}$. By using (J7), for any $u \in \partial B_{R_0} \cap P'$, we obtain

$$
\begin{aligned}
(\mathcal{A}u)_n &\le \sum_{i=1}^{N-1} G_1(n,i) f\left(i, \sum_{j=1}^{N-1} I_2(j) g(j,u_j)\right)\\
&\le \sum_{i=1}^{N-1} G_1(n,i) f\left(i, \sum_{j=1}^{N-1} I_2(j) g(j,R_0)\right) \le \sum_{i=1}^{N-1} G_1(n,i) f\left(i, m_0 \sum_{j=1}^{N-1} g(j,R_0)\right)\\
&< \frac{R_0}{m_0} \sum_{i=1}^{N-1} G_1(n,i) \le \frac{R_0}{m_0} \sum_{i=1}^{N-1} I_1(i) \le R_0, \quad \forall n = \overline{0,N}.
\end{aligned}
$$

So, $\|\mathcal{A}u\| < \|u\|$ for all $u \in \partial B_{R_0} \cap P'$.

By Theorem 1.3.1, we deduce that

$$
i(\mathcal{A}, B_{R_0} \cap P', P') = 1. \tag{3.32}
$$

On the other hand, from (J3), (J6), and the proofs of Theorems 3.2.1 and 3.2.2, we know that there exist a sufficiently large $L > R_0$ and a sufficiently small δ_2 with $0 < \delta_2 < R_0$ such that

$$
i(\mathcal{A}, B_L \cap P', P') = 0, \quad i(\mathcal{A}, B_{\delta_2} \cap P', P') = 0. \tag{3.33}
$$

From (3.32) and (3.33), we obtain

$$
i(\mathcal{A}, (B_L \setminus \bar{B}_{R_0}) \cap P', P') = -1, \quad i(\mathcal{A}, (B_{R_0} \setminus \bar{B}_{\delta_2}) \cap P', P') = 1.
$$

Then $\mathcal{A}$ has at least one fixed point u^1 in $(B_L \setminus \bar{B}_{R_0}) \cap P'$ and has at least one fixed point u^2 in $(B_{R_0} \setminus \bar{B}_{\delta_2}) \cap P'$. Therefore, problem (S′)–(BC) has two distinct positive solutions (u^1, v^1), $(u^2, v^2) \in P' \times P'$ with $\|u^i\| > 0$, $\|v^i\| > 0$ for $i = 1, 2$. The proof of Theorem 3.2.3 is completed. □

3.2.3 An example

We consider the system of second-order difference equations

$$
\begin{cases}
\Delta^2 u_{n-1} + a(v_n^\alpha + v_n^\beta) = 0, & n = \overline{1,29},\\
\Delta^2 v_{n-1} + b(u_n^\theta + u_n^\delta) = 0, & n = \overline{1,29},
\end{cases} \tag{$\overline{S'}$}
$$

with the multipoint boundary conditions

$$
\begin{cases}
u_0 = \dfrac{2}{3} u_{15}, \quad u_{30} = \dfrac{1}{3} u_8 + \dfrac{1}{6} u_{16} + \dfrac{1}{4} u_{24},\\
v_0 = \dfrac{1}{3} v_9 + \dfrac{1}{2} v_{20}, \quad v_{30} = \dfrac{1}{3} v_6 + \dfrac{1}{4} v_{18},
\end{cases} \tag{$\overline{BC'}$}
$$

where $\alpha > 1$, $\beta < 1$, $\theta > 2$, $\delta < 1$, and a, $b > 0$.

Here $N = 30, p = 1, q = 3, r = 2, l = 2, a_1 = 2/3, \xi_1 = 15, b_1 = 1/3, b_2 = 1/6$, $b_3 = 1/4$, $\eta_1 = 8$, $\eta_2 = 16$, $\eta_3 = 24$, $c_1 = 1/3$, $c_2 = 1/2$, $\zeta_1 = 9$, $\zeta_2 = 20$, $d_1 = 1/3, d_2 = 1/4, \rho_1 = 6, \rho_2 = 18, f(n,u) = a(u^\alpha + u^\beta)$, and $g(n,u) = b(u^\theta + u^\delta)$

for all $n = \overline{1,29}$, $u \in [0,\infty)$. We have $\sum_{i=1}^{1} a_i = 2/3 < 1$, $\sum_{i=1}^{3} b_i = 3/4 < 1$, $\sum_{i=1}^{2} c_i = 5/6 < 1$, $\sum_{i=1}^{2} d_i = 7/12 < 1$, $\Delta_1 = \frac{157}{18}$, and $\Delta_2 = \frac{28}{3}$. The functions I_1 and I_2 are given by

$$I_1(j) = \begin{cases} \frac{403j}{157} - \frac{j^2}{30}, & 1 \le j \le 7, \\ \frac{960}{157} + \frac{283j}{157} - \frac{j^2}{30}, & 8 \le j \le 15, \\ \frac{5280}{157} - \frac{j}{157} - \frac{j^2}{30}, & 16 \le j \le 23, \\ \frac{7440}{157} - \frac{91j}{157} - \frac{j^2}{30}, & 24 \le j \le 29, \end{cases}$$

$$I_2(j) = \begin{cases} \frac{19j}{7} - \frac{j^2}{30}, & 1 \le j \le 5, \\ \frac{27}{7} + \frac{29j}{14} - \frac{j^2}{30}, & 6 \le j \le 8, \\ \frac{639}{56} + \frac{69j}{56} - \frac{j^2}{30}, & 9 \le j \le 17, \\ \frac{1125}{56} + \frac{3j}{4} - \frac{j^2}{30}, & 18 \le j \le 19, \\ \frac{2535}{56} - \frac{57j}{112} - \frac{j^2}{30}, & 20 \le j \le 29. \end{cases}$$

We obtain $K_1 = \sum_{j=1}^{29} I_1(j) \approx 461.68046709$, $K_2 = \max_{j=\overline{1,29}} I_2(j) \approx 22.78928571$, and $m_0 = K_1$. The functions $f(n,u)$ and $g(n,u)$ are nondecreasing with respect to u for any $n \in \{1,\dots,29\}$, and for $c = 1$ and $p_1 = 1/2$, assumptions (J3) and (J6) are satisfied; we have

$$\tilde{f}_\infty^{\mathrm{i}} = \lim_{u\to\infty} \frac{a(u^\alpha + u^\beta)}{u^{1/2}} = \infty, \quad \tilde{g}_\infty^{\mathrm{i}} = \lim_{u\to\infty} \frac{b(u^\theta + u^\delta)}{u^2} = \infty,$$
$$\tilde{f}_0^{\mathrm{i}} = \lim_{u\to 0+} \frac{a(u^\alpha + u^\beta)}{u} = \infty, \quad \tilde{g}_0^{\mathrm{i}} = \lim_{u\to 0+} \frac{b(u^\theta + u^\delta)}{u} = \infty.$$

We take $R_0 = 1$ and then $\sum_{i=1}^{29} g(i,R_0) = 58b$ and $f\left(n, m_0 \sum_{i=1}^{29} g(i,1)\right) = f(n, 58bm_0)) = a[(58bm_0)^\alpha + (58bm_0)^\beta]$ for all $n = \overline{1,29}$. If $a[(58bm_0)^\alpha + (58bm_0)^\beta] < \frac{1}{m_0}$, assumption (J7) is satisfied. For example, if $\alpha = 3/2$, $\beta = 1/2$, $b = 1/(58m_0) \approx 3.734 \cdot 10^{-5}$, and $a < 1/(2m_0)$ ($a < 1.083 \cdot 10^{-3}$), then the above inequality is satisfied. By Theorem 3.2.3, we deduce that problem $(\overline{\mathrm{S}'})$–$(\overline{\mathrm{BC}'})$ has at least two positive solutions.

Remark 3.2.1. The results presented in this section were published in Henderson et al. (2014).

3.3 Remarks on some particular cases

In this section, we present briefly some existence and multiplicity results for the positive solutions for some particular cases of problem (S)–(BC) from Section 3.1 and problem (S′)–(BC) from Section 3.2.

3.3.1 Systems with parameters

In this section we study problem (S)–(BC) from Section 3.1 with $f(n,u,v)=\tilde{f}(u,v)$, $g(n,u,v) = \tilde{g}(u,v)$, $p = r = 1$, $\xi_1 = 1$, $\zeta_1 = 1$, $a_1 = \beta/(\alpha+\beta)$, and $c_1 = \delta/(\gamma+\delta)$.

Namely, we consider here the system of nonlinear second-order difference equations

$$\begin{cases} \Delta^2 u_{n-1} + \lambda c_n f(u_n, v_n) = 0, & n = \overline{1,N-1}, \\ \Delta^2 v_{n-1} + \mu d_n g(u_n, v_n) = 0, & n = \overline{1,N-1}, \end{cases} \tag{S$_0$}$$

with the multipoint boundary conditions

$$\begin{cases} \alpha u_0 - \beta \Delta u_0 = 0, & u_N = \sum_{i=1}^{m-2} a_i u_{\xi_i}, \\ \gamma v_0 - \delta \Delta v_0 = 0, & v_N = \sum_{i=1}^{p-2} b_i v_{\eta_i}, \end{cases} \tag{BC$_0$}$$

where $N, m, p \in \mathbb{N}$, $N \geq 2$, $m, p \geq 3$, $\alpha, \beta, \gamma, \delta, a_i \in \mathbb{R}$ for all $i = \overline{1,m-2}$, $b_i \in \mathbb{R}$ for all $i = \overline{1,p-2}$, $\xi_i \in \mathbb{N}$ for all $i = \overline{1,m-2}$, $\eta_i \in \mathbb{N}$ for all $i = \overline{1,p-2}$, $1 \leq \xi_1 < \cdots < \xi_{m-2} \leq N-1$, and $1 \leq \eta_1 < \cdots < \eta_{p-2} \leq N-1$.

By a positive solution of problem (S$_0$)–(BC$_0$) we mean a pair of sequences $((u_n)_{n=\overline{0,N}}, (v_n)_{n=\overline{0,N}})$ satisfying (S$_0$) and (BC$_0$) with $u_n \geq 0$ and $v_n \geq 0$ for all $n = \overline{0,N}$, and $u_n > 0$ for all $n = \overline{1,N}$ or $v_n > 0$ for all $n = \overline{1,N}$.

First, we consider the following second-order difference equation with $m+1$-point boundary conditions:

$$\Delta^2 u_{n-1} + y_n = 0, \quad n = \overline{1,N-1}, \tag{3.34}$$

$$\alpha u_0 - \beta \Delta u_0 = 0, \quad u_N = \sum_{i=1}^{m-2} a_i u_{\xi_i}, \tag{3.35}$$

where $m \in \mathbb{N}$, $m \geq 3$, $\alpha, \beta, a_i \in \mathbb{R}$ for all $i = \overline{1,m-2}$ and $\xi_i \in \mathbb{N}$ for all $i = \overline{1,m-2}$, $1 \leq \xi_1 < \cdots < \xi_{m-2} \leq N-1$.

Lemma 3.3.1 (Li and Sun, 2006; Luca, 2009). *If $\alpha, \beta, a_i \in \mathbb{R}$ for all $i = \overline{1,m-2}$, $\alpha+\beta \neq 0$, $\xi_i \in \mathbb{N}$ for all $i = \overline{1,m-2}$, $1 \leq \xi_1 < \cdots < \xi_{m-2} \leq N-1$, $d = \alpha\left(N - \sum_{i=1}^{m-2} a_i\xi_i\right) + \beta\left(1 - \sum_{i=1}^{m-2} a_i\right) \neq 0$, and $y_n \in \mathbb{R}$ for all $n = \overline{1,N-1}$, then the unique solution of problem* (3.34)–(3.35) *is given by*

$$u_n = \frac{\alpha n+\beta}{d}\sum_{i=1}^{N-1}(N-i)y_i - \frac{\alpha n+\beta}{d}\sum_{i=1}^{m-2} a_i \sum_{j=1}^{\xi_i-1}(\xi_i-j)y_j - \sum_{i=1}^{n-1}(n-i)y_i, \quad n = \overline{0,N}.$$

Here we consider that $\sum_{i=1}^{0} z_i = 0$ and $\sum_{i=1}^{-1} z_i = 0$.

Lemma 3.3.2 (Luca, 2009). *Under the assumptions of Lemma 3.3.1, the Green's function for problem* (3.34)–(3.35) *is given by*

$$G_1(n,i)=\begin{cases} \dfrac{\alpha n+\beta}{d}(N-i)-\dfrac{\alpha n+\beta}{d}\displaystyle\sum_{k=j}^{m-2}a_k(\xi_k-i)-(n-i), & if\ \ \xi_{j-1}\le i<\xi_j,\\ & n\ge i,\ \ j=\overline{1,m-2},\ \ \xi_0=1,\\ \dfrac{\alpha n+\beta}{d}(N-i)-\dfrac{\alpha n+\beta}{d}\displaystyle\sum_{k=j}^{m-2}a_k(\xi_k-i), & if\ \ \xi_{j-1}\le i<\xi_j,\ \ n\le i,\\ & j=\overline{1,m-2},\\ \dfrac{\alpha n+\beta}{d}(N-i)-(n-i), \ \ if\ \ \xi_{m-2}\le i\le n, & \\ \dfrac{\alpha n+\beta}{d}(N-i), \ \ if\ \ \xi_{m-2}\le i\le N-1,\ \ n\le i. & \end{cases}$$

Using the above Green's function, we can express the solution of problem (3.34)–(3.35) as $u_n=\sum_{i=1}^{N-1}G_1(n,i)y_i,\quad n=\overline{0,N}$.

Lemma 3.3.3 (Li and Sun, 2006; Luca, 2009). *If* $\alpha\ge 0,\quad \beta\ge 0,\ d>0,\ a_i\ge 0$ *for all* $i=\overline{1,m-2},\ \xi_i\in\mathbb{N}$ *for all* $i=\overline{1,m-2},\ 1\le\xi_1<\cdots<\xi_{m-2}\le N-1$, $N\ge\sum_{i=1}^{m-2}a_i\xi_i$, *and* $y_n\ge 0$ *for all* $n=\overline{1,N-1}$, *then the solution* $u_n,\ n=\overline{0,N}$ *of problem* (3.34)–(3.35) *satisfies* $u_n\ge 0$ *for all* $n=\overline{0,N}$.

Lemma 3.3.4 (Luca, 2009). *If* $\alpha\ge 0,\quad \beta\ge 0,\ d>0,\ a_i\ge 0$ *for all* $i=\overline{1,m-2},\ \xi_i\in\mathbb{N}$ *for all* $i=\overline{1,m-2},\ 1\le\xi_1<\cdots<\xi_{m-2}\le N-1,\ N\ge\sum_{i=1}^{m-2}a_i\xi_i$, *and* $y_n\ge 0$ *for all* $n=\overline{1,N-1}$, *then the solution of problem* (3.34)–(3.35) *satisfies*

$$u_n\le\frac{\alpha N+\beta}{d}\sum_{i=1}^{N-1}(N-i)y_i,\quad \forall n=\overline{0,N},$$

$$u_{\xi_j}\ge\frac{\alpha\xi_j+\beta}{d}\sum_{i=\xi_{m-2}}^{N-1}(N-i)y_i,\quad \forall j=\overline{1,m-2}.$$

Lemma 3.3.5 (Li and Sun, 2006). *Assume that* $\alpha\ge 0,\quad \beta\ge 0,\ d>0,\ a_i\ge 0$ *for all* $i=\overline{1,m-2},\ \xi_i\in\mathbb{N}$ *for all* $i=\overline{1,m-2},\ 1\le\xi_1<\cdots<\xi_{m-2}\le N-1$, $N>\sum_{i=1}^{m-2}a_i\xi_i>0$, *and* $y_n\ge 0$ *for all* $n=\overline{1,N-1}$. *Then the solution of problem* (3.34)–(3.35) *satisfies* $\min_{n=\overline{\xi_1,N}}u_n\ge r_1\max_{i=\overline{0,N}}u_i$, *where*

$$r_1=\min\left\{\frac{\xi_1}{N},\frac{\sum_{i=1}^{m-2}a_i(N-\xi_i)}{N-\sum_{i=1}^{m-2}a_i\xi_i},\frac{\sum_{i=1}^{m-2}a_i\xi_i}{N},\frac{\sum_{i=1}^{s-1}a_i\xi_i+\sum_{i=s}^{m-2}a_i(N-\xi_i)}{N-\sum_{i=s}^{m-2}a_i\xi_i},\right.$$
$$\left. s=2,\ldots,m-2\right\}.$$

We can also formulate results similar to those in Lemmas 3.3.1–3.3.5 for the discrete boundary value problem:

$$\Delta^2 v_{n-1}+h_n=0,\quad n=\overline{1,N-1},\tag{3.36}$$

$$\gamma v_0 - \delta \Delta v_0 = 0, \quad v_N = \sum_{i=1}^{p-2} b_i v_{\eta_i}, \tag{3.37}$$

where $p \in \mathbb{N}$, $p \geq 3$, γ, δ, $b_i \in \mathbb{R}$ for all $i = \overline{1,p-2}$, $\eta_i \in \mathbb{N}$ for all $i = \overline{1,p-2}$, $1 \leq \eta_1 < \cdots < \eta_{p-2} \leq N-1$, and $h_n \in \mathbb{R}$ for all $n = \overline{1,N-1}$. We denote by e, r_2, η_0, and G_2 the constants and the corresponding function for problem (3.36)–(3.37) defined in a similar manner as d, r_1, ξ_0, and G_1, respectively.

We present the assumptions that we shall use in the sequel:

($\widetilde{\text{A1}}$) $\alpha \geq 0$, $\beta \geq 0$, $\gamma \geq 0$, $\delta \geq 0$, $a_i \geq 0$ for all $i = \overline{1,m-2}$, $b_i \geq 0$ for all $i = \overline{1,p-2}$, $\xi_i \in \mathbb{N}$ for all $i = \overline{1,m-2}$, $1 \leq \xi_1 < \cdots < \xi_{m-2} \leq N-1$, $\eta_i \in \mathbb{N}$ for all $i = \overline{1,p-2}$, $1 \leq \eta_1 < \cdots < \eta_{p-2} \leq N-1$, $N > \sum_{i=1}^{m-2} a_i \xi_i > 0$, $N > \sum_{i=1}^{p-2} b_i \eta_i > 0$, $d = \alpha\left(N - \sum_{i=1}^{m-2} a_i \xi_i\right) + \beta\left(1 - \sum_{i=1}^{m-2} a_i\right) > 0$, $e = \gamma\left(N - \sum_{i=1}^{p-2} b_i \eta_i\right) + \delta\left(1 - \sum_{i=1}^{p-2} b_i\right) > 0$, $\xi_{m-2} \geq \eta_1$, and $\eta_{p-2} \geq \xi_1$.

($\widetilde{\text{A2}}$) The constants c_n, $d_n \geq 0$ for all $n = \overline{1,N-1}$, and there exists $i_0 \in \{\xi_{m-2}, \ldots, N-1\}$ and $j_0 \in \{\eta_{p-2}, \ldots, N-1\}$ such that $c_{i_0} > 0$, $d_{j_0} > 0$.

($\widetilde{\text{A3}}$) The functions $f, g : [0,\infty) \times [0,\infty) \to [0,\infty)$ are continuous.

We introduce the following extreme limits:

$$f_0^{\rm s} = \limsup_{u+v\to 0+} \frac{f(u,v)}{u+v}, \quad g_0^{\rm s} = \limsup_{u+v\to 0+} \frac{g(u,v)}{u+v},$$

$$f_0^{\rm i} = \liminf_{u+v\to 0+} \frac{f(u,v)}{u+v}, \quad g_0^{\rm i} = \liminf_{u+v\to 0+} \frac{g(u,v)}{u+v},$$

$$f_\infty^{\rm s} = \limsup_{u+v\to \infty} \frac{f(u,v)}{u+v}, \quad g_\infty^{\rm s} = \limsup_{u+v\to \infty} \frac{g(u,v)}{u+v},$$

$$f_\infty^{\rm i} = \liminf_{u+v\to \infty} \frac{f(u,v)}{u+v}, \quad g_\infty^{\rm i} = \liminf_{u+v\to \infty} \frac{g(u,v)}{u+v}.$$

In the definitions of the extreme limits above, the variables u and v are nonnegative.

We consider the Banach space $X = \mathbb{R}^{N+1} = \{u = (u_0, u_1, \ldots, u_N),\ u_i \in \mathbb{R},\ i = \overline{0,N}\}$ with the norm $\|u\| = \max_{n=\overline{0,N}} |u_n|$ and the Banach space $Y = X \times X$ with the norm $\|(u,v)\|_Y = \|u\| + \|v\|$.

We define the cone $C \subset Y$ by

$$C = \Bigg\{ (u,v) \in Y; \quad u = (u_n)_{n=\overline{0,N}}, \quad v = (v_n)_{n\in\overline{0,N}}, \quad u_n \geq 0, \quad v_n \geq 0,$$
$$\forall n = \overline{0,N} \quad \text{and} \quad \min_{n=\overline{\theta_0,N}} (u_n + v_n) \geq r\|(u,v)\|_Y \Bigg\},$$

where $\theta_0 = \max\{\xi_1, \eta_1\}$, $r = \min\{r_1, r_2\}$, and r_1 and r_2 are defined above.

Let $A_1, A_2 : Y \to X$ and $\mathcal{T} : Y \to Y$ be the operators defined by

$$A_1(u,v) = (A_1(u,v))_{n=\overline{0,N}}, \quad (A_1(u,v))_n = \lambda \sum_{i=1}^{N-1} G_1(n,i) c_i f(u_i, v_i), \quad n = \overline{0,N},$$

$$A_2(u,v) = (A_2(u,v))_{n=\overline{0,N}}, \quad (A_2(u,v))_n = \mu \sum_{i=1}^{N-1} G_2(n,i) d_i g(u_i, v_i), \quad n = \overline{0,N},$$

and $\mathcal{T}(u,v) = (A_1(u,v), A_2(u,v))$, $(u,v) = ((u_n)_{n=\overline{0,N}}, (v_n)_{n=\overline{0,N}}) \in Y$, where G_1 and G_2 are the Green's functions defined above.

The solutions of problem (S_0)–(BC_0) coincide with the fixed points of the operator $\mathcal{T}$ in the space Y. In addition, the operator $\mathcal{T}$ is a completely continuous operator.

We suppose first that $f_0^s, g_0^s, f_\infty^i, g_\infty^i \in (0,\infty)$, and for positive numbers $\alpha_1, \alpha_2 > 0$ such that $\alpha_1 + \alpha_2 = 1$, we define the positive numbers L_1, L_2, L_3, and L_4 by

$$L_1 = \alpha_1 \left(\frac{r(\alpha \xi_{m-2} + \beta)}{d} \sum_{j=\xi_{m-2}}^{N-1} (N-j) c_j f_\infty^i \right)^{-1},$$

$$L_2 = \alpha_1 \left(\frac{\alpha N + \beta}{d} \sum_{j=1}^{N-1} (N-j) c_j f_0^s \right)^{-1},$$

$$L_3 = \alpha_2 \left(\frac{r(\gamma \eta_{p-2} + \delta)}{e} \sum_{j=\eta_{p-2}}^{N-1} (N-j) d_j g_\infty^i \right)^{-1},$$

$$L_4 = \alpha_2 \left(\frac{\gamma N + \delta}{e} \sum_{j=1}^{N-1} (N-j) d_j g_0^s \right)^{-1}.$$

By using Lemmas 3.3.1–3.3.5, the operator $\mathcal{T}$, Theorem 1.1.1, and arguments similar to those used in the proofs of Theorems 3.1.1 and 3.1.2, we obtain the following results:

Theorem 3.3.1. *Assume that* ($\widetilde{A1}$)–($\widetilde{A3}$) *hold and* $\alpha_1, \alpha_2 > 0$ *are positive numbers such that* $\alpha_1 + \alpha_2 = 1$.

- ***(a)*** *If* $f_0^s, g_0^s, f_\infty^i, g_\infty^i \in (0,\infty)$, $L_1 < L_2$, *and* $L_3 < L_4$, *then for each* $\lambda \in (L_1, L_2)$ *and* $\mu \in (L_3, L_4)$ *there exists a positive solution* $((u_n)_{n=\overline{0,N}}, (v_n)_{n=\overline{0,N}})$ *for* (S_0)–(BC_0).
- ***(b)*** *If* $f_0^s = g_0^s = 0$ *and* $f_\infty^i, g_\infty^i \in (0,\infty)$, *then for each* $\lambda \in (L_1, \infty)$ *and* $\mu \in (L_3, \infty)$ *there exists a positive solution* $((u_n)_{n=\overline{0,N}}, (v_n)_{n=\overline{0,N}})$ *for* (S_0)–(BC_0).
- ***(c)*** *If* $f_0^s, g_0^s \in (0,\infty)$ *and* $f_\infty^i = g_\infty^i = \infty$, *then for each* $\lambda \in (0, L_2)$ *and* $\mu \in (0, L_4)$ *there exists a positive solution* $((u_n)_{n=\overline{0,N}}, (v_n)_{n=\overline{0,N}})$ *for* (S_0)–(BC_0).
- ***(d)*** *If* $f_0^s = g_0^s = 0$ *and* $f_\infty^i = g_\infty^i = \infty$, *then for each* $\lambda \in (0,\infty)$ *and* $\mu \in (0,\infty)$ *there exists a positive solution* $((u_n)_{n=\overline{0,N}}, (v_n)_{n=\overline{0,N}})$ *for* (S_0)–(BC_0).

We suppose now that $f_0^{\rm i}, g_0^{\rm i}, f_\infty^{\rm s}, g_\infty^{\rm s} \in (0, \infty)$, and for positive numbers $\alpha_1, \alpha_2 > 0$ such that $\alpha_1 + \alpha_2 = 1$, we define the positive numbers $\tilde{L}_1, \tilde{L}_2, \tilde{L}_3$, and $\tilde{L}_4$ by

$$\tilde{L}_1 = \alpha_1 \left(\frac{r(\alpha\xi_{m-2} + \beta)}{d} \sum_{j=\xi_{m-2}}^{N-1} (N-j) c_j f_0^{\rm i} \right)^{-1},$$

$$\tilde{L}_2 = \alpha_1 \left(\frac{\alpha N + \beta}{d} \sum_{j=1}^{N-1} (N-j) c_j f_\infty^{\rm s} \right)^{-1},$$

$$\tilde{L}_3 = \alpha_2 \left(\frac{r(\gamma\eta_{p-2} + \delta)}{e} \sum_{j=\eta_{p-2}}^{N-1} (N-j) d_j g_0^{\rm i} \right)^{-1},$$

$$\tilde{L}_4 = \alpha_2 \left(\frac{\gamma N + \delta}{e} \sum_{j=1}^{N-1} (N-j) d_j g_\infty^{\rm s} \right)^{-1}.$$

Theorem 3.3.2. *Assume that* $(\widetilde{\rm A1})$–$(\widetilde{\rm A3})$ *hold and* $\alpha_1, \alpha_2 > 0$ *are positive numbers such that* $\alpha_1 + \alpha_2 = 1$.

(a) *If* $f_0^{\rm i}, g_0^{\rm i}, f_\infty^{\rm s}, g_\infty^{\rm s} \in (0, \infty)$, $\tilde{L}_1 < \tilde{L}_2$, *and* $\tilde{L}_3 < \tilde{L}_4$, *then for each* $\lambda \in (\tilde{L}_1, \tilde{L}_2)$ *and* $\mu \in (\tilde{L}_3, \tilde{L}_4)$ *there exists a positive solution* $((u_n)_{n=\overline{0,N}}, (v_n)_{n=\overline{0,N}})$ *for* (S$_0$)–(BC$_0$).

(b) *If* $f_\infty^{\rm s} = g_\infty^{\rm s} = 0$ *and* $f_0^{\rm i}, g_0^{\rm i} \in (0, \infty)$, *then for each* $\lambda \in (\tilde{L}_1, \infty)$ *and* $\mu \in (\tilde{L}_3, \infty)$ *there exists a positive solution* $((u_n)_{n=\overline{0,N}}, (v_n)_{n=\overline{0,N}})$ *for* (S$_0$)–(BC$_0$).

(c) *If* $f_\infty^{\rm s}, g_\infty^{\rm s} \in (0, \infty)$ *and* $f_0^{\rm i} = g_0^{\rm i} = \infty$, *then for each* $\lambda \in (0, \tilde{L}_2)$ *and* $\mu \in (0, \tilde{L}_4)$ *there exists a positive solution* $((u_n)_{n=\overline{0,N}}, (v_n)_{n=\overline{0,N}})$ *for* (S$_0$)–(BC$_0$).

(d) *If* $f_\infty^{\rm s} = g_\infty^{\rm s} = 0$ *and* $f_0^{\rm i} = g_0^{\rm i} = \infty$, *then for each* $\lambda \in (0, \infty)$ *and* $\mu \in (0, \infty)$ *there exists a positive solution* $((u_n)_{n=\overline{0,N}}, (v_n)_{n=\overline{0,N}})$ *for* (S$_0$)–(BC$_0$).

In Theorems 3.3.1 and 3.3.2, we find that the operator $\mathcal{T}$ has at least one fixed point $(u, v) = ((u_n)_{n=\overline{0,N}}, (v_n)_{n=\overline{0,N}}) \in C$ such that $R_1 \le \|(u, v)\|_Y \le R_2$, with some $R_1, R_2 > 0$, $R_1 < R_2$. The last inequality implies that $\|u\| > 0$ or $\|v\| > 0$. If $\|u\| > 0$, then $u_n > 0$ for all $n = \overline{1, N}$. Because $\Delta^2 u_{n-1} = -\lambda c_n f(u_n, v_n) \le 0$ for all $n = \overline{1, N-1}$, we deduce that u is concave. This, together with the fact that $u_N \ge \min_{n=\overline{\xi_1, N}} u_n \ge r_1 \|u\| > 0$, implies that $u_n > 0$ for all $n = \overline{1, N}$. In a similar manner, if $\|v\| > 0$, then $v_n > 0$ for all $n = \overline{1, N}$.

Remark 3.3.1. The results presented in this section were published in Henderson and Luca (2012f).

3.3.2 Systems without parameters

In this section, we study system (S$'$)–(BC) from Section 3.2 with $a_i = 0$ for all $i = \overline{1, p}$ and $c_i = 0$ for all $i = \overline{1, r}$.

Namely, we consider here the system of nonlinear second-order difference equations

$$\begin{cases} \Delta^2 u_{n-1} + f(n, v_n) = 0, & n = \overline{1, N-1}, \\ \Delta^2 v_{n-1} + g(n, u_n) = 0, & n = \overline{1, N-1}, \end{cases} \qquad (\mathrm{S}_0')$$

with the multipoint boundary conditions

$$u_0 = 0, \quad u_N = \sum_{i=1}^{m-2} b_i u_{\xi_i}, \quad v_0 = 0, \quad v_N = \sum_{i=1}^{p-2} c_i v_{\eta_i}, \tag{BC$'_0$}$$

where $N, m, p \in \mathbb{N}$, $N \geq 2$, $m, p \geq 3$, $b_i \in \mathbb{R}$ for all $i = \overline{1, m-2}$, $c_i \in \mathbb{R}$ for all $i = \overline{1, p-2}$, $\xi_i \in \mathbb{N}$ for all $i = \overline{1, m-2}$, $\eta_i \in \mathbb{N}$ for all $i = \overline{1, p-2}$, $1 \leq \xi_1 < \cdots < \xi_{m-2} \leq N-1$, and $1 \leq \eta_1 < \cdots < \eta_{p-2} \leq N-1$.

By a positive solution of (S$'_0$)–(BC$'_0$), we mean a pair of sequences $((u_n)_{n=\overline{0,N}}, (v_n)_{n=\overline{0,N}})$ satisfying (S$'_0$) and (BC$'_0$) and $u_n > 0$, $v_n > 0$ for all $n = \overline{1, N}$.

We consider first the following nonlinear second-order difference equation with m-point boundary conditions:

$$\Delta^2 u_{n-1} + y_n = 0, \quad n = \overline{1, N-1}, \tag{3.38}$$

$$u_0 = 0, \quad u_N = \sum_{i=1}^{m-2} b_i u_{\xi_i}, \tag{3.39}$$

where $m \in \mathbb{N}$, $m \geq 3$, $b_i \in \mathbb{R}$ for all $i = \overline{1, m-2}$, $\xi_i \in \mathbb{N}$ for all $i = \overline{1, m-2}$, and $1 \leq \xi_1 < \cdots < \xi_{m-2} \leq N-1$.

Lemma 3.3.6 (Li and Sun, 2006; Luca, 2009). *If $b_i \in \mathbb{R}$ for all $i = \overline{1, m-2}$, $\xi_i \in \mathbb{N}$ for all $i = \overline{1, m-2}$, $1 \leq \xi_1 < \cdots < \xi_{m-2} \leq N-1$, $\tilde{d} = N - \sum_{i=1}^{m-2} b_i \xi_i \neq 0$, and $y_n \in \mathbb{R}$ for all $n = \overline{1, N-1}$, then the solution of problem* (3.38)–(3.39) *is given by*

$$u_n = \frac{n}{\tilde{d}} \sum_{i=1}^{N-1} (N-i) y_i - \frac{n}{\tilde{d}} \sum_{i=1}^{m-2} b_i \sum_{j=1}^{\xi_i - 1} (\xi_i - j) y_j - \sum_{i=1}^{n-1} (n-i) y_i, \quad n = \overline{0, N}.$$

We consider again that $\sum_{i=1}^{0} z_i = 0$ and $\sum_{i=1}^{-1} z_i = 0$.

Lemma 3.3.7 (Luca, 2009). *Under the assumptions of Lemma 3.3.6, the Green's function for the boundary value problem* (3.38)–(3.39) *is given by*

$$\tilde{G}_1(n,i) = \begin{cases} \dfrac{n(N-i)}{\tilde{d}} - \dfrac{n}{\tilde{d}} \displaystyle\sum_{k=j}^{m-2} b_k(\xi_k - i) - (n-i), & \text{if} \quad \xi_{j-1} \leq i < \xi_j, \\ \qquad n \geq i, \quad j = \overline{1, m-2}, \quad \xi_0 = 1, & \\ \dfrac{n(N-i)}{\tilde{d}} - \dfrac{n}{\tilde{d}} \displaystyle\sum_{k=j}^{m-2} b_k(\xi_k - i), & \text{if} \quad \xi_{j-1} \leq i < \xi_j, \quad n \leq i, \\ \qquad j = \overline{1, m-2}, & \\ \dfrac{n(N-i)}{\tilde{d}} - (n-i), & \text{if} \quad \xi_{m-2} \leq i \leq n, \\ \dfrac{n(N-i)}{\tilde{d}}, & \text{if} \quad \xi_{m-2} \leq i \leq N-1, \quad n \leq i. \end{cases}$$

Using the above Green's function, we can express the solution of problem (3.38)–(3.39) as $u_n = \sum_{i=1}^{N-1} \tilde{G}_1(n,i)y_i, \quad n = \overline{0,N}$.

Lemma 3.3.8 (Li and Sun, 2006; Luca, 2009). *If $b_i \geq 0$ for all $i = \overline{1,m-2}$, $\xi_i \in \mathbb{N}$ for all $i = \overline{1,m-2}$, $1 \leq \xi_1 < \cdots < \xi_{m-2} \leq N-1$, $\tilde{d} > 0$, and $y_n \geq 0$ for all $n = \overline{1,N-1}$, then the solution u_n, $n = \overline{0,N}$ of problem* (3.38)–(3.39) *satisfies $u_n \geq 0$ for all $n = \overline{0,N}$, and $\tilde{G}_1(n,i) \geq 0$ for all $n = \overline{0,N}$, $i = \overline{1,N-1}$.*

Lemma 3.3.9 (Li and Sun, 2006). *Assume that $b_i \geq 0$ for all $i = \overline{1,m-2}$, $\xi_i \in \mathbb{N}$ for all $i = \overline{1,m-2}$, $1 \leq \xi_1 < \cdots < \xi_{m-2} \leq N-1$, $\tilde{d} > 0$, $\sum_{i=1}^{m-2} b_i\xi_i > 0$, and $y_n \geq 0$ for all $n = \overline{1,N-1}$. Then the solution of problem* (3.38)–(3.39) *satisfies $\min_{n=\overline{\xi_1,N}} u_n \geq \gamma_1 \max_{i=\overline{0,N}} u_i$, where*

$$\gamma_1 = \min\left\{\frac{\xi_1}{N}, \frac{\sum_{k=1}^{m-2} b_k(N-\xi_k)}{N - \sum_{k=1}^{m-2} b_k\xi_k},\right.$$
$$\left.\frac{\sum_{k=1}^{m-2} b_k\xi_k}{N}, \frac{\sum_{k=1}^{s-1} b_k\xi_k + \sum_{k=s}^{m-2} b_k(N-\xi_k)}{N - \sum_{k=s}^{m-2} b_k\xi_k}, \quad s = 2,\ldots,m-2\right\}.$$

Lemma 3.3.10 (Henderson and Luca, 2013a). *If $b_i \geq 0$ for all $i = \overline{1,m-2}$, $\xi_i \in \mathbb{N}$ for all $i = \overline{1,m-2}$, $1 \leq \xi_1 < \cdots < \xi_{m-2} \leq N-1$, and $\tilde{d} > 0$, then the Green's function $\tilde{G}_1$ of problem* (3.38)–(3.39) *satisfies the inequalities*

$$\tilde{G}_1(n,i) \leq a_1 i(N-i), \quad \forall n = \overline{0,N}, \quad i = \overline{1,N-1},$$
$$\tilde{G}_1(n,i) \geq a_2 i(N-i), \quad \forall n = \overline{a,b}, \quad i = \overline{1,N-1}, \quad a \in \{1,\ldots,N-1\},$$
$$b \in \{\xi_{m-2},\ldots,N\}, \quad a < b,$$

where

$$a_1 = \max\left\{\frac{1}{\tilde{d}(N-\xi_1)}\sum_{k=1}^{m-2} b_k(N-\xi_k), \ \frac{1}{\tilde{d}}, \ \frac{1}{\tilde{d}\xi_{m-2}}\sum_{k=1}^{m-2} b_k\xi_k,\right.$$
$$\frac{1}{\tilde{d}\xi_{j-1}(N-\xi_j)}\left[\xi_{j-1}\left(N\sum_{k=j}^{m-2} b_k - \sum_{k=1}^{m-2} b_k\xi_k\right) + N\sum_{k=1}^{j-1} b_k\xi_k\right], \tag{3.40}$$
$$j = 2,\ldots,m-2,$$
$$\left.\frac{N}{\tilde{d}\xi_{j-1}(N-\xi_j)}\sum_{k=1}^{j-1} b_k\xi_k, \quad j = 2,\ldots,m-2\right\},$$

$$a_2 = \min\left\{\frac{1}{N}, \ \frac{1}{\tilde{d}N}\sum_{k=1}^{m-2} b_k\xi_k, \ \frac{a}{\tilde{d}N}, \ \frac{1}{\tilde{d}(N-\xi_{j-1})}\sum_{k=j}^{m-2}(b-\xi_k)b_k,\right.$$
$$j = 1,\ldots,m-2,$$

$$\frac{a}{\tilde{d}}\left[N-\sum_{k=j}^{m-2}b_k\xi_k+\left(\sum_{k=j}^{m-2}b_k-1\right)\xi_{j-1}\right]\cdot\frac{4}{N^2},\quad j=1,\ldots,m-2,$$
$$\left.\frac{a}{\tilde{d}}\left[N-\sum_{k=j}^{m-2}b_k\xi_k+\left(\sum_{k=j}^{m-2}b_k-1\right)\xi_j\right]\cdot\frac{4}{N^2},\quad j=1,\ldots,m-2\right\}. \tag{3.41}$$

We can also formulate results similar to those in Lemmas 3.3.6–3.3.10 for the following problem:

$$\Delta^2 v_{n-1}+h_n=0,\quad n=\overline{1,N-1}, \tag{3.42}$$

$$v_0=0,\quad v_N=\sum_{i=1}^{p-2}c_i v_{\eta_i}, \tag{3.43}$$

where $p\in\mathbb{N}$, $p\geq 3$, $c_i\in\mathbb{R}$ for all $i=\overline{1,p-2}$, $\eta_i\in\mathbb{N}$ for all $i=\overline{1,p-2}$, $1\leq\eta_1<\cdots<\eta_{p-2}\leq N-1$, and $h_n\in\mathbb{R}$ for all $n=\overline{1,N-1}$. We denote by $\tilde{e}$, γ_2, and $\tilde{G}_2$ the constants and the corresponding function for problem (3.42)–(3.43) defined in a similar manner as $\tilde{d}$, γ_1, and $\tilde{G}_1$, respectively.

Moreover, under assumptions similar to those from Lemma 3.3.10, we have the inequalities

$$\tilde{G}_2(n,i)\leq\tilde{a}_1 i(N-i),\quad\forall n=\overline{0,N},\quad i=\overline{1,N-1},$$
$$\tilde{G}_2(n,i)\geq\tilde{a}_2 i(N-i),\quad\forall n=\overline{a,b},\quad i=\overline{1,N-1},\quad a\in\{1,\ldots,N-1\},$$
$$b\in\{\eta_{p-2},\ldots,N\},\quad a<b,$$

where $\tilde{a}_1$ and $\tilde{a}_2$ are defined as a_1 and a_2 in (3.40) and (3.41) with m, b_i, and ξ_i replaced by p, c_i, and η_i, respectively.

In the rest of this section, we shall denote by a_1 and a_2 the constants from Lemma 3.3.10 for $\tilde{G}_1$ with $a=\xi_1$ and $b=N$, and by $\tilde{a}_1$ and $\tilde{a}_2$ the corresponding constants for $\tilde{G}_2$ with $a=\eta_1$ and $b=N$.

We present the assumptions that we shall use in the sequel:

($\widetilde{\mathrm{J1}}$) $\xi_i\in\mathbb{N}$ for all $i=\overline{1,m-2}$, $1\leq\xi_1<\cdots<\xi_{m-2}\leq N-1$, $\eta_i\in\mathbb{N}$ for all $i=\overline{1,p-2}$, $1\leq\eta_1<\cdots<\eta_{p-2}\leq N-1$, $b_i\geq 0$ for all $i=\overline{1,m-2}$, $c_i\geq 0$ for all $i=\overline{1,p-2}$, $\tilde{d}=N-\sum_{i=1}^{m-2}b_i\xi_i>0$, $\tilde{e}=N-\sum_{i=1}^{p-2}c_i\eta_i>0$, $\sum_{i=1}^{m-2}b_i\xi_i>0$, and $\sum_{i=1}^{p-2}c_i\eta_i>0$.

($\widetilde{\mathrm{J2}}$) The functions $f,g:\{1,\ldots,N-1\}\times[0,\infty)\to[0,\infty)$ are continuous, and $f(n,0)=0$, $g(n,0)=0$, for all $n=\overline{1,N-1}$.

The pair of sequences $\left((u_n)_{n=\overline{0,N}},(v_n)_{n=\overline{0,N}}\right)$ with $u_n\geq 0$, $v_n\geq 0$ for all $n=\overline{0,N}$, is a solution for problem (S_0')–(BC_0') if and only if $\left((u_n)_{n=\overline{0,N}},(v_n)_{n=\overline{0,N}}\right)$ with $u_n\geq 0$, $v_n\geq 0$ for all $n=\overline{0,N}$, is a solution for the nonlinear system

$$\begin{cases} u_n = \sum_{i=1}^{N-1} \tilde{G}_1(n,i) f\left(i, \sum_{j=1}^{N-1} \tilde{G}_2(i,j) g(j,u_j)\right), & n = \overline{0,N}, \\ v_n = \sum_{i=1}^{N-1} \tilde{G}_2(n,i) g(i,u_i), \quad n = \overline{0,N}. \end{cases}$$

We consider the Banach space $X = \mathbb{R}^{N+1} = \{u = (u_0, u_1, \ldots, u_N), \quad u_i \in \mathbb{R}, i = \overline{0,N}\}$ with the norm $\|u\| = \max_{i=\overline{0,N}} |u_i|$, for $u = (u_n)_{n=\overline{0,N}}$, and define the cone $C' \subset X$ by $C' = \{u \in X, \quad u = (u_n)_{n=\overline{0,N}}, \quad u_n \geq 0, \quad n = \overline{0,N}\}$.

We also define the operator $\mathcal{A} : C' \to X$ by

$$\mathcal{A}(u) = v, \quad u = (u_n)_{n=\overline{0,N}}, \quad v = (v_n)_{n=\overline{0,N}},$$

$$v_n = \sum_{i=1}^{N-1} \tilde{G}_1(n,i) f\left(i, \sum_{j=1}^{N-1} \tilde{G}_2(i,j) g(j,u_j)\right),$$

and the operators $\mathcal{B} : C' \to X$ and $\mathcal{C} : C' \to X$ by

$$\mathcal{B}(u) = v, \quad u = (u_n)_{n=\overline{0,N}}, \quad v = (v_n)_{n=\overline{0,N}}, \quad v_n = \sum_{i=1}^{N-1} \tilde{G}_1(n,i) u_i, \quad n = \overline{0,N},$$

$$\mathcal{C}(u) = v, \quad u = (u_n)_{n=\overline{0,N}}, \quad v = (v_n)_{n=\overline{0,N}}, \quad v_n = \sum_{i=1}^{N-1} \tilde{G}_2(n,i) u_i, \quad n = \overline{0,N}.$$

Under assumptions ($\widetilde{\text{J1}}$) and ($\widetilde{\text{J2}}$), using also Lemma 3.3.8, one can easily see that $\mathcal{A}$, $\mathcal{B}$, and $\mathcal{C}$ are completely continuous from C' to C'. Thus, the existence and multiplicity of positive solutions of system (S_0')–(BC_0') are equivalent to the existence and multiplicity of fixed points of the operator $\mathcal{A}$.

We also define the cone $C_0 = \{u \in C'; \quad \min_{n=\overline{\theta_0,N}} u_n \geq \gamma \|u\|\}$, where $\theta_0 = \max\{\xi_1, \eta_1\}$, $\gamma = \min\{\gamma_1, \gamma_2\}$, and γ_1 and $\gamma_2 > 0$ are defined above. From our assumptions and Lemma 3.3.9, we deduce that $\mathcal{B}(C') \subset C_0$ and $\mathcal{C}(C') \subset C_0$.

By using Lemmas 3.3.6–3.3.10, Theorems 1.3.1–1.3.3, and arguments similar to those used in the proofs of Theorems 3.2.1–3.2.3, we obtain the following results:

Theorem 3.3.3. *Assume that* ($\widetilde{\text{J1}}$) *and* ($\widetilde{\text{J2}}$) *hold. If the functions* f *and* g *also satisfy the following conditions* ($\widetilde{\text{J3}}$) *and* ($\widetilde{\text{J4}}$), *then problem* (S_0')–(BC_0') *has at least one positive solution* $((u_n)_{n=\overline{0,N}}, (v_n)_{n=\overline{0,N}})$:

($\widetilde{\text{J3}}$) *There exists a positive constant* $p_1 \in (0,1]$ *such that*

$$(1) \quad \tilde{f}_\infty^{\text{i}} = \liminf_{u\to\infty} \min_{n=\overline{1,N-1}} \frac{f(n,u)}{u^{p_1}} \in (0,\infty] \quad \textit{and}$$

$$(2) \quad \tilde{g}_\infty^{\text{i}} = \liminf_{u\to\infty} \min_{n=\overline{1,N-1}} \frac{g(n,u)}{u^{1/p_1}} = \infty.$$

($\widetilde{\text{J}}4$) *There exists a positive constant* $q_1 \in (0,\infty)$ *such that*

$$(1)\quad \tilde{f}_0^{s} = \limsup_{u\to 0^+} \max_{n=\overline{1,N-1}} \frac{f(n,u)}{u^{q_1}} \in [0,\infty) \quad \text{and}$$

$$(2)\quad \tilde{g}_0^{s} = \limsup_{u\to 0^+} \max_{n=\overline{1,N-1}} \frac{g(n,u)}{u^{1/q_1}} = 0.$$

Theorem 3.3.4. *Assume that* ($\widetilde{\text{J}}1$) *and* ($\widetilde{\text{J}}2$) *hold. If the functions* f *and* g *also satisfy the following conditions* ($\widetilde{\text{J}}5$) *and* ($\widetilde{\text{J}}6$), *then problem* (S_0')–(BC_0') *has at least one positive solution* $((u_n)_{n=\overline{0,N}}, (v_n)_{n=\overline{0,N}})$:

($\widetilde{\text{J}}5$) *There exists a positive constant* $r_1 \in (0,\infty)$ *such that*

$$(1)\quad \tilde{f}_\infty^{s} = \limsup_{u\to \infty} \max_{n=\overline{1,N-1}} \frac{f(n,u)}{u^{r_1}} \in [0,\infty) \quad \text{and} \quad (2)\quad \tilde{g}_\infty^{s} = \limsup_{u\to \infty} \max_{n=\overline{1,N-1}} \frac{g(n,u)}{u^{1/r_1}} = 0.$$

($\widetilde{\text{J}}6$) *The following conditions are satisfied:*

$$(1)\quad \tilde{f}_0^{i} = \liminf_{u\to 0^+} \min_{n=\overline{1,N-1}} \frac{f(n,u)}{u} \in (0,\infty] \quad \text{and} \quad (2)\quad \tilde{g}_0^{i} = \liminf_{u\to 0^+} \min_{n=\overline{1,N-1}} \frac{g(n,u)}{u} = \infty.$$

Theorem 3.3.5. *Assume that* ($\widetilde{\text{J}}1$) *and* ($\widetilde{\text{J}}2$) *hold. If the functions* f *and* g *also satisfy conditions* ($\widetilde{\text{J}}3$) *and* ($\widetilde{\text{J}}6$) *and the following condition* ($\widetilde{\text{J}}7$), *then problem* (S_0')–(BC_0') *has at least two positive solutions* (u^1, v^1) *and* (u^2, v^2):

($\widetilde{\text{J}}7$) *For each* $n = \overline{1, N-1}$, $f(n,u)$ *and* $g(n,u)$ *are nondecreasing with respect to* u, *and there exists a constant* $R_0 > 0$ *such that*

$$f\left(n, m_0 \sum_{i=1}^{N-1} g(i, R_0)\right) < \frac{R_0}{m_0}, \qquad \forall n = \overline{1, N-1},$$

where $m_0 = N^2/4 \max\{a_1(N-1), \tilde{a}_1\}$, *and* a_1 *and* $\tilde{a}_1$ *are defined above.*

In Theorems 3.3.3–3.3.5, we find that the operator $\mathcal{A}$ has at least one fixed point $u^1 \in (B_L \setminus \bar{B}_{\delta_1}) \cap C'$ (with some $\delta_1, L > 0$, $\delta_1 < L$), $u^1 = (u_n^1)_{n=\overline{0,N}}$—that is, $\delta_1 < \|u^1\| < L$ (or at least two fixed points). Let $v^1 = (v_n^1)_{n=\overline{0,N}}$, $v_n^1 = \sum_{i=1}^{N-1} \tilde{G}_2(n,i) g(i, u_i^1)$, $n = \overline{0,N}$. Then $(u^1, v^1) \in C' \times C'$ is a solution of (S_0')–(BC_0'). In addition, $u_n^1 > 0$, $v_n^1 > 0$ for all $n = \overline{1,N}$. Because $\Delta^2 u_{n-1}^1 = -f(n, v_n^1) \le 0$, we deduce that u^1 is concave. Combining this information with $u_N^1 \ge \min_{n=\overline{\xi_1,N}} u_n^1 \ge \gamma_1 \|u^1\| > 0$, we obtain $u_n^1 > 0$ for all $n = \overline{1,N}$. Because $u_n^1 = \sum_{i=1}^{N-1} \tilde{G}_1(n,i) f(i, v_i^1)$, $n = \overline{0,N}$ and $\|u^1\| > 0$, we deduce that $\|v^1\| > 0$. If we suppose that $v_n^1 = 0$, for all $n = \overline{0,N}$, then by using ($\widetilde{\text{J}}2$), we have $f(i, v_i^1) = f(i, 0) = 0$ for all $i = \overline{1, N-1}$. This implies $u_n^1 = 0$ for all $n = \overline{0,N}$, which contradicts $\|u^1\| > 0$. In a manner similar to that above, we obtain $v_n^1 > 0$ for all $n = \overline{1,N}$.

Remark 3.3.2. The results presented in this section were published in Henderson and Luca (2013a).

3.4 Boundary conditions with additional positive constants

In this section, we shall investigate the existence and nonexistence of positive solutions for a system of second-order difference equations with multipoint boundary conditions in which some positive constants appear.

3.4.1 Presentation of the problem

We consider the system of nonlinear second-order difference equations

$$\begin{cases} \Delta^2 u_{n-1} + s_n f(v_n) = 0, & n = \overline{1,N-1}, \\ \Delta^2 v_{n-1} + t_n g(u_n) = 0, & n = \overline{1,N-1}, \end{cases} \tag{S^0}$$

with the multipoint boundary conditions

$$u_0 = \sum_{i=1}^{p} a_i u_{\xi_i} + a_0, \quad u_N = \sum_{i=1}^{q} b_i u_{\eta_i}, \quad v_0 = \sum_{i=1}^{r} c_i v_{\zeta_i}, \quad v_N = \sum_{i=1}^{l} d_i v_{\rho_i} + b_0, \tag{BC^0}$$

where $N, p, q, r, l \in \mathbb{N}$, $N \geq 2$, $a_i \in \mathbb{R}$ for all $i = \overline{1,p}$, $b_i \in \mathbb{R}$ for all $i = \overline{1,q}$, $c_i \in \mathbb{R}$ for all $i = \overline{1,r}$, $d_i \in \mathbb{R}$ for all $i = \overline{1,l}$, $\xi_i \in \mathbb{N}$ for all $i = \overline{1,p}$, $\eta_i \in \mathbb{N}$ for all $i = \overline{1,q}$, $\zeta_i \in \mathbb{N}$ for all $i = \overline{1,r}$, $\rho_i \in \mathbb{N}$ for all $i = \overline{1,l}$, $1 \leq \xi_1 < \ldots < \xi_p \leq N-1$, $1 \leq \eta_1 < \cdots < \eta_q \leq N-1$, $1 \leq \zeta_1 < \cdots < \zeta_r \leq N-1$, $1 \leq \rho_1 < \cdots < \rho_l \leq N-1$, and a_0 and b_0 are positive constants.

By using the Schauder fixed point theorem (Theorem 1.6.1), we shall prove the existence of positive solutions of problem (S^0)–(BC^0). By a positive solution of (S^0)–(BC^0), we mean a pair of sequences $((u_n)_{n=\overline{0,N}}, (v_n)_{n=\overline{0,N}})$ satisfying (S^0) and (BC^0) with $u_n \geq 0$ and $v_n \geq 0$ for all $n = \overline{0,N}$, $u_n > 0$ for all $n = \overline{0,N-1}$, and $v_n > 0$ for all $n = \overline{1,N}$. We shall also give sufficient conditions for the nonexistence of positive solutions for this problem. System (S^0) with the multipoint boundary conditions $\alpha u_0 - \beta \Delta u_0 = 0$, $u_N = \sum_{i=1}^{m-2} a_i u_{\xi_i} + a_0$, $\gamma v_0 - \delta \Delta v_0 = 0$, $v_N = \sum_{i=1}^{p-2} b_i v_{\eta_i} + b_0$ ($a_0, b_0 > 0$) was investigated in Henderson and Luca (2013e).

We present the assumptions that we shall use in the sequel:

(H1) $\xi_i \in \mathbb{N}$ for all $i = \overline{1,p}$, $1 \leq \xi_1 < \cdots < \xi_p \leq N-1$, $a_i \geq 0$ for all $i = \overline{1,p}$, $\sum_{i=1}^{p} a_i < 1$, $\eta_i \in \mathbb{N}$ for all $i = \overline{1,q}$, $1 \leq \eta_1 < \cdots < \eta_q \leq N-1$, $b_i \geq 0$ for all $i = \overline{1,q}$, $\sum_{i=1}^{q} b_i < 1$, $\zeta_i \in \mathbb{N}$ for all $i = \overline{1,r}$, $1 \leq \zeta_1 < \cdots < \zeta_r \leq N-1$, $c_i \geq 0$ for all $i = \overline{1,r}$, $\sum_{i=1}^{r} c_i < 1$, $\rho_i \in \mathbb{N}$ for all $i = \overline{1,l}$, $1 \leq \rho_1 < \cdots < \rho_l \leq N-1$, $d_i \geq 0$ for all $i = \overline{1,l}$, $\sum_{i=1}^{l} d_i < 1$.

(H2) The constants $s_n, t_n \geq 0$ for all $n = \overline{1,N-1}$, and there exist $i_0, j_0 \in \{1, \ldots, N-1\}$ such that $s_{i_0} > 0$, $t_{j_0} > 0$.

(H3) $f, g : [0,\infty) \to [0,\infty)$ are continuous functions, and there exists $c_0 > 0$ such that $f(u) < \frac{c_0}{L}$, $g(u) < \frac{c_0}{L}$ for all $u \in [0, c_0]$, where $L = \max\{\sum_{i=1}^{N-1} s_i I_1(i), \sum_{i=1}^{N-1} t_i I_2(i)\}$, and I_1 and I_2 are defined in Section 3.1.2.

(H4) $f, g : [0,\infty) \to [0,\infty)$ are continuous functions and satisfy the conditions

$\lim_{u\to\infty} \frac{f(u)}{u} = \infty, \quad \lim_{u\to\infty} \frac{g(u)}{u} = \infty.$

Under assumption (H1), we have all auxiliary results in Lemmas 3.1.1–3.1.6 from Section 3.1.2. Besides, by (H2), we deduce that $\sum_{i=1}^{N-1} s_i I_1(i) > 0$ and $\sum_{i=1}^{N-1} t_i I_2(i) > 0$—that is, the constant L from (H3) is positive.

3.4.2 Main results

Our first theorem is the following existence result for problem (S^0)–(BC^0):

Theorem 3.4.1. *Assume that assumptions (H1)–(H3) hold. Then problem* (S^0)–(BC^0) *has at least one positive solution for $a_0 > 0$ and $b_0 > 0$ sufficiently small.*

Proof. We consider the problems

$$\Delta^2 h_{n-1} = 0, \quad n = \overline{1, N-1}, \quad h_0 = \sum_{i=1}^{p} a_i h_{\xi_i} + 1, \quad h_N = \sum_{i=1}^{q} b_i h_{\eta_i}, \tag{3.44}$$

$$\Delta^2 k_{n-1} = 0, \quad n = \overline{1, N-1}, \quad k_0 = \sum_{i=1}^{r} c_i k_{\zeta_i}, \quad k_N = \sum_{i=1}^{l} d_i k_{\rho_i} + 1. \tag{3.45}$$

Problems (3.44) and (3.45) have the solutions

$$h_n = \frac{1}{\Delta_1}\left[-n\left(1 - \sum_{i=1}^{q} b_i\right) + \left(N - \sum_{i=1}^{q} b_i \eta_i\right)\right], \quad n = \overline{0, N},$$
$$k_n = \frac{1}{\Delta_2}\left[n\left(1 - \sum_{i=1}^{r} c_i\right) + \sum_{i=1}^{r} c_i \zeta_i\right], \quad n = \overline{0, N}, \tag{3.46}$$

respectively, where Δ_1 and Δ_2 are defined in Section 3.1.2. By assumption (H1), we obtain $h_n > 0$ for all $n = \overline{0, N-1}$, and $k_n > 0$ for all $n = \overline{1, N}$.

We define the sequences $(x_n)_{n=\overline{0,N}}$ and $(y_n)_{n=\overline{0,N}}$ by

$$x_n = u_n - a_0 h_n, \quad y_n = v_n - b_0 k_n, \quad n = \overline{0, N},$$

where $((u_n)_{n=\overline{0,N}}, (v_n)_{n=\overline{0,N}})$ is a solution of (S^0)–(BC^0). Then (S^0)–(BC^0) can be equivalently written as

$$\begin{cases} \Delta^2 x_{n-1} + s_n f(y_n + b_0 k_n) = 0, & n = \overline{1, N-1}, \\ \Delta^2 y_{n-1} + t_n g(x_n + a_0 h_n) = 0, & n = \overline{1, N-1}, \end{cases} \tag{3.47}$$

with the boundary conditions

$$x_0 = \sum_{i=1}^{r} a_i x_{\xi_i}, \quad x_N = \sum_{i=1}^{q} b_i x_{\eta_i}, \quad y_0 = \sum_{i=1}^{r} c_i y_{\zeta_i}, \quad y_N = \sum_{i=1}^{l} d_i y_{\rho_i}. \tag{3.48}$$

Using the Green's functions G_1 and G_2 from Section 3.1.2, we find a pair $((x_n)_{n=\overline{0,N}}, (y_n)_{n=\overline{0,N}})$ is a solution of problem (3.47)–(3.48) if and only if it is a solution for the problem

$$\begin{cases} x_n = \sum_{i=1}^{N-1} G_1(n,i) s_i f\left(\sum_{j=1}^{N-1} G_2(i,j) t_j g(x_j + a_0 h_j) + b_0 k_i\right), \quad n = \overline{0,N}, \\ y_n = \sum_{i=1}^{N-1} G_2(n,i) t_i g(x_i + a_0 h_i), \quad n = \overline{0,N}, \end{cases} \tag{3.49}$$

where $(h_n)_{n=\overline{0,N}}$ and $(k_n)_{n=\overline{0,N}}$ are given in (3.46).

We consider the Banach space $X = \mathbb{R}^{N+1}$ with the norm $\|u\| = \max_{n=\overline{0,N}} |u_n|$, $u = (u_n)_{n=\overline{0,N}}$, and we define the set $M = \{(x_n)_{n=\overline{0,N}}, \quad 0 \le x_n \le c_0, \quad \forall n = \overline{0,N}\} \subset X$. We also define the operator $\mathcal{E} : M \to X$ by

$$\mathcal{E}(x) = \left(\sum_{i=1}^{N-1} G_1(n,i) s_i f\left(\sum_{j=1}^{N-1} G_2(i,j) t_j g(x_j + a_0 h_j) + b_0 k_i\right)\right)_{n=\overline{0,N}},$$
$$x = (x_n)_{n=\overline{0,N}} \in M.$$

For sufficiently small $a_0 > 0$ and $b_0 > 0$, by (H3), we deduce

$$f(y_n + b_0 k_n) \le \frac{c_0}{L}, \quad g(x_n + a_0 h_n) \le \frac{c_0}{L}, \quad \forall n = \overline{0,N}, \quad \forall (x_n)_n, \quad (y_n)_n \in M.$$

Then, by using Lemma 3.1.4, we obtain $\mathcal{E}(x)_n \ge 0$ for all $n = \overline{0,N}$ and $x = (x_n)_{n=\overline{0,N}} \in M$. By Lemma 3.1.5, for all $x \in M$, we have

$$\sum_{j=1}^{N-1} G_2(i,j) t_j g(x_j + a_0 h_j) \le \sum_{j=1}^{N-1} I_2(j) t_j g(x_j + a_0 h_j) \le \frac{c_0}{L} \sum_{j=1}^{N-1} t_j I_2(j) \le c_0,$$
$$\forall i = \overline{1, N-1},$$

and

$$\mathcal{E}(x)_n \le \sum_{i=1}^{N-1} I_1(i) s_i f\left(\sum_{j=1}^{N-1} G_2(i,j) t_j g(x_j + a_0 h_j) + b_0 k_i\right) \le \frac{c_0}{L} \sum_{i=1}^{N-1} s_i I_1(i) \le c_0,$$
$$\forall n = \overline{0,N}.$$

Therefore, $\mathcal{E}(M) \subset M$.

Using standard arguments, we deduce that $\mathcal{E}$ is completely continuous. By Theorem 1.6.1, we conclude that $\mathcal{E}$ has a fixed point $(x_n)_{n=\overline{0,N}} \in M$. This element together with $y = (y_n)_{n=\overline{0,N}}$ given by (3.49) represents a solution for problem (3.47)–(3.48). This shows that our problem (S^0)–(BC^0) has a positive solution $((u_n)_{n=\overline{0,N}}, (v_n)_{n=\overline{0,N}})$ with $u_n = x_n + a_0 h_n$, $v_n = y_n + b_0 k_n$, $n = \overline{0,N}$ ($u_n > 0$ for all $n = \overline{0, N-1}$, and $v_n > 0$ for all $n = \overline{1,N}$) for sufficiently small $a_0 > 0$ and $b_0 > 0$. □

In what follows, we present sufficient conditions for the nonexistence of positive solutions of (S^0)–(BC^0).

Theorem 3.4.2. *Assume that assumptions (H1), (H2), and (H4) hold. Then problem* (S^0)–(BC^0) *has no positive solution for* a_0 *and* b_0 *sufficiently large.*

Proof. We suppose that $((u_n)_{n=\overline{0,N}}, (v_n)_{n=\overline{0,N}})$ is a positive solution of (S^0)–(BC^0). Then $((x_n)_{n=\overline{0,N}}, (y_n)_{n=\overline{0,N}})$ with $x_n = u_n - a_0 h_n$, $y_n = v_n - b_0 k_n$, $n = \overline{0,N}$, is a solution for problem (3.47)–(3.48), where $(h_n)_{n=\overline{0,N}}$ and $(k_n)_{n=\overline{0,N}}$ are the solutions of problems (3.44) and (3.45), respectively, (given by (3.46)). By (H2), there exists $c \in \{1, 2, \dots, [\![N/2]\!]\}$ such that $i_0, j_0 \in \{c, ..., N-c\}$, and then $\sum_{i=c}^{N-c} s_i I_1(i) > 0$ and $\sum_{i=c}^{N-c} t_i I_2(i) > 0$. By using Lemma 3.1.4, we have $x_n \geq 0$, $y_n \geq 0$ for all $n = \overline{0,N}$, and by Lemma 3.1.6, we obtain $\min_{n=\overline{c,N-c}} x_n \geq \gamma_1 \|x\|$ and $\min_{n=\overline{c,N-c}} y_n \geq \gamma_2 \|y\|$, where γ_1 and γ_2 are defined in Section 3.1.2.

Using now (3.46), we deduce that

$$\min_{n=\overline{c,N-c}} h_n = h_{N-c} = \frac{h_{N-c}}{h_0}\|h\|, \qquad \min_{n=\overline{c,N-c}} k_n = k_c = \frac{k_c}{k_N}\|k\|.$$

Therefore, we obtain

$$\min_{n=\overline{c,N-c}} (x_n + a_0 h_n) \geq \gamma_1 \|x\| + \frac{a_0 h_{N-c}}{h_0}\|h\| \geq r_1(\|x\| + a_0\|h\|) \geq r_1\|x + a_0 h\|,$$

$$\min_{n=\overline{c,N-c}} (y_n + b_0 k_n) \geq \gamma_2 \|y\| + \frac{b_0 k_c}{k_N}\|k\| \geq r_2(\|y\| + b_0\|k\|) \geq r_2\|y + b_0 k\|,$$

where $r_1 = \min\{\gamma_1, h_{N-c}/h_0\}$, $r_2 = \min\{\gamma_2, k_c/k_N\}$.

We now consider $R = \left(\min\left\{\gamma_2 r_1 \sum_{i=c}^{N-c} t_i I_2(i),\ \gamma_1 r_2 \sum_{i=c}^{N-c} s_i I_1(i)\right\}\right)^{-1} > 0$.

By using (H4), for R defined above, we conclude that there exists $M_0 > 0$ such that $f(u) > 2Ru$, $g(u) > 2Ru$ for all $u \geq M_0$. We consider $a_0 > 0$ and $b_0 > 0$ sufficiently large such that $\min_{n=\overline{c,N-c}}(x_n + a_0 h_n) \geq M_0$ and $\min_{n=\overline{c,N-c}}(y_n + b_0 k_n) \geq M_0$. By (H2), (3.47), (3.48), and the above inequalities, we deduce that $\|x\| > 0$ and $\|y\| > 0$.

Now, by using Lemma 3.1.5 and the above considerations, we have

$$\begin{aligned} y_c &= \sum_{i=1}^{N-1} G_2(c,i) t_i g(x_i + a_0 h_i) \geq \gamma_2 \sum_{i=1}^{N-1} I_2(i) t_i g(x_i + a_0 h_i) \\ &\geq \gamma_2 \sum_{i=c}^{N-c} I_2(i) t_i g(x_i + a_0 h_i) \geq 2R\gamma_2 \sum_{i=c}^{N-c} I_2(i) t_i (x_i + a_0 h_i) \\ &\geq 2R\gamma_2 \sum_{i=c}^{N-c} I_2(i) t_i \min_{j=\overline{c,N-c}} \{x_j + a_0 h_j\} \geq 2R\gamma_2 r_1 \sum_{i=c}^{N-c} I_2(i) t_i \|x + a_0 h\| \\ &\geq 2\|x + a_0 h\| \geq 2\|x\|. \end{aligned}$$

Therefore, we obtain

$$\|x\| \leq y_c/2 \leq \|y\|/2. \tag{3.50}$$

In a similar manner, we deduce

$$\begin{aligned} x_c &= \sum_{i=1}^{N-1} G_1(c,i)s_i f(y_i + b_0 k_i) \geq \gamma_1 \sum_{i=1}^{N-1} I_1(i)s_i f(y_i + b_0 k_i) \\ &\geq \gamma_1 \sum_{i=c}^{N-c} I_1(i)s_i f(y_i + b_0 k_i) \geq 2R\gamma_1 \sum_{i=c}^{N-c} I_1(i)s_i(y_i + b_0 k_i) \\ &\geq 2R\gamma_1 \sum_{i=c}^{N-c} I_1(i)s_i \min_{j=c,N-c} \{y_j + b_0 k_j\} \geq 2R\gamma_1 r_2 \sum_{i=c}^{N-c} I_1(i)s_i \|y + b_0 k\| \\ &\geq 2\|y + b_0 k\| \geq 2\|y\|. \end{aligned}$$

So, we obtain

$$\|y\| \leq x_c/2 \leq \|x\|/2. \tag{3.51}$$

By (3.50) and (3.51), we obtain $\|x\| \leq \|y\|/2 \leq \|x\|/4$, which is a contradiction, because $\|x\| > 0$. Then, for a_0 and b_0 sufficiently large, our problem (S^0)–(BC^0) has no positive solution. □

Results similar to those in Theorems 3.4.1 and 3.4.2 can be obtained if instead of boundary conditions (BC^0) we have

$$u_0 = \sum_{i=1}^{p} a_i u_{\xi_i}, \quad u_N = \sum_{i=1}^{q} b_i u_{\eta_i} + a_0, \quad v_0 = \sum_{i=1}^{r} c_i v_{\zeta_i} + b_0, \quad v_N = \sum_{i=1}^{l} d_i v_{\rho_i}, \tag{BC_1^0}$$

or

$$u_0 = \sum_{i=1}^{p} a_i u_{\xi_i} + a_0, \quad u_N = \sum_{i=1}^{q} b_i u_{\eta_i}, \quad v_0 = \sum_{i=1}^{r} c_i v_{\zeta_i} + b_0, \quad v_N = \sum_{i=1}^{l} d_i v_{\rho_i}, \tag{BC_2^0}$$

or

$$u_0 = \sum_{i=1}^{p} a_i u_{\xi_i}, \quad u_N = \sum_{i=1}^{q} b_i u_{\eta_i} + a_0, \quad v_0 = \sum_{i=1}^{r} c_i v_{\zeta_i}, \quad v_N = \sum_{i=1}^{l} d_i v_{\rho_i} + b_0, \tag{BC_3^0}$$

where a_0 and b_0 are positive constants.

For problem (S^0)–(BC_1^0), instead of sequences $(h_n)_{n=\overline{0,N}}$ and $(k_n)_{n=\overline{0,N}}$ from the proof of Theorem 3.4.1, the solutions of the problems

$$\Delta^2 \tilde{h}_{n-1} = 0, \quad n = \overline{1, N-1}, \quad \tilde{h}_0 = \sum_{i=1}^{p} a_i \tilde{h}_{\xi_i}, \quad \tilde{h}_N = \sum_{i=1}^{q} b_i \tilde{h}_{\eta_i} + 1, \tag{3.52}$$

$$\Delta^2\tilde{k}_{n-1} = 0, \quad n = \overline{1, N-1}, \quad \tilde{k}_0 = \sum_{i=1}^{r} c_i \tilde{k}_{\zeta_i} + 1, \quad \tilde{k}_N = \sum_{i=1}^{l} d_i \tilde{k}_{\rho_i} \tag{3.53}$$

are

$$\tilde{h}_n = \frac{1}{\Delta_1}\left[n\left(1 - \sum_{i=1}^{p} a_i\right) + \sum_{i=1}^{p} a_i \xi_i\right], \quad n = \overline{0, N},$$

$$\tilde{k}_n = \frac{1}{\Delta_2}\left[-n\left(1 - \sum_{i=1}^{l} d_i\right) + \left(N - \sum_{i=1}^{l} d_i \rho_i\right)\right], \quad n = \overline{0, N},$$

respectively. By assumption (H1), we obtain $\tilde{h}_n > 0$ for all $n = \overline{1, N}$, and $\tilde{k}_n > 0$ for all $n = \overline{0, N-1}$.

For problem (S^0)–(BC_2^0), instead of sequences $(h_n)_{n=\overline{0,N}}$ and $(k_n)_{n=\overline{0,N}}$ from Theorem 3.4.1, the solutions of problems (3.44) and (3.53) are $(h_n)_{n=\overline{0,N}}$ and $(\tilde{k}_n)_{n=\overline{0,N}}$, respectively, which satisfy $h_n > 0$ for all $n = \overline{0, N-1}$, and $\tilde{k}_n > 0$ for all $n = \overline{0, N-1}$. For problem (S^0)–(BC_3^0), instead of sequences $(h_n)_{n=\overline{0,N}}$ and $(k_n)_{n=\overline{0,N}}$ from Theorem 3.4.1, the solutions of problems (3.52) and (3.45) are $(\tilde{h}_n)_{n=\overline{0,N}}$ and $(k_n)_{n=\overline{0,N}}$, respectively, which satisfy $\tilde{h}_n > 0$ for all $n = \overline{1, N}$, and $k_n > 0$ for all $n = \overline{1, N}$.

Therefore, we also obtain the following results:

Theorem 3.4.3. *Assume that assumptions (H1)–(H3) hold. Then problem* (S^0)–(BC_1^0) *has at least one positive solution* ($u_n > 0$ *for all* $n = \overline{1, N}$, *and* $v_n > 0$ *for all* $n = \overline{0, N-1}$) *for* $a_0 > 0$ *and* $b_0 > 0$ *sufficiently small.*

Theorem 3.4.4. *Assume that assumptions (H1), (H2), and (H4) hold. Then problem* (S^0)–(BC_1^0) *has no positive solution* ($u_n > 0$ *for all* $n = \overline{1, N}$, *and* $v_n > 0$ *for all* $n = \overline{0, N-1}$) *for* a_0 *and* b_0 *sufficiently large.*

Theorem 3.4.5. *Assume that assumptions (H1)–(H3) hold. Then problem* (S^0)–(BC_2^0) *has at least one positive solution* ($u_n > 0$ *for all* $n = \overline{0, N-1}$, *and* $v_n > 0$ *for all* $n = \overline{0, N-1}$) *for* $a_0 > 0$ *and* $b_0 > 0$ *sufficiently small.*

Theorem 3.4.6. *Assume that assumptions (H1), (H2), and (H4) hold. Then problem* (S^0)–(BC_2^0) *has no positive solution* ($u_n > 0$ *for all* $n = \overline{0, N-1}$, *and* $v_n > 0$ *for all* $n = \overline{0, N-1}$) *for* a_0 *and* b_0 *sufficiently large.*

Theorem 3.4.7. *Assume that assumptions (H1)–(H3) hold. Then problem* (S^0)–(BC_3^0) *has at least one positive solution* ($u_n > 0$ *for all* $n = \overline{1, N}$, *and* $v_n > 0$ *for all* $n = \overline{1, N}$) *for* $a_0 > 0$ *and* $b_0 > 0$ *sufficiently small.*

Theorem 3.4.8. *Assume that assumptions (H1), (H2), and (H4) hold. Then problem* (S^0)–(BC_3^0) *has no positive solution* ($u_n > 0$ *for all* $n = \overline{1, N}$, *and* $v_n > 0$ *for all* $n = \overline{1, N}$) *for* a_0 *and* b_0 *sufficiently large.*

3.4.3 An example

Example 3.4.1. We consider $N = 20$, $s_n = c/(n+1)$, $t_n = d/n$ for all $n = \overline{1, 19}$, $c > 0$, $d > 0$, $p = 2$, $q = 3$, $r = 1$, $l = 2$, $a_1 = 1/2$, $a_2 = 1/3$, $\xi_1 = 4$,

$\xi_2 = 16$, $b_1 = 1/3$, $b_2 = 1/4$, $b_3 = 1/5$, $\eta_1 = 5$, $\eta_2 = 10$, $\eta_3 = 15$, $c_1 = 3/4$, $\zeta_1 = 10$, $d_1 = 1/3$, $d_2 = 1/5$, $\rho_1 = 3$, and $\rho_2 = 18$. We also consider the functions $f, g : [0, \infty) \to [0, \infty)$, $f(x) = \frac{\tilde{a}x^{\alpha_1}}{2x+1}$, $g(x) = \frac{\tilde{b}x^{\alpha_2}}{3x+2}$ for all $x \in [0, \infty)$, with $\tilde{a}, \tilde{b} > 0$ and $\alpha_1, \alpha_2 > 2$. We have $\lim_{x\to\infty} f(x)/x = \lim_{x\to\infty} g(x)/x = \infty$.

Therefore, we consider the system of second-order difference equations

$$\begin{cases} \Delta^2 u_{n-1} + \dfrac{c\tilde{a}v_n^{\alpha_1}}{(n+1)(2v_n+1)} = 0, \quad n = \overline{1,19}, \\ \Delta^2 v_{n-1} + \dfrac{d\tilde{b}u_n^{\alpha_2}}{n(3u_n+2)} = 0, \quad n = \overline{1,19}, \end{cases} \qquad (\widetilde{S^0})$$

with the multipoint boundary conditions

$$\begin{cases} u_0 = \dfrac{1}{2}u_4 + \dfrac{1}{3}u_{16} + a_0, \quad u_{20} = \dfrac{1}{3}u_5 + \dfrac{1}{4}u_{10} + \dfrac{1}{5}u_{15}, \\ v_0 = \dfrac{3}{4}v_{10}, \quad v_{20} = \dfrac{1}{3}v_3 + \dfrac{1}{5}v_{18} + b_0. \end{cases} \qquad (\widetilde{BC^0})$$

By using problem (S_1)–(BC_1) from Section 3.1.4, we deduce that assumptions (H1), (H2), and (H4) are satisfied. In addition, by using the functions I_1 and I_2 from Section 3.1.4, we obtain $\tilde{A} = \sum_{i=1}^{19} I_1(i)/(i+1) \approx 30.1784002$ and $\tilde{B} = \sum_{i=1}^{19} I_2(i)/i \approx 23.63831254$, and then $L = \max\{c\tilde{A}, d\tilde{B}\}$. We choose $c_0 = 1$, and if we select $\tilde{a}$ and $\tilde{b}$ satisfying the conditions $\tilde{a} < \frac{3}{L} = 3\min\left\{\frac{1}{c\tilde{A}}, \frac{1}{d\tilde{B}}\right\}$ and $\tilde{b} < \frac{5}{L} = 5\min\left\{\frac{1}{c\tilde{A}}, \frac{1}{d\tilde{B}}\right\}$, then we conclude that $f(x) \le \tilde{a}/3 < 1/L$ and $g(x) \le \tilde{b}/5 < 1/L$ for all $x \in [0, 1]$. For example, if $c = 1$ and $d = 2$, then for $\tilde{a} \le 0.063$ and $\tilde{b} \le 0.105$ the above conditions for f and g are satisfied. So, assumption (H3) is also satisfied. By Theorems 3.4.1 and 3.4.2, we deduce that problem $(\widetilde{S^0})$–$(\widetilde{BC^0})$ has at least one positive solution (here $u_n > 0$ and $v_n > 0$ for all $n = \overline{0,20}$) for sufficiently small $a_0 > 0$ and $b_0 > 0$, and no positive solution for sufficiently large a_0 and b_0.

Systems of Riemann–Liouville fractional differential equations with uncoupled integral boundary conditions

4

4.1 Existence and nonexistence of positive solutions for systems with parameters and uncoupled boundary conditions

Fractional differential equations describe many phenomena in various fields of engineering and scientific disciplines such as physics, biophysics, chemistry, biology (such as blood flow phenomena), economics, control theory, signal and image processing, aerodynamics, viscoelasticity, and electromagnetics (Baleanu et al., 2012; Das, 2008; Kilbas et al., 2006; Podlubny, 1999; Sabatier et al., 2007; Samko et al., 1993). For some recent developments on the topic, see Agarwal et al. (2010a,b), Aghajani et al. (2012), Ahmad and Ntouyas (2012a,b), Bai (2010), Balachandran and Trujillo (2010), Baleanu et al. (2010), El-Shahed and Nieto (2010), Graef et al. (2012), Jiang and Yuan (2010), Lan and Lin (2011), Liang and Zhang (2009), Yuan (2010), and Yuan et al. (2012).

In this section, we shall investigate the existence and nonexistence of positive solutions for a system of nonlinear Riemann–Liouville fractional differential equations which contains some parameters and is subject to uncoupled Riemann–Stieltjes integral boundary conditions.

4.1.1 Presentation of the problem

We consider the system of nonlinear ordinary fractional differential equations

$$\begin{cases} D_{0+}^{\alpha} u(t) + \lambda f(t, u(t), v(t)) = 0, & t \in (0,1), \\ D_{0+}^{\beta} v(t) + \mu g(t, u(t), v(t)) = 0, & t \in (0,1), \end{cases} \tag{S}$$

with the uncoupled integral boundary conditions

$$\begin{cases} u(0) = u'(0) = \cdots = u^{(n-2)}(0) = 0, \quad u(1) = \displaystyle\int_0^1 u(s)\,\mathrm{d}H(s), \\ v(0) = v'(0) = \cdots = v^{(m-2)}(0) = 0, \quad v(1) = \displaystyle\int_0^1 v(s)\,\mathrm{d}K(s), \end{cases} \tag{BC}$$

Boundary Value Problems for Systems of Differential, Difference and Fractional Equations.
http://dx.doi.org/10.1016/B978-0-12-803652-5.00004-1

where $n-1<\alpha\le n$, $m-1<\beta\le m$, $n, m\in\mathbb{N}$, $n, m\ge 3$, D_{0+}^{α} and D_{0+}^{β} denote the Riemann–Liouville derivatives of orders α and β, respectively, and the integrals from (BC) are Riemann–Stieltjes integrals.

Under some assumptions on f and g, we give intervals for the parameters λ and μ such that positive solutions of (S)–(BC) exist. By a positive solution of problem (S)–(BC) we mean a pair of functions $(u, v)\in C([0,1])\times C([0,1])$ satisfying (S) and (BC) with $u(t)\ge 0$, $v(t)\ge 0$ for all $t\in[0,1]$ and $(u, v)\ne(0,0)$. The nonexistence of positive solutions for the above problem is also investigated. The case when H and K are step functions (i.e., where boundary conditions (BC) become multipoint boundary conditions) was investigated in Henderson and Luca (2013f), where we proved the existence of positive solutions of (S)–(BC) by using the Guo–Krasnosel'skii fixed point theorem. Problem (S)–(BC) studied by using some auxiliary results different from those presented in Section 4.1.2 was investigated in Henderson et al. (2015b).

4.1.2 Preliminaries and auxiliary results

In this section, we present the definitions, some lemmas from the theory of fractional calculus, and some auxiliary results that will be used to prove our main theorems.

Definition 4.1.1. The (left-sided) fractional integral of order $\alpha>0$ of a function $f:(0,\infty)\to\mathbb{R}$, denoted by $I_{0+}^{\alpha}f$, is given by

$$\left(I_{0+}^{\alpha}f\right)(t):=\frac{1}{\Gamma(\alpha)}\int_0^t (t-s)^{\alpha-1}f(s)\,\mathrm{d}s,\quad t>0,$$

provided the right-hand side is pointwise defined on $(0,\infty)$, where $\Gamma(\alpha)$ is the Euler gamma function defined by $\Gamma(\alpha)=\int_0^{\infty}t^{\alpha-1}e^{-t}\,\mathrm{d}t$, $\alpha>0$.

Definition 4.1.2. The Riemann–Liouville fractional derivative of order $\alpha\ge 0$ for a function $f:(0,\infty)\to\mathbb{R}$, denoted by $D_{0+}^{\alpha}f$, is given by

$$\left(D_{0+}^{\alpha}f\right)(t):=\left(\frac{\mathrm{d}}{\mathrm{d}t}\right)^n\left(I_{0+}^{n-\alpha}f\right)(t)=\frac{1}{\Gamma(n-\alpha)}\left(\frac{\mathrm{d}}{\mathrm{d}t}\right)^n\int_0^t\frac{f(s)}{(t-s)^{\alpha-n+1}}\,\mathrm{d}s,\quad t>0,$$

where $n=[\![\alpha]\!]+1$, provided that the right-hand side is pointwise defined on $(0,\infty)$.

The notation $[\![\alpha]\!]$ stands for the largest integer not greater than α. If $\alpha=m\in\mathbb{N}$, then $D_{0+}^{m}f(t)=f^{(m)}(t)$ for $t>0$, and if $\alpha=0$, then $D_{0+}^{0}f(t)=f(t)$ for $t>0$.

Lemma 4.1.1 (Kilbas et al., 2006).

(a) *If $\alpha>0$, $\beta>0$, and $f\in L^p(0,1)$ $(1\le p\le\infty)$, then the relation $(I_{0+}^{\alpha}I_{0+}^{\beta}f)(t)=(I_{0+}^{\alpha+\beta}f)(t)$ is satisfied at almost every point $t\in(0,1)$. If $\alpha+\beta>1$, then the above relation holds at any point of* $[0,1]$.

(b) *If $\alpha>0$ and $f\in L^p(0,1)$ $(1\le p\le\infty)$, then the relation $(D_{0+}^{\alpha}I_{0+}^{\alpha}f)(t)=f(t)$ holds almost everywhere on* $(0,1)$.

(c) *If $\alpha>\beta>0$ and $f\in L^p(0,1)$ $(1\le p\le\infty)$, then the relation $(D_{0+}^{\beta}I_{0+}^{\alpha}f)(t)=(I_{0+}^{\alpha-\beta}f)(t)$ holds almost everywhere on* $(0,1)$.

Lemma 4.1.2 (Kilbas et al., 2006). *Let $\alpha>0$ and $n=[\![\alpha]\!]+1$ for $\alpha\notin\mathbb{N}$ and $n=\alpha$ for $\alpha\in\mathbb{N}$; that is, n is the smallest integer greater than or equal to α.*

Then the solutions of the fractional differential equation $D_{0+}^{\alpha}u(t) = 0, 0 < t < 1$, *are*

$$u(t) = c_1 t^{\alpha-1} + c_2 t^{\alpha-2} + \cdots + c_n t^{\alpha-n}, \quad 0 < t < 1,$$

where $c_1, c_2, \ldots, c_n$ *are arbitrary real constants.*

Lemma 4.1.3. *Let* $\alpha > 0$, *n be the smallest integer greater than or equal to* α $(n-1 < \alpha \leq n)$, *and* $y \in L^1(0,1)$. *The solutions of the fractional equation* $D_{0+}^{\alpha}u(t) + y(t) = 0, \quad 0 < t < 1$, *are*

$$u(t) = -\frac{1}{\Gamma(\alpha)} \int_0^t (t-s)^{\alpha-1} y(s)\,\mathrm{d}s + c_1 t^{\alpha-1} + \cdots + c_n t^{\alpha-n}, \quad 0 < t < 1,$$

where $c_1, c_2, \ldots, c_n$ *are arbitrary real constants.*

Proof. By Lemma 4.1.1 (b), the equation $D_{0+}^{\alpha}u(t) + y(t) = 0$ can be written as

$$D_{0+}^{\alpha}u(t) + D_{0+}^{\alpha}(I_{0+}^{\alpha}y)(t) = 0 \quad \text{or} \quad D_{0+}^{\alpha}(u + I_{0+}^{\alpha}y)(t) = 0.$$

With use of Lemma 4.1.2, the solutions for the above equation are

$$\begin{aligned} u(t) + I_{0+}^{\alpha}y(t) = c_1 t^{\alpha-1} + \cdots + c_n t^{\alpha-n} \quad &\Leftrightarrow \\ u(t) &= -I_{0+}^{\alpha}y(t) + c_1 t^{\alpha-1} + \cdots + c_n t^{\alpha-n} \\ &= -\frac{1}{\Gamma(\alpha)} \int_0^t (t-s)^{\alpha-1} y(s)\,\mathrm{d}s + c_1 t^{\alpha-1} + \cdots + c_n t^{\alpha-n}, \quad 0 < t < 1, \end{aligned}$$

where $c_1, c_2, \ldots, c_n$ are arbitrary real constants. □

We consider now the fractional differential equation

$$D_{0+}^{\alpha}u(t) + y(t) = 0, \quad t \in (0,1), \tag{4.1}$$

with the integral boundary conditions

$$u(0) = u'(0) = \cdots = u^{(n-2)}(0) = 0, \quad u(1) = \int_0^1 u(s)\,\mathrm{d}H(s), \tag{4.2}$$

where $n - 1 < \alpha \leq n$, $n \in \mathbb{N}$, $n \geq 3$, and $H : [0,1] \to \mathbb{R}$ is a function of bounded variation.

Lemma 4.1.4. *If* $H : [0,1] \to \mathbb{R}$ *is a function of bounded variation,* $\Delta_1 = 1 - \int_0^1 s^{\alpha-1}\mathrm{d}H(s) \neq 0$, *and* $y \in C(0,1) \cap L^1(0,1)$, *then the unique solution of problem* (4.1)–(4.2) *is given by*

$$\begin{aligned} u(t) = &-\frac{1}{\Gamma(\alpha)} \int_0^t (t-s)^{\alpha-1} y(s)\,\mathrm{d}s + \frac{t^{\alpha-1}}{\Delta_1 \Gamma(\alpha)} \left[\int_0^1 (1-s)^{\alpha-1} y(s)\,\mathrm{d}s \right. \\ &\left. - \int_0^1 \left(\int_s^1 (\tau - s)^{\alpha-1}\,\mathrm{d}H(\tau) \right) y(s)\,\mathrm{d}s \right], \quad 0 \leq t \leq 1. \end{aligned} \tag{4.3}$$

Proof. By Lemma 4.1.3, the solutions of (4.1) are

$$u(t) = -\frac{1}{\Gamma(\alpha)} \int_0^t (t-s)^{\alpha-1} y(s)\,\mathrm{d}s + c_1 t^{\alpha-1} + \cdots + c_n t^{\alpha-n},$$

where $c_1, \ldots, c_n \in \mathbb{R}$. By using the conditions $u(0) = u'(0) = \cdots = u^{(n-2)}(0) = 0$, we obtain $c_2 = \cdots = c_n = 0$. Then we conclude

$$u(t) = c_1 t^{\alpha-1} - \frac{1}{\Gamma(\alpha)} \int_0^t (t-s)^{\alpha-1} y(s)\, ds.$$

Now, by condition $u(1) = \int_0^1 u(s)\, dH(s)$, we deduce

$$c_1 - \frac{1}{\Gamma(\alpha)} \int_0^1 (1-s)^{\alpha-1} y(s)\, ds = \int_0^1 \left[c_1 s^{\alpha-1} - \frac{1}{\Gamma(\alpha)} \int_0^s (s-\tau)^{\alpha-1} y(\tau)\, d\tau \right] dH(s),$$

or

$$c_1 \left(1 - \int_0^1 s^{\alpha-1} dH(s) \right) = \frac{1}{\Gamma(\alpha)} \int_0^1 (1-s)^{\alpha-1} y(s)\, ds - \frac{1}{\Gamma(\alpha)} \int_0^1 \left(\int_0^s (s-\tau)^{\alpha-1} y(\tau)\, d\tau \right) dH(s).$$

So, we obtain

$$\begin{aligned} c_1 &= \frac{1}{\Delta_1 \Gamma(\alpha)} \int_0^1 (1-s)^{\alpha-1} y(s)\, ds - \frac{1}{\Delta_1 \Gamma(\alpha)} \int_0^1 \left(\int_0^s (s-\tau)^{\alpha-1} y(\tau)\, d\tau \right) dH(s) \\ &= \frac{1}{\Delta_1 \Gamma(\alpha)} \int_0^1 (1-s)^{\alpha-1} y(s)\, ds - \frac{1}{\Delta_1 \Gamma(\alpha)} \int_0^1 \left(\int_\tau^1 (s-\tau)^{\alpha-1} dH(s) \right) y(\tau)\, d\tau \\ &= \frac{1}{\Delta_1 \Gamma(\alpha)} \int_0^1 (1-s)^{\alpha-1} y(s)\, ds - \frac{1}{\Delta_1 \Gamma(\alpha)} \int_0^1 \left(\int_s^1 (\tau-s)^{\alpha-1} dH(\tau) \right) y(s)\, ds. \end{aligned}$$

Therefore, we get the expression (4.3) for the solution of problem (4.1)–(4.2). □

Lemma 4.1.5. *Under the assumptions of Lemma 4.1.4, the Green's function for the boundary value problem* (4.1)–(4.2) *is given by*

$$G_1(t,s) = g_1(t,s) + \frac{t^{\alpha-1}}{\Delta_1} \int_0^1 g_1(\tau,s)\, dH(\tau), \quad (t,s) \in [0,1] \times [0,1], \tag{4.4}$$

where

$$g_1(t,s) = \frac{1}{\Gamma(\alpha)} \begin{cases} t^{\alpha-1}(1-s)^{\alpha-1} - (t-s)^{\alpha-1}, & 0 \le s \le t \le 1, \\ t^{\alpha-1}(1-s)^{\alpha-1}, & 0 \le t \le s \le 1. \end{cases} \tag{4.5}$$

Proof. By Lemma 4.1.4 and relation (4.3), we conclude

$$\begin{aligned} u(t) = \frac{1}{\Gamma(\alpha)} \Bigg\{ & \int_0^t [t^{\alpha-1}(1-s)^{\alpha-1} - (t-s)^{\alpha-1}] y(s)\, ds + \int_t^1 t^{\alpha-1}(1-s)^{\alpha-1} y(s)\, ds \\ & - \int_0^1 t^{\alpha-1}(1-s)^{\alpha-1} y(s)\, ds + \frac{t^{\alpha-1}}{\Delta_1} \Bigg[\int_0^1 (1-s)^{\alpha-1} y(s)\, ds \\ & - \int_0^1 \left(\int_s^1 (\tau-s)^{\alpha-1} dH(\tau) \right) y(s)\, ds \Bigg] \Bigg\} \end{aligned}$$

$$
\begin{aligned}
&= \frac{1}{\Gamma(\alpha)} \left\{ \int_0^t [t^{\alpha-1}(1-s)^{\alpha-1} - (t-s)^{\alpha-1}] y(s)\, ds + \int_t^1 t^{\alpha-1}(1-s)^{\alpha-1} y(s)\, ds \right. \\
&\quad - \frac{1}{\Delta_1} \left(1 - \int_0^1 \tau^{\alpha-1}\, dH(\tau)\right) \int_0^1 t^{\alpha-1}(1-s)^{\alpha-1} y(s)\, ds \\
&\quad \left. + \frac{t^{\alpha-1}}{\Delta_1} \left[\int_0^1 (1-s)^{\alpha-1} y(s)\, ds - \int_0^1 \left(\int_s^1 (\tau-s)^{\alpha-1}\, dH(\tau) \right) y(s)\, ds \right] \right\} \\
&= \frac{1}{\Gamma(\alpha)} \left\{ \int_0^t [t^{\alpha-1}(1-s)^{\alpha-1} - (t-s)^{\alpha-1}] y(s)\, ds + \int_t^1 t^{\alpha-1}(1-s)^{\alpha-1} y(s)\, ds \right. \\
&\quad + \frac{t^{\alpha-1}}{\Delta_1} \left[\int_0^1 \left(\int_0^1 \tau^{\alpha-1}(1-s)^{\alpha-1}\, dH(\tau) \right) y(s)\, ds \right. \\
&\quad \left.\left. - \int_0^1 \left(\int_s^1 (\tau-s)^{\alpha-1}\, dH(\tau) \right) y(s)\, ds \right] \right\} \\
&= \frac{1}{\Gamma(\alpha)} \left\{ \int_0^t [t^{\alpha-1}(1-s)^{\alpha-1} - (t-s)^{\alpha-1}] y(s)\, ds + \int_t^1 t^{\alpha-1}(1-s)^{\alpha-1} y(s)\, ds \right. \\
&\quad + \frac{t^{\alpha-1}}{\Delta_1} \left[\int_0^1 \left(\int_0^s \tau^{\alpha-1}(1-s)^{\alpha-1}\, dH(\tau) \right) y(s)\, ds + \int_0^1 \left(\int_s^1 \left[\tau^{\alpha-1}(1-s)^{\alpha-1} \right.\right.\right. \\
&\quad \left.\left.\left.\left. - (\tau-s)^{\alpha-1} \right] dH(\tau) \right) y(s)\, ds \right] \right\} = \int_0^1 g_1(t,s) y(s)\, ds \\
&\quad + \frac{t^{\alpha-1}}{\Delta_1} \int_0^1 \left(\int_0^1 g_1(\tau,s)\, dH(\tau) \right) y(s)\, ds = \int_0^1 G_1(t,s) y(s)\, ds, \quad \forall t \in [0,1],
\end{aligned}
$$

where g_1 and G_1 are given in (4.5) and (4.4), respectively. Hence, $u(t) = \int_0^1 G_1(t,s)$ $y(s)\, ds$ for all $t \in [0,1]$. □

Lemma 4.1.6. *The function g_1 given by* (4.5) *has the following properties;*

(a) $g_1 : [0,1] \times [0,1] \to \mathbb{R}_+$ *is a continuous function, and* $g_1(t,s) > 0$ *for all* $(t,s) \in (0,1) \times (0,1)$.

(b) $g_1(t,s) \le g_1(\theta_1(s), s)$ *for all* $(t,s) \in [0,1] \times [0,1]$.

(c) *For any* $c \in (0, 1/2)$,
$\min_{t\in[c,1-c]} g_1(t,s) \ge \gamma_1 g_1(\theta_1(s), s)$ *for all* $s \in [0,1]$,

where $\gamma_1 = c^{\alpha-1}$ *and* $\theta_1(s) = \begin{cases} \frac{s}{1-(1-s)^{\frac{\alpha-1}{\alpha-2}}}, & s \in (0,1], \\ \frac{\alpha-2}{\alpha-1}, & s = 0, \end{cases}$ *if* $n-1 < \alpha \le n$, $n \ge 3$.

The proof of Lemma 4.1.6 is similar to that of Lemma 3.3 from Ji and Guo (2009). We define θ_1 at $s = 0$ as $\frac{\alpha-2}{\alpha-1}$ so that θ_1 is a continuous function.

Lemma 4.1.7. *If* $H : [0,1] \to \mathbb{R}$ *is a nondecreasing function and* $\Delta_1 > 0$, *then the Green's function* G_1 *of problem* (4.1)–(4.2) *is continuous on* $[0,1] \times [0,1]$ *and satisfies* $G_1(t,s) \ge 0$ *for all* $(t,s) \in [0,1] \times [0,1]$. *Moreover, if* $y \in C(0,1) \cap L^1(0,1)$ *satisfies* $y(t) \ge 0$ *for all* $t \in (0,1)$, *then the solution* u *of problem* (4.1)–(4.2) *satisfies* $u(t) \ge 0$ *for all* $t \in [0,1]$.

Proof. By using the assumptions of this lemma, we have $G_1(t,s) \geq 0$ for all $(t,s) \in [0,1] \times [0,1]$, and so $u(t) \geq 0$ for all $t \in [0,1]$. □

Lemma 4.1.8. *Assume that $H : [0,1] \to \mathbb{R}$ is a nondecreasing function and $\Delta_1 > 0$. Then the Green's function G_1 of problem (4.1)–(4.2) satisfies the following inequalities:*

(a) $G_1(t,s) \leq J_1(s), \quad \forall (t,s) \in [0,1] \times [0,1]$, *where*

$$J_1(s) = g_1(\theta_1(s), s) + \frac{1}{\Delta_1} \int_0^1 g_1(\tau, s)\, dH(\tau), \quad \forall s \in [0,1].$$

(b) For every $c \in (0, 1/2)$, we have

$$\min_{t \in [c,1-c]} G_1(t,s) \geq \gamma_1 J_1(s) \geq \gamma_1 G_1(t', s), \quad \forall t', s \in [0,1].$$

Proof. The first inequality, (a), is evident. For the second inequality, (b), for $c \in (0,1/2)$ and $t \in [c, 1-c]$, t', $s \in [0,1]$, we deduce

$$\begin{aligned} G_1(t,s) &\geq c^{\alpha-1} g_1(\theta_1(s), s) + \frac{c^{\alpha-1}}{\Delta_1} \int_0^1 g_1(\tau, s)\, dH(\tau) \\ &= c^{\alpha-1} \left(g_1(\theta_1(s), s) + \frac{1}{\Delta_1} \int_0^1 g_1(\tau, s)\, dH(\tau) \right) = \gamma_1 J_1(s) \geq \gamma_1 G_1(t', s). \end{aligned}$$

Therefore, we obtain the inequalities (b) of this lemma. □

Lemma 4.1.9. *Assume that $H : [0,1] \to \mathbb{R}$ is a nondecreasing function, $\Delta_1 > 0$, $c \in (0,1/2)$, and $y \in C(0,1) \cap L^1(0,1)$, $y(t) \geq 0$ for all $t \in (0,1)$. Then the solution $u(t)$, $t \in [0,1]$ of problem (4.1)–(4.2) satisfies the inequality $\inf_{t \in [c,1-c]} u(t) \geq \gamma_1 \sup_{t' \in [0,1]} u(t')$.*

Proof. For $c \in (0,1/2)$, $t \in [c, 1-c]$, and $t' \in [0,1]$, we have

$$u(t) = \int_0^1 G_1(t,s) y(s)\, ds \geq \gamma_1 \int_0^1 J_1(s) y(s)\, ds \geq \gamma_1 \int_0^1 G_1(t', s) y(s)\, ds = \gamma_1 u(t').$$

Then we deduce the conclusion of this lemma. □

We can also formulate results similar to those in Lemmas 4.1.4–4.1.9 for the fractional differential equation

$$D_{0+}^{\beta} v(t) + h(t) = 0, \quad 0 < t < 1, \tag{4.6}$$

with the integral boundary conditions

$$v(0) = v'(0) = \cdots = v^{(m-2)}(0) = 0, \quad v(1) = \int_0^1 v(s)\, dK(s), \tag{4.7}$$

where $m - 1 < \beta \leq m$, $m \in \mathbb{N}$, $m \geq 3$, $K : [0,1] \to \mathbb{R}$ is a function of bounded variation, and $h \in C(0,1) \cap L^1(0,1)$. We denote by Δ_2, γ_2, g_2, θ_2, G_2, and J_2 the corresponding constants and functions for problem (4.6)–(4.7) defined in a similar manner as Δ_1, γ_1, g_1, θ_1, G_1, and J_1, respectively.

4.1.3 Main results

In this section, we give sufficient conditions on λ, μ, f, and g such that positive solutions with respect to a cone for our problem (S)–(BC) exist.

We present the assumptions that we shall use in the sequel:

(H1) $H, K : [0,1] \to \mathbb{R}$ are nondecreasing functions, $\Delta_1 = 1 - \int_0^1 s^{\alpha-1}\,\mathrm{d}H(s) > 0$, and $\Delta_2 = 1 - \int_0^1 s^{\beta-1}\,\mathrm{d}K(s) > 0$.

(H2) The functions $f, g : [0,1] \times [0,\infty) \times [0,\infty) \to [0,\infty)$ are continuous.

For $c \in (0, 1/2)$, we introduce the following extreme limits:

$$f_0^{\mathrm{s}} = \limsup_{u+v\to 0^+} \max_{t\in[0,1]} \frac{f(t,u,v)}{u+v}, \quad g_0^{\mathrm{s}} = \limsup_{u+v\to 0^+} \max_{t\in[0,1]} \frac{g(t,u,v)}{u+v},$$

$$f_0^{\mathrm{i}} = \liminf_{u+v\to 0^+} \min_{t\in[c,1-c]} \frac{f(t,u,v)}{u+v}, \quad g_0^{\mathrm{i}} = \liminf_{u+v\to 0^+} \min_{t\in[c,1-c]} \frac{g(t,u,v)}{u+v},$$

$$f_\infty^{\mathrm{s}} = \limsup_{u+v\to \infty} \max_{t\in[0,1]} \frac{f(t,u,v)}{u+v}, \quad g_\infty^{\mathrm{s}} = \limsup_{u+v\to \infty} \max_{t\in[0,1]} \frac{g(t,u,v)}{u+v},$$

$$f_\infty^{\mathrm{i}} = \liminf_{u+v\to \infty} \min_{t\in[c,1-c]} \frac{f(t,u,v)}{u+v}, \quad g_\infty^{\mathrm{i}} = \liminf_{u+v\to \infty} \min_{t\in[c,1-c]} \frac{g(t,u,v)}{u+v}.$$

In the definitions of the extreme limits above, the variables u and v are nonnegative.

By using the Green's functions G_1 and G_2 from Section 4.1.2 (Lemma 4.1.5), we can write our problem (S)–(BC) equivalently as the following nonlinear system of integral equations:

$$\begin{cases} u(t) = \lambda \displaystyle\int_0^1 G_1(t,s) f(s,u(s),v(s))\,\mathrm{d}s, & 0 \le t \le 1, \\ v(t) = \mu \displaystyle\int_0^1 G_2(t,s) g(s,u(s),v(s))\,\mathrm{d}s, & 0 \le t \le 1. \end{cases}$$

We consider the Banach space $X = C([0,1])$ with the supremum norm $\|\cdot\|$ and the Banach space $Y = X \times X$ with the norm $\|(u,v)\|_Y = \|u\| + \|v\|$. For $c \in (0,1/2)$, we define the cone $P \subset Y$ by

$$P = \{(u,v) \in Y;\ u(t) \ge 0,\ v(t) \ge 0,\ \forall t \in [0,1] \text{ and } \inf_{t\in[c,1-c]} (u(t)+v(t)) \ge \gamma \|(u,v)\|_Y\},$$

where $\gamma = \min\{\gamma_1, \gamma_2\}$, and γ_1 and γ_2 are defined in Section 4.1.2 (Lemma 4.1.6).

For $\lambda, \mu > 0$, we introduce the operators $Q_1, Q_2 : Y \to X$, and $\mathcal{Q} : Y \to Y$ defined by

$$Q_1(u,v)(t) = \lambda \int_0^1 G_1(t,s) f(s,u(s),v(s))\,\mathrm{d}s, \quad 0 \le t \le 1,$$

$$Q_2(u,v)(t) = \mu \int_0^1 G_2(t,s) g(s,u(s),v(s))\,\mathrm{d}s, \quad 0 \le t \le 1,$$

and $\mathcal{Q}(u,v) = (Q_1(u,v), Q_2(u,v))$, $(u,v) \in Y$. The solutions of our problem (S)–(BC) coincide with the fixed points of the operator $\mathcal{Q}$.

Lemma 4.1.10. *If (H1) and (H2) hold, and $c \in (0, 1/2)$, then $\mathcal{Q} : P \to P$ is a completely continuous operator.*

Proof. Let $(u, v) \in P$ be an arbitrary element. Because $Q_1(u, v)$ and $Q_2(u, v)$ satisfy problem (4.1)–(4.2) for $y(t) = \lambda f(t, u(t), v(t))$, $t \in [0, 1]$, and problem (4.6)–(4.7) for $h(t) = \mu g(t, u(t), v(t))$, $t \in [0, 1]$, respectively, then by Lemma 4.1.9, we obtain

$$\inf_{t\in[c,1-c]} Q_1(u, v)(t) \ge \gamma_1 \sup_{t'\in[0,1]} Q_1(u, v)(t') = \gamma_1 \|Q_1(u, v)\|,$$

$$\inf_{t\in[c,1-c]} Q_2(u, v)(t) \ge \gamma_2 \sup_{t'\in[0,1]} Q_2(u, v)(t') = \gamma_2 \|Q_2(u, v)\|.$$

Hence, we conclude

$$\begin{aligned}\inf_{t\in[c,1-c]} [Q_1(u, v)(t) + Q_2(u, v)(t)] &\ge \inf_{t\in[c,1-c]} Q_1(u, v)(t) + \inf_{t\in[c,1-c]} Q_2(u, v)(t)\\ &\ge \gamma_1 \|Q_1(u, v)\| + \gamma_2 \|Q_2(u, v)\|\\ &\ge \gamma \|(Q_1(u, v), Q_2(u, v))\|_Y = \gamma \|\mathcal{Q}(u, v)\|_Y.\end{aligned}$$

By Lemma 4.1.7, (H1), and (H2), we obtain $Q_1(u, v)(t) \ge 0$, $Q_2(u, v)(t) \ge 0$ for all $t \in [0, 1]$, and so we deduce that $\mathcal{Q}(u, v) \in P$. Hence, we get $\mathcal{Q}(P) \subset P$. By using standard arguments, we can easily show that Q_1 and Q_2 are completely continuous, and then $\mathcal{Q}$ is a completely continuous operator. □

For $c \in (0, 1/2)$, we denote $A = \int_c^{1-c} J_1(s)\,ds$, $B = \int_0^1 J_1(s)\,ds$, $C = \int_c^{1-c} J_2(s)\,ds$, and $D = \int_0^1 J_2(s)\,ds$, where J_1 and J_2 are defined in Section 4.1.2 (Lemma 4.1.8).

First, for $f_0^s, g_0^s, f_\infty^i, g_\infty^i \in (0, \infty)$ and numbers $\alpha_1, \alpha_2 \ge 0$, and $\tilde{\alpha}_1, \tilde{\alpha}_2 > 0$ such that $\alpha_1 + \alpha_2 = 1$ and $\tilde{\alpha}_1 + \tilde{\alpha}_2 = 1$, we define the numbers L_1, L_2, L_3, L_4, L_2', and L_4' by

$$L_1 = \frac{\alpha_1}{\gamma\gamma_1 f_\infty^i A},\quad L_2 = \frac{\tilde{\alpha}_1}{f_0^s B},\quad L_3 = \frac{\alpha_2}{\gamma\gamma_2 g_\infty^i C},\quad L_4 = \frac{\tilde{\alpha}_2}{g_0^s D},\quad L_2' = \frac{1}{f_0^s B},\quad L_4' = \frac{1}{g_0^s D}.$$

Theorem 4.1.1. *Assume that (H1) and (H2) hold, $c \in (0, 1/2)$, and $\alpha_1, \alpha_2 \ge 0$ and $\tilde{\alpha}_1, \tilde{\alpha}_2 > 0$ such that $\alpha_1 + \alpha_2 = 1$ and $\tilde{\alpha}_1 + \tilde{\alpha}_2 = 1$.*

(1) If $f_0^s, g_0^s, f_\infty^i, g_\infty^i \in (0, \infty)$, $L_1 < L_2$, and $L_3 < L_4$, then for each $\lambda \in (L_1, L_2)$ and $\mu \in (L_3, L_4)$ there exists a positive solution $(u(t), v(t))$, $t \in [0, 1]$ for (S)–(BC).

(2) If $f_0^s = 0$, $g_0^s, f_\infty^i, g_\infty^i \in (0, \infty)$, and $L_3 < L_4'$, then for each $\lambda \in (L_1, \infty)$ and $\mu \in (L_3, L_4')$ there exists a positive solution $(u(t), v(t))$, $t \in [0, 1]$ for (S)–(BC).

(3) If $g_0^s = 0$, $f_0^s, f_\infty^i, g_\infty^i \in (0, \infty)$, and $L_1 < L_2'$, then for each $\lambda \in (L_1, L_2')$ and $\mu \in (L_3, \infty)$ there exists a positive solution $(u(t), v(t))$, $t \in [0, 1]$ for (S)–(BC).

(4) If $f_0^s = g_0^s = 0$ and $f_\infty^i, g_\infty^i \in (0, \infty)$, then for each $\lambda \in (L_1, \infty)$ and $\mu \in (L_3, \infty)$ there exists a positive solution $(u(t), v(t))$, $t \in [0, 1]$ for (S)–(BC).

(5) If $\{f_0^s, g_0^s, f_\infty^i \in (0, \infty), g_\infty^i = \infty\}$ or $\{f_0^s, g_0^s, g_\infty^i \in (0, \infty), f_\infty^i = \infty\}$ or $\{f_0^s, g_0^s \in (0, \infty), f_\infty^i = g_\infty^i = \infty\}$, then for each $\lambda \in (0, L_2)$ and $\mu \in (0, L_4)$ there exists a positive solution $(u(t), v(t))$, $t \in [0, 1]$ for (S)–(BC).

(6) If $\{f_0^s = 0, g_0^s, f_\infty^i \in (0, \infty), g_\infty^i = \infty\}$ or $\{f_0^s = 0, f_\infty^i = \infty, g_0^s, g_\infty^i \in (0, \infty)\}$ or $\{f_0^s = 0, g_0^s \in (0, \infty), f_\infty^i = g_\infty^i = \infty\}$, then for each $\lambda \in (0, \infty)$ and $\mu \in (0, L_4')$ there exists a positive solution $(u(t), v(t))$, $t \in [0, 1]$ for (S)–(BC).

(7) If $\{f_0^s, f_\infty^i \in (0,\infty),\ g_0^s = 0,\ g_\infty^i = \infty\}$ or $\{f_0^s, g_\infty^i \in (0,\infty),\ g_0^s = 0,\ f_\infty^i = \infty\}$ or $\{f_0^s \in (0,\infty),\ g_0^s = 0,\ f_\infty^i = g_\infty^i = \infty\}$, then for each $\lambda \in (0, L_2')$ and $\mu \in (0,\infty)$ there exists a positive solution $(u(t), v(t))$, $t \in [0,1]$ for (S)–(BC).

(8) If $\{f_0^s = g_0^s = 0,\ f_\infty^i \in (0,\infty),\ g_\infty^i = \infty\}$ or $\{f_0^s = g_0^s = 0,\ f_\infty^i = \infty,\ g_\infty^i \in (0,\infty)\}$ or $\{f_0^s = g_0^s = 0,\ f_\infty^i = g_\infty^i = \infty\}$, then for each $\lambda \in (0,\infty)$ and $\mu \in (0,\infty)$ there exists a positive solution $(u(t), v(t))$, $t \in [0,1]$ for (S)–(BC).

Proof. We consider the above cone $P \subset Y$ and the operators Q_1, Q_2, and $\mathcal{Q}$. Because the proofs of the above cases are similar, in what follows we shall prove one of them—namely, the second case of (5). So, we suppose $f_0^s, g_0^s, g_\infty^i \in (0,\infty)$ and $f_\infty^i = \infty$. Let $\lambda \in (0, L_2)$ and $\mu \in (0, L_4)$—that is, $\lambda \in (0, \frac{\tilde{\alpha}_1}{f_0^s B})$ and $\mu \in (0, \frac{\tilde{\alpha}_2}{g_0^s D})$. We choose $\alpha_2' > 0$, $\alpha_2' < \min\{\mu\gamma\gamma_2 g_\infty^i C, 1\}$. Let $\alpha_1' = 1 - \alpha_2'$ and let $\varepsilon > 0$ be a positive number such that $\varepsilon < g_\infty^i$ and

$$\frac{\alpha_1'\varepsilon}{\gamma\gamma_1 A} \le \lambda, \quad \frac{\alpha_2'}{\gamma\gamma_2(g_\infty^i - \varepsilon)C} \le \mu, \quad \frac{\tilde{\alpha}_1}{(f_0^s + \varepsilon)B} \ge \lambda, \quad \frac{\tilde{\alpha}_2}{(g_0^s + \varepsilon)D} \ge \mu.$$

By using (H2) and the definitions of f_0^s and g_0^s, we deduce that there exists $R_1 > 0$ such that $f(t,u,v) \le (f_0^s + \varepsilon)(u+v)$ and $g(t,u,v) \le (g_0^s + \varepsilon)(u+v)$ for all $t \in [0,1]$ and $u, v \ge 0$, with $0 \le u + v \le R_1$. We define the set $\Omega_1 = \{(u,v) \in Y,\ \ \|(u,v)\|_Y < R_1\}$. Now let $(u,v) \in P \cap \partial\Omega_1$—that is, $(u,v) \in P$ with $\|(u,v)\|_Y = R_1$ or equivalently $\|u\| + \|v\| = R_1$. Then $u(t) + v(t) \le R_1$ for all $t \in [0,1]$, and by Lemma 4.1.8, we obtain

$$\begin{aligned} Q_1(u,v)(t) &\le \lambda \int_0^1 J_1(s) f(s, u(s), v(s))\, ds \le \lambda \int_0^1 J_1(s)(f_0^s + \varepsilon)(u(s) + v(s))\, ds \\ &\le \lambda (f_0^s + \varepsilon) \int_0^1 J_1(s)(\|u\| + \|v\|)\, ds \\ &= \lambda (f_0^s + \varepsilon) B \|(u,v)\|_Y \le \tilde{\alpha}_1 \|(u,v)\|_Y, \quad \forall t \in [0,1]. \end{aligned}$$

Therefore, $\|Q_1(u,v)\| \le \tilde{\alpha}_1 \|(u,v)\|_Y$. In a similar manner, we conclude

$$\begin{aligned} Q_2(u,v)(t) &\le \mu \int_0^1 J_2(s) g(s, u(s), v(s))\, ds \le \mu \int_0^1 J_2(s)(g_0^s + \varepsilon)(u(s) + v(s))\, ds \\ &\le \mu (g_0^s + \varepsilon) \int_0^1 J_2(s)(\|u\| + \|v\|)\, ds \\ &= \mu (g_0^s + \varepsilon) D \|(u,v)\|_Y \le \tilde{\alpha}_2 \|(u,v)\|_Y, \quad \forall t \in [0,1]. \end{aligned}$$

Therefore, $\|Q_2(u,v)\| \le \tilde{\alpha}_2 \|(u,v)\|_Y$.

Then for $(u,v) \in P \cap \partial\Omega_1$, we deduce

$$\|\mathcal{Q}(u,v)\|_Y = \|Q_1(u,v)\| + \|Q_2(u,v)\| \le \tilde{\alpha}_1 \|(u,v)\|_Y + \tilde{\alpha}_2 \|(u,v)\|_Y = \|(u,v)\|_Y. \tag{4.8}$$

By the definitions of f_∞^i and g_∞^i, there exists $\bar{R}_2 > 0$ such that $f(t,u,v) \ge \frac{1}{\varepsilon}(u+v)$ and $g(t,u,v) \ge (g_\infty^i - \varepsilon)(u+v)$, for all $u,\ v \ge 0$, with $u+v \ge \bar{R}_2$, and $t \in [c, 1-c]$. We consider $R_2 = \max\{2R_1, \bar{R}_2/\gamma\}$, and we define $\Omega_2 = \{(u,v) \in Y, \quad \|(u,v)\|_Y < R_2\}$. Then for $(u,v) \in P$ with $\|(u,v)\|_Y = R_2$, we obtain

$$u(t) + v(t) \geq \inf_{t\in[c,1-c]} (u(t) + v(t)) \geq \gamma \|(u,v)\|_Y = \gamma R_2 \geq \bar{R}_2, \quad \forall t \in [c, 1-c].$$

Then, by Lemma 4.1.8, we conclude

$$\begin{aligned} Q_1(u,v)(c) &\geq \lambda\gamma_1 \int_0^1 J_1(s)f(s,u(s),v(s))\,ds \geq \lambda\gamma_1 \int_c^{1-c} J_1(s)f(s,u(s),v(s))\,ds \\ &\geq \lambda\gamma_1 \int_c^{1-c} J_1(s)\frac{1}{\varepsilon}(u(s)+v(s))\,ds \\ &\geq \frac{\lambda\gamma_1}{\varepsilon}\int_c^{1-c} J_1(s)\gamma\|(u,v)\|_Y\,ds \geq \alpha_1'\|(u,v)\|_Y. \end{aligned}$$

So, $\|Q_1(u,v)\| \geq Q_1(u,v)(c) \geq \alpha_1'\|(u,v)\|_Y$.

In a similar manner, we deduce

$$\begin{aligned} Q_2(u,v)(c) &\geq \mu\gamma_2 \int_0^1 J_2(s)g(s,u(s),v(s))\,ds \geq \mu\gamma_2 \int_c^{1-c} J_2(s)g(s,u(s),v(s))\,ds \\ &\geq \mu\gamma_2 \int_c^{1-c} J_2(s)(g_\infty^{\rm i} - \varepsilon)(u(s)+v(s))\,ds \\ &\geq \mu\gamma_2(g_\infty^{\rm i} - \varepsilon)\int_c^{1-c} J_2(s)\gamma\|(u,v)\|_Y\,ds \geq \alpha_2'\|(u,v)\|_Y. \end{aligned}$$

So, $\|Q_2(u,v)\| \geq Q_2(u,v)(c) \geq \alpha_2'\|(u,v)\|_Y$.

Hence, for $(u,v) \in P \cap \partial\Omega_2$, we obtain

$$\|Q(u,v)\|_Y = \|Q_1(u,v)\| + \|Q_2(u,v)\| \geq (\alpha_1' + \alpha_2')\|(u,v)\|_Y = \|(u,v)\|_Y. \quad (4.9)$$

By using (4.8), (4.9), Lemma 4.1.10, and Theorem 1.1.1 (1), we conclude that Q has a fixed point $(u,v) \in P \cap (\bar{\Omega}_2 \setminus \Omega_1)$ such that $R_1 \leq \|u\| + \|v\| \leq R_2$. □

In what follows, for $f_0^{\rm i}, g_0^{\rm i}, f_\infty^{\rm s}, g_\infty^{\rm s} \in (0,\infty)$ and numbers $\alpha_1, \alpha_2 \geq 0$ and $\tilde{\alpha}_1, \tilde{\alpha}_2 > 0$ such that $\alpha_1 + \alpha_2 = 1$ and $\tilde{\alpha}_1 + \tilde{\alpha}_2 = 1$, we define the numbers $\tilde{L}_1, \tilde{L}_2, \tilde{L}_3, \tilde{L}_4, \tilde{L}_2'$, and $\tilde{L}_4'$ by

$$\tilde{L}_1 = \frac{\alpha_1}{\gamma\gamma_1 f_0^{\rm i} A}, \quad \tilde{L}_2 = \frac{\tilde{\alpha}_1}{f_\infty^{\rm s} B}, \quad \tilde{L}_3 = \frac{\alpha_2}{\gamma\gamma_2 g_0^{\rm i} C}, \quad \tilde{L}_4 = \frac{\tilde{\alpha}_2}{g_\infty^{\rm s} D}, \quad \tilde{L}_2' = \frac{1}{f_\infty^{\rm s} B}, \quad \tilde{L}_4' = \frac{1}{g_\infty^{\rm s} D}.$$

Theorem 4.1.2. *Assume that (H1) and (H2) hold, $c \in (0,1/2)$, and $\alpha_1, \alpha_2 \geq 0$ and $\tilde{\alpha}_1, \tilde{\alpha}_2 > 0$ such that $\alpha_1 + \alpha_2 = 1$ and $\tilde{\alpha}_1 + \tilde{\alpha}_2 = 1$.*

(1) If $f_0^{\rm i}, g_0^{\rm i}, f_\infty^{\rm s}, g_\infty^{\rm s} \in (0,\infty)$, $\tilde{L}_1 < \tilde{L}_2$, and $\tilde{L}_3 < \tilde{L}_4$, then for each $\lambda \in (\tilde{L}_1, \tilde{L}_2)$ and $\mu \in (\tilde{L}_3, \tilde{L}_4)$ there exists a positive solution $(u(t), v(t))$, $t \in [0,1]$ for (S)–(BC).

(2) If $f_0^{\rm i}, g_0^{\rm i}, f_\infty^{\rm s} \in (0,\infty)$, $g_\infty^{\rm s} = 0$, and $\tilde{L}_1 < \tilde{L}_2'$, then for each $\lambda \in (\tilde{L}_1, \tilde{L}_2')$ and $\mu \in (\tilde{L}_3, \infty)$ there exists a positive solution $(u(t), v(t))$, $t \in [0,1]$ for (S)–(BC).

(3) If $f_0^{\rm i}, g_0^{\rm i}, g_\infty^{\rm s} \in (0,\infty)$, $f_\infty^{\rm s} = 0$, and $\tilde{L}_3 < \tilde{L}_4'$, then for each $\lambda \in (\tilde{L}_1, \infty)$ and $\mu \in (\tilde{L}_3, \tilde{L}_4')$ there exists a positive solution $(u(t), v(t))$, $t \in [0,1]$ for (S)–(BC).

(4) If $f_0^{\rm i}$, $g_0^{\rm i} \in (0,\infty)$ and $f_\infty^{\rm s} = g_\infty^{\rm s} = 0$, then for each $\lambda \in (\tilde{L}_1,\infty)$ and $\mu \in (\tilde{L}_3,\infty)$ there exists a positive solution $(u(t), v(t))$, $t \in [0,1]$ for (S)–(BC).

(5) If $\{f_0^{\rm i} = \infty,\ g_0^{\rm i}, f_\infty^{\rm s}, g_\infty^{\rm s} \in (0,\infty)\}$ or $\{f_0^{\rm i}, f_\infty^{\rm s}, g_\infty^{\rm s} \in (0,\infty),\ g_0^{\rm i} = \infty\}$ or $\{f_0^{\rm i} = g_0^{\rm i} = \infty,$ $f_\infty^{\rm s}, g_\infty^{\rm s} \in (0,\infty)\}$, then for each $\lambda \in (0,\tilde{L}_2)$ and $\mu \in (0,\tilde{L}_4)$ there exists a positive solution $(u(t), v(t))$, $t \in [0,1]$ for (S)–(BC).

(6) If $\{f_0^{\rm i} = \infty,\ g_0^{\rm i}, f_\infty^{\rm s} \in (0,\infty),\ g_\infty^{\rm s} = 0\}$ or $\{f_0^{\rm i}, f_\infty^{\rm s} \in (0,\infty),\ g_0^{\rm i} = \infty,\ g_\infty^{\rm s} = 0\}$ or $\{f_0^{\rm i} = g_0^{\rm i} = \infty,\ f_\infty^{\rm s} \in (0,\infty),\ g_\infty^{\rm s} = 0\}$, then for each $\lambda \in (0,\tilde{L}_2')$ and $\mu \in (0,\infty)$ there exists a positive solution $(u(t), v(t))$, $t \in [0,1]$ for (S)–(BC).

(7) If $\{f_0^{\rm i} = \infty,\ g_0^{\rm i}, g_\infty^{\rm s} \in (0,\infty),\ f_\infty^{\rm s} = 0\}$ or $\{f_0^{\rm i}, g_\infty^{\rm s} \in (0,\infty),\ g_0^{\rm i} = \infty,\ f_\infty^{\rm s} = 0\}$ or $\{f_0^{\rm i} = g_0^{\rm i} = \infty,\ f_\infty^{\rm s} = 0,\ g_\infty^{\rm s} \in (0,\infty)\}$, then for each $\lambda \in (0,\infty)$ and $\mu \in (0,\tilde{L}_4')$ there exists a positive solution $(u(t), v(t))$, $t \in [0,1]$ for (S)–(BC).

(8) If $\{f_0^{\rm i} = \infty,\ g_0^{\rm i} \in (0,\infty),\ f_\infty^{\rm s} = g_\infty^{\rm s} = 0\}$ or $\{f_0^{\rm i} \in (0,\infty),\ g_0^{\rm i} = \infty,\ f_\infty^{\rm s} = g_\infty^{\rm s} = 0\}$ or $\{f_0^{\rm i} = g_0^{\rm i} = \infty,\ f_\infty^{\rm s} = g_\infty^{\rm s} = 0\}$, then for each $\lambda \in (0,\infty)$ and $\mu \in (0,\infty)$ there exists a positive solution $(u(t), v(t))$, $t \in [0,1]$ for (S)–(BC).

Proof. We consider the above cone $P \subset Y$ and the operators Q_1, Q_2, and $\mathcal{Q}$. Because the proofs of the above cases are similar, in what follows we shall prove one of them—namely, the third case of (6). So, we suppose $f_0^{\rm i} = g_0^{\rm i} = \infty$, $f_\infty^{\rm s} \in (0,\infty)$, and $g_\infty^{\rm s} = 0$. Let $\lambda \in (0,\tilde{L}_2')$—that is, $\lambda \in (0, \frac{1}{f_\infty^{\rm s} B})$—and $\mu \in (0,\infty)$. We choose $\tilde{\alpha}_1' \in (\lambda f_\infty^{\rm s} B, 1)$ and $\alpha_1' \in (0,1)$. Let $\tilde{\alpha}_2' = 1 - \tilde{\alpha}_1'$ and $\alpha_2' = 1 - \alpha_1'$, and let $\varepsilon > 0$ be a positive number such that

$$\frac{\alpha_1'\varepsilon}{\gamma\gamma_1 A} \leq \lambda, \quad \frac{\alpha_2'\varepsilon}{\gamma\gamma_2 C} \leq \mu, \quad \frac{\tilde{\alpha}_1'}{(f_\infty^{\rm s} + \varepsilon)B} \geq \lambda, \quad \frac{\tilde{\alpha}_2'}{\varepsilon D} \geq \mu.$$

By using (H2) and the definitions of $f_0^{\rm i}$ and $g_0^{\rm i}$, we deduce that there exists $R_3 > 0$ such that $f(t,u,v) \geq \frac{1}{\varepsilon}(u+v)$, $g(t,u,v) \geq \frac{1}{\varepsilon}(u+v)$ for all u, $v \geq 0$, with $0 \leq u+v \leq R_3$, and $t \in [c, 1-c]$. We denote $\Omega_3 = \{(u,v) \in Y,\quad \|(u,v)\|_Y < R_3\}$. Let $(u,v) \in P$ with $\|(u,v)\|_Y = R_3$—that is, $\|u\| + \|v\| = R_3$. Because $u(t)+v(t) \leq \|u\|+\|v\| = R_3$ for all $t \in [0,1]$, then by using Lemma 4.1.8, we obtain

$$\begin{aligned} Q_1(u,v)(c) &\geq \lambda\gamma_1 \int_c^{1-c} J_1(s) f(s,u(s),v(s))\,{\rm d}s \geq \lambda\gamma_1 \int_c^{1-c} J_1(s)\frac{1}{\varepsilon}(u(s)+v(s))\,{\rm d}s \\ &\geq \frac{\lambda\gamma\gamma_1}{\varepsilon} \int_c^{1-c} J_1(s)(\|u\|+\|v\|)\,{\rm d}s \geq \alpha_1'\|(u,v)\|_Y. \end{aligned}$$

Therefore, $\|Q_1(u,v)\| \geq Q_1(u,v)(c) \geq \alpha_1'\|(u,v)\|_Y$. In a similar manner, we conclude

$$\begin{aligned} Q_2(u,v)(c) &\geq \mu\gamma_2 \int_c^{1-c} J_2(s) g(s,u(s),v(s))\,{\rm d}s \geq \mu\gamma_2 \int_c^{1-c} J_2(s)\frac{1}{\varepsilon}(u(s)+v(s))\,{\rm d}s \\ &\geq \frac{\mu\gamma\gamma_2}{\varepsilon} \int_c^{1-c} J_2(s)(\|u\|+\|v\|)\,{\rm d}s \geq \alpha_2'\|(u,v)\|_Y. \end{aligned}$$

So, $\|Q_2(u,v)\| \geq Q_2(u,v)(c) \geq \alpha_2'\|(u,v)\|_Y$.

Thus, for an arbitrary element $(u,v) \in P \cap \partial\Omega_3$, we deduce

$$\|\mathcal{Q}(u,v)\|_Y \geq (\alpha_1' + \alpha_2')\|(u,v)\|_Y = \|(u,v)\|_Y. \tag{4.10}$$

Now, we define the functions $f^*, g^* : [0,1] \times \mathbb{R}_+ \to \mathbb{R}_+, f^*(t,x) = \max_{0\le u+v\le x} f(t,u,v), g^*(t,x) = \max_{0\le u+v\le x} g(t,u,v)$ for all $t \in [0,1]$ and $x \ge 0$. Then $f(t,u,v) \le f^*(t,x)$, $g(t,u,v) \le g^*(t,x)$ for all $t \in [0,1]$, $u, v \ge 0$, and $u+v \le x$. The functions $f^*(t,\cdot)$ and $g^*(t,\cdot)$ are nondecreasing for every $t \in [0,1]$, and they satisfy the conditions

$$\limsup_{x\to\infty} \max_{t\in[0,1]} \frac{f^*(t,x)}{x} \le f_\infty^s, \quad \lim_{x\to\infty} \max_{t\in[0,1]} \frac{g^*(t,x)}{x} = 0.$$

Therefore, for $\varepsilon > 0$, there exists $\bar{R}_4 > 0$ such that for all $x \ge \bar{R}_4$ and $t \in [0,1]$, we have

$$\frac{f^*(t,x)}{x} \le \limsup_{x\to\infty} \max_{t\in[0,1]} \frac{f^*(t,x)}{x} + \varepsilon \le f_\infty^s + \varepsilon, \quad \frac{g^*(t,x)}{x} \le \lim_{x\to\infty} \max_{t\in[0,1]} \frac{g^*(t,x)}{x} + \varepsilon = \varepsilon,$$

and so $f^*(t,x) \le (f_\infty^s + \varepsilon)x$ and $g^*(t,x) \le \varepsilon x$.

We consider $R_4 = \max\{2R_3, \bar{R}_4\}$, and we denote $\Omega_4 = \{(u,v) \in Y, \quad \|(u,v)\|_Y < R_4\}$. Let $(u,v) \in P \cap \partial\Omega_4$. By the definitions of f^* and g^*, we conclude

$$f(t,u(t),v(t)) \le f^*(t, \|(u,v)\|_Y), \quad g(t,u(t),v(t)) \le g^*(t, \|(u,v)\|_Y), \quad \forall t \in [0,1].$$

Then for all $t \in [0,1]$, we obtain

$$\begin{aligned} Q_1(u,v)(t) &\le \lambda \int_0^1 J_1(s) f(s,(u(s),v(s)))\,ds \le \lambda \int_0^1 J_1(s) f^*(s, \|(u,v)\|_Y)\,ds \\ &\le \lambda (f_\infty^s + \varepsilon) \int_0^1 J_1(s) \|(u,v)\|_Y\,ds \le \tilde{\alpha}_1' \|(u,v)\|_Y, \end{aligned}$$

and so $\|Q_1(u,v)\| \le \tilde{\alpha}_1' \|(u,v)\|_Y$.

In a similar manner, we deduce

$$\begin{aligned} Q_2(u,v)(t) &\le \mu \int_0^1 J_2(s) g(s,u(s),v(s))\,ds \le \mu \int_0^1 J_2(s) g^*(s, \|(u,v)\|_Y)\,ds \\ &\le \mu\varepsilon \int_0^1 J_2(s) \|(u,v)\|_Y\,ds \le \tilde{\alpha}_2' \|(u,v)\|_Y, \end{aligned}$$

and so $\|Q_2(u,v)\| \le \tilde{\alpha}_2' \|(u,v)\|_Y$.

Therefore, for $(u,v) \in P \cap \partial\Omega_4$, it follows that

$$\|\mathcal{Q}(u,v)\|_Y \le (\tilde{\alpha}_1' + \tilde{\alpha}_2') \|(u,v)\|_Y = \|(u,v)\|_Y. \tag{4.11}$$

By using (4.10), (4.11), Lemma 4.1.10, and Theorem 1.1.1 (2), we conclude that $\mathcal{Q}$ has a fixed point $(u,v) \in P \cap (\bar{\Omega}_4 \setminus \Omega_3)$ such that $R_3 \le \|(u,v)\|_Y \le R_4$. □

We present now intervals for λ and μ for which there exists no positive solution of problem (S)–(BC).

Theorem 4.1.3. *Assume that (H1) and (H2) hold. If $f_0^s, f_\infty^s, g_0^s, g_\infty^s < \infty$, then there exist positive constants λ_0 and μ_0 such that for every $\lambda \in (0,\lambda_0)$ and $\mu \in (0,\mu_0)$, the boundary value problem (S)–(BC) has no positive solution.*

Proof. From the definitions of $f_0^s, f_\infty^s, g_0^s, g_\infty^s$, we deduce that there exist M_1, $M_2 > 0$ such that $f(t,u,v) \le M_1(u+v)$, $g(t,u,v) \le M_2(u+v)$, for all $t \in [0,1]$ and $u, v \ge 0$. As in the proof of Theorem 1.2.1, one can prove that $\lambda_0 = \frac{1}{2M_1B}$ and $\mu_0 = \frac{1}{2M_2D}$ satisfy our theorem. □

Theorem 4.1.4. *Assume that (H1) and (H2) hold.*

- **(a)** *If $f_0^i, f_\infty^i > 0$ and $f(t,u,v) > 0$ for all $t \in [c, 1-c]$, $u \ge 0$, $v \ge 0$, $u+v > 0$, then there exists a positive constant $\tilde{\lambda}_0$ such that for every $\lambda > \tilde{\lambda}_0$ and $\mu > 0$ the boundary value problem (S)–(BC) has no positive solution.*
- **(b)** *If $g_0^i, g_\infty^i > 0$ and $g(t,u,v) > 0$ for all $t \in [c, 1-c]$, $u \ge 0$, $v \ge 0$, $u+v > 0$, then there exists a positive constant $\tilde{\mu}_0$ such that for every $\mu > \tilde{\mu}_0$ and $\lambda > 0$ the boundary value problem (S)–(BC) has no positive solution.*
- **(c)** *If $f_0^i, f_\infty^i, g_0^i, g_\infty^i > 0$ and $f(t,u,v) > 0$, $g(t,u,v) > 0$ for all $t \in [c, 1-c]$, $u \ge 0$, $v \ge 0$, $u+v > 0$, then there exist positive constants $\hat{\lambda}_0$ and $\hat{\mu}_0$ such that for every $\lambda > \hat{\lambda}_0$ and $\mu > \hat{\mu}_0$ the boundary value problem (S)–(BC) has no positive solution.*

Proof. **(a)** From the assumptions of the theorem, we deduce that there exists $m_1 > 0$ such that
$f(t,u,v) \ge m_1(u+v)$, for all $t \in [c, 1-c]$, $u, v \ge 0$. As in the proof of Theorem 1.2.2, one can prove that $\tilde{\lambda}_0 = \frac{1}{\gamma\gamma_1 m_1 A}$ satisfies our theorem (a).

(b) From the assumptions of the theorem, we deduce that there exists $m_2 > 0$ such that $g(t,u,v) \ge m_2(u+v)$, for all $t \in [c, 1-c]$, $u, v \ge 0$. Then one can show that $\tilde{\mu}_0 = \frac{1}{\gamma\gamma_2 m_2 C}$ satisfies our theorem (b).

(c) We define $\hat{\lambda}_0 = \frac{\tilde{\lambda}_0}{2}$ and $\hat{\mu}_0 = \frac{\tilde{\mu}_0}{2}$, and one can show that they satisfy our theorem (c). □

4.1.4 Examples

Let $\alpha = \frac{5}{2}$ $(n=3)$, $\beta = \frac{10}{3}$ $(m=4)$, $H(t) = \begin{cases} 0, & t \in [0, 1/4), \\ 3, & t \in [1/4, 3/4), \\ 7/2, & t \in [3/4, 1] \end{cases}$ and $K(t) = t^4$ for all $t \in [0,1]$. Then $\int_0^1 u(s)\,dH(s) = 3u\left(\frac{1}{4}\right) + \frac{1}{2}u\left(\frac{3}{4}\right)$ and $\int_0^1 v(s)\,dK(s) = 4\int_0^1 s^3 v(s)\,ds$.

We consider the system of fractional differential equations

$$\begin{cases} D_{0+}^{5/2}u(t) + \lambda f(t,u(t),v(t)) = 0, & 0 < t < 1, \\ D_{0+}^{10/3}v(t) + \mu g(t,u(t),v(t)) = 0, & 0 < t < 1, \end{cases} \tag{S$_1$}$$

with the boundary conditions

$$\begin{cases} u(0) = u'(0) = 0, \quad u(1) = 3u\left(\frac{1}{4}\right) + \frac{1}{2}u\left(\frac{3}{4}\right), \\ v(0) = v'(0) = v''(0) = 0, \quad v(1) = 4\int_0^1 s^3 v(s)\,ds. \end{cases} \tag{BC$_1$}$$

Then we obtain $\Delta_1 = 1 - \int_0^1 s^{3/2}\,dH(s) = 1 - 3\left(\frac{1}{4}\right)^{3/2} - \frac{1}{2}\left(\frac{3}{4}\right)^{3/2} = \frac{10-3\sqrt{3}}{16} \approx 0.3002 > 0$ and $\Delta_2 = 1 - \int_0^1 s^{7/3}\,dK(s) = 1 - 4\int_0^1 s^{16/3}\,ds = \frac{7}{19} \approx 0.3684 > 0$.

We also deduce

$$g_1(t,s)=\frac{4}{3\sqrt{\pi}}\begin{cases} t^{3/2}(1-s)^{3/2}-(t-s)^{3/2}, & 0\le s\le t\le 1,\\ t^{3/2}(1-s)^{3/2}, & 0\le t\le s\le 1,\end{cases}$$

$$g_2(t,s)=\frac{1}{\Gamma(10/3)}\begin{cases} t^{7/3}(1-s)^{7/3}-(t-s)^{7/3}, & 0\le s\le t\le 1,\\ t^{7/3}(1-s)^{7/3}, & 0\le t\le s\le 1,\end{cases}$$

$\theta_1(s)=\frac{1}{3-3s+s^2}$ for all $s\in[0,1]$, $\theta_2(s)=\frac{s}{1-(1-s)^{7/4}}$ for all $s\in(0,1]$, $\theta_2(0)=4/7$. For the functions J_1 and J_2, we obtain

$$J_1(s)=\begin{cases} \frac{4}{3\sqrt{\pi}}\left\{\frac{s(1-s)^{3/2}}{(3-3s+s^2)^{1/2}}+\frac{1}{10-3\sqrt{3}}\left\{6\left[(1-s)^{3/2}-(1-4s)^{3/2}\right]\right.\right.\\ \quad\left.\left.+\left[3\sqrt{3}(1-s)^{3/2}-(3-4s)^{3/2}\right]\right\}\right\}, & 0\le s<\frac{1}{4},\\ \frac{4}{3\sqrt{\pi}}\left\{\frac{s(1-s)^{3/2}}{(3-3s+s^2)^{1/2}}+\frac{1}{10-3\sqrt{3}}\left\{6(1-s)^{3/2}\right.\right.\\ \quad\left.\left.+\left[3\sqrt{3}(1-s)^{3/2}-(3-4s)^{3/2}\right]\right\}\right\}, & \frac{1}{4}\le s<\frac{3}{4},\\ \frac{4}{3\sqrt{\pi}}\left[\frac{s(1-s)^{3/2}}{(3-3s+s^2)^{1/2}}+\frac{3(2+\sqrt{3})(1-s)^{3/2}}{10-3\sqrt{3}}\right], & \frac{3}{4}\le s\le 1,\end{cases}$$

and

$$J_2(s)=\begin{cases} \frac{1}{\Gamma(10/3)}\left\{\frac{s^{7/3}(1-s)^{7/3}}{[1-(1-s)^{7/4}]^{4/3}}+\frac{76}{7}\left[\frac{3}{19}(1-s)^{7/3}-\frac{3}{19}(1-s)^{19/3}\right.\right.\\ \quad\left.\left.-\frac{9}{16}s(1-s)^{16/3}-\frac{9}{13}s^2(1-s)^{13/3}-\frac{3}{10}s^3(1-s)^{10/3}\right]\right\}, & 0<s\le 1,\\ 0, \quad s=0.\end{cases}$$

For $c=\frac{1}{4}$, we deduce $\gamma_1=\frac{1}{8}$ and $\gamma=\gamma_2=\frac{1}{16\sqrt[3]{4}}$. After some computations, we conclude $A=\int_{1/4}^{3/4}J_1(s)\,ds\approx 0.27639507$, $B=\int_0^1 J_1(s)\,ds\approx 0.42677595$, $C=\int_{1/4}^{3/4}J_2(s)\,ds\approx 0.02808183$, and $D=\int_0^1 J_2(s)\,ds\approx 0.04007233$.

Example 4.1.1. We consider the functions

$$f(t,u,v)=\frac{\sqrt{1-t}\,[p_1(u+v)+1](u+v)(q_1+\sin v)}{u+v+1},$$

$$g(t,u,v)=\frac{\sqrt[3]{1-t}\,[p_2(u+v)+1](u+v)(q_2+\cos u)}{u+v+1},$$

for all $t\in[0,1]$, $u,\ v\in[0,\infty)$, where $p_1,\ p_2>0$ and $q_1,\ q_2>1$.

We have $f_0^s=q_1$, $g_0^s=q_2+1$, $f_\infty^i=p_1(q_1-1)/2$, and $g_\infty^i=p_2(q_2-1)/\sqrt[3]{4}$. For $\alpha_1,\ \alpha_2>0$ with $\alpha_1+\alpha_2=1$, we consider $\tilde{\alpha}_1=\alpha_1$ and $\tilde{\alpha}_2=\alpha_2$. Then we obtain

$$L_1=\frac{256\sqrt[3]{4}\alpha_1}{p_1(q_1-1)A},\quad L_2=\frac{\alpha_1}{q_1B},\quad L_3=\frac{1024\alpha_2}{p_2(q_2-1)C},\quad L_4=\frac{\alpha_2}{(q_2+1)D}.$$

The conditions $L_1 < L_2$ and $L_3 < L_4$ become

$$\frac{p_1(q_1-1)}{q_1} > \frac{256\sqrt[3]{4}B}{A}, \quad \frac{p_2(q_2-1)}{q_2+1} > \frac{1024D}{C}.$$

If $\frac{p_1(q_1-1)}{q_1} \geq 628$ and $\frac{p_2(q_2-1)}{q_2+1} \geq 1462$, then the above conditions are satisfied. For example, if $\alpha_1 = \alpha_2 = 1/2$, $q_1 = q_2 = 2$, $p_1 = 1256$, and $p_2 = 4386$, we obtain $L_1 \approx 0.585298$, $L_2 \approx 0.585787$, $L_3 \approx 4.156961$, and $L_4 \approx 4.159145$. Therefore, by Theorem 4.1.1 (1), for each $\lambda \in (L_1, L_2)$ and $\mu \in (L_3, L_4)$, there exists a positive solution $(u(t), v(t))$, $t \in [0,1]$ for problem (S_1)–(BC_1).

Because $f_0^s = q_1$, $f_\infty^s = p_1(q_1+1)$, $g_0^s = q_2+1$, and $g_\infty^s = p_2(q_2+1)$ are finite, we can apply Theorem 4.1.3. For $q_1 = q_2 = 2$, $p_1 = 1256$, and $p_2 = 4386$, we deduce

$$M_1 = \sup_{u,v\geq 0} \frac{[p_1(u+v)+1](q_1+\sin v)}{u+v+1} \approx 3768,$$

$$M_2 = \sup_{u,v\geq 0} \frac{[p_2(u+v)+1](q_2+\cos u)}{u+v+1} \approx 13{,}158.$$

Then we obtain $\lambda_0 = 1/(2M_1B) \approx 3.1 \cdot 10^{-4}$ and $\mu_0 = 1/(2M_2D) \approx 9.5 \cdot 10^{-4}$. Therefore, by Theorem 4.1.3, we conclude that for any $\lambda \in (0, \lambda_0)$ and $\mu \in (0, \mu_0)$ problem (S_1)–(BC_1) has no positive solution.

Example 4.1.2. We consider the functions

$$f(t,u,v) = \frac{a_0 t(u+v)^2(q_1+\sin v)}{u+v+1}, \quad g(t,u,v) = \frac{b_0 t^2(u+v)^2(q_2+\cos u)}{u+v+1},$$

for all $t \in [0,1]$, $u, v \in [0,\infty)$, where $a_0, b_0 > 0$, $q_1, q_2 > 1$.

We have $f_0^s = g_0^s = 0$, $f_\infty^i = a_0(q_1-1)/4$, and $g_\infty^i = b_0(q_2-1)/(16)$. Then, for $\alpha_1, \alpha_2 \geq 0$ with $\alpha_1 + \alpha_2 = 1$, we conclude $L_1 = \frac{512\sqrt[3]{4}\alpha_1}{a_0(q_1-1)A}$ and $L_3 = \frac{4096\sqrt[3]{16}\alpha_2}{b_0(q_2-1)C}$. For example, if $a_0 = b_0 = 1$, $q_1 = q_2 = 2$, and $\alpha_1 = \alpha_2 = 1/2$, we obtain $L_1 \approx 1470.26744$ and $L_3 \approx 183{,}771.40707$. Then, by Theorem 4.1.1 (4), we deduce that for each $\lambda \in (L_1, \infty)$ and $\mu \in (L_3, \infty)$ there exists a positive solution $(u(t), v(t))$, $t \in [0,1]$ for problem (S_1)–(BC_1).

Example 4.1.3. We consider the functions

$$f(t,u,v) = \frac{t}{1+t^2}[p_1(u^2+v^2)+q_1(u+v)], \quad g(t,u,v) = \frac{t^2}{1+t}[p_2(u^2+v^2)+q_2(u+v)],$$

for all $t \in [0,1]$, $u, v \in [0,\infty)$, where $p_1, p_2, q_1, q_2 > 0$.

We have $f_0^s = q_1/2$, $g_0^s = q_2/2$, and $f_\infty^i = g_\infty^i = \infty$. Then, for $\tilde{\alpha}_1, \tilde{\alpha}_2 > 0$ with $\tilde{\alpha}_1 + \tilde{\alpha}_2 = 1$, we obtain $L_2 = \frac{2\tilde{\alpha}_1}{q_1 B}$ and $L_4 = \frac{2\tilde{\alpha}_2}{q_2 D}$. For example, if $\tilde{\alpha}_1 = \tilde{\alpha}_2 = 1/2$ and $q_1 = q_2 = 1$, we have $L_2 \approx 2.34315$ and $L_2 \approx 24.95487$. Then, by Theorem 4.1.1 (5), we deduce that for each $\lambda \in (0, L_2)$ and $\mu \in (0, L_4)$ there exists a positive solution $(u(t), v(t))$, $t \in [0,1]$ for problem (S_1)–(BC_1).

Example 4.1.4. We consider the functions

$$f(t,u,v) = p_1 t^a (u^2 + v^2), \quad g(t,u,v) = p_2(1-t)^b (e^{u+v} - 1), \quad \forall t \in [0,1], \quad u, v \in [0,\infty),$$

where a, b, p_1, $p_2 > 0$.

We have $f_0^s = 0$, $g_0^s = p_2$, $f_\infty^i = g_\infty^i = \infty$, and $L_4' = \frac{1}{p_2 D}$. For example, if $p_2 = 2$, we obtain $L_4' \approx 12.47744$. Then, by Theorem 4.1.1 (6), we conclude that for each $\lambda \in (0,\infty)$ and $\mu \in (0, L_4')$ there exists a positive solution $(u(t), v(t))$, $t \in [0,1]$ for problem (S_1)–(BC_1).

Example 4.1.5. We consider the functions

$$f(t,u,v) = a(u+v)^{p_1}, \quad g(t,u,v) = b(u+v)^{p_2}, \quad \forall t \in [0,1], \quad u, v \in [0,\infty),$$

where a, $b > 0$ and p_1, $p_2 > 1$.

We have $f_0^s = g_0^s = 0$ and $f_\infty^i = g_\infty^i = \infty$. Then, by Theorem 4.1.1 (8), for each $\lambda \in (0,\infty)$ and $\mu \in (0,\infty)$ there exists a positive solution $(u(t), v(t))$, $t \in [0,1]$ for problem (S_1)–(BC_1).

Example 4.1.6. We consider the functions

$$f(t,u,v) = a(u+v)^{q_1}, \quad g(t,u,v) = b(u+v)^{q_2}, \quad \forall t \in [0,1], \quad u, v \in [0,\infty),$$

where a, $b > 0$ and q_1, $q_2 \in (0,1)$.

We have $f_0^i = g_0^i = \infty$ and $f_\infty^s = g_\infty^s = 0$. Then, by Theorem 4.1.2 (8), for each $\lambda \in (0,\infty)$ and $\mu \in (0,\infty)$ there exists a positive solution $(u(t), v(t))$, $t \in [0,1]$ for problem (S_1)–(BC_1).

4.2 Existence and multiplicity of positive solutions for systems without parameters and uncoupled boundary conditions

In this section, we investigate the existence and multiplicity of positive solutions for problem (S)–(BC) from Section 4.1 with $\lambda = \mu = 1$, and f and g dependent only on t and v, and t and u, respectively. The nonlinearities f and g are nonsingular functions, or singular functions at $t = 0$ and/or $t = 1$.

4.2.1 Nonsingular nonlinearities

We consider the system of nonlinear ordinary fractional differential equations

$$\begin{cases} D_{0+}^{\alpha} u(t) + f(t, v(t)) = 0, & t \in (0,1), \\ D_{0+}^{\beta} v(t) + g(t, u(t)) = 0, & t \in (0,1), \end{cases} \tag{S'}$$

with the uncoupled integral boundary conditions

$$\begin{cases} u(0) = u'(0) = \cdots = u^{(n-2)}(0) = 0, & u(1) = \displaystyle\int_0^1 u(s)\,dH(s), \\ v(0) = v'(0) = \cdots = v^{(m-2)}(0) = 0, & v(1) = \displaystyle\int_0^1 v(s)\,dK(s), \end{cases} \tag{BC}$$

where $n-1<\alpha\le n$, $m-1<\beta\le m$, $n, m\in\mathbb{N}$, $n, m\ge 3$, D_{0+}^{α} and D_{0+}^{β} denote the Riemann–Liouville derivatives of orders α and β, respectively, and the integrals from (BC) are Riemann–Stieltjes integrals.

Under sufficient conditions on functions f and g, we prove the existence and multiplicity of positive solutions of the above problem, by applying the fixed point index theory. By a positive solution of problem (S′)–(BC), we mean a pair of functions $(u,v)\in C([0,1])\times C([0,1])$ satisfying (S′) and (BC) with $u(t)\ge 0$, $v(t)\ge 0$ for all $t\in[0,1]$ and $\sup_{t\in[0,1]}u(t)>0$, $\sup_{t\in[0,1]}v(t)>0$.

We present the basic assumptions that we shall use in the sequel:

(I1) $H, K:[0,1]\to\mathbb{R}$ are nondecreasing functions, $\Delta_1=1-\int_0^1 s^{\alpha-1}\,\mathrm{d}H(s)>0$, and $\Delta_2=1-\int_0^1 s^{\beta-1}\,\mathrm{d}K(s)>0$.

(I2) The functions $f, g:[0,1]\times[0,\infty)\to[0,\infty)$ are continuous, and $f(t,0)=g(t,0)=0$ for all $t\in[0,1]$.

Under assumption (I1), we have all auxiliary results in Lemmas 4.1.4–4.1.9 from Section 4.1.2.

The pair of functions $(u,v)\in C([0,1])\times C([0,1])$ is a solution for our problem (S′)–(BC) if and only if $(u,v)\in C([0,1])\times C([0,1])$ is a solution for the nonlinear integral system

$$\begin{cases} u(t)=\displaystyle\int_0^1 G_1(t,s)f\left(s,\int_0^1 G_2(s,\tau)g(\tau,u(\tau))\,\mathrm{d}\tau\right)\mathrm{d}s, & t\in[0,1],\\ v(t)=\displaystyle\int_0^1 G_2(t,s)g(s,u(s))\,\mathrm{d}s, \quad t\in[0,1]. \end{cases}$$

We consider the Banach space $X=C([0,1])$ with the supremum norm $\|\cdot\|$, and define the cone $P'\subset X$ by $P'=\{u\in X,\ u(t)\ge 0,\ \forall t\in[0,1]\}$.

We also define the operator $\mathcal{A}:P'\to X$ by

$$(\mathcal{A}u)(t)=\int_0^1 G_1(t,s)f\left(s,\int_0^1 G_2(s,\tau)g(\tau,u(\tau))\,\mathrm{d}\tau\right)\mathrm{d}s,\quad t\in[0,1],$$

and the operators $\mathcal{B}:P'\to X$ and $\mathcal{C}:P'\to X$ by

$$(\mathcal{B}u)(t)=\int_0^1 G_1(t,s)u(s)\,\mathrm{d}s,\quad (\mathcal{C}u)(t)=\int_0^1 G_2(t,s)u(s)\,\mathrm{d}s,\quad t\in[0,1].$$

Under assumptions (I1) and (I2), using also Lemma 4.1.7, one can easily see that $\mathcal{A}$, $\mathcal{B}$, and $\mathcal{C}$ are completely continuous from P' to P'. Thus, the existence and multiplicity of positive solutions of system (S′)–(BC) are equivalent to the existence and multiplicity of fixed points of the operator $\mathcal{A}$.

Theorem 4.2.1. *Assume that (I1) and (I2) hold. If the functions f and g also satisfy the following conditions (I3) and (I4), then problem* (S′)–(BC) *has at least one positive solution* $(u(t),v(t))$, $t\in[0,1]$*:*

(I3) *There exist positive constants $p\in(0,1]$ and $c\in(0,1/2)$ such that*

$$(1)\ \tilde{f}_\infty^{\mathrm{i}}=\liminf_{u\to\infty}\inf_{t\in[c,1-c]}\frac{f(t,u)}{u^p}\in(0,\infty]\quad \textit{and}\quad (2)\ \tilde{g}_\infty^{\mathrm{i}}=\liminf_{u\to\infty}\inf_{t\in[c,1-c]}\frac{g(t,u)}{u^{1/p}}=\infty.$$

(I4) There exists a positive constant $q \in (0, \infty)$ such that

$$(1)\ \tilde{f}_0^s = \limsup_{u \to 0+} \sup_{t \in [0,1]} \frac{f(t,u)}{u^q} \in [0, \infty) \quad \textit{and} \quad (2)\ \tilde{g}_0^s = \limsup_{u \to 0+} \sup_{t \in [0,1]} \frac{g(t,u)}{u^{1/q}} = 0.$$

Proof. Because the proof of the theorem is similar to that of Theorem 2.2.1, we shall sketch only some parts of it. From (1) of assumption (I3), we deduce that there exist C_1, $C_2 > 0$ such that

$$f(t,u) \geq C_1 u^p - C_2, \quad \forall (t,u) \in [c, 1-c] \times [0, \infty). \tag{4.12}$$

Then for $u \in P'$, by using (4.12) and Lemmas 4.1.7 and 4.1.8, we obtain after some computations

$$(\mathcal{A}u)(t) \geq C_1 \int_c^{1-c} G_1(t,s) \left(\int_c^{1-c} (G_2(s,\tau))^p (g(\tau, u(\tau)))^p \, d\tau \right) ds - C_3, \quad \forall t \in [0,1], \tag{4.13}$$

where $C_3 = C_2 \int_c^{1-c} J_1(s)\, ds$.

For c given in (I3), we define the cone $P_0 = \{u \in P', \quad \inf_{t \in [c,1-c]} u(t) \geq \gamma \|u\|\}$, where $\gamma = \min\{\gamma_1, \gamma_2\}$, and γ_1 and γ_2 are defined in Section 4.1.2 (Lemma 4.1.6). From our assumptions and Lemma 4.1.9, for any $y \in P'$ we can easily show that $u = \mathcal{B}y \in P_0$ and $v = \mathcal{C}y \in P_0$—that is, $\mathcal{B}(P') \subset P_0$ and $\mathcal{C}(P') \subset P_0$.

We now consider the function $u_0(t) = \int_0^1 G_1(t,s)\, ds = (\mathcal{B}y_0)(t) \geq 0$, $t \in [0,1]$, with $y_0(t) = 1$ for all $t \in [0,1]$. We define the set

$$M = \{u \in P'; \quad \text{there exists} \quad \lambda \geq 0 \quad \text{such that} \quad u = \mathcal{A}u + \lambda u_0\}.$$

We shall show that $M \subset P_0$ and that M is a bounded subset of X. If $u \in M$, then there exists $\lambda \geq 0$ such that $u(t) = (\mathcal{A}u)(t) + \lambda u_0(t)$, $t \in [0,1]$. From the definition of u_0, we have

$$u(t) = (\mathcal{A}u)(t) + \lambda(\mathcal{B}y_0)(t) = \mathcal{B}(Fu(t)) + \lambda(\mathcal{B}y_0)(t) = \mathcal{B}(Fu(t) + \lambda y_0(t)) \in P_0,$$

where $F : P' \to P'$ is defined by $(Fu)(t) = f\left(t, \int_0^1 G_2(t,s) g(s, u(s))\, ds\right)$. Therefore, $M \subset P_0$, and from the definition of P_0, we get

$$\|u\| \leq \frac{1}{\gamma} \inf_{t \in [c,1-c]} u(t), \quad \forall u \in M. \tag{4.14}$$

From (2) of assumption (I3), we conclude that for $\varepsilon_0 = (2/(C_1 m_1 m_2 \gamma_1 \gamma_2^p))^{1/p} > 0$ there exists $C_4 > 0$ such that

$$(g(t,u))^p \geq \varepsilon_0^p u - C_4, \quad \forall (t,u) \in [c, 1-c] \times [0, \infty), \tag{4.15}$$

where $m_1 = \int_c^{1-c} J_1(\tau)\, d\tau > 0$, $m_2 = \int_c^{1-c} (J_2(\tau))^p\, d\tau > 0$.

For $u \in M$ and $t \in [c, 1-c]$, by using Lemma 4.1.8 and relations (4.13) and (4.15), we obtain

$$
\begin{aligned}
u(t) &= (\mathcal{A}u)(t) + \lambda u_0(t) \geq (\mathcal{A}u)(t) \\
&\geq C_1\gamma_1\gamma_2^p \int_c^{1-c} J_1(s)\left(\int_c^{1-c} (J_2(\tau))^p \left(\varepsilon_0^p u(\tau) - C_4\right) \mathrm{d}\tau\right) \mathrm{d}s - C_3 \\
&\geq C_1\varepsilon_0^p\gamma_1\gamma_2^p \left(\int_c^{1-c} J_1(s)\,\mathrm{d}s\right)\left(\int_c^{1-c} (J_2(\tau))^p u(\tau)\,\mathrm{d}\tau\right) - C_5 \\
&\geq C_1\varepsilon_0^p\gamma_1\gamma_2^p \left(\int_c^{1-c} J_1(s)\,\mathrm{d}s\right)\left(\int_c^{1-c} (J_2(\tau))^p\,\mathrm{d}\tau\right) \cdot \inf_{\tau\in[c,1-c]} u(\tau) - C_5 \\
&= 2 \inf_{\tau\in[c,1-c]} u(\tau) - C_5,
\end{aligned}
$$

where $C_5 = C_3 + C_1C_4m_1m_2\gamma_1\gamma_2^p > 0$.

Hence, $\inf_{t\in[c,1-c]} u(t) \geq 2\inf_{t\in[c,1-c]} u(t) - C_5$, and so

$$
\inf_{t\in[c,1-c]} u(t) \leq C_5, \quad \forall u \in M. \tag{4.16}
$$

Now from relations (4.14) and (4.16), we obtain $\|u\| \leq (\inf_{t\in[c,1-c]} u(t))/\gamma \leq C_5/\gamma$ for all $u \in M$—that is, M is a bounded subset of X.

Besides, there exists a sufficiently large $L > 0$ such that

$$
u \neq \mathcal{A}u + \lambda u_0, \quad \forall u \in \partial B_L \cap P, \quad \forall \lambda \geq 0.
$$

From Theorem 1.3.2, we deduce that the fixed point index of the operator $\mathcal{A}$ over $B_L \cap P'$ with respect to P' is

$$
i(\mathcal{A}, B_L \cap P', P') = 0. \tag{4.17}
$$

Next, from assumption (I4), we conclude that there exist $M_0 > 0$ and $\delta_1 \in (0,1)$ such that

$$
f(t,u) \leq M_0u^q, \ \forall (t,u) \in [0,1]\times[0,1]; \quad g(t,u) \leq \varepsilon_1 u^{1/q}, \ \forall (t,u) \in [0,1]\times[0,\delta_1], \tag{4.18}
$$

where $\varepsilon_1 = \min\{1/M_2, (1/(2M_0M_1M_2^q))^{1/q}\} > 0$, $M_1 = \int_0^1 J_1(s)\,\mathrm{d}s > 0$, $M_2 = \int_0^1 J_2(s)\,\mathrm{d}s > 0$. Hence, for any $u \in \bar{B}_{\delta_1} \cap P'$ and $t \in [0,1]$ we obtain

$$
\int_0^1 G_2(t,s)g(s,u(s))\,\mathrm{d}s \leq \varepsilon_1 \int_0^1 J_2(s)(u(s))^{1/q}\,\mathrm{d}s \leq \varepsilon_1 M_2 \|u\|^{1/q} \leq 1. \tag{4.19}
$$

Therefore, by (4.18) and (4.19), we deduce that for any $u \in \bar{B}_{\delta_1} \cap P'$ and $t \in [0,1]$

$$
\begin{aligned}
(\mathcal{A}u)(t) &\leq M_0 \int_0^1 G_1(t,s)\left(\int_0^1 G_2(s,\tau)g(\tau,u(\tau))\,\mathrm{d}\tau\right)^q \mathrm{d}s \\
&\leq M_0\varepsilon_1^q M_2^q \|u\| \int_0^1 J_1(s)\,\mathrm{d}s = M_0\varepsilon_1^q M_1 M_2^q \|u\| \leq \frac{1}{2}\|u\|.
\end{aligned}
$$

This implies that $\|\mathcal{A}u\| \le \|u\|/2$ for all $u \in \partial B_{\delta_1} \cap P'$. From Theorem 1.3.1, we conclude that the fixed point index of the operator $\mathcal{A}$ over $B_{\delta_1} \cap P'$ with respect to P' is

$$i(\mathcal{A}, B_{\delta_1} \cap P', P') = 1. \tag{4.20}$$

Combining (4.17) and (4.20), we obtain

$$i(\mathcal{A}, (B_L \setminus \bar{B}_{\delta_1}) \cap P', P') = i(\mathcal{A}, B_L \cap P', P') - i(\mathcal{A}, B_{\delta_1} \cap P', P') = -1.$$

We deduce that $\mathcal{A}$ has at least one fixed point $u_1 \in (B_L \setminus \bar{B}_{\delta_1}) \cap P'$—that is, $\delta_1 < \|u_1\| < L$.

Let $v_1(t) = \int_0^1 G_2(t,s)g(s,u_1(s))\,ds$. Then $(u_1, v_1) \in P' \times P'$ is a solution of (S′)–(BC). In addition, $\|v_1\| > 0$. If we suppose that $v_1(t) = 0$, for all $t \in [0,1]$, then by using (I2), we have $f(s, v_1(s)) = f(s,0) = 0$ for all $s \in [0,1]$. This implies $u_1(t) = \int_0^1 G_1(t,s)f(s,v_1(s))\,ds = 0$ for all $t \in [0,1]$, which contradicts $\|u_1\| > 0$. The proof of Theorem 4.2.1 is completed. □

Using arguments similar to those used in the proofs of Theorems 2.2.2 and 2.2.3, we also obtain the following results for our problem (S′)–(BC):

Theorem 4.2.2. *Assume that (I1) and (I2) hold. If the functions f and g also satisfy the following conditions (I5) and (I6), then problem* (S′)–(BC) *has at least one positive solution* $(u(t), v(t))$, $t \in [0,1]$:

(I5) *There exists a positive constant* $r \in (0,\infty)$ *such that*

$$(1)\ \tilde{f}^{\rm s}_\infty = \limsup_{u\to\infty} \sup_{t\in[0,1]} \frac{f(t,u)}{u^r} \in [0,\infty) \quad \textit{and} \quad (2)\ \tilde{g}^{\rm s}_\infty = \limsup_{u\to\infty} \sup_{t\in[0,1]} \frac{g(t,u)}{u^{1/r}} = 0.$$

(I6) *There exists* $c \in (0,1/2)$ *such that*

$$(1)\ \tilde{f}^{\rm i}_0 = \liminf_{u\to 0+} \inf_{t\in[c,1-c]} \frac{f(t,u)}{u} \in (0,\infty] \quad \textit{and} \quad (2)\ \tilde{g}^{\rm i}_0 = \liminf_{u\to 0+} \inf_{t\in[c,1-c]} \frac{g(t,u)}{u} = \infty.$$

Theorem 4.2.3. *Assume that (I1)–(I3) and (I6) hold. If the functions f and g also satisfy the following condition (I7), then problem* (S′)–(BC) *has at least two positive solutions* $(u_1(t), v_1(t))$, $(u_2(t), v_2(t))$, $t \in [0,1]$:

(I7) *For each* $t \in [0,1]$, $f(t,u)$ *and* $g(t,u)$ *are nondecreasing with respect to u, and there exists a constant* $N > 0$ *such that*

$$f\left(t, m_0 \int_0^1 g(s,N)\,ds\right) < \frac{N}{m_0}, \quad \forall t \in [0,1],$$

where $m_0 = \max\{K_1, K_2\}$, $K_1 = \max_{s\in[0,1]} J_1(s)$ *and* $K_2 = \max_{s\in[0,1]} J_2(s)$.

We present now an example which illustrates our results above.

Example 4.2.1. Let $\alpha = 5/2$, $(n = 3)$, $\beta = 7/3$, $(m = 3)$, $H(t) = \begin{cases} 0, & t \in [0,1/4), \\ 2, & t \in [1/4,3/4), \\ 3, & t \in [3/4,1], \end{cases}$

and $K(t) = t^3$ for all $t \in [0,1]$. Then, $\int_0^1 u(s)\,dH(s) = 2u\left(\frac{1}{4}\right) + u\left(\frac{3}{4}\right)$ and $\int_0^1 v(s)\,dK(s) = 3\int_0^1 s^2 v(s)\,ds$.

We consider the system of fractional differential equations

$$\begin{cases} D_{0+}^{5/2}u(t) + f(t, v(t)) = 0, & t \in (0,1), \\ D_{0+}^{7/3}v(t) + g(t, u(t)) = 0, & t \in (0,1), \end{cases} \tag{S$_2$}$$

with the boundary conditions

$$\begin{cases} u(0) = u'(0) = 0, & u(1) = 2u\left(\frac{1}{4}\right) + u\left(\frac{3}{4}\right), \\ v(0) = v'(0) = 0, & v(1) = 3\int_0^1 s^2 v(s)\,ds. \end{cases} \tag{BC$_2$}$$

Here we consider $f(t,u) = a(u^{\alpha_0} + u^{\beta_0})$ and $g(t,u) = b(u^{\gamma_0} + u^{\delta_0})$ for all $t \in [0,1]$, $u \in [0,\infty)$, where $\alpha_0 > 1$, $0 < \beta_0 < 1$, $\gamma_0 > 2$, $0 < \delta_0 < 1$, and $a, b > 0$.

Then, we obtain $\Delta_1 = 1 - \int_0^1 s^{\alpha-1}\,dH(s) = 1 - 2\left(\frac{1}{4}\right)^{3/2} - \left(\frac{3}{4}\right)^{3/2} = \frac{3(2-\sqrt{3})}{8} \approx 0.1005 > 0$ and $\Delta_2 = 1 - \int_0^1 s^{\beta-1}\,dK(s) = 1 - 3\int_0^1 s^{10/3}\,ds = \frac{4}{13} \approx 0.3077 > 0$. We also deduce

$$g_1(t,s) = \frac{4}{3\sqrt{\pi}} \begin{cases} t^{3/2}(1-s)^{3/2} - (t-s)^{3/2}, & 0 \le s \le t \le 1, \\ t^{3/2}(1-s)^{3/2}, & 0 \le t \le s \le 1, \end{cases}$$

$$g_2(t,s) = \frac{1}{\Gamma(7/3)} \begin{cases} t^{4/3}(1-s)^{4/3} - (t-s)^{4/3}, & 0 \le s \le t \le 1, \\ t^{4/3}(1-s)^{4/3}, & 0 \le t \le s \le 1, \end{cases}$$

$\theta_1(s) = \frac{1}{3-3s+s^2}$, and $\theta_2(s) = \frac{1}{4-6s+4s^2-s^3}$, for all $s \in [0,1]$.

For the functions J_1 and J_2, we obtain

$$\begin{aligned} J_1(s) &= g_1(\theta_1(s), s) + \frac{1}{\Delta_1}\int_0^1 g_1(\tau, s)\,dH(\tau) \\ &= g_1\left(\frac{1}{3-3s+s^2}, s\right) + \frac{1}{\Delta_1}\left[2g_1\left(\frac{1}{4}, s\right) + g_1\left(\frac{3}{4}, s\right)\right] \\ &= \begin{cases} \frac{4}{3\sqrt{\pi}}\left\{\frac{s(1-s)^{3/2}}{(3-3s+s^2)^{1/2}} + \frac{1}{3(2-\sqrt{3})}\left[2\left((1-s)^{3/2} - (1-4s)^{3/2}\right)\right.\right. \\ \qquad \left.\left. + \left(3\sqrt{3}(1-s)^{3/2} - (3-4s)^{3/2}\right)\right]\right\}, & 0 \le s < \frac{1}{4}, \\ \frac{4}{3\sqrt{\pi}}\left\{\frac{s(1-s)^{3/2}}{(3-3s+s^2)^{1/2}} + \frac{1}{3(2-\sqrt{3})}\left[2(1-s)^{3/2}\right.\right. \\ \qquad \left.\left. + \left(3\sqrt{3}(1-s)^{3/2} - (3-4s)^{3/2}\right)\right]\right\}, & \frac{1}{4} \le s < \frac{3}{4}, \\ \frac{4}{3\sqrt{\pi}}\left[\frac{s(1-s)^{3/2}}{(3-3s+s^2)^{1/2}} + \frac{(2+3\sqrt{3})(1-s)^{3/2}}{3(2-\sqrt{3})}\right], & \frac{3}{4} \le s \le 1, \end{cases} \end{aligned}$$

and

$$\begin{aligned} J_2(s) &= g_2(\theta_2(s), s) + \frac{1}{\Delta_2}\int_0^1 g_2(\tau, s)\,dK(\tau) = g_2\left(\frac{1}{4-6s+4s^2-s^3}, s\right) \\ &\quad + \frac{3}{\Delta_2}\int_0^1 \tau^2 g_2(\tau, s)\,d\tau \end{aligned}$$

$$= \frac{1}{\Gamma(7/3)} \left[\frac{s(1-s)^{4/3}}{(4-6s+4s^2-s^3)^{1/3}} + \frac{9}{4}(1-s)^{4/3} - \frac{117}{20}s(1-s)^{10/3} - \frac{117}{28}s^2(1-s)^{7/3} - \frac{9}{4}(1-s)^{13/3} \right], \quad s \in [0,1].$$

We have $K_1 = \max_{s\in[0,1]} J_1(s) \approx 1.8076209$ and $K_2 = \max_{s\in[0,1]} J_2(s) \approx 0.37377618$. Then $m_0 = \max\{K_1, K_2\} = K_1$. The functions $f(t,u)$ and $g(t,u)$ are nondecreasing with respect to u for any $t \in [0,1]$, and for $p = 1/2$ and $c \in (0,1/2)$ assumptions (I3) and (I6) are satisfied; we obtain $\tilde{f}_\infty^{\rm i} = \infty$, $\tilde{g}_\infty^{\rm i} = \infty$, $\tilde{f}_0^{\rm i} = \infty$, and $\tilde{g}_0^{\rm i} = \infty$. We take $N = 1$, and then $\int_0^1 g(s,1)\,ds = 2b$ and $f(t, 2bm_0) = a[(2bm_0)^{\alpha_0} + (2bm_0)^{\beta_0}]$. If $a[(2bm_0)^{\alpha_0} + (2bm_0)^{\beta_0}] < \frac{1}{m_0} \Leftrightarrow a\left[m_0^{\alpha_0+1}(2b)^{\alpha_0} + m_0^{\beta_0+1}(2b)^{\beta_0}\right] < 1$, then assumption (I7) is satisfied. For example, if $\alpha_0 = 3/2$, $\beta_0 = 1/3$, $b = 1/2$, and $a < \frac{1}{m_0^{5/2} + m_0^{4/3}}$ (e.g., $a \le 0.1516$), then the above inequality is satisfied. By Theorem 4.2.3, we deduce that problem (S_2)–(BC_2) has at least two positive solutions.

4.2.2 *Singular nonlinearities*

We consider the system of nonlinear ordinary fractional differential equations

$$\begin{cases} D_{0+}^{\alpha} u(t) + f(t, v(t)) = 0, & t \in (0,1), \\ D_{0+}^{\beta} v(t) + g(t, u(t)) = 0, & t \in (0,1), \end{cases} \tag{S'}$$

with the uncoupled integral boundary conditions

$$\begin{cases} u(0) = u'(0) = \cdots = u^{(n-2)}(0) = 0, & u(1) = \displaystyle\int_0^1 u(s)\,dH(s), \\ v(0) = v'(0) = \cdots = v^{(m-2)}(0) = 0, & v(1) = \displaystyle\int_0^1 v(s)\,dK(s), \end{cases} \tag{BC}$$

where $n-1 < \alpha \le n$, $m-1 < \beta \le m$, $n, m \in \mathbb{N}$, $n, m \ge 3$, D_{0+}^{α} and D_{0+}^{β} denote the Riemann–Liouville derivatives of orders α and β, respectively, and the integrals from (BC) are Riemann–Stieltjes integrals.

We present some weaker assumptions on f and g, which do not possess any sublinear or superlinear growth conditions and may be singular at $t = 0$ and/or $t = 1$, such that positive solutions for problem (S′)–(BC) exist. By a positive solution of problem (S′)–(BC), we mean a pair of functions $(u,v) \in C([0,1]) \times C([0,1])$ satisfying (S′) and (BC) with $u(t) \ge 0$, $v(t) \ge 0$ for all $t \in [0,1]$ and $\sup_{t\in[0,1]} u(t) > 0$, $\sup_{t\in[0,1]} v(t) > 0$.

We present the basic assumptions that we shall use in the sequel:

(L1) $H, K : [0,1] \to \mathbb{R}$ are nondecreasing functions, $\Delta_1 = 1 - \int_0^1 s^{\alpha-1}\,dH(s) > 0$, and $\Delta_2 = 1 - \int_0^1 s^{\beta-1}\,dK(s) > 0$.

(L2) The functions $f, g \in C((0,1) \times \mathbb{R}_+, \mathbb{R}_+)$, and there exist $p_i \in C((0,1), \mathbb{R}_+)$ and $q_i \in C(\mathbb{R}_+, \mathbb{R}_+)$, $i = 1, 2$, with $0 < \int_0^1 p_i(t)\,dt < \infty$, $i = 1, 2$, $q_1(0) = 0$, $q_2(0) = 0$ such that

$$f(t,x) \le p_1(t)q_1(x), \quad g(t,x) \le p_2(t)q_2(x), \quad \forall t \in (0,1), \quad x \in [0,\infty).$$

Under assumption (L1), we have all auxiliary results in Lemmas 4.1.4–4.1.9 from Section 4.1.2.

The pair of functions $(u, v) \in C([0, 1]) \times C([0, 1])$ is a solution for our problem (S′)–(BC) if and only if $(u, v) \in C([0, 1]) \times C([0, 1])$ is a solution for the nonlinear integral system

$$\begin{cases} u(t) = \displaystyle\int_0^1 G_1(t,s)f\left(s, \int_0^1 G_2(s,\tau)g(\tau,u(\tau))\,d\tau\right) ds, & t \in [0,1], \\ v(t) = \displaystyle\int_0^1 G_2(t,s)g(s,u(s))\,ds, \quad t \in [0,1]. \end{cases}$$

We consider again the Banach space $X = C([0, 1])$ with the supremum norm $\|\cdot\|$ and define the cone $P' \subset X$ by $P' = \{u \in X, \quad u(t) \ge 0, \quad \forall t \in [0, 1]\}$.

We also define the operator $\mathcal{D} : P' \to X$ by

$$(\mathcal{D}u)(t) = \int_0^1 G_1(t,s)f\left(s, \int_0^1 G_2(s,\tau)g(\tau,u(\tau))\,d\tau\right) ds, \quad t \in [0,1].$$

Lemma 4.2.1. *Assume that (L1) and (L2) hold. Then $\mathcal{D} : P' \to P'$ is completely continuous.*

Proof. We denote $\tilde{\alpha} = \int_0^1 J_1(s)p_1(s)\,ds$ and $\tilde{\beta} = \int_0^1 J_2(s)p_2(s)\,ds$. Using (L2), we deduce that $0 < \tilde{\alpha} < \infty$ and $0 < \tilde{\beta} < \infty$. By Lemma 4.1.7 and the corresponding lemma for G_2, we get that $\mathcal{D}$ maps P' into P'.

We shall prove that $\mathcal{D}$ maps bounded sets into relatively compact sets. Suppose $E \subset P'$ is an arbitrary bounded set. Then there exists $\overline{M}_1 > 0$ such that $\|u\| \le \overline{M}_1$ for all $u \in E$. By using (L2) and Lemma 4.1.8, we obtain $\|\mathcal{D}u\| \le \tilde{\alpha}\overline{M}_3$ for all $u \in E$, where $\overline{M}_3 = \sup_{x\in[0,\tilde{\beta}\overline{M}_2]} q_1(x)$ and $\overline{M}_2 = \sup_{x\in[0,\overline{M}_1]} q_2(x)$. In what follows, we shall prove that $\mathcal{D}(E)$ is equicontinuous. By using Lemma 4.1.5, we have

$$\begin{aligned}
(\mathcal{D}u)(t) &= \int_0^1 G_1(t,s)f\left(s, \int_0^1 G_2(s,\tau)g(\tau,u(\tau))\,d\tau\right) ds \\
&= \int_0^1 \left[g_1(t,s) + \frac{t^{\alpha-1}}{\Delta_1}\int_0^1 g_1(\tau,s)\,dH(\tau)\right] f\left(s, \int_0^1 G_2(s,\tau)g(\tau,u(\tau))\,d\tau\right) ds \\
&= \frac{1}{\Gamma(\alpha)}\int_0^t [t^{\alpha-1}(1-s)^{\alpha-1} - (t-s)^{\alpha-1}] f\left(s, \int_0^1 G_2(s,\tau)g(\tau,u(\tau))\,d\tau\right) ds \\
&\quad + \frac{1}{\Gamma(\alpha)}\int_t^1 t^{\alpha-1}(1-s)^{\alpha-1} f\left(s, \int_0^1 G_2(s,\tau)g(\tau,u(\tau))\,d\tau\right) ds \\
&\quad + \frac{t^{\alpha-1}}{\Delta_1}\int_0^1 \left(\int_0^1 g_1(\tau,s)\,dH(\tau)\right) f\left(s, \int_0^1 G_2(s,\tau)g(\tau,u(\tau))\,d\tau\right) ds, \\
&\quad \forall t \in [0,1].
\end{aligned}$$

Therefore, for any $t \in (0,1)$, we obtain

$$
\begin{aligned}
(\mathcal{D}u)'(t) &= \frac{1}{\Gamma(\alpha)} \int_0^t [(\alpha-1)t^{\alpha-2}(1-s)^{\alpha-1} - (\alpha-1)(t-s)^{\alpha-2}] \\
&\quad \times f\left(s, \int_0^1 G_2(s,\tau) g(\tau, u(\tau))\,\mathrm{d}\tau\right) \mathrm{d}s \\
&\quad + \frac{1}{\Gamma(\alpha)} \int_t^1 (\alpha-1)t^{\alpha-2}(1-s)^{\alpha-1} f\left(s, \int_0^1 G_2(s,\tau) g(\tau, u(\tau))\,\mathrm{d}\tau\right) \mathrm{d}s \\
&\quad + \frac{(\alpha-1)t^{\alpha-2}}{\Delta_1} \int_0^1 \left(\int_0^1 g_1(\tau,s)\,\mathrm{d}H(\tau)\right) f\left(s, \int_0^1 G_2(s,\tau) g(\tau, u(\tau))\,\mathrm{d}\tau\right) \mathrm{d}s.
\end{aligned}
$$

So, for any $t \in (0,1)$, we deduce

$$
\begin{aligned}
|(\mathcal{D}u)'(t)| &\le \frac{1}{\Gamma(\alpha-1)} \int_0^t [t^{\alpha-2}(1-s)^{\alpha-1} + (t-s)^{\alpha-2}] p_1(s) \\
&\quad \times q_1\left(\int_0^1 G_2(s,\tau) g(\tau, u(\tau))\,\mathrm{d}\tau\right) \mathrm{d}s + \frac{1}{\Gamma(\alpha-1)} \int_t^1 t^{\alpha-2}(1-s)^{\alpha-1} p_1(s) \\
&\quad \times q_1\left(\int_0^1 G_2(s,\tau) g(\tau, u(\tau))\,\mathrm{d}\tau\right) \mathrm{d}s + \frac{(\alpha-1)t^{\alpha-2}}{\Delta_1} \\
&\quad \times \int_0^1 \left(\int_0^1 g_1(\tau,s)\,\mathrm{d}H(\tau)\right) p_1(s) q_1\left(\int_0^1 G_2(s,\tau) g(\tau, u(\tau))\,\mathrm{d}\tau\right) \mathrm{d}s \\
&\le \overline{M}_3 \left(\frac{1}{\Gamma(\alpha-1)} \int_0^t [t^{\alpha-2}(1-s)^{\alpha-1} + (t-s)^{\alpha-2}] p_1(s)\,\mathrm{d}s \right. \\
&\quad + \frac{1}{\Gamma(\alpha-1)} \int_t^1 t^{\alpha-2}(1-s)^{\alpha-1} p_1(s)\,\mathrm{d}s \\
&\quad \left. + \frac{(\alpha-1)t^{\alpha-2}}{\Delta_1} \int_0^1 \left(\int_0^1 g_1(\tau,s)\,\mathrm{d}H(\tau)\right) p_1(s)\,\mathrm{d}s \right). \qquad (4.21)
\end{aligned}
$$

We denote

$$
\begin{aligned}
h(t) &= \frac{1}{\Gamma(\alpha-1)} \int_0^t [t^{\alpha-2}(1-s)^{\alpha-1} + (t-s)^{\alpha-2}] p_1(s)\,\mathrm{d}s \\
&\quad + \frac{1}{\Gamma(\alpha-1)} \int_t^1 t^{\alpha-2}(1-s)^{\alpha-1} p_1(s)\,\mathrm{d}s, \quad t \in (0,1), \\
\mu(t) &= h(t) + \frac{(\alpha-1)t^{\alpha-2}}{\Delta_1} \int_0^1 \left(\int_0^1 g_1(\tau,s)\,\mathrm{d}H(\tau)\right) p_1(s)\,\mathrm{d}s, \quad t \in (0,1).
\end{aligned}
$$

For the integral of the function h, by exchanging the order of integration, we obtain

$$
\begin{aligned}
\int_0^1 h(t)\,\mathrm{d}t &= \frac{1}{\Gamma(\alpha-1)} \int_0^1 \left(\int_0^t [t^{\alpha-2}(1-s)^{\alpha-1} + (t-s)^{\alpha-2}] p_1(s)\,\mathrm{d}s\right) \mathrm{d}t \\
&\quad + \frac{1}{\Gamma(\alpha-1)} \int_0^1 \left(\int_t^1 t^{\alpha-2}(1-s)^{\alpha-1} p_1(s)\,\mathrm{d}s\right) \mathrm{d}t
\end{aligned}
$$

$$= \frac{1}{\Gamma(\alpha-1)}\int_0^1\left(\int_s^1 [t^{\alpha-2}(1-s)^{\alpha-1}+(t-s)^{\alpha-2}]p_1(s)\,dt\right)ds$$
$$+\frac{1}{\Gamma(\alpha-1)}\int_0^1\left(\int_0^s t^{\alpha-2}(1-s)^{\alpha-1}p_1(s)\,dt\right)ds$$
$$= \frac{2}{\Gamma(\alpha)}\int_0^1 (1-s)^{\alpha-1}p_1(s)\,ds < \infty.$$

For the integral of the function μ, we have

$$\int_0^1 \mu(t)\,dt = \int_0^1 h(t)\,dt + \frac{\alpha-1}{\Delta_1}\left(\int_0^1\left(\int_0^1 g_1(\tau,s)\,dH(\tau)\right)p_1(s)\,ds\right)\left(\int_0^1 t^{\alpha-2}\,dt\right)$$
$$\le \frac{2}{\Gamma(\alpha)}\int_0^1 (1-s)^{\alpha-1}p_1(s)\,ds + \frac{H(1)-H(0)}{\Delta_1}\int_0^1 g_1(\theta_1(s),s)p_1(s)\,ds$$
$$\le \frac{1}{\Gamma(\alpha)}\left(2+\frac{H(1)-H(0)}{\Delta_1}\right)\int_0^1 (1-s)^{\alpha-1}p_1(s)\,ds < \infty. \tag{4.22}$$

We deduce that $\mu \in L^1(0,1)$. Thus, for any given $t_1, t_2 \in [0,1]$ with $t_1 \le t_2$ and $u \in E$, by (4.21), we conclude

$$|(\mathcal{D}u)(t_1)-(\mathcal{D}u)(t_2)| = \left|\int_{t_1}^{t_2} (\mathcal{D}u)'(t)\,dt\right| \le \overline{M}_3\int_{t_1}^{t_2}\mu(t)\,dt. \tag{4.23}$$

From (4.22), (4.23), and the absolute continuity of the integral function, we find that $\mathcal{D}(E)$ is equicontinuous. By the Ascoli-Arzelà theorem, we deduce that $\mathcal{D}(E)$ is relatively compact; therefore, $\mathcal{D}$ is a compact operator. Besides, we can easily show that $\mathcal{D}$ is continuous on P'; hence, $\mathcal{D}: P' \to P'$ is completely continuous. □

Theorem 4.2.4. *Assume that (L1) and (L2) hold. If the functions f and g also satisfy the following conditions (L3) and (L4), then problem* (S′)–(BC) *has at least one positive solution $(u(t), v(t))$, $t \in [0,1]$:*

(L3) *There exist $\alpha_1, \alpha_2 \in (0,\infty)$ with $\alpha_1\alpha_2 \le 1$ such that*

(1) $q_{1\infty}^s = \limsup_{x\to\infty} \frac{q_1(x)}{x^{\alpha_1}} \in [0,\infty)$ *and* (2) $q_{2\infty}^s = \limsup_{x\to\infty} \frac{q_2(x)}{x^{\alpha_2}} = 0.$

(L4) *There exist $\beta_1, \beta_2 \in (0,\infty)$ with $\beta_1\beta_2 \le 1$ and $c \in \left(0,\frac{1}{2}\right)$ such that*

(1) $\hat{f}_0^i = \liminf_{x\to0^+} \inf_{t\in[c,1-c]} \frac{f(t,x)}{x^{\beta_1}} \in (0,\infty]$ *and* (2) $\hat{g}_0^i = \liminf_{x\to0^+} \inf_{t\in[c,1-c]} \frac{g(t,x)}{x^{\beta_2}} = \infty.$

Proof. Because the proof of this theorem is similar to that of Theorem 1.4.2, we shall sketch only some parts of it. For c given in (L4), we consider the cone $P_0 = \{u \in X,\ u(t) \ge 0,\ \forall t \in [0,1],\ \inf_{t\in[c,1-c]} u(t) \ge \gamma\|u\|\}$, where $\gamma = \min\{\gamma_1,\gamma_2\}$. Under assumptions (L1) and (L2), we obtain $\mathcal{D}(P') \subset P_0$. By (L3), we deduce that there exist $C_6, C_7, C_8 > 0$ and $\varepsilon_2 \in (0, (2^{\alpha_1}C_6\tilde{\alpha}\tilde{\beta}^{\alpha_1})^{-1/\alpha_1})$ such that

$$q_1(x) \le C_6x^{\alpha_1}+C_7,\quad q_2(x) \le \varepsilon_2x^{\alpha_2}+C_8,\quad \forall x \in [0,\infty). \tag{4.24}$$

By using (4.24) and (L2), for any $u \in P_0$, we conclude

$$
\begin{aligned}
(\mathcal{D}u)(t) &\le \int_0^1 G_1(t,s)p_1(s)q_1\left(\int_0^1 G_2(s,\tau)g(\tau,u(\tau))\,d\tau\right) ds \\
&\le C_6\int_0^1 G_1(t,s)p_1(s)\left(\int_0^1 G_2(s,\tau)g(\tau,u(\tau))\,d\tau\right)^{\alpha_1} ds + C_7\int_0^1 J_1(s)p_1(s)\,ds \\
&\le C_6\int_0^1 J_1(s)p_1(s)\left[\int_0^1 G_2(s,\tau)p_2(\tau)\left(\varepsilon_2(u(\tau))^{\alpha_2}+C_8\right)d\tau\right]^{\alpha_1} ds + \tilde{\alpha}C_7 \\
&\le C_6\left(\int_0^1 J_1(s)p_1(s)\,ds\right)\left(\int_0^1 J_2(\tau)p_2(\tau)\,d\tau\right)^{\alpha_1}\left(\varepsilon_2\|u\|^{\alpha_2}+C_8\right)^{\alpha_1} + \tilde{\alpha}C_7 \\
&\le C_6 2^{\alpha_1}\varepsilon_2^{\alpha_1}\tilde{\alpha}\tilde{\beta}^{\alpha_1}\|u\|^{\alpha_1\alpha_2} + C_6 2^{\alpha_1}\tilde{\alpha}\tilde{\beta}^{\alpha_1}C_8^{\alpha_1} + \tilde{\alpha}C_7, \quad \forall t \in [0,1].
\end{aligned}
$$

By the definition of ε_2, we can choose sufficiently large $R_1 > 0$ such that

$$
\|\mathcal{D}u\| \le \|u\|, \quad \forall u \in \partial B_{R_1} \cap P_0. \tag{4.25}
$$

From (L4), we deduce that there exist positive constants $C_9 > 0$, $x_1 > 0$ and $\varepsilon_3 \ge (1/(C_9\gamma_1\gamma_2^{\beta_1}\gamma^{\beta_1\beta_2}\overline{\theta}_1\overline{\theta}_2^{\beta_1}))^{1/\beta_1}$ such that

$$
f(t,x) \ge C_9 x^{\beta_1}, \quad g(t,x) \ge \varepsilon_3 x^{\beta_2}, \quad \forall (t,x) \in [c,1-c]\times[0,x_1], \tag{4.26}
$$

where $\overline{\theta}_1 = \int_c^{1-c} J_1(s)\,ds$ and $\overline{\theta}_2 = \int_c^{1-c} J_2(s)\,ds$. From the assumption $q_2(0) = 0$ and the continuity of q_2, we conclude that there exists sufficiently small $\varepsilon_4 \in (0,\min\{x_1,1\})$ such that $q_2(x) \le \tilde{\beta}^{-1}x_1$ for all $x \in [0,\varepsilon_4]$, where $\tilde{\beta} = \int_0^1 J_2(s)p_2(s)\,ds$. Therefore, for any $u \in \partial B_{\varepsilon_4} \cap P_0$ and $s \in [0,1]$ we have

$$
\int_0^1 G_2(s,\tau)g(\tau,u(\tau))\,d\tau \le \int_0^1 J_2(\tau)p_2(\tau)q_2(u(\tau))\,d\tau \le x_1. \tag{4.27}
$$

By (4.26), (4.27), and Lemmas 4.1.8 and 4.1.9, for any $u \in \partial B_{\varepsilon_4} \cap P_0$ and $t \in [c,1-c]$ we obtain

$$
\begin{aligned}
(\mathcal{D}u)(t) &\ge C_9\int_c^{1-c} G_1(t,s)\left(\int_c^{1-c} G_2(s,\tau)g(\tau,u(\tau))\,d\tau\right)^{\beta_1} ds \\
&\ge C_9\int_c^{1-c} G_1(t,s)\left(\varepsilon_3\int_c^{1-c} G_2(s,\tau)(u(\tau))^{\beta_2}\,d\tau\right)^{\beta_1} ds \\
&\ge C_9\gamma_1\int_c^{1-c} J_1(s)\left((\varepsilon_3\gamma_2)^{\beta_1}\left(\int_c^{1-c} J_2(\tau)(u(\tau))^{\beta_2}\,d\tau\right)^{\beta_1}\right) ds \\
&\ge C_9\gamma_1\gamma_2^{\beta_1}\varepsilon_3^{\beta_1}\gamma^{\beta_1\beta_2}\overline{\theta}_1\overline{\theta}_2^{\beta_1}\|u\|^{\beta_1\beta_2} \ge \|u\|^{\beta_1\beta_2} \ge \|u\|.
\end{aligned}
$$

Therefore,

$$
\|\mathcal{D}u\| \ge \|u\|, \quad \forall u \in \partial B_{\varepsilon_4} \cap P_0. \tag{4.28}
$$

By (4.25), (4.28), Lemma 4.2.1, and Theorem 1.1.1, we deduce that $\mathcal{D}$ has at least one fixed point $u_2 \in (\bar{B}_{R_1} \setminus B_{\varepsilon_4}) \cap P_0$. Then our problem (S′)–(BC) has at least one positive solution $(u_2, v_2) \in P_0 \times P_0$, where $v_2(t) = \int_0^1 G_2(t,s)g(s,u_2(s))\,ds$. The proof of Theorem 4.2.4 is completed. □

Using arguments similar to those used in the proof of Theorem 1.4.1, we also obtain the following result for our problem (S′)–(BC):

Theorem 4.2.5. *Assume that (L1) and (L2) hold. If the functions f and g also satisfy the following conditions (L5) and (L6), then problem* (S′)–(BC) *has at least one positive solution $(u(t), v(t))$, $t \in [0,1]$:*

(L5) *There exist $r_1, r_2 \in (0,\infty)$ with $r_1 r_2 \geq 1$ such that*

$$(1)\quad q_{10}^s = \limsup_{x\to 0^+} \frac{q_1(x)}{x^{r_1}} \in [0,\infty) \quad \textit{and} \quad (2)\quad q_{20}^s = \limsup_{x\to 0^+} \frac{q_2(x)}{x^{r_2}} = 0.$$

(L6) *There exist $l_1, l_2 \in (0,\infty)$ with $l_1 l_2 \geq 1$ and $c \in \left(0,\frac{1}{2}\right)$ such that*

$$(1)\ \hat{f}_\infty^i = \liminf_{x\to\infty} \inf_{t\in[c,1-c]} \frac{f(t,x)}{x^{l_1}} \in (0,\infty] \quad \textit{and} \quad (2)\ \hat{g}_\infty^i = \liminf_{x\to\infty} \inf_{t\in[c,1-c]} \frac{g(t,x)}{x^{l_2}} = \infty.$$

We now present an example for the above results.

Example 4.2.2. We consider problem (S_2)–(BC_2) from Example 4.2.1, where $f(t,x) = \frac{x^a}{t^{\zeta_1}(1-t)^{\rho_1}}$ and $g(t,x) = \frac{x^b}{t^{\zeta_2}(1-t)^{\rho_2}}$, for all $t \in (0,1)$, $x \in [0,\infty)$, with $a,\ b > 1$ and $\zeta_1, \rho_1, \zeta_2, \rho_2 \in (0,1)$. Here $f(t,x) = p_1(t)q_1(x)$ and $g(t,x) = p_2(t)q_2(x)$, where $p_1(t) = \frac{1}{t^{\zeta_1}(1-t)^{\rho_1}}$ and $p_2(t) = \frac{1}{t^{\zeta_2}(1-t)^{\rho_2}}$ for all $t \in (0,1)$, and $q_1(x) = x^a$ and $q_2(x) = x^b$ for all $x \in [0,\infty)$. We have $0 < \int_0^1 p_1(s)\,ds < \infty$, $0 < \int_0^1 p_2(s)\,ds < \infty$.

In (L5), for $r_1 < a,\ r_2 < b$, and $r_1 r_2 \geq 1$, we obtain $q_{10}^s = 0$ and $q_{20}^s = 0$. In (L6), for $l_1 < a,\ l_2 < b,\ l_1 l_2 \geq 1$, and $c \in \left(0,\frac{1}{2}\right)$, we have $\hat{f}_\infty^i = \infty$ and $\hat{g}_\infty^i = \infty$.

For example, if $a = 3/2$, $b = 2$, $r_1 = 1$, $r_2 = 3/2$, $l_1 = 1$, and $l_2 = 3/2$, the above conditions are satisfied. Then, by Theorem 4.2.5, we deduce that problem (S_2)–(BC_2) has at least one positive solution.

Remark 4.2.1. The results presented in this section were published in Henderson and Luca (2014a).

4.3 Uncoupled boundary conditions with additional positive constants

In this section, we shall investigate the existence and nonexistence of positive solutions for a system of nonlinear ordinary fractional differential equations with uncoupled integral boundary conditions in which some positive constants appear.

4.3.1 Presentation of the problem

We consider the system of nonlinear ordinary fractional differential equations

$$\begin{cases} D_{0+}^{\alpha} u(t) + a(t)f(v(t)) = 0, & t \in (0,1), \\ D_{0+}^{\beta} v(t) + b(t)g(u(t)) = 0, & t \in (0,1), \end{cases} \tag{S^0}$$

with the uncoupled integral boundary conditions

$$\begin{cases} u(0) = u'(0) = \cdots = u^{(n-2)}(0) = 0, & u(1) = \displaystyle\int_0^1 u(s)\,\mathrm{d}H(s) + a_0, \\ v(0) = v'(0) = \cdots = v^{(m-2)}(0) = 0, & v(1) = \displaystyle\int_0^1 v(s)\,\mathrm{d}K(s) + b_0, \end{cases} \tag{BC^0}$$

where $n - 1 < \alpha \le n$, $m - 1 < \beta \le m$, $n, m \in \mathbb{N}$, $n, m \ge 3$, D_{0+}^{α} and D_{0+}^{β} denote the Riemann–Liouville derivatives of orders α and β, respectively, the integrals from (BC^0) are Riemann–Stieltjes integrals, and a_0 and b_0 are positive constants.

By using the Schauder fixed point theorem (Theorem 1.6.1), we shall prove the existence of positive solutions of problem (S^0)–(BC^0). By a positive solution of (S^0)–(BC^0), we mean a pair of functions $(u, v) \in C([0, 1]; \mathbb{R}_+) \times C([0, 1]; \mathbb{R}_+)$ satisfying (S^0) and (BC^0) with $u(t) > 0$, $v(t) > 0$ for all $t \in (0, 1]$. We shall also give sufficient conditions for the nonexistence of positive solutions for this problem.

We present the assumptions that we shall use in the sequel:

(J1) $H, K : [0, 1] \to \mathbb{R}$ are nondecreasing functions, $\Delta_1 = 1 - \int_0^1 s^{\alpha-1}\,\mathrm{d}H(s) > 0$, and $\Delta_2 = 1 - \int_0^1 s^{\beta-1}\,\mathrm{d}K(s) > 0$.

(J2) The functions $a, b : [0, 1] \to [0, \infty)$ are continuous, and there exist $t_1, t_2 \in (0, 1)$ such that $a(t_1) > 0$, $b(t_2) > 0$.

(J3) $f, g : [0, \infty) \to [0, \infty)$ are continuous functions, and there exists $c_0 > 0$ such that $f(u) < \frac{c_0}{L}$, $g(u) < \frac{c_0}{L}$ for all $u \in [0, c_0]$, where $L = \max\{\int_0^1 a(s)J_1(s)\,\mathrm{d}s, \int_0^1 b(s)J_2(s)\,\mathrm{d}s\}$, and J_1 and J_2 are defined in Section 4.1.2.

(J4) $f, g : [0, \infty) \to [0, \infty)$ are continuous functions and satisfy the conditions $\lim_{u\to\infty} \frac{f(u)}{u} = \infty$, $\lim_{u\to\infty} \frac{g(u)}{u} = \infty$.

Under assumption (J1), we have all auxiliary results in Lemmas 4.1.4–4.1.9 from Section 4.1.2. Besides, by (J2) we deduce that $\int_0^1 a(s)J_1(s)\,\mathrm{d}s > 0$ and $\int_0^1 b(s)J_2(s)\,\mathrm{d}s > 0$—that is, the constant L from (J3) is positive.

4.3.2 Main results

Our first theorem is the following existence result for problem (S^0)–(BC^0):

Theorem 4.3.1. *Assume that assumptions (J1)–(J3) hold. Then problem* (S^0)–(BC^0) *has at least one positive solution for $a_0 > 0$ and $b_0 > 0$ sufficiently small.*

Proof. We consider the problems

$$\begin{cases} D_{0+}^{\alpha} h(t) = 0, & t \in (0, 1), \\ h(0) = h'(0) = \cdots = h^{(n-2)}(0) = 0, & h(1) = \displaystyle\int_0^1 h(s)\,\mathrm{d}H(s) + 1, \end{cases} \tag{4.29}$$

$$\begin{cases} D_{0+}^{\beta} k(t) = 0, & t \in (0, 1), \\ k(0) = k'(0) = \cdots = k^{(m-2)}(0) = 0, & k(1) = \displaystyle\int_0^1 k(s)\,\mathrm{d}K(s) + 1. \end{cases} \tag{4.30}$$

Problems (4.29) and (4.30) have the solutions

$$h(t) = \frac{t^{\alpha-1}}{\Delta_1}, \quad k(t) = \frac{t^{\beta-1}}{\Delta_2}, \quad t \in [0,1], \tag{4.31}$$

respectively, where Δ_1 and Δ_2 are defined in (J1). By assumption (J1), we obtain $h(t) > 0$ and $k(t) > 0$ for all $t \in (0,1]$.

We define the functions $x(t)$ and $y(t)$ for all $t \in [0,1]$ by

$$x(t) = u(t) - a_0 h(t), \quad y(t) = v(t) - b_0 k(t), \quad t \in [0,1],$$

where (u, v) is a solution of (S^0)–(BC^0). Then (S^0)–(BC^0) can be equivalently written as

$$\begin{cases} D_{0+}^{\alpha} x(t) + a(t) f(y(t) + b_0 k(t)) = 0, & t \in (0,1), \\ D_{0+}^{\beta} y(t) + b(t) g(x(t) + a_0 h(t)) = 0, & t \in (0,1), \end{cases} \tag{4.32}$$

with the boundary conditions

$$\begin{cases} x(0) = x'(0) = \cdots = x^{(n-2)}(0) = 0, & x(1) = \int_0^1 x(s)\, \mathrm{d}H(s), \\ y(0) = y'(0) = \cdots = y^{(m-2)}(0) = 0, & y(1) = \int_0^1 y(s)\, \mathrm{d}K(s). \end{cases} \tag{4.33}$$

Using the Green's functions G_1 and G_2 from Section 4.1.2, we find a pair (x, y) is a solution of problem (4.32)–(4.33) if and only if (x, y) is a solution for the nonlinear integral equations

$$\begin{cases} x(t) = \int_0^1 G_1(t,s) a(s) f\left(\int_0^1 G_2(s,\tau) b(\tau) g(x(\tau) + a_0 h(\tau))\, \mathrm{d}\tau + b_0 k(s)\right) \mathrm{d}s, \ t \in [0,1], \\ y(t) = \int_0^1 G_2(t,s) b(s) g(x(s) + a_0 h(s))\, \mathrm{d}s, \quad t \in [0,1], \end{cases} \tag{4.34}$$

where $h(t)$ and $k(t)$ for $t \in [0,1]$ are given by (4.31).

We consider the Banach space $X = C([0,1])$ with the supremum norm $\|\cdot\|$, and define the set

$$E = \{x \in C([0,1]), \quad 0 \le x(t) \le c_0, \quad \forall t \in [0,1]\} \subset X.$$

We also define the operator $\mathcal{S} : E \to X$ by

$$(\mathcal{S}x)(t) = \int_0^1 G_1(t,s) a(s) f\left(\int_0^1 G_2(s,\tau) b(\tau) g(x(\tau) + a_0 h(\tau))\, \mathrm{d}\tau + b_0 k(s)\right) \mathrm{d}s,$$
$$0 \le t \le 1, \quad x \in E.$$

For sufficiently small $a_0 > 0$ and $b_0 > 0$, by (J3), we deduce

$$f(y(t) + b_0 k(t)) \le \frac{c_0}{L}, \quad g(x(t) + a_0 h(t)) \le \frac{c_0}{L}, \quad \forall t \in [0,1], \quad \forall x, y \in E.$$

Then, by using Lemma 4.1.7, we obtain $(\mathcal{S}x)(t) \geq 0$ for all $t \in [0,1]$ and $x \in E$. By Lemma 4.1.8, for all $x \in E$, we have

$$\int_0^1 G_2(s,\tau)b(\tau)g(x(\tau)+a_0h(\tau))\,\mathrm{d}\tau \leq \int_0^1 J_2(\tau)b(\tau)g(x(\tau)+a_0h(\tau))\,\mathrm{d}\tau$$
$$\leq \frac{c_0}{L}\int_0^1 J_2(\tau)b(\tau)\,\mathrm{d}\tau \leq c_0, \quad \forall s \in [0,1],$$

and

$$(\mathcal{S}x)(t) \leq \int_0^1 J_1(s)a(s)f\left(\int_0^1 G_2(s,\tau)b(\tau)g(x(\tau)+a_0h(\tau))\,\mathrm{d}\tau + b_0k(s)\right)\mathrm{d}s$$
$$\leq \frac{c_0}{L}\int_0^1 J_1(s)a(s)\,\mathrm{d}s \leq c_0, \quad \forall t \in [0,1].$$

Therefore, $\mathcal{S}(E) \subset E$.

Using standard arguments, we deduce that $\mathcal{S}$ is completely continuous. By Theorem 1.6.1, we conclude that $\mathcal{S}$ has a fixed point $x \in E$. This element together with y given by (4.34) represents a solution for (4.32) and (4.33). This shows that our problem (S^0)–(BC^0) has a positive solution (u,v) with $u = x + a_0h$, $v = y + b_0k$ for sufficiently small $a_0 > 0$ and $b_0 > 0$. □

In what follows, we present sufficient conditions for the nonexistence of positive solutions of (S^0)–(BC^0).

Theorem 4.3.2. *Assume that assumptions (J1), (J2), and (J4) hold. Then, problem* (S^0)–(BC^0) *has no positive solution for a_0 and b_0 sufficiently large.*

Proof. We suppose that (u,v) is a positive solution of (S^0)–(BC^0). Then (x,y) with $x = u - a_0h$, $y = v - b_0k$ is a solution for problem (4.32)–(4.33), where h and k are the solutions of problems (4.29) and (4.30), respectively, (given by (4.31)). By (J2), there exists $c \in (0,1/2)$ such that $t_1, t_2 \in (c, 1-c)$, and then $\int_c^{1-c} a(s)J_1(s)\,\mathrm{d}s > 0$ and $\int_c^{1-c} b(s)J_2(s)\,\mathrm{d}s > 0$. Now by using Lemma 4.1.7, we have $x(t) \geq 0$, $y(t) \geq 0$ for all $t \in [0,1]$, and by Lemma 4.1.9 we obtain $\inf_{t\in[c,1-c]} x(t) \geq \gamma_1\|x\|$ and $\inf_{t\in[c,1-c]} y(t) \geq \gamma_2\|y\|$.

Using now (4.31), we deduce that

$$\inf_{t\in[c,1-c]} h(t) = h(c) = \frac{h(c)}{h(1)}\|h\| = \gamma_1\|h\|, \quad \inf_{t\in[c,1-c]} k(t) = k(c) = \frac{k(c)}{k(1)}\|k\| = \gamma_2\|k\|.$$

Therefore, we obtain

$$\inf_{t\in[c,1-c]} (x(t)+a_0h(t)) \geq \gamma_1\|x\| + a_0\gamma_1\|h\| \geq \gamma_1\|x+a_0h\|,$$

$$\inf_{t\in[c,1-c]} (y(t)+b_0k(t)) \geq \gamma_2\|y\| + b_0\gamma_2\|k\| \geq \gamma_2\|y+b_0k\|.$$

We now consider $R = \left(\min\left\{\gamma_1\gamma_2\int_c^{1-c} a(s)J_1(s)\mathrm{d}s, \gamma_1\gamma_2\int_c^{1-c} b(s)J_2(s)\mathrm{d}s\right\}\right)^{-1} > 0.$

By using (J4), for R defined above, we conclude that there exists $M > 0$ such that $f(u) > 2Ru$ and $g(u) > 2Ru$ for all $u \geq M$. We consider $a_0 > 0$ and $b_0 > 0$ sufficiently large such that $\inf_{t\in[c,1-c]}(x(t)+a_0h(t)) \geq M$ and $\inf_{t\in[c,1-c]}(y(t)$

$+b_0k(t)) \geq M$. By (J2), (4.32), (4.33), and the above inequalities, we deduce that $\|x\| > 0$ and $\|y\| > 0$.

Now by using Lemma 4.1.8 and the above considerations, we have

$$\begin{aligned}
y(c) &= \int_0^1 G_2(c,s)b(s)g(x(s)+a_0h(s))\,\mathrm{d}s \geq \gamma_2 \int_0^1 J_2(s)b(s)g(x(s)+a_0h(s))\,\mathrm{d}s \\
&\geq \gamma_2 \int_c^{1-c} J_2(s)b(s)g(x(s)+a_0h(s))\,\mathrm{d}s \geq 2R\gamma_2 \int_c^{1-c} J_2(s)b(s)(x(s)+a_0h(s))\,\mathrm{d}s \\
&\geq 2R\gamma_2 \int_c^{1-c} J_2(s)b(s) \inf_{\tau\in[c,1-c]}(x(\tau)+a_0h(\tau))\,\mathrm{d}s \\
&\geq 2R\gamma_1\gamma_2 \int_c^{1-c} J_2(s)b(s)\|x+a_0h\|\,\mathrm{d}s \geq 2\|x+a_0h\| \geq 2\|x\|.
\end{aligned}$$

Therefore, we obtain

$$\|x\| \leq y(c)/2 \leq \|y\|/2. \tag{4.35}$$

In a similar manner, we deduce

$$\begin{aligned}
x(c) &= \int_0^1 G_1(c,s)a(s)f(y(s)+b_0k(s))\,\mathrm{d}s \geq \gamma_1 \int_0^1 J_1(s)a(s)f(y(s)+b_0k(s))\,\mathrm{d}s \\
&\geq \gamma_1 \int_c^{1-c} J_1(s)a(s)f(y(s)+b_0k(s))\,\mathrm{d}s \geq 2R\gamma_1 \int_c^{1-c} J_1(s)a(s)(y(s)+b_0k(s))\,\mathrm{d}s \\
&\geq 2R\gamma_1 \int_c^{1-c} J_1(s)a(s) \inf_{\tau\in[c,1-c]}(y(\tau)+b_0k(\tau))\,\mathrm{d}s \\
&\geq 2R\gamma_1\gamma_2 \int_c^{1-c} J_1(s)a(s)\|y+b_0k\|\,\mathrm{d}s \geq 2\|y+b_0k\| \geq 2\|y\|.
\end{aligned}$$

So, we obtain

$$\|y\| \leq x(c)/2 \leq \|x\|/2. \tag{4.36}$$

By (4.35) and (4.36), we conclude that $\|x\| \leq \|y\|/2 \leq \|x\|/4$, which is a contradiction, because $\|x\| > 0$. Then, for a_0 and b_0 sufficiently large, our problem (S^0)–(BC^0) has no positive solution. □

4.3.3 An example

Example 4.3.1. We consider $a(t) = 1$ and $b(t) = 1$ for all $t \in [0,1]$, $\alpha = \frac{5}{2}$ $(n = 3)$, $\beta = \frac{10}{3}$ $(m = 4)$, $H(t) = \begin{cases} 0, & t \in [0,1/4), \\ 3, & t \in [1/4,3/4), \\ 7/2, & t \in [3/4,1], \end{cases}$

and $K(t) = t^4$ for all $t \in [0,1]$. Then $\int_0^1 u(s)\,\mathrm{d}H(s) = 3u\left(\frac{1}{4}\right) + \frac{1}{2}u\left(\frac{3}{4}\right)$ and $\int_0^1 v(s)\,\mathrm{d}K(s) = 4\int_0^1 s^3 v(s)\,\mathrm{d}s$. We also consider the functions $f, g : [0,\infty) \to [0,\infty)$, $f(x) = \frac{\tilde{a}x^{\alpha_0}}{x^{\beta_0}+\tilde{c}}$, $g(x) = \frac{\tilde{b}x^{\gamma_0}}{x^{\delta_0}+\tilde{d}}$ for all $x \in [0,\infty)$, with $\tilde{a}, \tilde{b}, \tilde{c}, \tilde{d} > 0$,

$\alpha_0, \beta_0, \gamma_0, \delta_0 > 0$, $\alpha_0 > \beta_0 + 1$, and $\gamma_0 > \delta_0 + 1$. We have $\lim_{x\to\infty} f(x)/x = \lim_{x\to\infty} g(x)/x = \infty$.

Therefore, we consider the system of fractional differential equations

$$\begin{cases} D_{0+}^{5/2} u(t) + \dfrac{\tilde{a} v^{\alpha_0}(t)}{v^{\beta_0}(t) + \tilde{c}} = 0, \quad t \in (0,1), \\ D_{0+}^{10/3} v(t) + \dfrac{\tilde{b} u^{\gamma_0}(t)}{u^{\delta_0}(t) + \tilde{d}} = 0, \quad t \in (0,1), \end{cases} \tag{S_1^0}$$

with the boundary conditions

$$\begin{cases} u(0) = u'(0) = 0, \quad u(1) = 3u\left(\dfrac{1}{4}\right) + \dfrac{1}{2} u\left(\dfrac{3}{4}\right) + a_0, \\ v(0) = v'(0) = v''(0) = 0, \quad v(1) = 4\displaystyle\int_0^1 s^3 v(s)\, ds + b_0. \end{cases} \tag{BC_1^0}$$

By using also Δ_1 and Δ_2 from Section 4.1.4, we deduce that assumptions (J1), (J2), and (J4) are satisfied. In addition, by using functions J_1 and J_2 from Section 4.1.4, we obtain $\tilde{A} = \int_0^1 J_1(s)\, ds \approx 0.42677595$, $\tilde{B} = \int_0^1 J_2(s)\, ds \approx 0.04007233$, and then $L = \tilde{A}$. We choose $c_0 = 1$, and if we select $\tilde{a}, \tilde{b}, \tilde{c}, \tilde{d}$ satisfying the conditions $\tilde{a} < \frac{1+\tilde{c}}{L} = \frac{1+\tilde{c}}{\tilde{A}}$ and $\tilde{b} < \frac{1+\tilde{d}}{L} = \frac{1+\tilde{d}}{\tilde{A}}$, then we conclude that $f(x) \le \frac{\tilde{a}}{1+\tilde{c}} < \frac{1}{L}$ and $g(x) \le \frac{\tilde{b}}{1+\tilde{d}} < \frac{1}{L}$ for all $x \in [0,1]$. For example, if $\tilde{c} = \tilde{d} = 1$, then for $\tilde{a} \le 4.68$ and $\tilde{b} \le 4.68$ the above conditions for f and g are satisfied. So, assumption (J3) is also satisfied. By Theorems 4.3.1 and 4.3.2, we deduce that problem (S_1^0)–(BC_1^0) has at least one positive solution for sufficiently small $a_0 > 0$ and $b_0 > 0$, and no positive solution for sufficiently large a_0 and b_0.

4.4 A system of semipositone fractional boundary value problems

In this section, we investigate the existence of positive solutions for a system of nonlinear ordinary fractional differential equations with sign-changing nonlinearities subject to uncoupled integral boundary conditions.

4.4.1 Presentation of the problem

We consider the system of nonlinear ordinary fractional differential equations

$$\begin{cases} D_{0+}^{\alpha} u(t) + \lambda f(t, u(t), v(t)) = 0, \quad t \in (0,1), \\ D_{0+}^{\beta} v(t) + \mu g(t, u(t), v(t)) = 0, \quad t \in (0,1), \end{cases} \tag{$\tilde{S}$}$$

with the uncoupled integral boundary conditions

$$\begin{cases} u(0) = u'(0) = \cdots = u^{(n-2)}(0) = 0, \quad u(1) = \displaystyle\int_0^1 u(s)\,\mathrm{d}H(s), \\ v(0) = v'(0) = \cdots = v^{(m-2)}(0) = 0, \quad v(1) = \displaystyle\int_0^1 v(s)\,\mathrm{d}K(s), \end{cases} \tag{$\widetilde{\mathrm{BC}}$}$$

where $n-1 < \alpha \le n$, $m-1 < \beta \le m$, $n, m \in \mathbb{N}$, $n, m \ge 3$, D_{0+}^{α} and D_{0+}^{β} denote the Riemann–Liouville derivatives of orders α and β, respectively, the integrals from $(\widetilde{\mathrm{BC}})$ are Riemann–Stieltjes integrals and f and g are sign-changing continuous functions (i.e., we have a so-called system of semipositone boundary value problems).

By using the nonlinear alternative of Leray–Schauder type (Theorem 2.5.1), we present intervals for parameters λ and μ such that problem $(\tilde{\mathrm{S}})$–$(\widetilde{\mathrm{BC}})$ has at least one positive solution. By a positive solution of problem $(\tilde{\mathrm{S}})$–$(\widetilde{\mathrm{BC}})$ we mean a pair of functions $(u, v) \in C([0,1]; \mathbb{R}_+) \times C([0,1]; \mathbb{R}_+)$ satisfying $(\tilde{\mathrm{S}})$ and $(\widetilde{\mathrm{BC}})$ with $u(t) > 0$, $v(t) > 0$ for all $t \in (0,1)$. For the case when f and g are nonnegative functions, the existence of positive solutions of the above problem ($u(t) \ge 0$, $v(t) \ge 0$ for all $t \in [0,1]$, $(u, v) \ne (0,0)$) was studied in Section 4.1 by using the Guo–Krasnosel'skii fixed point theorem. The positive solutions ($u(t) \ge 0$, $v(t) \ge 0$ for all $t \in [0,1]$, $\sup_{t\in[0,1]} u(t) > 0$, $\sup_{t\in[0,1]} v(t) > 0$) of system $(\tilde{\mathrm{S}})$ with $\lambda = \mu = 1$ and with $f(t,u,v)$ and $g(t,u,v)$ replaced by $\tilde{f}(t,v)$ and $\tilde{g}(t,u)$, respectively ($\tilde{f}$ and $\tilde{g}$ are nonnegative functions), with the boundary conditions $(\widetilde{\mathrm{BC}})$, were investigated in Section 4.2 (the nonsingular and singular cases) by applying some theorems from the fixed point index theory and the Guo–Krasnosel'skii fixed point theorem.

4.4.2 Auxiliary results

In this section, we present some auxiliary results related to the following fractional differential equation

$$D_{0+}^{\alpha} u(t) + z(t) = 0, \quad 0 < t < 1, \tag{4.37}$$

with the integral boundary conditions

$$u(0) = u'(0) = \cdots = u^{(n-2)}(0) = 0, \quad u(1) = \int_0^1 u(s)\,\mathrm{d}H(s), \tag{4.38}$$

where $n-1 < \alpha \le n$, $n \in \mathbb{N}$, $n \ge 3$, and $H : [0,1] \to \mathbb{R}$ is a function of bounded variation.

Lemma 4.4.1. *If $H : [0,1] \to \mathbb{R}$ is a function of bounded variation, $\Delta_1 = 1 - \int_0^1 s^{\alpha-1}\mathrm{d}H(s) \ne 0$, and $z \in C(0,1) \cap L^1(0,1)$, then the unique solution of problem (4.37)–(4.38) is $u(t) = \int_0^1 G_1(t,s)z(s)\,\mathrm{d}s$, where*

$$G_1(t,s) = g_1(t,s) + \frac{t^{\alpha-1}}{\Delta_1}\int_0^1 g_1(\tau,s)\,\mathrm{d}H(\tau), \quad (t,s) \in [0,1] \times [0,1], \tag{4.39}$$

$$g_1(t,s) = \frac{1}{\Gamma(\alpha)} \begin{cases} t^{\alpha-1}(1-s)^{\alpha-1} - (t-s)^{\alpha-1}, & 0 \le s \le t \le 1, \\ t^{\alpha-1}(1-s)^{\alpha-1}, & 0 \le t \le s \le 1. \end{cases} \tag{4.40}$$

For the proof of the above lemma, see Lemmas 4.1.4 and 4.1.5 in Section 4.1.2.

Lemma 4.4.2. *The function g_1 given by* (4.40) *has the following properties:*

(a) $g_1 : [0,1] \times [0,1] \to \mathbb{R}_+$ *is a continuous function, and* $g_1(t,s) > 0$ *for all* $(t,s) \in (0,1) \times (0,1)$.

(b) $g_1(t,s) \le h_1(s)$ *for all* $(t,s) \in [0,1] \times [0,1]$, *where* $h_1(s) = \frac{s(1-s)^{\alpha-1}}{\Gamma(\alpha-1)}$.

(c) $g_1(t,s) \ge k_1(t)h_1(s)$ *for all* $(t,s) \in [0,1] \times [0,1]$, *where*

$$k_1(t) = \min\left\{\frac{(1-t)t^{\alpha-2}}{\alpha-1}, \frac{t^{\alpha-1}}{\alpha-1}\right\} = \begin{cases} \dfrac{t^{\alpha-1}}{\alpha-1}, & 0 \le t \le \frac{1}{2}, \\ \dfrac{(1-t)t^{\alpha-2}}{\alpha-1}, & \frac{1}{2} \le t \le 1. \end{cases} \tag{4.41}$$

Proof. The first part, (a), is evident. For the second part, (b), see Yuan (2010). For part (c), for $s \le t$, we obtain

$$\begin{aligned} g_1(t,s) &= \frac{1}{\Gamma(\alpha)}[t^{\alpha-1}(1-s)^{\alpha-1} - (t-s)^{\alpha-1}] = \frac{1}{\Gamma(\alpha)}[(t-ts)^{\alpha-1} - (t-s)^{\alpha-1}] \\ &= \frac{1}{\Gamma(\alpha)}[(t-ts)^{\alpha-1} - (t-s)^{\alpha-2}(t-s)] \\ &\ge \frac{1}{\Gamma(\alpha)}[(t-ts)^{\alpha-2}(t-ts) - (t-ts)^{\alpha-2}(t-s)] \\ &= \frac{1}{\Gamma(\alpha)}t^{\alpha-2}(1-t)s(1-s)^{\alpha-2} \ge \frac{t^{\alpha-2}(1-t)}{\alpha-1} \cdot \frac{s(1-s)^{\alpha-1}}{\Gamma(\alpha-1)}. \end{aligned}$$

If $s \ge t$, we have

$$g_1(t,s) = \frac{1}{\Gamma(\alpha)}t^{\alpha-1}(1-s)^{\alpha-1} \ge \frac{t^{\alpha-1}}{\alpha-1} \cdot \frac{s(1-s)^{\alpha-1}}{\Gamma(\alpha-1)}.$$

Therefore, we deduce that $g_1(t,s) \ge k_1(t)h_1(s)$ for all $(t,s) \in [0,1] \times [0,1]$, where $k_1(t)$, $t \in [0,1]$ is defined in (4.41). □

Lemma 4.4.3. *If $H : [0,1] \to \mathbb{R}$ is a nondecreasing function and $\Delta_1 > 0$, then the Green's function G_1 of problem* (4.37)–(4.38), *given by* (4.39), *is continuous on* $[0,1] \times [0,1]$ *and satisfies* $G_1(t,s) \ge 0$ *for all* $(t,s) \in [0,1] \times [0,1]$, *and* $G_1(t,s) > 0$ *for all* $(t,s) \in (0,1) \times (0,1)$. *Moreover, if* $z \in C(0,1) \cap L^1(0,1)$ *satisfies* $z(t) \ge 0$ *for all* $t \in (0,1)$, *then the solution u of problem* (4.37)–(4.38) *satisfies* $u(t) \ge 0$ *for all* $t \in [0,1]$.

For the proof of the above lemma, see Lemma 4.1.7.

Lemma 4.4.4. *Assume that $H : [0,1] \to \mathbb{R}$ is a nondecreasing function and $\Delta_1 > 0$. Then the Green's function G_1 of problem* (4.37)–(4.38) *satisfies the following inequalities:*

(a) $G_1(t,s) \le \tilde{J}_1(s), \quad \forall (t,s) \in [0,1] \times [0,1]$, *where*

$$\tilde{J}_1(s) = \tau_1 h_1(s), \quad \tau_1 = 1 + \frac{1}{\Delta_1}\int_0^1 dH(\tau) = 1 + \frac{1}{\Delta_1}(H(1) - H(0)).$$

(b) $G_1(t,s) \geq \tilde{\gamma}_1(t)\tilde{J}_1(s), \quad \forall\,(t,s) \in [0,1] \times [0,1]$, *where*

$$\tilde{\gamma}_1(t) = \frac{1}{\tau_1}\left(k_1(t) + \frac{t^{\alpha-1}}{\Delta_1}\int_0^1 k_1(\tau)\,\mathrm{d}H(\tau)\right).$$

Proof.

(a) For all $(t,s) \in [0,1] \times [0,1]$, we have

$$\begin{aligned} G_1(t,s) &= g_1(t,s) + \frac{t^{\alpha-1}}{\Delta_1}\int_0^1 g_1(\tau,s)\,\mathrm{d}H(\tau) \\ &\leq h_1(s) + \frac{1}{\Delta_1}\int_0^1 h_1(s)\,\mathrm{d}H(\tau) = h_1(s)\left(1 + \frac{1}{\Delta_1}\int_0^1 \mathrm{d}H(\tau)\right) = \tilde{J}_1(s). \end{aligned}$$

(b) For the second inequality, for all $(t,s) \in [0,1] \times [0,1]$, we obtain

$$\begin{aligned} G_1(t,s) &\geq k_1(t)h_1(s) + \frac{t^{\alpha-1}}{\Delta_1}\int_0^1 k_1(\tau)h_1(s)\,\mathrm{d}H(\tau) \\ &= \frac{1}{\tau_1}(\tau_1 h_1(s))\left(k_1(t) + \frac{t^{\alpha-1}}{\Delta_1}\int_0^1 k_1(\tau)\,\mathrm{d}H(\tau)\right) = \tilde{\gamma}_1(t)\tilde{J}_1(s). \end{aligned}$$

We observe that $\tilde{\gamma}_1(t) > 0$ for all $t \in (0,1)$, and if $H \not\equiv$ const., then $\tilde{\gamma}_1(1) > 0$.

□

Lemma 4.4.5. *Assume that* $H : [0,1] \to \mathbb{R}$ *is a nondecreasing function,* $\Delta_1 > 0$, *and* $z \in C(0,1) \cap L^1(0,1)$, $z(t) \geq 0$ *for all* $t \in (0,1)$. *Then the solution* $u(t)$, $t \in [0,1]$ *of problem* (4.37)–(4.38) *satisfies the inequality* $u(t) \geq \tilde{\gamma}_1(t)\sup_{t' \in [0,1]} u(t')$ *for all* $t \in [0,1]$.

Proof. For all $t,\ t' \in [0,1]$, we obtain

$$\begin{aligned} u(t) &= \int_0^1 G_1(t,s)z(s)\,\mathrm{d}s \geq \int_0^1 \tilde{\gamma}_1(t)\tilde{J}_1(s)z(s)\,\mathrm{d}s = \tilde{\gamma}_1(t)\int_0^1 \tilde{J}_1(s)z(s)\,\mathrm{d}s \\ &\geq \tilde{\gamma}_1(t)\int_0^1 G_1(t',s)z(s)\,\mathrm{d}s = \tilde{\gamma}_1(t)u(t'). \end{aligned}$$

Therefore, we deduce that $u(t) \geq \tilde{\gamma}_1(t)\sup_{t' \in [0,1]} u(t')$ for all $t \in [0,1]$. □

We can also formulate results similar to those in Lemmas 4.4.1–4.4.5 for the fractional differential equation

$$D_{0+}^{\beta} v(t) + \tilde{z}(t) = 0, \quad 0 < t < 1, \tag{4.42}$$

with the integral boundary conditions

$$v(0) = v'(0) = \cdots = v^{(m-2)}(0) = 0, \quad v(1) = \int_0^1 v(s)\,\mathrm{d}K(s), \tag{4.43}$$

where $m - 1 < \beta \leq m$, $m \in \mathbb{N}$, $m \geq 3$, $K : [0,1] \to \mathbb{R}$ is a nondecreasing function, and $\tilde{z} \in C(0,1) \cap L^1(0,1)$. We denote by Δ_2, g_2, G_2, h_2, k_2, τ_2, $\tilde{J}_2$, and $\tilde{\gamma}_2$ the corresponding constants and functions for problem (4.42)–(4.43) defined in a similar manner as Δ_1, g_1, G_1, h_1, k_1, τ_1, $\tilde{J}_1$, and $\tilde{\gamma}_1$, respectively.

4.4.3 Main result

In this section, we investigate the existence of positive solutions for our problem $(\tilde{\text{S}})$–$(\widetilde{\text{BC}})$. We present now the assumptions that we shall use in the sequel:

$(\widehat{\text{H1}})$ $H, K : [0,1] \to \mathbb{R}$ are nondecreasing functions, $\Delta_1 = 1 - \int_0^1 s^{\alpha-1}\,dH(s) > 0$, and $\Delta_2 = 1 - \int_0^1 s^{\beta-1}\,dK(s) > 0$.

$(\widehat{\text{H2}})$ The functions $f, g \in C([0,1] \times [0,\infty) \times [0,\infty), (-\infty,+\infty))$, and there exist functions $p_1, p_2 \in C([0,1], (0,\infty))$ such that $f(t,u,v) \geq -p_1(t)$ and $g(t,u,v) \geq -p_2(t)$ for any $t \in [0,1]$ and $u, v \in [0,\infty)$.

$(\widehat{\text{H3}})$ $f(t,0,0) > 0$, $g(t,0,0) > 0$ for all $t \in [0,1]$.

We consider the system of nonlinear fractional differential equations

$$\begin{cases} D_{0+}^{\alpha}x(t) + \lambda(f(t,[x(t)-q_1(t)]^*,[y(t)-q_2(t)]^*) + p_1(t)) = 0, & 0 < t < 1, \\ D_{0+}^{\beta}y(t) + \mu(g(t,[x(t)-q_1(t)]^*,[y(t)-q_2(t)]^*) + p_2(t)) = 0, & 0 < t < 1, \end{cases} \tag{4.44}$$

with the integral boundary conditions

$$\begin{cases} x(0) = x'(0) = \cdots = x^{(n-2)}(0) = 0, \quad x(1) = \displaystyle\int_0^1 x(s)\,dH(s), \\ y(0) = y'(0) = \cdots = y^{(m-2)}(0) = 0, \quad y(1) = \displaystyle\int_0^1 y(s)\,dK(s), \end{cases} \tag{4.45}$$

where $z(t)^* = z(t)$ if $z(t) \geq 0$ and $z(t)^* = 0$ if $z(t) < 0$. Here q_1 and q_2 are given by $q_1(t) = \lambda\int_0^1 G_1(t,s)p_1(s)\,ds$ and $q_2(t) = \mu\int_0^1 G_2(t,s)p_2(s)\,ds$—that is, they are the solutions of the problems

$$\begin{cases} D_{0+}^{\alpha}q_1(t) + \lambda p_1(t) = 0, \quad t \in (0,1), \\ q_1(0) = q_1'(0) = \cdots = q_1^{(n-2)}(0) = 0, \quad q_1(1) = \displaystyle\int_0^1 q_1(s)\,dH(s), \end{cases} \tag{4.46}$$

and

$$\begin{cases} D_{0+}^{\beta}q_2(t) + \mu p_2(t) = 0, \quad t \in (0,1), \\ q_2(0) = q_2'(0) = \cdots = q_2^{(m-2)}(0) = 0, \quad q_2(1) = \displaystyle\int_0^1 q_2(s)\,dK(s), \end{cases} \tag{4.47}$$

respectively. By $(\hat{\text{H}}1)$, $(\hat{\text{H}}2)$, and Lemma 4.4.3, we have $q_1(t) \geq 0$, $q_2(t) \geq 0$ for all $t \in [0,1]$, and $q_1(t) > 0$, $q_2(t) > 0$ for all $t \in (0,1)$.

We shall prove that there exists a solution (x,y) for the boundary value problem (4.44)–(4.45) with $x(t) \geq q_1(t)$ and $y(t) \geq q_2(t)$ for all $t \in [0,1]$. In this case, the functions $u(t) = x(t) - q_1(t)$ and $v(t) = y(t) - q_2(t)$, $t \in [0,1]$ represent a positive solution of the boundary value problem $(\tilde{\text{S}})$–$(\widetilde{\text{BC}})$. By (4.44)–(4.45) and (4.46)–(4.47), we have

$$\begin{aligned} D_{0+}^{\alpha}u(t) = D_{0+}^{\alpha}x(t) - D_{0+}^{\alpha}q_1(t) &= -\lambda f(t,[x(t)-q_1(t)]^*,[y(t)-q_2(t)]^*) \\ &\quad - \lambda p_1(t) + \lambda p_1(t) = -\lambda f(t,u(t),v(t)), \quad \forall t \in (0,1), \end{aligned}$$

$$D_{0+}^{\beta}v(t) = D_{0+}^{\beta}y(t) - D_{0+}^{\beta}q_2(t) = -\mu g(t,[x(t)-q_1(t)]^*,[y(t)-q_2(t)]^*)$$
$$- \mu p_2(t) + \mu p_2(t) = -\mu g(t,u(t),v(t)), \quad \forall t \in (0,1),$$

and

$$u(0) = x(0) - q_1(0) = 0, \dots, u^{(n-2)}(0) = x^{(n-2)}(0) - q_1^{(n-2)}(0) = 0,$$
$$v(0) = y(0) - q_2(0) = 0, \dots, v^{(m-2)}(0) = y^{(m-2)}(0) - q_2^{(m-2)}(0) = 0,$$
$$u(1) = x(1) - q_1(1) = \int_0^1 x(s)\,\mathrm{d}H(s) - \int_0^1 q_1(s)\,\mathrm{d}H(s) = \int_0^1 u(s)\,\mathrm{d}H(s),$$
$$v(1) = y(1) - q_2(1) = \int_0^1 y(s)\,\mathrm{d}K(s) - \int_0^1 q_2(s)\,\mathrm{d}K(s) = \int_0^1 v(s)\,\mathrm{d}K(s).$$

Therefore, in what follows, we shall investigate the boundary value problem (4.44)–(4.45).

By using Lemma 4.4.1, we find problem (4.44)–(4.45) is equivalent to the system

$$\begin{cases} x(t) = \lambda \displaystyle\int_0^1 G_1(t,s)\left(f(s,[x(s)-q_1(s)]^*,[y(s)-q_2(s)]^*) + p_1(s)\right)\mathrm{d}s, & t \in [0,1], \\ y(t) = \mu \displaystyle\int_0^1 G_2(t,s)\left(g(s,[x(s)-q_1(s)]^*,[y(s)-q_2(s)]^*) + p_2(s)\right)\mathrm{d}s, & t \in [0,1]. \end{cases}$$

We consider the Banach space $X = C([0,1])$ with the supremum norm $\|\cdot\|$ and the Banach space $Y = X \times X$ with the norm $\|(x,y)\|_Y = \|x\| + \|y\|$. We also define the cones

$$\tilde{P}_1 = \{x \in X, \quad x(t) \geq \tilde{\gamma}_1(t)\|x\|, \quad \forall t \in [0,1]\} \subset X,$$
$$\tilde{P}_2 = \{y \in X, \quad y(t) \geq \tilde{\gamma}_2(t)\|y\|, \quad \forall t \in [0,1]\} \subset X,$$

and $\tilde{P} = \tilde{P}_1 \times \tilde{P}_2 \subset Y$.

For $\lambda, \mu > 0$, we define now the operator $\tilde{Q} : \tilde{P} \to Y$ by $\tilde{Q}(x,y) = (\tilde{Q}_1(x,y), \tilde{Q}_2(x,y))$ with

$$\tilde{Q}_1(x,y)(t) = \lambda \int_0^1 G_1(t,s)$$
$$\times \left(f(s,[x(s)-q_1(s)]^*,[y(s)-q_2(s)]^*) + p_1(s)\right)\mathrm{d}s, \quad 0 \leq t \leq 1,$$
$$\tilde{Q}_2(x,y)(t) = \mu \int_0^1 G_2(t,s)$$
$$\times \left(g(s,[x(s)-q_1(s)]^*,[y(s)-q_2(s)]^*) + p_2(s)\right)\mathrm{d}s, \quad 0 \leq t \leq 1.$$

Lemma 4.4.6. *If $(\widehat{H1})$ and $(\widehat{H2})$ hold, then the operator $\tilde{Q} : \tilde{P} \to \tilde{P}$ is a completely continuous operator.*

Proof. The operators $\tilde{Q}_1$ and $\tilde{Q}_2$ are well defined. For every $(x,y) \in \tilde{P}$, by Lemma 4.4.4, we have $\tilde{Q}_1(x,y)(t) < \infty$ and $\tilde{Q}_2(x,y)(t) < \infty$ for all $t \in [0,1]$. Then, by Lemma 4.4.5, we obtain

$$\tilde{Q}_1(x,y)(t) \geq \tilde{\gamma}_1(t) \sup_{t' \in [0,1]} \tilde{Q}_1(x,y)(t'), \quad \tilde{Q}_2(x,y)(t) \geq \tilde{\gamma}_2(t) \sup_{t' \in [0,1]} \tilde{Q}_2(x,y)(t'),$$

for all $t \in [0,1]$. Therefore, we conclude

$$\tilde{Q}_1(x,y)(t) \geq \tilde{\gamma}_1(t)\|\tilde{Q}_1(x,y)\|, \quad \tilde{Q}_2(x,y)(t) \geq \tilde{\gamma}_2(t)\|\tilde{Q}_2(x,y)\|, \quad \forall t \in [0,1],$$

and $\tilde{Q}(x,y) = (\tilde{Q}_1(x,y), \tilde{Q}_2(x,y)) \in \tilde{P}$.

By using standard arguments, we deduce that the operator $\tilde{Q} : \tilde{P} \to \tilde{P}$ is a completely continuous operator. □

It is clear that $(x,y) \in \tilde{P}$ is a solution of problem (4.44)–(4.45) if and only if (x,y) is a fixed point of $\tilde{Q}$.

Theorem 4.4.1. *Assume that $(\widehat{H1})$–$(\widehat{H3})$ hold. Then there exist constants $\lambda_0 > 0$ and $\mu_0 > 0$ such that for any $\lambda \in (0,\lambda_0]$ and $\mu \in (0,\mu_0]$ the boundary value problem $(\tilde{S})$–$(\widetilde{BC})$ has at least one positive solution.*

Proof. Let $\delta \in (0,1)$ be fixed. From $(\hat{H3})$, there exists $R_0 > 0$ such that

$$f(t,u,v) \geq \delta f(t,0,0) > 0, \;\; g(t,u,v) \geq \delta g(t,0,0) > 0, \;\; \forall t \in [0,1], \;\; u, v \in [0,R_0]. \tag{4.48}$$

We define

$$\bar{f}(R_0) = \max_{0\leq t\leq 1,\, 0\leq u,v\leq R_0} \{f(t,u,v) + p_1(t)\} \geq \max_{0\leq t\leq 1} \{\delta f(t,0,0) + p_1(t)\} > 0,$$

$$\bar{g}(R_0) = \max_{0\leq t\leq 1,\, 0\leq u,v\leq R_0} \{g(t,u,v) + p_2(t)\} \geq \max_{0\leq t\leq 1} \{\delta g(t,0,0) + p_2(t)\} > 0,$$

$$c_1 = \int_0^1 \tilde{J}_1(s)\,ds > 0, \quad c_2 = \int_0^1 \tilde{J}_2(s)\,ds > 0, \quad \lambda_0 = \frac{R_0}{4c_1\bar{f}(R_0)} > 0,$$

$$\mu_0 = \frac{R_0}{4c_2\bar{g}(R_0)} > 0.$$

We shall show that for any $\lambda \in (0,\lambda_0]$ and $\mu \in (0,\mu_0]$, problem (4.44)–(4.45) has at least one positive solution.

So, let $\lambda \in (0,\lambda_0]$ and $\mu \in (0,\mu_0]$ be arbitrary, but fixed for the moment. We define the set $U = \{(x,y) \in \tilde{P}, \quad \|(x,y)\|_Y < R_0\}$. We suppose that there exist $(x_0,y_0) \in \partial U$, $(\|(x_0,y_0)\|_Y = R_0$ or $\|x_0\|+\|y_0\| = R_0)$, and $\nu \in (0,1)$ such that $(x_0,y_0) = \nu\tilde{Q}(x_0,y_0)$ or $x_0 = \nu\tilde{Q}_1(x_0,y_0), \quad y_0 = \nu\tilde{Q}_2(x_0,y_0)$.

We deduce that

$$[x_0(t) - q_1(t)]^* = x_0(t) - q_1(t) \leq x_0(t) \leq R_0, \quad \text{if} \quad x_0(t) - q_1(t) \geq 0,$$
$$[x_0(t) - q_1(t)]^* = 0, \quad \text{for} \quad x_0(t) - q_1(t) < 0, \quad \forall t \in [0,1],$$
$$[y_0(t) - q_2(t)]^* = y_0(t) - q_2(t) \leq y_0(t) \leq R_0, \quad \text{if} \quad y_0(t) - q_2(t) \geq 0,$$
$$[y_0(t) - q_2(t)]^* = 0, \quad \text{for} \quad y_0(t) - q_2(t) < 0, \quad \forall t \in [0,1].$$

Then for all $t \in [0,1]$, we obtain

$$
\begin{aligned}
x_0(t) &= \nu\tilde{Q}_1(x_0,y_0)(t) \le \tilde{Q}_1(x_0,y_0)(t) \\
&= \lambda \int_0^1 G_1(t,s)\left(f(s,[x_0(s)-q_1(s)]^*,[y_0(s)-q_2(s)]^*)+p_1(s)\right)\mathrm{d}s \\
&\le \lambda \int_0^1 G_1(t,s)\bar{f}(R_0)\,\mathrm{d}s \le \lambda \int_0^1 \tilde{J}_1(s)\bar{f}(R_0)\,\mathrm{d}s \le \lambda_0 c_1 \bar{f}(R_0) = R_0/4, \\
y_0(t) &= \nu\tilde{Q}_2(x_0,y_0)(t) \le \tilde{Q}_2(x_0,y_0)(t) \\
&= \mu \int_0^1 G_2(t,s)\left(g(s,[x_0(s)-q_1(s)]^*,[y_0(s)-q_2(s)]^*)+p_2(s)\right)\mathrm{d}s \\
&\le \mu \int_0^1 G_2(t,s)\bar{g}(R_0)\,\mathrm{d}s \le \mu \int_0^1 \tilde{J}_2(s)\bar{g}(R_0)\,\mathrm{d}s \le \mu_0 c_2 \bar{g}(R_0) = R_0/4.
\end{aligned}
$$

Hence, $\|x_0\| \le R_0/4$ and $\|y_0\| \le R_0/4$. Then $R_0 = \|(x_0,y_0)\|_Y = \|x_0\| + \|y_0\| \le \frac{R_0}{4} + \frac{R_0}{4} = \frac{R_0}{2}$, which is a contradiction.

Therefore, by Lemma 4.4.6 and Theorem 2.5.1, we deduce that $\tilde{Q}$ has a fixed point $(x,y) \in \bar{U} \cap \tilde{P}$. That is, $(x,y) = \tilde{Q}(x,y) \Leftrightarrow x = \tilde{Q}_1(x,y), \quad y = \tilde{Q}_2(x,y)$, and $\|x\| + \|y\| \le R_0$, with $x(t) \ge \tilde{\gamma}_1(t)\|x\| \ge 0$ and $y(t) \ge \tilde{\gamma}_2(t)\|y\| \ge 0$ for all $t \in [0,1]$.

Moreover, by (4.48), we obtain

$$
\begin{aligned}
x(t) &= \tilde{Q}_1(x,y)(t) = \lambda \int_0^1 G_1(t,s)\left(f(s,[x(s)-q_1(s)]^*,[y(s)-q_2(s)]^*)\right. \\
&\quad \left. + p_1(s)\right)\mathrm{d}s \\
&\ge \lambda \int_0^1 G_1(t,s)(\delta f(s,0,0)+p_1(s))\,\mathrm{d}s > \lambda \int_0^1 G_1(t,s)p_1(s)\,\mathrm{d}s = q_1(t) > 0, \\
&\quad \forall\, t \in (0,1), \\
y(t) &= \tilde{Q}_2(x,y)(t) = \mu \int_0^1 G_2(t,s)\left(g(s,[x(s)-q_1(s)]^*,[y(s)-q_2(s)]^*)\right. \\
&\quad \left. + p_2(s)\right)\mathrm{d}s \\
&\ge \mu \int_0^1 G_2(t,s)(\delta g(s,0,0)+p_2(s))\,\mathrm{d}s > \mu \int_0^1 G_2(t,s)p_2(s)\,\mathrm{d}s = q_2(t) > 0, \\
&\quad \forall\, t \in (0,1).
\end{aligned}
$$

Let $u(t) = x(t) - q_1(t) \ge 0$ and $v(t) = y(t) - q_2(t) \ge 0$ for all $t \in [0,1]$, with $u(t) > 0$, $v(t) > 0$ on $(0,1)$. Then (u,v) is a positive solution of the boundary value problem $(\tilde{\mathrm{S}})$–$(\widetilde{\mathrm{BC}})$. □

4.4.4 Examples

Let $\alpha = \frac{5}{2}$ $(n = 3)$, $\beta = \frac{10}{3}$ $(m = 4)$, $H(t) = \begin{cases} 0, & t \in [0, 1/4), \\ 3, & t \in [1/4, 3/4), \\ 7/2, & t \in [3/4, 1], \end{cases}$

and $K(t) = t^4$ for all $t \in [0, 1]$. Then $\int_0^1 u(s)\,dH(s) = 3u\left(\frac{1}{4}\right) + \frac{1}{2}u\left(\frac{3}{4}\right)$ and $\int_0^1 v(s)\,dK(s) = 4\int_0^1 s^3 v(s)\,ds$.

We consider the system of fractional differential equations

$$\begin{cases} D_{0+}^{5/2}u(t) + \lambda f(t, u(t), v(t)) = 0, & 0 < t < 1, \\ D_{0+}^{10/3}v(t) + \mu g(t, u(t), v(t)) = 0, & 0 < t < 1, \end{cases} \qquad (\tilde{S}_0)$$

with the boundary conditions

$$\begin{cases} u(0) = u'(0) = 0, \quad u(1) = 3u\left(\dfrac{1}{4}\right) + \dfrac{1}{2}u\left(\dfrac{3}{4}\right), \\ v(0) = v'(0) = v''(0) = 0, \quad v(1) = 4\displaystyle\int_0^1 s^3 v(s)\,ds. \end{cases} \qquad (\widetilde{BC}_0)$$

Then we obtain $\Delta_1 = 1 - \int_0^1 s^{3/2}\,dH(s) = 1 - 3\left(\frac{1}{4}\right)^{3/2} - \frac{1}{2}\left(\frac{3}{4}\right)^{3/2} = \frac{10-3\sqrt{3}}{16} \approx 0.3002 > 0$ and $\Delta_2 = 1 - \int_0^1 s^{7/3}\,dK(s) = 1 - 4\int_0^1 s^{16/3}\,ds = \frac{7}{19} \approx 0.3684 > 0$.

We also deduce

$$g_1(t, s) = \frac{4}{3\sqrt{\pi}} \begin{cases} t^{3/2}(1-s)^{3/2} - (t-s)^{3/2}, & 0 \le s \le t \le 1, \\ t^{3/2}(1-s)^{3/2}, & 0 \le t \le s \le 1, \end{cases}$$

$$g_2(t, s) = \frac{1}{\Gamma(10/3)} \begin{cases} t^{7/3}(1-s)^{7/3} - (t-s)^{7/3}, & 0 \le s \le t \le 1, \\ t^{7/3}(1-s)^{7/3}, & 0 \le t \le s \le 1, \end{cases}$$

$\tau_1 = \frac{633+168\sqrt{3}}{73}$, $h_1(s) = \frac{2}{\sqrt{\pi}}s(1-s)^{3/2}$, $\tilde{J}_1(s) = \tau_1 h_1(s) = \frac{1266+336\sqrt{3}}{73\sqrt{\pi}} \times s(1-s)^{3/2}$, $s \in [0, 1]$, $\tau_2 = \frac{26}{7}$, $h_2(s) = \frac{1}{\Gamma(7/3)}s(1-s)^{7/3}$, $\tilde{J}_2(s) = \tau_2 h_2(s) = \frac{26}{7\Gamma(7/3)}s(1-s)^{7/3}$, $s \in [0, 1]$, $c_1 = \int_0^1 \tilde{J}_1(s)\,ds \approx 1.63225815$, and $c_2 = \int_0^1 \tilde{J}_2(s)\,ds \approx 0.2159704$.

Example 4.4.1. We consider the functions

$$f(t, u, v) = (u-a)(u-b) + \cos(\theta_1 v), \quad g(t, u, v) = (v-c)(v-d) + \sin(\theta_2 u),$$

for all $t \in [0, 1]$, $u, v \in [0, \infty)$, where $b > a > 0$, $d > c > 0$, and $\theta_1, \theta_2 > 0$.

There exists $M_0 > 0$ such that $f(t, u, v) + M_0 \ge 0$ and $g(t, u, v) + M_0 \ge 0$ for all $t \in [0, 1]$ and $u, v \in [0, \infty)$ $(p_1(t) = p_2(t) = M_0$, for all $t \in [0, 1])$. $M_0 = \max\left\{\frac{(b-a)^2}{4} + 1, \frac{(d-c)^2}{4} + 1\right\}$ satisfies the above inequalities.

Let $\delta = \min\left\{\frac{a(2b-a)}{4(ab+1)}, \frac{c(2d-c)}{4cd}\right\} < 1$ and $R_0 = \min\left\{\frac{a}{2}, \frac{c}{2}, \frac{\pi}{2\theta_1}, \frac{\pi}{2\theta_2}\right\}$. Then

$$f(t, u, v) \ge \delta f(t, 0, 0) = \delta(ab+1), \quad g(t, u, v) \ge \delta g(t, 0, 0) = \delta cd,$$

for all $t \in [0,1]$ and $u,\ v \in [0, R_0]$. Besides,

$$\bar{f}(R_0) = \max_{0\le t\le 1,\ 0\le u,\, v\le R_0} \{f(t,u,v) + p_1(t)\} = ab + 1 + M_0,$$

$$\bar{g}(R_0) = \max_{0\le t\le 1,\ 0\le u,\, v\le R_0} \{g(t,u,v) + p_2(t)\} = cd + \sin(\theta_2 R_0) + M_0.$$

Then $\lambda_0 = \frac{R_0}{4c_1(ab+1+M_0)}$ and $\mu_0 = \frac{R_0}{4c_2(cd+\sin(\theta_2 R_0)+M_0)}$. For example, if $a = 1$, $b = 2$, $c = 3$, $d = 4$, and $\theta_1 = \theta_2 = 1$, then $R_0 = 1/2$, $\delta = 1/4$, $M_0 = 5/4$, $\bar{f}(R_0) = 4.25$, $\bar{g}(R_0) \approx 13.72942553$, $\lambda_0 = \frac{1}{34c_1} \approx 0.01801906$, and $\mu_0 = \frac{1}{8(12+\sin(1/2)+5/4)c_2} \approx 0.04215638$.

By Theorem 4.4.1, for any $\lambda \in (0, \lambda_0]$ and $\mu \in (0, \mu_0]$, we deduce that problem $(\tilde{S}_0)$–$(\widetilde{BC}_0)$ has a positive solution (u, v).

Example 4.4.2. We consider the functions

$$f(t,u,v) = v^a + \cos(\theta_1 u), \quad g(t,u,v) = u^b + \cos(\theta_2 v), \quad \forall t \in [0,1],\ u,\ v \in [0,\infty),$$

where $a,\ b,\ \theta_1,\ \theta_2 > 0$.

There exists $M_0 > 0$ ($M_0 = 1$) such that $f(t,u,v) + M_0 \ge 0$ and $g(t,u,v) + M_0 \ge 0$ for all $t \in [0,1]$ and $u,\ v \in [0,\infty)$ ($p_1(t) = p_2(t) = M_0$ for all $t \in [0,1]$).

Let $\delta = \frac{1}{2} < 1$ and $R_0 = \min\left\{\frac{\pi}{3\theta_1}, \frac{\pi}{3\theta_2}\right\}$. Then

$$f(t,u,v) \ge \delta f(t,0,0) = \frac{1}{2}, \quad g(t,u,v) \ge \delta g(t,0,0) = \frac{1}{2}, \quad \forall t \in [0,1],\ u,\ v \in [0,R_0].$$

Besides,

$$\bar{f}(R_0) = \max_{0\le t\le 1,\ 0\le u,\, v\le R_0} \{f(t,u,v) + p_1(t)\} = R_0^a + 2,$$

$$\bar{g}(R_0) = \max_{0\le t\le 1,\ 0\le u,\, v\le R_0} \{g(t,u,v) + p_2(t)\} = R_0^b + 2.$$

Then $\lambda_0 = \frac{R_0}{4c_1(R_0^a+2)}$ and $\mu_0 = \frac{R_0}{4c_2(R_0^b+2)}$. For example, if $a = 2$, $b = 1/2$, and $\theta_1 = \theta_2 = 1$, then $R_0 = \frac{\pi}{3}$, $\bar{f}(R_0) = \frac{\pi^2}{9} + 2$, $\bar{g}(R_0) = \sqrt{\frac{\pi}{3}} + 2$, $\lambda_0 = \frac{3\pi}{4(\pi^2+18)c_1} \approx 0.05179543$, and $\mu_0 = \frac{\pi}{4(\sqrt{3\pi}+6)c_2} \approx 0.40094916$.

By Theorem 4.4.1, for any $\lambda \in (0, \lambda_0]$ and $\mu \in (0, \mu_0]$, we deduce that problem $(\tilde{S}_0)$–$(\widetilde{BC}_0)$ has a positive solution (u, v).

Remark 4.4.1. The results presented in this section were published in Luca and Tudorache (2014).

Systems of Riemann–Liouville fractional differential equations with coupled integral boundary conditions

5

5.1 Existence of positive solutions for systems with parameters and coupled boundary conditions

In this section, we shall investigate the existence of positive solutions for a system of nonlinear Riemann–Liouville fractional differential equations which contains some parameters and is subject to coupled Riemann–Stieltjes integral boundary conditions. Coupled boundary conditions appear in the study of reaction-diffusion equations and Sturm–Liouville problems, and have applications in many fields of sciences and engineering, such as thermal conduction and mathematical biology (see, e.g., Amann 1986, 1988; Aronson 1978; Deng 1995, 1996; Lin and Xie 1998; Pedersen and Lin 2001).

5.1.1 Presentation of the problem

We consider the system of nonlinear ordinary fractional differential equations

$$\begin{cases} D_{0+}^{\alpha}u(t) + \lambda f(t, u(t), v(t)) = 0, & t \in (0,1), \\ D_{0+}^{\beta}v(t) + \mu g(t, u(t), v(t)) = 0, & t \in (0,1), \end{cases} \tag{S}$$

with the coupled integral boundary conditions

$$\begin{cases} u(0) = u'(0) = \cdots = u^{(n-2)}(0) = 0, \quad u(1) = \displaystyle\int_0^1 v(s)\,\mathrm{d}H(s), \\ v(0) = v'(0) = \cdots = v^{(m-2)}(0) = 0, \quad v(1) = \displaystyle\int_0^1 u(s)\,\mathrm{d}K(s), \end{cases} \tag{BC}$$

where $n-1 < \alpha \le n$, $m-1 < \beta \le m$, $n, m \in \mathbb{N}$, $n, m \ge 3$, D_{0+}^{α} and D_{0+}^{β} denote the Riemann–Liouville derivatives of orders α and β, respectively, and the integrals from (BC) are Riemann–Stieltjes integrals.

We shall give sufficient conditions on λ, μ, f, and g such that positive solutions of (S)–(BC) exist. By a positive solution of problem (S)–(BC), we mean a pair of functions $(u, v) \in C([0,1]) \times C([0,1])$ satisfying (S) and (BC) with $u(t) \ge 0$, $v(t) \ge 0$ for all $t \in [0,1]$ and $(u, v) \ne (0,0)$. System (S) with $\alpha = \beta$, $\lambda = \mu$, and the coupled

Boundary Value Problems for Systems of Differential, Difference and Fractional Equations.
http://dx.doi.org/10.1016/B978-0-12-803652-5.00005-3

boundary conditions $u^{(i)}(0) = v^{(i)}(0) = 0$ for $i = 0, 1, \ldots, n-2$, $u(1) = av(\xi)$ and $v(1) = bu(\eta)$ with $\xi, \eta \in (0,1)$ and $0 < ab\xi\eta < 1$ was investigated in Yuan et al. (2012). In that paper, the authors proved the existence of multiple positive solutions, where the functions f and g are continuous and semipositone.

5.1.2 Auxiliary results

In this section, we present some auxiliary results that will be used to prove our main theorems.

We consider the fractional differential system

$$\begin{cases} D_{0+}^{\alpha} u(t) + x(t) = 0, & t \in (0,1), \\ D_{0+}^{\beta} v(t) + y(t) = 0, & t \in (0,1), \end{cases} \tag{5.1}$$

with the coupled integral boundary conditions

$$\begin{cases} u(0) = u'(0) = \cdots = u^{(n-2)}(0) = 0, \quad u(1) = \displaystyle\int_0^1 v(s)\,\mathrm{d}H(s), \\ v(0) = v'(0) = \cdots = v^{(m-2)}(0) = 0, \quad v(1) = \displaystyle\int_0^1 u(s)\,\mathrm{d}K(s), \end{cases} \tag{5.2}$$

where $n-1 < \alpha \le n$, $m-1 < \beta \le m$, $n, m \in \mathbb{N}$, $n, m \ge 3$, and $H, K : [0,1] \to \mathbb{R}$ are functions of bounded variation.

Lemma 5.1.1. *If $H, K : [0,1] \to \mathbb{R}$ are functions of bounded variation, $\Delta = 1 - \left(\int_0^1 \tau^{\alpha-1}\,\mathrm{d}K(\tau)\right)\left(\int_0^1 \tau^{\beta-1}\,\mathrm{d}H(\tau)\right) \ne 0$, and $x, y \in C(0,1) \cap L^1(0,1)$, then the unique solution of problem* (5.1)–(5.2) *is given by*

$$\begin{aligned} u(t) = & -\frac{1}{\Gamma(\alpha)} \int_0^t (t-s)^{\alpha-1} x(s)\,\mathrm{d}s + \frac{t^{\alpha-1}}{\Delta}\left[\frac{1}{\Gamma(\alpha)}\int_0^1 (1-s)^{\alpha-1} x(s)\,\mathrm{d}s \right. \\ & \left. -\frac{1}{\Gamma(\alpha)}\left(\int_0^1 \tau^{\beta-1}\,\mathrm{d}H(\tau)\right)\left(\int_0^1\left(\int_s^1 (\tau-s)^{\alpha-1}\,\mathrm{d}K(\tau)\right) x(s)\,\mathrm{d}s\right)\right] \\ & + \frac{t^{\alpha-1}}{\Delta}\left[-\frac{1}{\Gamma(\beta)}\int_0^1\left(\int_s^1 (\tau-s)^{\beta-1}\,\mathrm{d}H(\tau)\right) y(s)\,\mathrm{d}s \right. \\ & \left. + \frac{1}{\Gamma(\beta)}\left(\int_0^1 \tau^{\beta-1}\,\mathrm{d}H(\tau)\right)\left(\int_0^1 (1-s)^{\beta-1} y(s)\,\mathrm{d}s\right)\right], \\ v(t) = & -\frac{1}{\Gamma(\beta)} \int_0^t (t-s)^{\beta-1} y(s)\,\mathrm{d}s + \frac{t^{\beta-1}}{\Delta}\left[\frac{1}{\Gamma(\beta)}\int_0^1 (1-s)^{\beta-1} y(s)\,\mathrm{d}s \right. \\ & \left. -\frac{1}{\Gamma(\beta)}\left(\int_0^1 \tau^{\alpha-1}\,\mathrm{d}K(\tau)\right)\left(\int_0^1\left(\int_s^1 (\tau-s)^{\beta-1}\,\mathrm{d}H(\tau)\right) y(s)\,\mathrm{d}s\right)\right] \\ & + \frac{t^{\beta-1}}{\Delta}\left[-\frac{1}{\Gamma(\alpha)}\int_0^1\left(\int_s^1 (\tau-s)^{\alpha-1}\,\mathrm{d}K(\tau)\right) x(s)\,\mathrm{d}s \right. \\ & \left. + \frac{1}{\Gamma(\alpha)}\left(\int_0^1 \tau^{\alpha-1}\,\mathrm{d}K(\tau)\right)\left(\int_0^1 (1-s)^{\alpha-1} x(s)\,\mathrm{d}s\right)\right]. \end{aligned} \tag{5.3}$$

Proof. By Lemma 4.1.3, the solutions of system (5.1) are

$$\begin{cases} u(t) = -\dfrac{1}{\Gamma(\alpha)}\displaystyle\int_0^t (t-s)^{\alpha-1}x(s)\,\mathrm{d}s + c_1 t^{\alpha-1} + \cdots + c_n t^{\alpha-n}, & t \in [0,1], \\ v(t) = -\dfrac{1}{\Gamma(\beta)}\displaystyle\int_0^t (t-s)^{\beta-1}y(s)\,\mathrm{d}s + d_1 t^{\beta-1} + \cdots + d_m t^{\beta-m}, & t \in [0,1], \end{cases}$$

where $c_1, \dots, c_n, d_1, \dots, d_m \in \mathbb{R}$. By using the conditions $u(0) = u'(0) = \cdots = u^{(n-2)}(0) = 0$ and $v(0) = v'(0) = \cdots = v^{(m-2)}(0) = 0$, we obtain $c_2 = c_3 = \cdots = c_n = 0$ and $d_2 = d_3 = \cdots = d_m = 0$. Then we conclude

$$\begin{cases} u(t) = c_1 t^{\alpha-1} - \dfrac{1}{\Gamma(\alpha)}\displaystyle\int_0^t (t-s)^{\alpha-1}x(s)\,\mathrm{d}s, & t \in [0,1], \\ v(t) = d_1 t^{\beta-1} - \dfrac{1}{\Gamma(\beta)}\displaystyle\int_0^t (t-s)^{\beta-1}y(s)\,\mathrm{d}s, & t \in [0,1]. \end{cases}$$

Now, by conditions $u(1) = \int_0^1 v(s)\,\mathrm{d}H(s)$ and $v(1) = \int_0^1 u(s)\,\mathrm{d}K(s)$, we deduce

$$\begin{cases} c_1 - \dfrac{1}{\Gamma(\alpha)}\displaystyle\int_0^1 (1-s)^{\alpha-1}x(s)\,\mathrm{d}s = \int_0^1 \left(d_1 s^{\beta-1} - \dfrac{1}{\Gamma(\beta)}\int_0^s (s-\tau)^{\beta-1}y(\tau)\,\mathrm{d}\tau\right)\mathrm{d}H(s), \\ d_1 - \dfrac{1}{\Gamma(\beta)}\displaystyle\int_0^1 (1-s)^{\beta-1}y(s)\,\mathrm{d}s = \int_0^1 \left(c_1 s^{\alpha-1} - \dfrac{1}{\Gamma(\alpha)}\int_0^s (s-\tau)^{\alpha-1}x(\tau)\,\mathrm{d}\tau\right)\mathrm{d}K(s), \end{cases}$$

or equivalently

$$\begin{cases} c_1 - d_1\displaystyle\int_0^1 s^{\beta-1}\,\mathrm{d}H(s) = \dfrac{1}{\Gamma(\alpha)}\int_0^1 (1-s)^{\alpha-1}x(s)\,\mathrm{d}s \\ \qquad -\dfrac{1}{\Gamma(\beta)}\displaystyle\int_0^1 \left(\int_s^1 (\tau-s)^{\beta-1}\,\mathrm{d}H(\tau)\right) y(s)\,\mathrm{d}s, \\ -c_1\displaystyle\int_0^1 s^{\alpha-1}\,\mathrm{d}K(s) + d_1 = \dfrac{1}{\Gamma(\beta)}\int_0^1 (1-s)^{\beta-1}y(s)\,\mathrm{d}s \\ \qquad -\dfrac{1}{\Gamma(\alpha)}\displaystyle\int_0^1 \left(\int_s^1 (\tau-s)^{\alpha-1}\,\mathrm{d}K(\tau)\right) x(s)\,\mathrm{d}s. \end{cases}$$

The above system in the unknowns c_1 and d_1 has the determinant

$$\Delta = \begin{vmatrix} 1 & -\displaystyle\int_0^1 s^{\beta-1}\mathrm{d}H(s) \\ -\displaystyle\int_0^1 s^{\alpha-1}\mathrm{d}K(s) & 1 \end{vmatrix} = 1 - \left(\int_0^1 s^{\alpha-1}\mathrm{d}K(s)\right)\left(\int_0^1 s^{\beta-1}\mathrm{d}H(s)\right).$$

By the assumptions of this lemma, we have $\Delta \neq 0$.

So, we obtain

$$c_1 = \frac{1}{\Delta}\left[\frac{1}{\Gamma(\alpha)}\int_0^1 (1-s)^{\alpha-1}x(s)\,\mathrm{d}s\right.$$
$$-\frac{1}{\Gamma(\alpha)}\left(\int_0^1 \tau^{\beta-1}\,\mathrm{d}H(\tau)\right)\left(\int_0^1 \left(\int_s^1 (\tau-s)^{\alpha-1}\,\mathrm{d}K(\tau)\right) x(s)\,\mathrm{d}s\right)$$

$$- \frac{1}{\Gamma(\beta)} \int_0^1 \left(\int_s^1 (\tau - s)^{\beta-1} \, dH(\tau) \right) y(s) \, ds$$
$$+ \frac{1}{\Gamma(\beta)} \left(\int_0^1 \tau^{\beta-1} \, dH(\tau) \right) \left(\int_0^1 (1-s)^{\beta-1} y(s) \, ds \right) \Bigg],$$

and

$$d_1 = \frac{1}{\Delta} \Bigg[- \frac{1}{\Gamma(\alpha)} \int_0^1 \left(\int_s^1 (\tau - s)^{\alpha-1} \, dK(\tau) \right) x(s) \, ds$$
$$+ \frac{1}{\Gamma(\alpha)} \left(\int_0^1 \tau^{\alpha-1} \, dK(\tau) \right) \left(\int_0^1 (1-s)^{\alpha-1} x(s) \, ds \right)$$
$$+ \frac{1}{\Gamma(\beta)} \int_0^1 (1-s)^{\beta-1} y(s) \, ds$$
$$- \frac{1}{\Gamma(\beta)} \left(\int_0^1 \tau^{\alpha-1} \, dK(\tau) \right) \left(\int_0^1 \left(\int_s^1 (\tau - s)^{\beta-1} \, dH(\tau) \right) y(s) \, ds \right) \Bigg].$$

Therefore, we deduce expression (5.3) for the solution of problem (5.1)–(5.2). □

Lemma 5.1.2. *Under the assumptions of Lemma 5.1.1, the solution* (u, v) *of problem* (5.1)–(5.2) *can be written as*

$$\begin{cases} u(t) = \displaystyle\int_0^1 G_1(t,s) x(s) \, ds + \int_0^1 G_2(t,s) y(s) \, ds, & t \in [0,1], \\ v(t) = \displaystyle\int_0^1 G_3(t,s) y(s) \, ds + \int_0^1 G_4(t,s) x(s) \, ds, & t \in [0,1], \end{cases} \tag{5.4}$$

where

$$\begin{cases} G_1(t,s) = g_1(t,s) + \dfrac{t^{\alpha-1}}{\Delta} \left(\displaystyle\int_0^1 \tau^{\beta-1} dH(\tau) \right) \left(\int_0^1 g_1(\tau,s) \, dK(\tau) \right), \\ G_2(t,s) = \dfrac{t^{\alpha-1}}{\Delta} \displaystyle\int_0^1 g_2(\tau,s) \, dH(\tau), \\ G_3(t,s) = g_2(t,s) + \dfrac{t^{\beta-1}}{\Delta} \left(\displaystyle\int_0^1 \tau^{\alpha-1} dK(\tau) \right) \left(\int_0^1 g_2(\tau,s) \, dH(\tau) \right), \\ G_4(t,s) = \dfrac{t^{\beta-1}}{\Delta} \displaystyle\int_0^1 g_1(\tau,s) \, dK(\tau), \quad \forall t, s \in [0,1], \end{cases} \tag{5.5}$$

and

$$\begin{cases} g_1(t,s) = \dfrac{1}{\Gamma(\alpha)} \begin{cases} t^{\alpha-1}(1-s)^{\alpha-1} - (t-s)^{\alpha-1}, & 0 \le s \le t \le 1, \\ t^{\alpha-1}(1-s)^{\alpha-1}, & 0 \le t \le s \le 1, \end{cases} \\ g_2(t,s) = \dfrac{1}{\Gamma(\beta)} \begin{cases} t^{\beta-1}(1-s)^{\beta-1} - (t-s)^{\beta-1}, & 0 \le s \le t \le 1, \\ t^{\beta-1}(1-s)^{\beta-1}, & 0 \le t \le s \le 1. \end{cases} \end{cases} \tag{5.6}$$

Proof. By Lemma 5.1.1 and relation (5.3), we conclude

$$
\begin{aligned}
u(t) &= \frac{1}{\Gamma(\alpha)}\left[\int_0^t [t^{\alpha-1}(1-s)^{\alpha-1}-(t-s)^{\alpha-1}]x(s)\,\mathrm{d}s + \int_t^1 t^{\alpha-1}(1-s)^{\alpha-1}x(s)\,\mathrm{d}s\right.\\
&\quad \left. - \int_0^1 t^{\alpha-1}(1-s)^{\alpha-1}x(s)\,\mathrm{d}s\right] + \frac{t^{\alpha-1}}{\Delta\Gamma(\alpha)}\int_0^1 (1-s)^{\alpha-1}x(s)\,\mathrm{d}s\\
&\quad - \frac{t^{\alpha-1}}{\Delta\Gamma(\alpha)}\left(\int_0^1 \tau^{\beta-1}\,\mathrm{d}H(\tau)\right)\left(\int_0^1\left(\int_s^1 (\tau-s)^{\alpha-1}\,\mathrm{d}K(\tau)\right)x(s)\,\mathrm{d}s\right)\\
&\quad + \frac{t^{\alpha-1}}{\Delta\Gamma(\beta)}\left[\int_0^1\left(\int_0^1 \tau^{\beta-1}(1-s)^{\beta-1}\,\mathrm{d}H(\tau)\right)y(s)\,\mathrm{d}s\right.\\
&\quad \left. - \int_0^1\left(\int_s^1 (\tau-s)^{\beta-1}\,\mathrm{d}H(\tau)\right)y(s)\,\mathrm{d}s\right]\\
&= \frac{1}{\Gamma(\alpha)}\left[\int_0^t [t^{\alpha-1}(1-s)^{\alpha-1}-(t-s)^{\alpha-1}]x(s)\,\mathrm{d}s + \int_t^1 t^{\alpha-1}(1-s)^{\alpha-1}x(s)\,\mathrm{d}s\right.\\
&\quad - \frac{1}{\Delta}\int_0^1 t^{\alpha-1}(1-s)^{\alpha-1}x(s)\,\mathrm{d}s + \frac{1}{\Delta}\left(\int_0^1 \tau^{\alpha-1}\mathrm{d}K(\tau)\right)\left(\int_0^1 \tau^{\beta-1}\,\mathrm{d}H(\tau)\right)\\
&\quad \times\left(\int_0^1 t^{\alpha-1}(1-s)^{\alpha-1}x(s)\,\mathrm{d}s\right) + \frac{1}{\Delta}\int_0^1 t^{\alpha-1}(1-s)^{\alpha-1}x(s)\,\mathrm{d}s\\
&\quad \left. - \frac{t^{\alpha-1}}{\Delta}\left(\int_0^1 \tau^{\beta-1}\mathrm{d}H(\tau)\right)\left(\int_0^1\left(\int_s^1 (\tau-s)^{\alpha-1}\mathrm{d}K(\tau)\right)x(s)\,\mathrm{d}s\right)\right]\\
&\quad + \frac{t^{\alpha-1}}{\Delta\Gamma(\beta)}\left[\int_0^1\left(\int_0^1 \tau^{\beta-1}(1-s)^{\beta-1}\mathrm{d}H(\tau)\right)y(s)\,\mathrm{d}s\right.\\
&\quad \left. - \int_0^1\left(\int_s^1 (\tau-s)^{\beta-1}\mathrm{d}H(\tau)\right)y(s)\,\mathrm{d}s\right]\\
&= \frac{1}{\Gamma(\alpha)}\left\{\int_0^t [t^{\alpha-1}(1-s)^{\alpha-1}-(t-s)^{\alpha-1}]x(s)\,\mathrm{d}s + \int_t^1 t^{\alpha-1}(1-s)^{\alpha-1}x(s)\,\mathrm{d}s\right.\\
&\quad + \frac{t^{\alpha-1}}{\Delta}\left(\int_0^1 \tau^{\beta-1}\mathrm{d}H(\tau)\right)\left[\int_0^1\left(\int_0^1 \tau^{\alpha-1}(1-s)^{\alpha-1}\mathrm{d}K(\tau)\right)x(s)\,\mathrm{d}s\right.\\
&\quad \left.\left. - \int_0^1\left(\int_s^1 (\tau-s)^{\alpha-1}\mathrm{d}K(\tau)\right)x(s)\,\mathrm{d}s\right]\right\}\\
&\quad + \frac{t^{\alpha-1}}{\Delta\Gamma(\beta)}\left[\int_0^1\left(\int_0^1 \tau^{\beta-1}(1-s)^{\beta-1}\mathrm{d}H(\tau)\right)y(s)\,\mathrm{d}s\right.\\
&\quad \left. - \int_0^1\left(\int_s^1 (\tau-s)^{\beta-1}\mathrm{d}H(\tau)\right)y(s)\,\mathrm{d}s\right].
\end{aligned}
$$

Therefore, we obtain

$$
\begin{aligned}
u(t) &= \frac{1}{\Gamma(\alpha)}\left\{\int_0^t [t^{\alpha-1}(1-s)^{\alpha-1} - (t-s)^{\alpha-1}]x(s)\,\mathrm{d}s + \int_t^1 t^{\alpha-1}(1-s)^{\alpha-1}x(s)\,\mathrm{d}s\right.\\
&\quad + \frac{t^{\alpha-1}}{\Delta}\left(\int_0^1 \tau^{\beta-1}\mathrm{d}H(\tau)\right)\left[\int_0^1\left(\int_0^s \tau^{\alpha-1}(1-s)^{\alpha-1}\mathrm{d}K(\tau)\right)x(s)\,\mathrm{d}s\right.\\
&\quad \left.\left.+ \int_0^1\left(\int_s^1 \tau^{\alpha-1}(1-s)^{\alpha-1}\mathrm{d}K(\tau)\right)x(s)\,\mathrm{d}s - \int_0^1\left(\int_s^1 (\tau-s)^{\alpha-1}\mathrm{d}K(\tau)\right)x(s)\,\mathrm{d}s\right]\right\}\\
&\quad + \frac{t^{\alpha-1}}{\Delta\Gamma(\beta)}\left\{\int_0^1\left(\int_0^s \tau^{\beta-1}(1-s)^{\beta-1}\mathrm{d}H(\tau)\right)y(s)\,\mathrm{d}s\right.\\
&\quad \left.+ \int_0^1\left(\int_s^1 \tau^{\beta-1}(1-s)^{\beta-1}\mathrm{d}H(\tau)\right)y(s)\,\mathrm{d}s - \int_0^1\left(\int_s^1 (\tau-s)^{\beta-1}\mathrm{d}H(\tau)\right)y(s)\,\mathrm{d}s\right\}\\
&= \frac{1}{\Gamma(\alpha)}\left\{\int_0^t [t^{\alpha-1}(1-s)^{\alpha-1} - (t-s)^{\alpha-1}]x(s)\,\mathrm{d}s + \int_t^1 t^{\alpha-1}(1-s)^{\alpha-1}x(s)\,\mathrm{d}s\right.\\
&\quad + \frac{t^{\alpha-1}}{\Delta}\left(\int_0^1 \tau^{\beta-1}\mathrm{d}H(\tau)\right)\left[\int_0^1\left(\int_0^s \tau^{\alpha-1}(1-s)^{\alpha-1}\mathrm{d}K(\tau)\right)x(s)\,\mathrm{d}s\right.\\
&\quad \left.\left.+ \int_0^1\left(\int_s^1 [\tau^{\alpha-1}(1-s)^{\alpha-1} - (\tau-s)^{\alpha-1}]\,\mathrm{d}K(\tau)\right)x(s)\,\mathrm{d}s\right]\right\}\\
&\quad + \frac{t^{\alpha-1}}{\Delta\Gamma(\beta)}\left\{\int_0^1\left(\int_0^s \tau^{\beta-1}(1-s)^{\beta-1}\mathrm{d}H(\tau)\right)y(s)\,\mathrm{d}s\right.\\
&\quad \left.+ \int_0^1\left(\int_s^1 [\tau^{\beta-1}(1-s)^{\beta-1} - (\tau-s)^{\beta-1}]\,\mathrm{d}H(\tau)\right)y(s)\,\mathrm{d}s\right\}\\
&= \int_0^1 g_1(t,s)x(s)\,\mathrm{d}s + \frac{t^{\alpha-1}}{\Delta}\left(\int_0^1 \tau^{\beta-1}\mathrm{d}H(\tau)\right)\left(\int_0^1\left(\int_0^1 g_1(\tau,s)\,\mathrm{d}K(\tau)\right)x(s)\,\mathrm{d}s\right)\\
&\quad + \frac{t^{\alpha-1}}{\Delta}\left(\int_0^1\left(\int_0^1 g_2(\tau,s)\,\mathrm{d}H(\tau)\right)y(s)\,\mathrm{d}s\right) = \int_0^1 G_1(t,s)x(s)\,\mathrm{d}s + \int_0^1 G_2(t,s)y(s)\,\mathrm{d}s.
\end{aligned}
$$

In a similar manner, we deduce

$$
\begin{aligned}
v(t) &= \frac{1}{\Gamma(\beta)}\left[\int_0^t [t^{\beta-1}(1-s)^{\beta-1} - (t-s)^{\beta-1}]y(s)\,\mathrm{d}s + \int_t^1 t^{\beta-1}(1-s)^{\beta-1}y(s)\,\mathrm{d}s\right.\\
&\quad \left.- \int_0^1 t^{\beta-1}(1-s)^{\beta-1}y(s)\,\mathrm{d}s\right] + \frac{t^{\beta-1}}{\Delta\Gamma(\beta)}\int_0^1 (1-s)^{\beta-1}y(s)\,\mathrm{d}s\\
&\quad - \frac{t^{\beta-1}}{\Delta\Gamma(\beta)}\left(\int_0^1 \tau^{\alpha-1}\,\mathrm{d}K(\tau)\right)\left(\int_0^1\left(\int_s^1 (\tau-s)^{\beta-1}\,\mathrm{d}H(\tau)\right)y(s)\,\mathrm{d}s\right)\\
&\quad + \frac{t^{\beta-1}}{\Delta\Gamma(\alpha)}\left[\int_0^1\left(\int_0^1 \tau^{\alpha-1}(1-s)^{\alpha-1}\,\mathrm{d}K(\tau)\right)x(s)\,\mathrm{d}s\right.
\end{aligned}
$$

$$
\begin{aligned}
&- \int_0^1 \left(\int_s^1 (\tau - s)^{\alpha-1} \, \mathrm{d}K(\tau) \right) x(s) \, \mathrm{d}s \Bigg] \\
&= \frac{1}{\Gamma(\beta)} \Bigg[\int_0^t [t^{\beta-1}(1-s)^{\beta-1} - (t-s)^{\beta-1}] y(s) \, \mathrm{d}s + \int_t^1 t^{\beta-1}(1-s)^{\beta-1} y(s) \, \mathrm{d}s \\
&\quad - \frac{1}{\Delta} \int_0^1 t^{\beta-1}(1-s)^{\beta-1} y(s) \, \mathrm{d}s + \frac{1}{\Delta} \left(\int_0^1 \tau^{\alpha-1} \mathrm{d}K(\tau) \right) \left(\int_0^1 \tau^{\beta-1} \, \mathrm{d}H(\tau) \right) \\
&\quad \times \left(\int_0^1 t^{\beta-1}(1-s)^{\beta-1} y(s) \, \mathrm{d}s \right) + \frac{1}{\Delta} \int_0^1 t^{\beta-1}(1-s)^{\beta-1} y(s) \, \mathrm{d}s \\
&\quad - \frac{t^{\beta-1}}{\Delta} \left(\int_0^1 \tau^{\alpha-1} \mathrm{d}K(\tau) \right) \left(\int_0^1 \left(\int_s^1 (\tau - s)^{\beta-1} \mathrm{d}H(\tau) \right) y(s) \, \mathrm{d}s \right) \Bigg] \\
&\quad + \frac{t^{\beta-1}}{\Delta \Gamma(\alpha)} \Bigg[\int_0^1 \left(\int_0^1 \tau^{\alpha-1}(1-s)^{\alpha-1} \mathrm{d}K(\tau) \right) x(s) \, \mathrm{d}s \\
&\quad - \int_0^1 \left(\int_s^1 (\tau - s)^{\alpha-1} \mathrm{d}K(\tau) \right) x(s) \, \mathrm{d}s \Bigg] \\
&= \frac{1}{\Gamma(\beta)} \Bigg\{ \int_0^t [t^{\beta-1}(1-s)^{\beta-1} - (t-s)^{\beta-1}] y(s) \, \mathrm{d}s + \int_t^1 t^{\beta-1}(1-s)^{\beta-1} y(s) \, \mathrm{d}s \\
&\quad + \frac{t^{\beta-1}}{\Delta} \left(\int_0^1 \tau^{\alpha-1} \mathrm{d}K(\tau) \right) \Bigg[\int_0^1 \left(\int_0^1 \tau^{\beta-1}(1-s)^{\beta-1} \mathrm{d}H(\tau) \right) y(s) \, \mathrm{d}s \\
&\quad - \int_0^1 \left(\int_s^1 (\tau - s)^{\beta-1} \mathrm{d}H(\tau) \right) y(s) \, \mathrm{d}s \Bigg] \Bigg\} \\
&\quad + \frac{t^{\beta-1}}{\Delta \Gamma(\alpha)} \Bigg[\int_0^1 \left(\int_0^1 \tau^{\alpha-1}(1-s)^{\alpha-1} \mathrm{d}K(\tau) \right) x(s) \, \mathrm{d}s \\
&\quad - \int_0^1 \left(\int_s^1 (\tau - s)^{\alpha-1} \mathrm{d}K(\tau) \right) x(s) \, \mathrm{d}s \Bigg].
\end{aligned}
$$

So, we conclude

$$
\begin{aligned}
v(t) &= \frac{1}{\Gamma(\beta)} \Bigg\{ \int_0^t [t^{\beta-1}(1-s)^{\beta-1} - (t-s)^{\beta-1}] y(s) \, \mathrm{d}s + \int_t^1 t^{\beta-1}(1-s)^{\beta-1} y(s) \, \mathrm{d}s \\
&\quad + \frac{t^{\beta-1}}{\Delta} \left(\int_0^1 \tau^{\alpha-1} \mathrm{d}K(\tau) \right) \Bigg[\int_0^1 \left(\int_0^s \tau^{\beta-1}(1-s)^{\beta-1} \mathrm{d}H(\tau) \right) y(s) \, \mathrm{d}s \\
&\quad + \int_0^1 \left(\int_s^1 \tau^{\beta-1}(1-s)^{\beta-1} \mathrm{d}H(\tau) \right) y(s) \, \mathrm{d}s - \int_0^1 \left(\int_s^1 (\tau - s)^{\beta-1} \mathrm{d}H(\tau) \right) y(s) \, \mathrm{d}s \Bigg] \Bigg\} \\
&\quad + \frac{t^{\beta-1}}{\Delta \Gamma(\alpha)} \Bigg\{ \int_0^1 \left(\int_0^s \tau^{\alpha-1}(1-s)^{\alpha-1} \mathrm{d}K(\tau) \right) x(s) \, \mathrm{d}s \\
&\quad + \int_0^1 \left(\int_s^1 \tau^{\alpha-1}(1-s)^{\alpha-1} \mathrm{d}K(\tau) \right) x(s) \, \mathrm{d}s - \int_0^1 \left(\int_s^1 (\tau - s)^{\alpha-1} \mathrm{d}K(\tau) \right) x(s) \, \mathrm{d}s \Bigg\}
\end{aligned}
$$

$$= \frac{1}{\Gamma(\beta)}\left\{\int_0^t [t^{\beta-1}(1-s)^{\beta-1} - (t-s)^{\beta-1}]y(s)\,ds + \int_t^1 t^{\beta-1}(1-s)^{\beta-1}y(s)\,ds\right.$$

$$+ \frac{t^{\beta-1}}{\Delta}\left(\int_0^1 \tau^{\alpha-1}dK(\tau)\right)\left[\int_0^1\left(\int_0^s \tau^{\beta-1}(1-s)^{\beta-1}dH(\tau)\right)y(s)\,ds\right.$$

$$\left.\left. + \int_0^1\left(\int_s^1 [\tau^{\beta-1}(1-s)^{\beta-1} - (\tau-s)^{\beta-1}]\,dH(\tau)\right)y(s)\,ds\right]\right\}$$

$$+ \frac{t^{\beta-1}}{\Delta\Gamma(\alpha)}\left\{\int_0^1\left(\int_0^s \tau^{\alpha-1}(1-s)^{\alpha-1}dK(\tau)\right)x(s)\,ds\right.$$

$$\left. + \int_0^1\left(\int_s^1 [\tau^{\alpha-1}(1-s)^{\alpha-1} - (\tau-s)^{\alpha-1}]\,dK(\tau)\right)x(s)\,ds\right\}$$

$$= \int_0^1 g_2(t,s)y(s)\,ds + \frac{t^{\beta-1}}{\Delta}\left(\int_0^1 \tau^{\alpha-1}dK(\tau)\right)\left(\int_0^1\left(\int_0^1 g_2(\tau,s)\,dH(\tau)\right)y(s)\,ds\right)$$

$$+ \frac{t^{\beta-1}}{\Delta}\left(\int_0^1\left(\int_0^1 g_1(\tau,s)\,dK(\tau)\right)x(s)\,ds\right) = \int_0^1 G_3(t,s)y(s)\,ds + \int_0^1 G_4(t,s)x(s)\,ds.$$

Therefore, we obtain expression (5.4) for the solution of problem (5.1)–(5.2). □

Lemma 5.1.3. *The functions g_1 and g_2 given by* (5.6) *have the following properties:*

(a) $g_1, g_2 : [0,1]\times[0,1] \to \mathbb{R}_+$ *are continuous functions, and* $g_1(t,s) > 0$, $g_2(t,s) > 0$ *for all* $(t,s) \in (0,1)\times(0,1)$*;*

(b) $g_1(t,s) \le g_1(\theta_1(s),s)$ *and* $g_2(t,s) \le g_2(\theta_2(s),s)$ *for all* $(t,s) \in [0,1]\times[0,1]$*;*

(c) *For any* $c \in (0,1/2)$, *we have* $\min_{t\in[c,1-c]} g_1(t,s) \ge \gamma_1 g_1(\theta_1(s),s)$ *and* $\min_{t\in[c,1-c]} g_2(t,s) \ge \gamma_2 g_2(\theta_2(s),s)$ *for all* $s \in [0,1]$,

where $\gamma_1 = c^{\alpha-1}$, $\gamma_2 = c^{\beta-1}$, $\theta_1(s) = \begin{cases} \dfrac{s}{1-(1-s)^{\frac{\alpha-1}{\alpha-2}}}, & s \in (0,1], \\ \dfrac{\alpha-2}{\alpha-1}, & s = 0, \end{cases}$ *if* $n-1 < \alpha \le n$,

$n \ge 3$, *and* $\theta_2(s) = \begin{cases} \dfrac{s}{1-(1-s)^{\frac{\beta-1}{\beta-2}}}, & s \in (0,1], \\ \dfrac{\beta-2}{\beta-1}, & s = 0. \end{cases}$ *if* $m-1 < \beta \le m$, $m \ge 3$.

The proof of Lemma 5.1.3 is similar to that of Lemma 3.3 from Ji and Guo (2009). We define θ_1 at $s = 0$ as $\frac{\alpha-2}{\alpha-1}$ so that θ_1 is a continuous function.

Lemma 5.1.4. *If* $H, K : [0,1] \to \mathbb{R}$ *are nondecreasing functions, and* $\Delta > 0$, *then* G_i, $i = 1,\ldots,4$ *given by* (5.5) *are continuous functions on* $[0,1]\times[0,1]$ *and satisfy* $G_i(t,s) \ge 0$ *for all* $(t,s) \in [0,1]\times[0,1]$, $i = 1,\ldots,4$. *Moreover, if* $x, y \in C(0,1)\cap L^1(0,1)$ *satisfy* $x(t) \ge 0$, $y(t) \ge 0$ *for all* $t \in (0,1)$, *then the solution* (u,v) *of problem* (5.1)–(5.2) *satisfies* $u(t) \ge 0$, $v(t) \ge 0$ *for all* $t \in [0,1]$.

Proof. By using the assumptions of this lemma, we have $G_i(t,s) \ge 0$ for all $(t,s) \in [0,1]\times[0,1]$, $i = 1,\ldots,4$, and so $u(t) \ge 0$, $v(t) \ge 0$ for all $t \in [0,1]$. □

Lemma 5.1.5. *Assume that* $H, K : [0,1] \to \mathbb{R}$ *are nondecreasing functions, and* $\Delta > 0$. *Then the functions* G_i, $i = 1,\ldots,4$ *satisfy the following inequalities:*

(a_1) $G_1(t,s) \le J_1(s), \quad \forall (t,s) \in [0,1] \times [0,1]$, *where*

$$J_1(s) = g_1(\theta_1(s),s) + \frac{1}{\Delta}\left(\int_0^1 \tau^{\beta-1} \mathrm{d}H(\tau)\right)\left(\int_0^1 g_1(\tau,s)\,\mathrm{d}K(\tau)\right).$$

(a_2) *For every* $c \in (0,1/2)$, *we have*

$$\min_{t\in[c,1-c]} G_1(t,s) \ge \gamma_1 J_1(s) \ge \gamma_1 G_1(t',s), \quad \forall t',\, s \in [0,1].$$

(b_1) $G_2(t,s) \le J_2(s), \quad \forall (t,s) \in [0,1]\times[0,1]$, *where* $J_2(s) = \frac{1}{\Delta}\int_0^1 g_2(\tau,s)\,\mathrm{d}H(\tau)$.
(b_2) *For every* $c \in (0,1/2)$, *we have*

$$\min_{t\in[c,1-c]} G_2(t,s) \ge \gamma_1 J_2(s) \ge \gamma_1 G_2(t',s), \quad \forall t',\, s \in [0,1].$$

(c_1) $G_3(t,s) \le J_3(s), \quad \forall (t,s) \in [0,1] \times [0,1]$, *where*

$$J_3(s) = g_2(\theta_2(s),s) + \frac{1}{\Delta}\left(\int_0^1 \tau^{\alpha-1} \mathrm{d}K(\tau)\right)\left(\int_0^1 g_2(\tau,s)\,\mathrm{d}H(\tau)\right).$$

(c_2) *For every* $c \in (0,1/2)$, *we have*

$$\min_{t\in[c,1-c]} G_3(t,s) \ge \gamma_2 J_3(s) \ge \gamma_2 G_3(t',s), \quad \forall t',\, s \in [0,1].$$

(d_1) $G_4(t,s) \le J_4(s), \quad \forall (t,s) \in [0,1]\times[0,1]$, *where* $J_4(s) = \frac{1}{\Delta}\int_0^1 g_1(\tau,s)\,\mathrm{d}K(\tau)$.
(d_2) *For every* $c \in (0,1/2)$, *we have*

$$\min_{t\in[c,1-c]} G_4(t,s) \ge \gamma_2 J_4(s) \ge \gamma_2 G_4(t',s), \quad \forall t',\, s \in [0,1].$$

Proof. Inequalities (a_1), (b_1), (c_1), and (d_1) are evident. For the other inequalities, for $c \in (0,1/2)$ and $t \in [c,1-c]$, $t', s \in [0,1]$, we deduce

$$\begin{aligned}
G_1(t,s) &\ge \gamma_1 g_1(\theta_1(s),s) + \frac{c^{\alpha-1}}{\Delta}\left(\int_0^1 \tau^{\beta-1}\mathrm{d}H(\tau)\right)\left(\int_0^1 g_1(\tau,s)\,\mathrm{d}K(\tau)\right) \\
&= \gamma_1 J_1(s) \ge \gamma_1 G_1(t',s), \\
G_2(t,s) &\ge \frac{c^{\alpha-1}}{\Delta}\int_0^1 g_2(\tau,s)\,\mathrm{d}H(\tau) = \gamma_1 J_2(s) \ge \gamma_1 G_2(t',s), \\
G_3(t,s) &\ge \gamma_2 g_2(\theta_2(s),s) + \frac{c^{\beta-1}}{\Delta}\left(\int_0^1 \tau^{\alpha-1}\mathrm{d}K(\tau)\right)\left(\int_0^1 g_2(\tau,s)\,\mathrm{d}H(\tau)\right) \\
&= \gamma_2 J_3(s) \ge \gamma_2 G_3(t',s), \\
G_4(t,s) &\ge \frac{c^{\beta-1}}{\Delta}\int_0^1 g_1(\tau,s)\,\mathrm{d}K(\tau) = \gamma_2 J_4(s) \ge \gamma_2 G_4(t',s).
\end{aligned}$$

Therefore, we obtain the inequalities (a_2), (b_2), (c_2), and (d_2) of this lemma. □

Lemma 5.1.6. *Assume that* $H, K : [0,1] \to \mathbb{R}$ *are nondecreasing functions,* $\Delta > 0$, $c \in (0,1/2)$, *and* $x, y \in C(0,1) \cap L^1(0,1)$, $x(t) \ge 0$, $y(t) \ge 0$ *for all*

$t \in (0,1)$. *Then the solution* $(u(t), v(t))$, $t \in [0,1]$ *of problem* (5.1)–(5.2) *satisfies the inequalities*

$$\inf_{t\in[c,1-c]} u(t) \geq \gamma_1 \sup_{t'\in[0,1]} u(t'), \qquad \inf_{t\in[c,1-c]} v(t) \geq \gamma_2 \sup_{t'\in[0,1]} v(t').$$

Proof. For $c \in (0, 1/2)$, $t \in [c, 1-c]$, and $t' \in [0,1]$, we have

$$\begin{aligned}
u(t) &= \int_0^1 G_1(t,s)x(s)\,ds + \int_0^1 G_2(t,s)y(s)\,ds \geq \gamma_1 \int_0^1 J_1(s)x(s)\,ds \\
&\quad + \gamma_1 \int_0^1 J_2(s)y(s)\,ds \geq \gamma_1 \int_0^1 G_1(t',s)x(s)\,ds \\
&\quad + \gamma_1 \int_0^1 G_2(t',s)y(s)\,ds = \gamma_1 u(t'), \\
v(t) &= \int_0^1 G_3(t,s)y(s)\,ds + \int_0^1 G_4(t,s)x(s)\,ds \geq \gamma_2 \int_0^1 J_3(s)y(s)\,ds \\
&\quad + \gamma_2 \int_0^1 J_4(s)x(s)\,ds \geq \gamma_2 \int_0^1 G_3(t',s)y(s)\,ds \\
&\quad + \gamma_2 \int_0^1 G_4(t',s)x(s)\,ds = \gamma_2 v(t').
\end{aligned}$$

Then we deduce the conclusion of this lemma. □

5.1.3 Main results

In this section, we shall give sufficient conditions on λ, μ, f, and g such that positive solutions with respect to a cone for our problem (S)–(BC) exist.

We present the assumptions that we shall use in the sequel:

(H1) $H, K : [0,1] \to \mathbb{R}$ are nondecreasing functions, and $\Delta = 1 - \left(\int_0^1 \tau^{\alpha-1}\,dK(\tau)\right)\left(\int_0^1 \tau^{\beta-1}\,dH(\tau)\right) > 0$.

(H2) The functions $f, g : [0,1] \times [0,\infty) \times [0,\infty) \to [0,\infty)$ are continuous.

For $c \in (0, 1/2)$, we introduce the following extreme limits:

$$\begin{aligned}
f_0^s &= \limsup_{u+v\to 0^+} \max_{t\in[0,1]} \frac{f(t,u,v)}{u+v}, & g_0^s &= \limsup_{u+v\to 0^+} \max_{t\in[0,1]} \frac{g(t,u,v)}{u+v}, \\
f_0^i &= \liminf_{u+v\to 0^+} \min_{t\in[c,1-c]} \frac{f(t,u,v)}{u+v}, & g_0^i &= \liminf_{u+v\to 0^+} \min_{t\in[c,1-c]} \frac{g(t,u,v)}{u+v}, \\
f_\infty^s &= \limsup_{u+v\to\infty} \max_{t\in[0,1]} \frac{f(t,u,v)}{u+v}, & g_\infty^s &= \limsup_{u+v\to\infty} \max_{t\in[0,1]} \frac{g(t,u,v)}{u+v}, \\
f_\infty^i &= \liminf_{u+v\to\infty} \min_{t\in[c,1-c]} \frac{f(t,u,v)}{u+v}, & g_\infty^i &= \liminf_{u+v\to\infty} \min_{t\in[c,1-c]} \frac{g(t,u,v)}{u+v}.
\end{aligned}$$

In the definitions of the extreme limits above, the variables u and v are nonnegative.

By using the functions $G_i, i = 1, \ldots, 4$ from Section 5.1.2 (Lemma 5.1.2), we can write problem (S)–(BC) equivalently as the following nonlinear system of integral equations:

$$\begin{cases} u(t) = \lambda \int_0^1 G_1(t,s)f(s,u(s),v(s))\,\mathrm{d}s + \mu \int_0^1 G_2(t,s)g(s,u(s),v(s))\,\mathrm{d}s, & t \in [0,1], \\ v(t) = \mu \int_0^1 G_3(t,s)g(s,u(s),v(s))\,\mathrm{d}s + \lambda \int_0^1 G_4(t,s)f(s,u(s),v(s))\,\mathrm{d}s, & t \in [0,1]. \end{cases}$$

We consider the Banach space $X = C([0,1])$ with the supremum norm $\|\cdot\|$ and the Banach space $Y = X \times X$ with the norm $\|(u,v)\|_Y = \|u\| + \|v\|$. We define the cone $P \subset Y$ by

$$P = \{(u,v) \in Y; \quad u(t) \geq 0, \quad v(t) \geq 0, \quad \forall t \in [0,1] \quad \text{and} \\ \inf_{t \in [c,1-c]} (u(t) + v(t)) \geq \gamma \|(u,v)\|_Y\},$$

where $\gamma = \min\{\gamma_1, \gamma_2\}$, and γ_1 and γ_2 are defined in Section 5.1.2.

For λ, $\mu > 0$, we introduce the operators T_1, $T_2 : Y \to X$ and $\mathcal{T} : Y \to Y$ defined by

$$T_1(u,v)(t) = \lambda \int_0^1 G_1(t,s)f(s,u(s),v(s))\,\mathrm{d}s + \mu \int_0^1 G_2(t,s)g(s,u(s),v(s))\,\mathrm{d}s, \quad 0 \leq t \leq 1,$$

$$T_2(u,v)(t) = \mu \int_0^1 G_3(t,s)g(s,u(s),v(s))\,\mathrm{d}s + \lambda \int_0^1 G_4(t,s)f(s,u(s),v(s))\,\mathrm{d}s, \quad 0 \leq t \leq 1,$$

and $\mathcal{T}(u,v) = (T_1(u,v), T_2(u,v)), (u,v) \in Y$. The positive solutions of our problem (S)–(BC) coincide with the fixed points of the operator $\mathcal{T}$.

Lemma 5.1.7. *If (H1) and (H2) hold and $c \in (0,1/2)$, then $\mathcal{T} : P \to P$ is a completely continuous operator.*

Proof. Let $(u,v) \in P$ be an arbitrary element. Because $T_1(u,v)$ and $T_2(u,v)$ satisfy problem (5.1)–(5.2) for $x(t) = \lambda f(t,u(t),v(t))$, $t \in [0,1]$, and $y(t) = \mu g(t,u(t),v(t))$, $t \in [0,1]$, then by Lemma 5.1.6, we obtain

$$\inf_{t \in [c,1-c]} T_1(u,v)(t) \geq \gamma_1 \sup_{t' \in [0,1]} T_1(u,v)(t') = \gamma_1 \|T_1(u,v)\|,$$

$$\inf_{t \in [c,1-c]} T_2(u,v)(t) \geq \gamma_2 \sup_{t' \in [0,1]} T_2(u,v)(t') = \gamma_2 \|T_2(u,v)\|.$$

Hence, we conclude

$$\inf_{t \in [c,1-c]} [T_1(u,v)(t) + T_2(u,v)(t)] \geq \inf_{t \in [c,1-c]} T_1(u,v)(t) + \inf_{t \in [c,1-c]} T_2(u,v)(t) \\ \geq \gamma_1 \|T_1(u,v)\| + \gamma_2 \|T_2(u,v)\| \geq \gamma \|(T_1(u,v), T_2(u,v))\|_Y = \gamma \|\mathcal{T}(u,v)\|_Y.$$

By Lemma 5.1.4, (H1), and (H2), we obtain $T_1(u,v)(t) \geq 0$ and $T_2(u,v)(t) \geq 0$ for all $t \in [0,1]$, and so we deduce that $\mathcal{T}(u,v) \in P$. Hence, we get $\mathcal{T}(P) \subset P$. By using standard arguments, we can easily show that T_1 and T_2 are completely continuous, and then $\mathcal{T}$ is a completely continuous operator. □

For $c \in (0,1/2)$, we denote $A = \int_0^1 J_1(s)\,ds$, $B = \int_0^1 J_2(s)\,ds$, $C = \int_0^1 J_3(s)\,ds$, $D = \int_0^1 J_4(s)\,ds$, $\tilde{A} = \int_c^{1-c} J_1(s)\,ds$, $\tilde{B} = \int_c^{1-c} J_2(s)\,ds$, $\tilde{C} = \int_c^{1-c} J_3(s)\,ds$, and $\tilde{D} = \int_c^{1-c} J_4(s)\,ds$, where $J_i, i = 1, \dots, 4$ are defined in Section 5.1.2 (Lemma 5.1.5).

For $f_0^s, g_0^s, f_\infty^i, g_\infty^i \in (0,\infty)$ and numbers $\alpha_1, \alpha_2 \in [0,1]$, $\alpha_3, \alpha_4 \in (0,1)$, $a \in [0,1]$, and $b \in (0,1)$, we define the numbers

$$L_1 = \max\left\{\frac{a\alpha_1}{\gamma\gamma_1 f_\infty^i \tilde{A}}, \frac{(1-a)\alpha_2}{\gamma\gamma_2 f_\infty^i \tilde{D}}\right\}, \quad L_2 = \min\left\{\frac{b\alpha_3}{f_0^s A}, \frac{(1-b)\alpha_4}{f_0^s D}\right\},$$

$$L_3 = \max\left\{\frac{a(1-\alpha_1)}{\gamma\gamma_1 g_\infty^i \tilde{B}}, \frac{(1-a)(1-\alpha_2)}{\gamma\gamma_2 g_\infty^i \tilde{C}}\right\}, \quad L_4 = \min\left\{\frac{b(1-\alpha_3)}{g_0^s B}, \frac{(1-b)(1-\alpha_4)}{g_0^s C}\right\}.$$

Theorem 5.1.1. *Assume that (H1) and (H2) hold, $c \in (0,1/2)$, $f_0^s, g_0^s, f_\infty^i, g_\infty^i \in (0,\infty)$, $\alpha_1, \alpha_2 \in [0,1]$, $\alpha_3, \alpha_4 \in (0,1)$, $a \in [0,1]$, $b \in (0,1)$, $L_1 < L_2$, and $L_3 < L_4$. Then for each $\lambda \in (L_1, L_2)$ and $\mu \in (L_3, L_4)$ there exists a positive solution $(u(t), v(t))$, $t \in [0,1]$ for (S)–(BC).*

Proof. For c given in the theorem, we consider the above cone $P \subset Y$ and the operators T_1, T_2, and $\mathcal{T}$. Let $\lambda \in (L_1, L_2)$ and $\mu \in (L_3, L_4)$, and let $\varepsilon > 0$ a positive number such that $\varepsilon < f_\infty^i$, $\varepsilon < g_\infty^i$, and

$$\frac{a\alpha_1}{\gamma\gamma_1(f_\infty^i-\varepsilon)\tilde{A}} \le \lambda, \quad \frac{a(1-\alpha_1)}{\gamma\gamma_1(g_\infty^i-\varepsilon)\tilde{B}} \le \mu, \quad \frac{(1-a)\alpha_2}{\gamma\gamma_2(f_\infty^i-\varepsilon)\tilde{D}} \le \lambda, \quad \frac{(1-a)(1-\alpha_2)}{\gamma\gamma_2(g_\infty^i-\varepsilon)\tilde{C}} \le \mu,$$
$$\frac{b\alpha_3}{(f_0^s+\varepsilon)A} \ge \lambda, \quad \frac{b(1-\alpha_3)}{(g_0^s+\varepsilon)B} \ge \mu, \quad \frac{(1-b)\alpha_4}{(f_0^s+\varepsilon)D} \ge \lambda, \quad \frac{(1-b)(1-\alpha_4)}{(g_0^s+\varepsilon)C} \ge \mu.$$

By using (H2) and the definitions of f_0^s and g_0^s, we deduce that there exists $R_1 > 0$ such that $f(t,u,v) \le (f_0^s+\varepsilon)(u+v)$ and $g(t,u,v) \le (g_0^s+\varepsilon)(u+v)$ for all $t \in [0,1]$ and $u, v \ge 0$ with $0 \le u+v \le R_1$. We define the set $\Omega_1 = \{(u,v) \in Y, \|(u,v)\|_Y < R_1\}$. Now let $(u,v) \in P \cap \partial\Omega_1$—that is, $(u,v) \in P$ with $\|(u,v)\|_Y = R_1$ or equivalently $\|u\| + \|v\| = R_1$. Then $u(t) + v(t) \le R_1$ for all $t \in [0,1]$, and by Lemma 5.1.5, we obtain

$$\begin{aligned}
T_1(u,v)(t) &= \lambda \int_0^1 G_1(t,s) f(s,u(s),v(s))\,ds + \mu \int_0^1 G_2(t,s) g(s,u(s),v(s))\,ds \\
&\le \lambda \int_0^1 J_1(s) f(s,u(s),v(s))\,ds + \mu \int_0^1 J_2(s) g(s,u(s),v(s))\,ds \\
&\le \lambda \int_0^1 J_1(s)(f_0^s+\varepsilon)(u(s)+v(s))\,ds + \mu \int_0^1 J_2(s)(g_0^s+\varepsilon)(u(s)+v(s))\,ds \\
&\le \lambda(f_0^s+\varepsilon) \int_0^1 J_1(s)(\|u\|+\|v\|)\,ds + \mu(g_0^s+\varepsilon) \int_0^1 J_2(s)(\|u\|+\|v\|)\,ds \\
&= [\lambda(f_0^s+\varepsilon)A + \mu(g_0^s+\varepsilon)B]\|(u,v)\|_Y \\
&\le [b\alpha_3 + b(1-\alpha_3)]\|(u,v)\|_Y = b\|(u,v)\|_Y, \quad t \in [0,1].
\end{aligned}$$

Therefore, $\|T_1(u,v)\| \le b\|(u,v)\|_Y$.

In a similar manner, we conclude

$$\begin{aligned}
T_2(u,v)(t) &= \mu \int_0^1 G_3(t,s)g(s,u(s),v(s))\,ds + \lambda \int_0^1 G_4(t,s)f(s,u(s),v(s))\,ds \\
&\le \mu \int_0^1 J_3(s)g(s,u(s),v(s))\,ds + \lambda \int_0^1 J_4(s)f(s,u(s),v(s))\,ds \\
&\le \mu \int_0^1 J_3(s)(g_0^s+\varepsilon)(u(s)+v(s))\,ds + \lambda \int_0^1 J_4(s)(f_0^s+\varepsilon)(u(s)+v(s))\,ds \\
&\le \mu(g_0^s+\varepsilon) \int_0^1 J_3(s)(\|u\|+\|v\|)\,ds + \lambda(f_0^s+\varepsilon) \int_0^1 J_4(s)(\|u\|+\|v\|)\,ds \\
&= [\mu(g_0^s+\varepsilon)C + \lambda(f_0^s+\varepsilon)D]\|(u,v)\|_Y \\
&\le [(1-b)(1-\alpha_4) + (1-b)\alpha_4]\|(u,v)\|_Y = (1-b)\|(u,v)\|_Y, \quad t\in[0,1].
\end{aligned}$$

Hence, $\|T_2(u,v)\| \le (1-b)\|(u,v)\|_Y$.

Then for $(u,v)\in P\cap\partial\Omega_1$, we deduce

$$\|\mathcal{T}(u,v)\|_Y = \|T_1(u,v)\| + \|T_2(u,v)\| \le b\|(u,v)\|_Y + (1-b)\|(u,v)\|_Y = \|(u,v)\|_Y.$$

By the definitions of f_∞^i and g_∞^i, there exists $\overline{R}_2>0$ such that $f(t,u,v)\ge(f_\infty^i-\varepsilon)(u+v)$ and $g(t,u,v)\ge(g_\infty^i-\varepsilon)(u+v)$ for all $u, v\ge 0$, with $u+v\ge\overline{R}_2$, and $t\in[c,1-c]$. We consider $R_2=\max\{2R_1,\overline{R}_2/\gamma\}$, and we define $\Omega_2=\{(u,v)\in Y, \|(u,v)\|_Y<R_2\}$. Then for $(u,v)\in P$ with $\|(u,v)\|_Y=R_2$, we obtain

$$u(t)+v(t) \ge \inf_{t\in[c,1-c]}(u(t)+v(t)) \ge \gamma\|(u,v)\|_Y = \gamma R_2 \ge \overline{R}_2, \quad \forall t\in[c,1-c].$$

Then, by Lemma 5.1.5, we conclude

$$\begin{aligned}
T_1(u,v)(c) &\ge \lambda\gamma_1 \int_0^1 J_1(s)f(s,u(s),v(s))\,ds + \mu\gamma_1 \int_0^1 J_2(s)g(s,u(s),v(s))\,ds \\
&\ge \lambda\gamma_1 \int_c^{1-c} J_1(s)f(s,u(s),v(s))\,ds + \mu\gamma_1 \int_c^{1-c} J_2(c)g(s,u(s),v(s))\,ds \\
&\ge \lambda\gamma_1 \int_c^{1-c} J_1(s)(f_\infty^i-\varepsilon)(u(s)+v(s))\,ds + \mu\gamma_1 \int_c^{1-c} J_2(s)(g_\infty^i-\varepsilon)(u(s)+v(s))\,ds \\
&\ge \lambda\gamma\gamma_1(f_\infty^i-\varepsilon) \int_c^{1-c} J_1(s)\|(u,v)\|_Y\,ds + \mu\gamma\gamma_1(g_\infty^i-\varepsilon) \int_c^{1-c} J_2(s)\|(u,v)\|_Y\,ds \\
&= [\lambda\gamma\gamma_1(f_\infty^i-\varepsilon)\tilde{A} + \mu\gamma\gamma_1(g_\infty^i-\varepsilon)\tilde{B}]\|(u,v)\|_Y \\
&\ge [a\alpha_1 + a(1-\alpha_1)]\|(u,v)\|_Y = a\|(u,v)\|_Y.
\end{aligned}$$

So, $\|T_1(u,v)\| \ge T_1(u,v)(c) \ge a\|(u,v)\|_Y$.

In a similar manner, we deduce

$$\begin{aligned}
T_2(u,v)(c) &\ge \mu\gamma_2 \int_0^1 J_3(s)g(s,u(s),v(s))\,ds + \lambda\gamma_2 \int_0^1 J_4(s)f(s,u(s),v(s))\,ds \\
&\ge \mu\gamma_2 \int_c^{1-c} J_3(s)g(s,u(s),v(s))\,ds + \lambda\gamma_2 \int_c^{1-c} J_4(s)f(s,u(s),v(s))\,ds \\
&\ge \mu\gamma_2 \int_c^{1-c} J_3(s)(g_\infty^i-\varepsilon)(u(s)+v(s))\,ds + \lambda\gamma_2 \int_c^{1-c} J_4(s)(f_\infty^i-\varepsilon)(u(s)+v(s))\,ds
\end{aligned}$$

$$\geq \mu\gamma\gamma_2(g_\infty^{\rm i} - \varepsilon)\int_c^{1-c} J_3(s)\|(u,v)\|_Y\,{\rm d}s + \lambda\gamma\gamma_2(f_\infty^{\rm i} - \varepsilon)\int_c^{1-c} J_4(s)\|(u,v)\|_Y\,{\rm d}s$$
$$= [\mu\gamma\gamma_2(g_\infty^{\rm i} - \varepsilon)\tilde{C} + \lambda\gamma\gamma_2(f_\infty^{\rm i} - \varepsilon)\tilde{D}]\|(u,v)\|_Y$$
$$\geq [(1-a)(1-\alpha_2) + (1-a)\alpha_2]\|(u,v)\|_Y = (1-a)\|(u,v)\|_Y.$$

So, $\|T_2(u,v)\| \geq T_2(u,v)(c) \geq (1-a)\|(u,v)\|_Y$.

Hence, for $(u,v) \in P \cap \partial\Omega_2$, we obtain

$$\|\mathcal{T}(u,v)\|_Y = \|T_1(u,v)\| + \|T_2(u,v)\| \geq a\|(u,v)\|_Y + (1-a)\|(u,v)\|_Y = \|(u,v)\|_Y.$$

By using Lemma 5.1.7 and Theorem 1.1.1, we conclude that $\mathcal{T}$ has a fixed point $(u,v) \in P \cap (\overline{\Omega}_2 \setminus \Omega_1)$ such that $R_1 \leq \|u\| + \|v\| \leq R_2$. □

We investigate now the extreme cases, where $f_0^{\rm s}$, $g_0^{\rm s}$ could be 0, or $f_\infty^{\rm i}$, $g_\infty^{\rm i}$ could be ∞. If we denote $L_2' = \min\left\{\frac{b}{f_0^{\rm s}A}, \frac{1-b}{f_0^{\rm s}D}\right\}$ and $L_4' = \min\left\{\frac{b}{g_0^{\rm s}B}, \frac{1-b}{g_0^{\rm s}C}\right\}$, then by using arguments similar to those used in the proof of Theorem 5.1.1, we obtain the following results:

Theorem 5.1.2. *Assume that (H1) and (H2) hold,* $c \in (0,1/2)$, $f_0^{\rm s} = 0$, $g_0^{\rm s}, f_\infty^{\rm i}, g_\infty^{\rm i} \in (0,\infty)$, $\alpha_1, \alpha_2 \in [0,1]$, $a \in [0,1]$, $b \in (0,1)$, *and* $L_3 < L_4'$, *then for each* $\lambda \in (L_1,\infty)$ *and* $\mu \in (L_3, L_4')$ *there exists a positive solution* $(u(t), v(t)), t \in [0,1]$ *for (S)–(BC).*

Theorem 5.1.3. *Assume that (H1) and (H2) hold,* $c \in (0,1/2)$, $g_0^{\rm s} = 0$, $f_0^{\rm s}, f_\infty^{\rm i}, g_\infty^{\rm i} \in (0,\infty)$, $\alpha_1, \alpha_2 \in [0,1]$, $a \in [0,1]$, $b \in (0,1)$, *and* $L_1 < L_2'$, *then for each* $\lambda \in (L_1, L_2')$ *and* $\mu \in (L_3,\infty)$ *there exists a positive solution* $(u(t), v(t)), t \in [0,1]$ *for (S)–(BC).*

Theorem 5.1.4. *Assume that (H1) and (H2) hold,* $c \in (0,1/2)$, $f_0^{\rm s} = g_0^{\rm s} = 0$, $f_\infty^{\rm i}, g_\infty^{\rm i} \in (0,\infty)$, $\alpha_1, \alpha_2 \in [0,1]$, *and* $a \in [0,1]$, *then for each* $\lambda \in (L_1,\infty)$ *and* $\mu \in (L_3,\infty)$ *there exists a positive solution* $(u(t), v(t)), t \in [0,1]$ *for (S)–(BC).*

Theorem 5.1.5. *Assume that (H1) and (H2) hold,* $c \in (0,1/2)$, $\{f_0^{\rm s}, g_0^{\rm s}, f_\infty^{\rm i} \in (0,\infty), g_\infty^{\rm i} = \infty\}$ *or* $\{f_0^{\rm s}, g_0^{\rm s}, g_\infty^{\rm i} \in (0,\infty), f_\infty^{\rm i} = \infty\}$ *or* $\{f_0^{\rm s}, g_0^{\rm s} \in (0,\infty), f_\infty^{\rm i} = g_\infty^{\rm i} = \infty\}$, $\alpha_3, \alpha_4 \in (0,1)$, *and* $b \in (0,1)$, *then for each* $\lambda \in (0, L_2)$ *and* $\mu \in (0, L_4)$ *there exists a positive solution* $(u(t), v(t)), t \in [0,1]$ *for (S)–(BC).*

Theorem 5.1.6. *Assume that (H1) and (H2) hold,* $c \in (0,1/2)$, $\{f_0^{\rm s} = 0, g_0^{\rm s}, f_\infty^{\rm i} \in (0,\infty), g_\infty^{\rm i} = \infty\}$ *or* $\{f_0^{\rm s} = 0, f_\infty^{\rm i} = \infty, g_0^{\rm s}, g_\infty^{\rm i} \in (0,\infty)\}$ *or* $\{f_0^{\rm s} = 0, g_0^{\rm s} \in (0,\infty), f_\infty^{\rm i} = g_\infty^{\rm i} = \infty\}$, *and* $b \in (0,1)$, *then for each* $\lambda \in (0,\infty)$ *and* $\mu \in (0, L_4')$ *there exists a positive solution* $(u(t), v(t)), t \in [0,1]$ *for (S)–(BC).*

Theorem 5.1.7. *Assume that (H1) and (H2) hold,* $c \in (0,1/2)$, $\{f_0^{\rm s}, f_\infty^{\rm i} \in (0,\infty), g_0^{\rm s} = 0, g_\infty^{\rm i} = \infty\}$ *or* $\{f_0^{\rm s}, g_\infty^{\rm i} \in (0,\infty), g_0^{\rm s} = 0, f_\infty^{\rm i} = \infty\}$ *or* $\{f_0^{\rm s} \in (0,\infty), g_0^{\rm s} = 0, f_\infty^{\rm i} = g_\infty^{\rm i} = \infty\}$, *and* $b \in (0,1)$, *then for each* $\lambda \in (0, L_2')$ *and* $\mu \in (0,\infty)$, *there exists a positive solution* $(u(t), v(t)), t \in [0,1]$ *for (S)–(BC).*

Theorem 5.1.8. *Assume that (H1) and (H2) hold,* $c \in (0,1/2)$, $\{f_0^{\rm s} = g_0^{\rm s} = 0, f_\infty^{\rm i} \in (0,\infty), g_\infty^{\rm i} = \infty\}$ *or* $\{f_0^{\rm s} = g_0^{\rm s} = 0, f_\infty^{\rm i} = \infty, g_\infty^{\rm i} \in (0,\infty)\}$ *or* $\{f_0^{\rm s} = g_0^{\rm s} = 0, f_\infty^{\rm i} = g_\infty^{\rm i} = \infty\}$, *then for each* $\lambda \in (0,\infty)$ *and* $\mu \in (0,\infty)$ *there exists a positive solution* $(u(t), v(t)), t \in [0,1]$ *for (S)–(BC).*

Remark 5.1.1. Theorems 5.1.1–5.1.4 each contains nine subcases because α_1, α_2 can be 0, or 1, or between 0 and 1.

In what follows, for $f_0^{\rm i}$, $g_0^{\rm i}$, $f_\infty^{\rm s}$, $g_\infty^{\rm s} \in (0,\infty)$ and numbers $\alpha_1, \alpha_2 \in [0,1]$, $\alpha_3, \alpha_4 \in (0,1)$, $a \in [0,1]$, and $b \in (0,1)$, we define the numbers

$$\tilde{L}_1 = \max\left\{\frac{a\alpha_1}{\gamma\gamma_1 f_0^{\rm i}\tilde{A}}, \frac{(1-a)\alpha_2}{\gamma\gamma_2 f_0^{\rm i}\tilde{D}}\right\}, \quad \tilde{L}_2 = \min\left\{\frac{b\alpha_3}{f_\infty^{\rm s}A}, \frac{(1-b)\alpha_4}{f_\infty^{\rm s}D}\right\},$$

$$\tilde{L}_3 = \max\left\{\frac{a(1-\alpha_1)}{\gamma\gamma_1 g_0^{\rm i}\tilde{B}}, \frac{(1-a)(1-\alpha_2)}{\gamma\gamma_2 g_0^{\rm i}\tilde{C}}\right\}, \quad \tilde{L}_4 = \min\left\{\frac{b(1-\alpha_3)}{g_\infty^{\rm s}B}, \frac{(1-b)(1-\alpha_4)}{g_\infty^{\rm s}C}\right\}.$$

Theorem 5.1.9. *Assume that (H1) and (H2) hold, $c \in (0,1/2)$, $f_0^{\rm i}$, $g_0^{\rm i}$, $f_\infty^{\rm s}$, $g_\infty^{\rm s} \in (0,\infty)$, $\alpha_1, \alpha_2 \in [0,1]$, $\alpha_3, \alpha_4 \in (0,1)$, $a \in [0,1]$, $b \in (0,1)$, $\tilde{L}_1 < \tilde{L}_2$, and $\tilde{L}_3 < \tilde{L}_4$. Then for each $\lambda \in (\tilde{L}_1, \tilde{L}_2)$ and $\mu \in (\tilde{L}_3, \tilde{L}_4)$ there exists a positive solution $(u(t), v(t))$, $t \in [0,1]$ for (S)–(BC).*

Proof. For c given in the theorem, we consider again the above cone $P \subset Y$ and the operators T_1, T_2, and $\mathcal{T}$. Let $\lambda \in (\tilde{L}_1, \tilde{L}_2)$ and $\mu \in (\tilde{L}_3, \tilde{L}_4)$, and let $\varepsilon > 0$ be a positive number such that $\varepsilon < f_0^{\rm i}$, $\varepsilon < g_0^{\rm i}$, and

$$\frac{a\alpha_1}{\gamma\gamma_1(f_0^{\rm i}-\varepsilon)\tilde{A}} \le \lambda, \quad \frac{a(1-\alpha_1)}{\gamma\gamma_1(g_0^{\rm i}-\varepsilon)\tilde{B}} \le \mu, \quad \frac{(1-a)\alpha_2}{\gamma\gamma_2(f_0^{\rm i}-\varepsilon)\tilde{D}} \le \lambda, \quad \frac{(1-a)(1-\alpha_2)}{\gamma\gamma_2(g_0^{\rm i}-\varepsilon)\tilde{C}} \le \mu,$$
$$\frac{b\alpha_3}{(f_\infty^{\rm s}+\varepsilon)A} \ge \lambda, \quad \frac{b(1-\alpha_3)}{(g_\infty^{\rm s}+\varepsilon)B} \ge \mu, \quad \frac{(1-b)\alpha_4}{(f_\infty^{\rm s}+\varepsilon)D} \ge \lambda, \quad \frac{(1-b)(1-\alpha_4)}{(g_\infty^{\rm s}+\varepsilon)C} \ge \mu.$$

By using (H2) and the definitions of $f_0^{\rm i}$ and $g_0^{\rm i}$, we deduce that there exists $R_3 > 0$ such that $f(t,u,v) \ge (f_0^{\rm i}-\varepsilon)(u+v)$ and $g(t,u,v) \ge (g_0^{\rm i}-\varepsilon)(u+v)$ for all $u, v \ge 0$, with $0 \le u+v \le R_3$, and $t \in [c, 1-c]$. We denote $\Omega_3 = \{(u,v) \in Y, \|(u,v)\|_Y < R_3\}$. Let $(u,v) \in P$ with $\|(u,v)\|_Y = R_3$—that is, $\|u\| + \|v\| = R_3$. Because $u(t)+v(t) \le \|u\|+\|v\| = R_3$ for all $t \in [0,1]$, by using Lemma 5.1.5, we obtain

$$\begin{aligned}
T_1(u,v)(c) &\ge \lambda\gamma_1 \int_0^1 J_1(s)f(s,u(s),v(s))\,{\rm d}s + \mu\gamma_1 \int_0^1 J_2(s)g(s,u(s),v(s))\,{\rm d}s \\
&\ge \lambda\gamma_1 \int_c^{1-c} J_1(s)f(s,u(s),v(s))\,{\rm d}s + \mu\gamma_1 \int_c^{1-c} J_2(s)g(s,u(s),v(s))\,{\rm d}s \\
&\ge \lambda\gamma_1 \int_c^{1-c} J_1(s)(f_0^{\rm i}-\varepsilon)(u(s)+v(s))\,{\rm d}s + \mu\gamma_1 \int_c^{1-c} J_2(s)(g_0^{\rm i}-\varepsilon)(u(s)+v(s))\,{\rm d}s \\
&\ge \lambda\gamma\gamma_1(f_0^{\rm i}-\varepsilon) \int_c^{1-c} J_1(s)(\|u\|+\|v\|)\,{\rm d}s + \mu\gamma\gamma_1(g_0^{\rm i}-\varepsilon) \int_c^{1-c} J_2(s)(\|u\|+\|v\|)\,{\rm d}s \\
&= \lambda\gamma\gamma_1(f_0^{\rm i}-\varepsilon)\tilde{A}\|(u,v)\|_Y + \mu\gamma\gamma_1(g_0^{\rm i}-\varepsilon)\tilde{B}\|(u,v)\|_Y \\
&\ge [a\alpha_1 + a(1-\alpha_1)]\|(u,v)\|_Y = a\|(u,v)\|_Y.
\end{aligned}$$

Therefore, $\|T_1(u,v)\| \ge T_1(u,v)(c) \ge a\|(u,v)\|_Y$.

In a similar manner, we conclude

$$\begin{aligned}
T_2(u,v)(c) &\ge \mu\gamma_2 \int_0^1 J_3(s)g(s,u(s),v(s))\,{\rm d}s + \lambda\gamma_2 \int_0^1 J_4(s)f(s,u(s),v(s))\,{\rm d}s \\
&\ge \mu\gamma_2 \int_c^{1-c} J_3(s)g(s,u(s),v(s))\,{\rm d}s + \lambda\gamma_2 \int_c^{1-c} J_4(s)f(s,u(s),v(s))\,{\rm d}s
\end{aligned}$$

$$\geq \mu\gamma_2 \int_c^{1-c} J_3(s)(g_0^{\rm i} - \varepsilon)(u(s) + v(s))\,{\rm d}s + \lambda\gamma_2 \int_c^{1-c} J_4(s)(f_0^{\rm i} - \varepsilon)(u(s) + v(s))\,{\rm d}s$$
$$\geq \mu\gamma\gamma_2(g_0^{\rm i} - \varepsilon) \int_c^{1-c} J_3(s)(\|u\| + \|v\|)\,{\rm d}s + \lambda\gamma\gamma_2(f_0^{\rm i} - \varepsilon) \int_c^{1-c} J_4(s)(\|u\| + \|v\|)\,{\rm d}s$$
$$= \mu\gamma\gamma_2(g_0^{\rm i} - \varepsilon)\tilde{C}\|(u,v)\|_Y + \lambda\gamma\gamma_2(f_0^{\rm i} - \varepsilon)\tilde{D}\|(u,v)\|_Y$$
$$\geq [(1-a)(1-\alpha_2) + (1-a)\alpha_2]\|(u,v)\|_Y = (1-a)\|(u,v)\|_Y.$$

So, $\|T_2(u,v)\| \geq T_2(u,v)(c) \geq (1-a)\|(u,v)\|_Y$.

Thus, for an arbitrary element $(u,v) \in P \cap \partial\Omega_3$, we deduce

$$\|\mathcal{T}(u,v)\|_Y = \|T_1(u,v)\| + \|T_2(u,v)\| \geq a\|(u,v)\|_Y + (1-a)\|(u,v)\|_Y = \|(u,v)\|_Y.$$

Now, we define the functions $f^*, g^* : [0,1] \times \mathbb{R}_+ \to \mathbb{R}_+, f^*(t,x) = \max_{0 \leq u+v \leq x} f(t,u,v)$, $g^*(t,x) = \max_{0 \leq u+v \leq x} g(t,u,v)$, $t \in [0,1]$, $x \in \mathbb{R}_+$. Then $f(t,u,v) \leq f^*(t,x)$ and $g(t,u,v) \leq g^*(t,x)$ for all $t \in [0,1]$, $u, v \geq 0$, and $u + v \leq x$. The functions $f^*(t,\cdot)$ and $g^*(t,\cdot)$ are nondecreasing for every $t \in [0,1]$, and they satisfy the conditions

$$\limsup_{x\to\infty} \max_{t\in[0,1]} \frac{f^*(t,x)}{x} \leq f_\infty^{\rm s}, \quad \limsup_{x\to\infty} \max_{t\in[0,1]} \frac{g^*(t,x)}{x} \leq g_\infty^{\rm s}.$$

Therefore, for $\varepsilon > 0$, there exists $\bar{R}_4 > 0$ such that for all $x \geq \bar{R}_4$ and $t \in [0,1]$, we have

$$\frac{f^*(t,x)}{x} \leq \limsup_{x\to\infty} \max_{t\in[0,1]} \frac{f^*(t,x)}{x} + \varepsilon \leq f_\infty^{\rm s} + \varepsilon,$$
$$\frac{g^*(t,x)}{x} \leq \limsup_{x\to\infty} \max_{t\in[0,1]} \frac{g^*(t,x)}{x} + \varepsilon \leq g_\infty^{\rm s} + \varepsilon,$$

and so $f^*(t,x) \leq (f_\infty^{\rm s} + \varepsilon)x$ and $g^*(t,x) \leq (g_\infty^{\rm s} + \varepsilon)x$.

We consider $R_4 = \max\{2R_3, \bar{R}_4\}$ and denote $\Omega_4 = \{(u,v) \in Y, \|(u,v)\|_Y < R_4\}$. Let $(u,v) \in P \cap \partial\Omega_4$. By the definitions of f^* and g^*, we obtain

$$f(t,u(t),v(t)) \leq f^*(t,\|(u,v)\|_Y), \quad g(t,u(t),v(t)) \leq g^*(t,\|(u,v)\|_Y), \quad \forall t \in [0,1].$$

Then for all $t \in [0,1]$, we conclude

$$T_1(u,v)(t) \leq \lambda \int_0^1 J_1(s)f(s,u(s),v(s))\,{\rm d}s + \mu \int_0^1 J_2(s)g(s,u(s),v(s))\,{\rm d}s$$
$$\leq \lambda \int_0^1 J_1(s)f^*(s,\|(u,v)\|_Y)\,{\rm d}s + \mu \int_0^1 J_2(s)g^*(s,\|(u,v)\|_Y)\,{\rm d}s$$
$$\leq \lambda(f_\infty^{\rm s} + \varepsilon) \int_0^1 J_1(s)\|(u,v)\|_Y\,{\rm d}s + \mu(g_\infty^{\rm s} + \varepsilon) \int_0^1 J_2(s)\|(u,v)\|_Y\,{\rm d}s$$
$$= [\lambda(f_\infty^{\rm s} + \varepsilon)A + \mu(g_\infty^{\rm s} + \varepsilon)B]\|(u,v)\|_Y$$
$$\leq [b\alpha_3 + b(1-\alpha_3)]\|(u,v)\|_Y = b\|(u,v)\|_Y.$$

Therefore, $\|T_1(u,v)\| \leq b\|(u,v)\|_Y$.

In a similar manner, we deduce

$$\begin{aligned}T_2(u,v)(t) &\le \mu\int_0^1 J_3(s)g(s,u(s),v(s))\,ds + \lambda\int_0^1 J_4(s)f(s,u(s),v(s))\,ds\\&\le \mu\int_0^1 J_3(s)g^*(s,\|(u,v)\|_Y)\,ds + \lambda\int_0^1 J_4(s)f^*(s,\|(u,v)\|_Y)\,ds\\&\le \mu(g_\infty^s+\varepsilon)\int_0^1 J_3(s)\|(u,v)\|_Y\,ds + \lambda(f_\infty^s+\varepsilon)\int_0^1 J_4(s)\|(u,v)\|_Y\,ds\\&= [\mu(g_\infty^s+\varepsilon)C + \lambda(f_\infty^s+\varepsilon)D]\|(u,v)\|_Y\\&\le [(1-b)(1-\alpha_4)+(1-b)\alpha_4]\|(u,v)\|_Y = (1-b)\|(u,v)\|_Y.\end{aligned}$$

So, $\|T_2(u,v)\| \le (1-b)\|(u,v)\|_Y$.

Then for $(u,v) \in P\cap \partial\Omega_4$, it follows that

$$\|\mathcal{T}(u,v)\|_Y = \|T_1(u,v)\|_Y + \|T_2(u,v)\|_Y \le b\|(u,v)\|_Y + (1-b)\|(u,v)\|_Y = \|(u,v)\|_Y.$$

By using Lemma 5.1.7 and Theorem 1.1.1, we conclude that $\mathcal{T}$ has a fixed point $(u,v) \in P\cap(\overline{\Omega}_4\setminus\Omega_3)$ such that $R_3 \le \|(u,v)\|_Y \le R_4$. □

We investigate now the extreme cases, where f_∞^s, g_∞^s could be 0, or f_0^i, g_0^i could be ∞. If we denote $\tilde{L}_2' = \min\left\{\frac{b}{f_\infty^s A}, \frac{1-b}{f_\infty^s D}\right\}$ and $\tilde{L}_4' = \min\left\{\frac{b}{g_\infty^s B}, \frac{1-b}{g_\infty^s C}\right\}$, then by using arguments similar to those used in the proof of Theorem 5.1.9, we obtain the following results:

Theorem 5.1.10. *Assume that (H1) and (H2) hold,* $c \in (0,1/2)$, $f_0^i, g_0^i, f_\infty^s \in (0,\infty)$, $g_\infty^s = 0$, $\alpha_1,\alpha_2 \in [0,1]$, $a \in [0,1]$, $b \in (0,1)$, *and* $\tilde{L}_1 < \tilde{L}_2'$, *then for each* $\lambda \in (\tilde{L}_1,\tilde{L}_2')$ *and* $\mu \in (\tilde{L}_3,\infty)$ *there exists a positive solution* $(u(t),v(t)), t \in [0,1]$ *for (S)–(BC).*

Theorem 5.1.11. *Assume that (H1) and (H2) hold,* $c \in (0,1/2)$, $f_0^i, g_0^i, g_\infty^s \in (0,\infty)$, $f_\infty^s = 0$, $\alpha_1,\alpha_2 \in [0,1]$, $a \in [0,1]$, $b \in (0,1)$, *and* $\tilde{L}_3 < \tilde{L}_4'$, *then for each* $\lambda \in (\tilde{L}_1,\infty)$ *and* $\mu \in (\tilde{L}_3,\tilde{L}_4')$ *there exists a positive solution* $(u(t),v(t)), t \in [0,1]$ *for (S)–(BC).*

Theorem 5.1.12. *Assume that (H1) and (H2) hold,* $c \in (0,1/2)$, $f_0^i, g_0^i \in (0,\infty)$, $f_\infty^s = g_\infty^s = 0$, $\alpha_1,\alpha_2 \in [0,1]$, *and* $a \in [0,1]$, *then for each* $\lambda \in (\tilde{L}_1,\infty)$ *and* $\mu \in (\tilde{L}_3,\infty)$ *there exists a positive solution* $(u(t),v(t)), t \in [0,1]$ *for (S)–(BC).*

Theorem 5.1.13. *Assume that (H1) and (H2) hold,* $c \in (0,1/2)$, $\{f_0^i = \infty, g_0^i, f_\infty^s, g_\infty^s \in (0,\infty)\}$ *or* $\{f_0^i, f_\infty^s, g_\infty^s \in (0,\infty), g_0^i = \infty\}$ *or* $\{f_0^i = g_0^i = \infty, f_\infty^s, g_\infty^s \in (0,\infty)\}$, $\alpha_3,\alpha_4 \in (0,1)$, *and* $b \in (0,1)$, *then for each* $\lambda \in (0,\tilde{L}_2)$ *and* $\mu \in (0,\tilde{L}_4)$ *there exists a positive solution* $(u(t),v(t)), t \in [0,1]$ *for (S)–(BC).*

Theorem 5.1.14. *Assume that (H1) and (H2) hold,* $c \in (0,1/2)$, $\{f_0^i = \infty, g_0^i, f_\infty^s \in (0,\infty), g_\infty^s = 0\}$ *or* $\{f_0^i, f_\infty^s \in (0,\infty), g_0^i = \infty, g_\infty^s = 0\}$ *or* $\{f_0^i = g_0^i = \infty, f_\infty^s \in (0,\infty), g_\infty^s = 0\}$, *and* $b \in (0,1)$, *then for each* $\lambda \in (0,\tilde{L}_2')$ *and* $\mu \in (0,\infty)$ *there exists a positive solution* $(u(t),v(t)), t \in [0,1]$ *for (S)–(BC).*

Theorem 5.1.15. *Assume that (H1) and (H2) hold, $c \in (0, 1/2)$, $\{f_0^{\rm i} = \infty$, $g_0^{\rm i}$, $g_\infty^{\rm s} \in (0, \infty)$, $f_\infty^{\rm s} = 0\}$ or $\{f_0^{\rm i}$, $g_\infty^{\rm s} \in (0, \infty)$, $g_0^{\rm i} = \infty$, $f_\infty^{\rm s} = 0\}$ or $\{f_0^{\rm i} = g_0^{\rm i} = \infty$, $f_\infty^{\rm s} = 0$, $g_\infty^{\rm s} \in (0, \infty)\}$, and $b \in (0, 1)$, then for each $\lambda \in (0, \infty)$ and $\mu \in (0, \tilde{L}_4')$ there exists a positive solution $(u(t), v(t))$, $t \in [0, 1]$ for (S)–(BC).*

Theorem 5.1.16. *Assume that (H1) and (H2) hold, $c \in (0, 1/2)$, and $\{f_0^{\rm i} = \infty$, $g_0^{\rm i} \in (0, \infty)$, $f_\infty^{\rm s} = g_\infty^{\rm s} = 0\}$ or $\{f_0^{\rm i} \in (0, \infty)$, $g_0^{\rm i} = \infty$, $f_\infty^{\rm s} = g_\infty^{\rm s} = 0\}$ or $\{f_0^{\rm i} = g_0^{\rm i} = \infty$, $f_\infty^{\rm s} = g_\infty^{\rm s} = 0\}$, then for each $\lambda \in (0, \infty)$ and $\mu \in (0, \infty)$ there exists a positive solution $(u(t), v(t))$, $t \in [0, 1]$ for (S)–(BC).*

5.1.4 Examples

Let $\alpha = 7/3\ (n = 3)$, $\beta = 5/2\ (m = 3)$, $H(t) = t^2$, and $K(t) = \begin{cases} 0, & t \in [0, 1/3), \\ 1, & t \in [1/3, 2/3), \\ 3/2, & t \in [2/3, 1], \end{cases}$

for all $t \in [0, 1]$. Then $\int_0^1 v(s)\,{\rm d}H(s) = 2\int_0^1 sv(s)\,{\rm d}s$ and $\int_0^1 u(s)\,{\rm d}K(s) = u\left(\frac{1}{3}\right) + \frac{1}{2}u\left(\frac{2}{3}\right)$.

We consider the system of fractional differential equations

$$\begin{cases} D_{0+}^{7/3}u(t) + \lambda f(t, u(t), v(t)) = 0, & t \in (0, 1), \\ D_{0+}^{5/2}v(t) + \mu g(t, u(t), v(t)) = 0, & t \in (0, 1), \end{cases} \qquad (S_1)$$

with the boundary conditions

$$\begin{cases} u(0) = u'(0) = 0, & u(1) = 2\displaystyle\int_0^1 sv(s)\,{\rm d}s, \\ v(0) = v'(0) = 0, & v(1) = u\left(\dfrac{1}{3}\right) + \dfrac{1}{2}u\left(\dfrac{2}{3}\right). \end{cases} \qquad (BC_1)$$

Then we obtain $\Delta = 1 - \left(\int_0^1 \tau^{4/3}{\rm d}K(\tau)\right)\left(\int_0^1 \tau^{3/2}{\rm d}H(\tau)\right) = 1 - \left(\left(\frac{1}{3}\right)^{4/3} + \frac{1}{2} \times \left(\frac{2}{3}\right)^{4/3}\right)\left(2\int_0^1 \tau^{5/2}\,{\rm d}\tau\right) = \frac{21\sqrt[3]{3} - 4 - 4\sqrt[3]{2}}{21\sqrt[3]{3}} \approx 0.70153491 > 0$. We also deduce

$$g_1(t, s) = \frac{1}{\Gamma(7/3)}\begin{cases} t^{4/3}(1-s)^{4/3} - (t-s)^{4/3}, & 0 \le s \le t \le 1, \\ t^{4/3}(1-s)^{4/3}, & 0 \le t \le s \le 1, \end{cases}$$

$$g_2(t, s) = \frac{4}{3\sqrt{\pi}}\begin{cases} t^{3/2}(1-s)^{3/2} - (t-s)^{3/2}, & 0 \le s \le t \le 1, \\ t^{3/2}(1-s)^{3/2}, & 0 \le t \le s \le 1, \end{cases}$$

$\theta_1(s) = \frac{1}{4-6s+4s^2-s^3}$, and $\theta_2(s) = \frac{1}{3-3s+s^2}$ for all $s \in [0, 1]$. For the functions J_i, $i = 1, \ldots, 4$, we obtain

$$J_1(s)=\begin{cases}\dfrac{1}{\Gamma(7/3)}\left\{\dfrac{s(1-s)^{4/3}}{(4-6s+4s^2-s^3)^{1/3}}+\dfrac{2}{21\sqrt[3]{3}\Delta}\left[2(1-s)^{4/3}-2(1-3s)^{4/3}\right.\right.\\ \qquad \left.\left.+\,(2-2s)^{4/3}-(2-3s)^{4/3}\right]\right\}, \quad 0\le s<1/3,\\ \dfrac{1}{\Gamma(7/3)}\left\{\dfrac{s(1-s)^{4/3}}{(4-6s+4s^2-s^3)^{1/3}}+\dfrac{2}{21\sqrt[3]{3}\Delta}\left[2(1-s)^{4/3}+(2-2s)^{4/3}\right.\right.\\ \qquad \left.\left.-\,(2-3s)^{4/3}\right]\right\}, \quad 1/3\le s<2/3,\\ \dfrac{1}{\Gamma(7/3)}\left\{\dfrac{s(1-s)^{4/3}}{(4-6s+4s^2-s^3)^{1/3}}+\dfrac{2}{21\sqrt[3]{3}\Delta}\left[2(1-s)^{4/3}+(2-2s)^{4/3}\right]\right\},\\ \qquad 2/3\le s\le 1,\end{cases}$$

$$J_2(s)=\frac{16}{3\sqrt{\pi}\Delta}\left\{\frac{1}{7}(1-s)^{3/2}-\frac{1}{7}(1-s)^{7/2}-\frac{1}{5}s(1-s)^{5/2}\right\}, \quad s\in[0,1],$$

$$J_3(s)=\frac{4}{3\sqrt{\pi}}\left\{\frac{s(1-s)^{3/2}}{(3-3s+s^2)^{1/2}}+\frac{4(1+\sqrt[3]{2})}{3\sqrt[3]{3}\Delta}\left[\frac{1}{7}(1-s)^{3/2}-\frac{1}{7}(1-s)^{7/2}-\frac{1}{5}s(1-s)^{5/2}\right]\right\},$$
$s\in[0,1]$,

$$J_4(s)=\begin{cases}\dfrac{1}{6\sqrt[3]{3}\Delta\Gamma(7/3)}\left[2(1-s)^{4/3}-2(1-3s)^{4/3}+(2-2s)^{4/3}-(2-3s)^{4/3}\right], & 0\le s<1/3,\\ \dfrac{1}{6\sqrt[3]{3}\Delta\Gamma(7/3)}\left[2(1-s)^{4/3}+(2-2s)^{4/3}-(2-3s)^{4/3}\right], & 1/3\le s<2/3,\\ \dfrac{1}{6\sqrt[3]{3}\Delta\Gamma(7/3)}\left[2(1-s)^{4/3}+(2-2s)^{4/3}\right], & 2/3\le s\le 1.\end{cases}$$

For $c=1/4$, we deduce $\gamma_1=4^{-4/3}\approx 0.15749013$ and $\gamma_2=\frac{1}{8}$, $\gamma=\gamma_2$. After some computations, we conclude $A=\int_0^1 J_1(s)\,ds\approx 0.15972386$, $\tilde{A}=\int_{1/4}^{3/4} J_1(s)\,ds\approx 0.11335535$, $B=\int_0^1 J_2(s)\,ds\approx 0.05446581$, $\tilde{B}=\int_{1/4}^{3/4} J_2(s)\,ds\approx 0.03892266$, $C=\int_0^1 J_3(s)\,ds\approx 0.09198682$, $\tilde{C}=\int_{1/4}^{3/4} J_3(s)\,ds\approx 0.06559293$, $D=\int_0^1 J_4(s)\,ds\approx 0.12885992$, and $\tilde{D}=\int_{1/4}^{3/4} J_4(s)\,ds\approx 0.09158825$.

Example 5.1.1. We consider the functions

$$f(t,u,v)=\frac{\sqrt[3]{t}\,[p_1(u+v)+1](u+v)(q_1+\sin v)}{u+v+1},$$
$$g(t,u,v)=\frac{\sqrt{1-t}\,[p_2(u+v)+1](u+v)(q_2+\cos u)}{u+v+1},$$

for all $t\in[0,1]$ and $u,\ v\in[0,\infty)$, where $p_1,\ p_2>0$ and $q_1,\ q_2>1$.

We have $f_0^{\rm s}=q_1$, $g_0^{\rm s}=q_2+1$, $f_\infty^{\rm i}=4^{-1/3}p_1(q_1-1)$, and $g_\infty^{\rm i}=\frac{1}{2}p_2(q_2-1)$. We take $a=b=\alpha_1=\alpha_2=\alpha_3=\alpha_4=1/2$; then we obtain

$$L_1=\frac{16\sqrt[3]{4}}{p_1(q_1-1)\tilde{D}},\quad L_2=\frac{1}{4q_1A},\quad L_3=\frac{16\sqrt[3]{4}}{p_2(q_2-1)\tilde{B}},\quad L_4=\frac{1}{4(q_2+1)C}.$$

The conditions $L_1<L_2$ and $L_3<L_4$ become

$$\frac{p_1(q_1-1)}{q_1}>\frac{64\sqrt[3]{4}A}{\tilde{D}},\quad \frac{p_2(q_2-1)}{q_2+1}>\frac{64\sqrt[3]{4}C}{\tilde{B}}.$$

If $\frac{p_1(q_1-1)}{q_1} \geq 178$ and $\frac{p_2(q_2-1)}{q_2+1} \geq 241$, then the above conditions are satisfied. Therefore, by Theorem 5.1.1, for each $\lambda \in (L_1, L_2)$ and $\mu \in (L_3, L_4)$ there exists a positive solution $(u(t), v(t))$, $t \in [0,1]$ for problem (S$_1$)–(BC$_1$). For example, if $q_1 = 2$, $q_2 = 3$, $p_1 = 356$, and $p_2 = 482$, then we obtain $L_1 \approx 0.77896311$, $L_2 \approx 0.78260064$, $L_3 \approx 0.67690395$, and $L_4 \approx 0.67944517$.

Example 5.1.2. We consider the functions

$$f(t,u,v) = p_1 t^{\tilde{a}}(u^2+v^2), \quad g(t,u,v) = p_2(1-t)^{\tilde{b}}(e^{u+v}-1), \quad \forall t \in [0,1], \quad u,\, v \in [0,\infty),$$

where $\tilde{a}$, $\tilde{b}$, p_1, $p_2 > 0$.

We have $f_0^{\rm s} = 0$, $g_0^{\rm s} = p_2$, and $f_\infty^{\rm i} = g_\infty^{\rm i} = \infty$. For $b = 1/2$, we obtain $L_4' = \frac{1}{2p_2C}$. Then, by Theorem 5.1.6, we conclude that for each $\lambda \in (0,\infty)$ and $\mu \in (0, L_4')$ there exists a positive solution $(u(t), v(t))$, $t \in [0,1]$ for problem (S$_1$)–(BC$_1$). For example, if $p_2 = 1$, we obtain $L_4' \approx 5.4355614$.

Remark 5.1.2. The results presented in this section were published in Henderson and Luca (2014c).

5.2 Existence and multiplicity of positive solutions for systems without parameters and coupled boundary conditions

In this section, we investigate the existence and multiplicity of positive solutions for problem (S)–(BC) from Section 5.1 with $\lambda = \mu = 1$, and f and g dependent only on t and v, and t and u, respectively. The nonlinearities f and g are nonsingular functions, or singular functions at $t = 0$ and/or $t = 1$.

5.2.1 *Nonsingular nonlinearities*

We consider the system of nonlinear ordinary fractional differential equations

$$\begin{cases} D_{0+}^{\alpha} u(t) + f(t, v(t)) = 0, & t \in (0,1), \\ D_{0+}^{\beta} v(t) + g(t, u(t)) = 0, & t \in (0,1), \end{cases} \tag{S$'$}$$

with the coupled integral boundary conditions

$$\begin{cases} u(0) = u'(0) = \cdots = u^{(n-2)}(0) = 0, & u(1) = \displaystyle\int_0^1 v(s)\, dH(s), \\ v(0) = v'(0) = \cdots = v^{(m-2)}(0) = 0, & v(1) = \displaystyle\int_0^1 u(s)\, dK(s), \end{cases} \tag{BC}$$

where $n-1 < \alpha \leq n$, $m-1 < \beta \leq m$, $n,\, m \in \mathbb{N}$, $n,\, m \geq 3$, D_{0+}^{α} and D_{0+}^{β} denote the Riemann–Liouville derivatives of orders α and β, respectively, and the integrals from (BC) are Riemann–Stieltjes integrals.

Under sufficient conditions on the functions f and g, we prove the existence and multiplicity of positive solutions of the above problem, by applying the fixed point index theory. By a positive solution of problem (S$'$)–(BC) we mean a pair of functions

$(u, v) \in C([0, 1]) \times C([0, 1])$ satisfying (S′) and (BC) with $u(t) \geq 0$, $v(t) \geq 0$ for all $t \in [0, 1]$ and $(u, v) \neq (0, 0)$.

We present the basic assumptions that we shall use in the sequel:

(I1) $H, K : [0, 1] \to \mathbb{R}$ are nondecreasing functions, and $\Delta = 1 - \left(\int_0^1 \tau^{\alpha-1}\, dK(\tau)\right) \times \left(\int_0^1 \tau^{\beta-1}\, dH(\tau)\right) > 0$.

(I2) The functions $f, g : [0, 1] \times [0, \infty) \to [0, \infty)$ are continuous.

Under assumption (I1), we have all auxiliary results in Lemmas 5.1.1–5.1.6 from Section 5.1.2.

By using the functions G_i, $i = 1, \dots, 4$ from Section 5.1.2 (Lemma 5.1.2), we can write our problem (S′)–(BC) equivalently as the following nonlinear system of integral equations:

$$\begin{cases} u(t) = \int_0^1 G_1(t, s) f(s, v(s))\, ds + \int_0^1 G_2(t, s) g(s, u(s))\, ds, & t \in [0, 1], \\ v(t) = \int_0^1 G_3(t, s) g(s, u(s))\, ds + \int_0^1 G_4(t, s) f(s, v(s))\, ds, & t \in [0, 1]. \end{cases}$$

We consider the Banach space $X = C([0, 1])$ with the supremum norm $\|\cdot\|$ and the Banach space $Y = X \times X$ with the norm $\|(u, v)\|_Y = \|u\| + \|v\|$. We define the cone $P' \subset Y$ by

$$P' = \{(u, v) \in Y, \quad u(t) \geq 0, \quad v(t) \geq 0 \quad \text{for all} \quad t \in [0, 1]\}.$$

We introduce the operators $Q_1, Q_2 : Y \to X$ and $Q : Y \to Y$ defined by

$$Q_1(u, v)(t) = \int_0^1 G_1(t, s) f(s, v(s))\, ds + \int_0^1 G_2(t, s) g(s, u(s))\, ds, \quad 0 \leq t \leq 1,$$

$$Q_2(u, v)(t) = \int_0^1 G_3(t, s) g(s, u(s))\, ds + \int_0^1 G_4(t, s) f(s, v(s))\, ds, \quad 0 \leq t \leq 1,$$

and $Q(u, v) = (Q_1(u, v), Q_2(u, v))$, $(u, v) \in Y$.

Under assumptions (I1) and (I2), it is easy to see that the operator $Q : P' \to P'$ is completely continuous. Thus, the existence and multiplicity of positive solutions of problem (S′)–(BC) are equivalent to the existence and multiplicity of fixed points of the operator Q.

Theorem 5.2.1. *Assume that (I1) and (I2) hold. If the functions f and g also satisfy the following conditions (I3) and (I4), then problem* (S′)–(BC) *has at least one positive solution* $(u(t), v(t))$, $t \in [0, 1]$*:*

(I3) *There exists $c \in (0, 1/2)$ such that*

$$(1)\quad \tilde{f}_\infty^{\mathrm{i}} = \lim_{u\to\infty} \inf_{t\in[c,1-c]} \frac{f(t, u)}{u} = \infty \quad \textit{and} \quad (2)\quad \tilde{g}_\infty^{\mathrm{i}} = \lim_{u\to\infty} \inf_{t\in[c,1-c]} \frac{g(t, u)}{u} = \infty.$$

(I4) *There exist $p \geq 1$ and $q \geq 1$ such that*

$$(1)\quad \tilde{f}_0^{\mathrm{s}} = \lim_{u\to0^+} \sup_{t\in[0,1]} \frac{f(t, u)}{u^p} = 0 \quad \textit{and} \quad (2)\quad \tilde{g}_0^{\mathrm{s}} = \lim_{u\to0^+} \sup_{t\in[0,1]} \frac{g(t, u)}{u^q} = 0.$$

Proof. For c given in (I3), we define the cone

$$P_0 = \{(u,v) \in P', \quad \inf_{t\in[c,1-c]} u(t) \geq \gamma_1 \|u\|, \quad \inf_{t\in[c,1-c]} v(t) \geq \gamma_2 \|v\|\},$$

where γ_1 and γ_2 are defined in Section 5.1.2 (Lemma 5.1.3).

From our assumptions and Lemma 5.1.6, for any $(u,v) \in P'$, we deduce that $\mathcal{Q}(u,v) = (Q_1(u,v), Q_2(u,v)) \in P_0$—that is, $\mathcal{Q}(P') \subset P_0$.

We consider the functions $u_0, v_0 : [0,1] \to \mathbb{R}$ defined by

$$\begin{cases} u_0(t) = \displaystyle\int_0^1 G_1(t,s)\,ds + \int_0^1 G_2(t,s)\,ds, & 0 \leq t \leq 1, \\ v_0(t) = \displaystyle\int_0^1 G_3(t,s)\,ds + \int_0^1 G_4(t,s)\,ds, & 0 \leq t \leq 1; \end{cases}$$

that is, (u_0, v_0) is the solution of problem (5.1)–(5.2) with $x(t) = x_0(t)$, $y(t) = y_0(t)$, $x_0(t) = 1$, and $y_0(t) = 1$ for all $t \in [0,1]$. Hence, $(u_0, v_0) = \mathcal{Q}(x_0, y_0) \in P_0$.

We define the set

$$M = \{(u,v) \in P', \quad \text{there exists} \quad \lambda \geq 0 \quad \text{such that} \quad (u,v) = \mathcal{Q}(u,v) + \lambda(u_0, v_0)\}.$$

We shall show that $M \subset P_0$ and that M is a bounded set of Y. If $(u,v) \in M$, then there exists $\lambda \geq 0$ such that $(u,v) = \mathcal{Q}(u,v) + \lambda(u_0, v_0)$, or equivalently

$$\begin{cases} u(t) = \displaystyle\int_0^1 G_1(t,s)(f(s,v(s)) + \lambda)\,ds + \int_0^1 G_2(t,s)(g(s,u(s)) + \lambda)\,ds, & 0 \leq t \leq 1, \\ v(t) = \displaystyle\int_0^1 G_3(t,s)(g(s,u(s)) + \lambda)\,ds + \int_0^1 G_4(t,s)(f(s,v(s)) + \lambda)\,ds, & 0 \leq t \leq 1. \end{cases}$$

By Lemma 5.1.6, we obtain $(u,v) \in P_0$, so $M \subset P_0$, and

$$\|u\| \leq \frac{1}{\gamma_1} \inf_{t\in[c,1-c]} u(t), \quad \|v\| \leq \frac{1}{\gamma_2} \inf_{t\in[c,1-c]} v(t), \quad \forall\, (u,v) \in M. \tag{5.7}$$

From (I3), we conclude that for $\varepsilon_1 = \frac{2}{\gamma_2 m_4} > 0$ and $\varepsilon_2 = \frac{2}{\gamma_1 m_2} > 0$ there exist $C_1, C_2 > 0$ such that

$$f(t,u) \geq \varepsilon_1 u - C_1, \quad g(t,u) \geq \varepsilon_2 u - C_2, \quad \forall\, (t,u) \in [c, 1-c] \times [0,\infty), \tag{5.8}$$

where $m_i = \int_c^{1-c} J_i(s)\,ds$, $i = 2,4$ and J_i, $i = 2,4$ are defined in Lemma 5.1.5.

For $(u,v) \in M$ and $t \in [c, 1-c]$, by using Lemma 5.1.5 and relations (5.8), we obtain

$$\begin{aligned} u(t) &= Q_1(u,v)(t) + \lambda u_0(t) \geq Q_1(u,v) \\ &= \int_0^1 G_1(t,s) f(s,v(s))\,ds + \int_0^1 G_2(t,s) g(s,u(s))\,ds \\ &\geq \int_c^{1-c} G_2(t,s) g(s,u(s))\,ds \geq \gamma_1 \int_c^{1-c} J_2(s)(\varepsilon_2 u(s) - C_2)\,ds \\ &\geq \gamma_1 \varepsilon_2 m_2 \inf_{s\in[c,1-c]} u(s) - \gamma_1 m_2 C_2 = 2 \inf_{s\in[c,1-c]} u(s) - C_3, \quad C_3 = \gamma_1 m_2 C_2, \end{aligned}$$

$$
\begin{aligned}
v(t) &= Q_2(u,v)(t) + \lambda v_0(t) \geq Q_2(u,v) \\
&= \int_0^1 G_3(t,s) g(s,u(s))\,\mathrm{d}s + \int_0^1 G_4(t,s) f(s,v(s))\,\mathrm{d}s \\
&\geq \int_c^{1-c} G_4(t,s) f(s,v(s))\,\mathrm{d}s \geq \gamma_2 \int_c^{1-c} J_4(s)(\varepsilon_1 v(s) - C_1)\,\mathrm{d}s \\
&\geq \gamma_2 \varepsilon_1 m_4 \inf_{s\in[c,1-c]} v(s) - \gamma_2 m_4 C_1 = 2 \inf_{s\in[c,1-c]} v(s) - C_4, \quad C_4 = \gamma_2 m_4 C_1.
\end{aligned}
$$

Therefore, we deduce

$$
\inf_{t\in[c,1-c]} u(t) \leq C_3, \qquad \inf_{t\in[c,1-c]} v(t) \leq C_4, \quad \forall\,(u,v) \in M. \tag{5.9}
$$

Now from relations (5.7) and (5.9), we obtain

$$
\|u\| \leq \frac{1}{\gamma_1} C_3, \quad \|v\| \leq \frac{1}{\gamma_2} C_4, \quad \text{and} \quad \|(u,v)\|_Y = \|u\| + \|v\| \leq \frac{C_3}{\gamma_1} + \frac{C_4}{\gamma_2} = C_5
$$

for all $(u,v) \in M$—that is, M is a bounded set of Y.

Besides, there exists a sufficiently large $R_1 > 1$ such that

$$
(u,v) \neq \mathcal{Q}(u,v) + \lambda(u_0,v_0), \quad \forall\,(u,v) \in \partial B_{R_1} \cap P', \quad \forall \lambda \geq 0.
$$

From Theorem 1.3.2, we deduce that the fixed point index of the operator $\mathcal{Q}$ over $B_{R_1} \cap P'$ with respect to P' is

$$
i(\mathcal{Q}, B_{R_1} \cap P', P') = 0. \tag{5.10}
$$

Next, from assumption (I4), we conclude that for $\varepsilon_3 = \min\left\{\frac{1}{4M_1}, \frac{1}{4M_4}\right\}$ and $\varepsilon_4 = \min\left\{\frac{1}{4M_2}, \frac{1}{4M_3}\right\}$ there exists $r_1 \in (0,1]$ such that

$$
f(t,u) \leq \varepsilon_3 u^p, \quad g(t,u) \leq \varepsilon_4 u^q, \quad \forall\,(t,u) \in [0,1] \times [0,r_1], \tag{5.11}
$$

where $M_i = \int_0^1 J_i(s)\,\mathrm{d}s$, $i = 1,\ldots,4$.

By using Lemma 5.1.5 and relations (5.11), we deduce that for all $(u,v) \in \bar{B}_{r_1} \cap P'$ and $t \in [0,1]$

$$
\begin{aligned}
Q_1(u,v)(t) &\leq \int_0^1 J_1(s)\varepsilon_3 (v(s))^p\,\mathrm{d}s + \int_0^1 J_2(s)\varepsilon_4 (u(s))^q\,\mathrm{d}s \\
&\leq \varepsilon_3 M_1 \|v\|^p + \varepsilon_4 M_2 \|u\|^q \leq \frac{1}{4}\|v\| + \frac{1}{4}\|u\| = \frac{1}{4}\|(u,v)\|_Y, \\
Q_2(u,v)(t) &\leq \int_0^1 J_3(s)\varepsilon_4 (u(s))^q\,\mathrm{d}s + \int_0^1 J_4(s)\varepsilon_3 (v(s))^p\,\mathrm{d}s \\
&\leq \varepsilon_4 M_3 \|u\|^q + \varepsilon_3 M_4 \|v\|^p \leq \frac{1}{4}\|u\| + \frac{1}{4}\|v\| = \frac{1}{4}\|(u,v)\|_Y.
\end{aligned}
$$

These imply that

$$
\|Q_1(u,v)\| \leq \frac{1}{4}\|(u,v)\|_Y, \quad \|Q_2(u,v)\| \leq \frac{1}{4}\|(u,v)\|_Y,
$$

$$\|\mathcal{Q}(u,v)\|_Y = \|\mathcal{Q}_1(u,v)\| + \|\mathcal{Q}_2(u,v)\| \le \frac{1}{2}\|(u,v)\|_Y, \quad \forall\,(u,v) \in \partial B_{r_1} \cap P'.$$

From Theorem 1.3.1, we conclude that the fixed point index of the operator $\mathcal{Q}$ over $B_{r_1} \cap P'$ with respect to P' is

$$i(\mathcal{Q}, B_{r_1} \cap P', P') = 1. \tag{5.12}$$

Combining (5.10) and (5.12), we obtain

$$i(\mathcal{Q}, (B_{R_1} \setminus \bar{B}_{r_1}) \cap P', P') = i(\mathcal{Q}, B_{R_1} \cap P', P') - i(\mathcal{Q}, B_{r_1} \cap P', P') = -1.$$

We deduce that $\mathcal{Q}$ has at least one fixed point $(u,v) \in (B_{R_1} \setminus \bar{B}_{r_1}) \cap P'$—that is, $r_1 < \|(u,v)\|_Y < R_1$. Thus, problem (S′)–(BC) has at least one positive solution $(u,v) \in P'$. The proof of Theorem 5.2.1 is completed. □

Theorem 5.2.2. *Assume that (I1) and (I2) hold. If the functions f and g also satisfy the following conditions (I5) and (I6), then problem* (S′)–(BC) *has at least one positive solution $(u(t), v(t)), t \in [0,1]$:*

(I5) (1) $\tilde{f}^{\rm s}_\infty = \lim\limits_{u\to\infty} \sup\limits_{t\in[0,1]} \dfrac{f(t,u)}{u} = 0$ *and* (2) $\tilde{g}^{\rm s}_\infty = \lim\limits_{u\to\infty} \sup\limits_{t\in[0,1]} \dfrac{g(t,u)}{u} = 0.$

(I6) *There exist $c \in (0,1/2)$, $\hat{p} \in (0,1]$, and $\hat{q} \in (0,1]$ such that*

(1) $\tilde{f}^{\rm i}_0 = \lim\limits_{u\to 0^+} \inf\limits_{t\in[c,1-c]} \dfrac{f(t,u)}{u^{\hat{p}}} = \infty$ *and* (2) $\tilde{g}^{\rm i}_0 = \lim\limits_{u\to 0^+} \inf\limits_{t\in[c,1-c]} \dfrac{g(t,u)}{u^{\hat{q}}} = \infty.$

Proof. From assumption (I5), we deduce that for $\varepsilon_5 = \min\left\{\frac{1}{4M_1}, \frac{1}{4M_4}\right\}$ and $\varepsilon_6 = \min\left\{\frac{1}{4M_2}, \frac{1}{4M_3}\right\}$ there exist C_6, $C_7 > 0$ such that

$$f(t,u) \le \varepsilon_5 u + C_6, \quad g(t,u) \le \varepsilon_6 u + C_7, \quad \forall\,(t,u) \in [0,1] \times [0,\infty). \tag{5.13}$$

Hence, for $(u,v) \in P'$, by using Lemma 5.1.5 and relations (5.13), we obtain

$$\begin{aligned}
\mathcal{Q}_1(u,v)(t) &\le \int_0^1 J_1(s)(\varepsilon_5 v(s) + C_6)\,ds + \int_0^1 J_2(s)(\varepsilon_6 u(s) + C_7)\,ds \\
&\le \varepsilon_5\|v\| \int_0^1 J_1(s)\,ds + C_6 \int_0^1 J_1(s)\,ds + \varepsilon_6\|u\| \int_0^1 J_2(s)\,ds \\
&\quad + C_7 \int_0^1 J_2(s)\,ds \\
&= \varepsilon_5\|v\|M_1 + C_6 M_1 + \varepsilon_6\|u\|M_2 + C_7 M_2 \\
&\le \frac{1}{4}\|v\| + \frac{1}{4}\|u\| + C_8 = \frac{1}{4}\|(u,v)\|_Y + C_8, \quad \forall\, t \in [0,1], \\
&\quad C_8 = C_6 M_1 + C_7 M_2,
\end{aligned}$$

$$\mathcal{Q}_2(u,v)(t) \le \int_0^1 J_3(s)(\varepsilon_6 u(s) + C_7)\,ds + \int_0^1 J_4(s)(\varepsilon_5 v(s) + C_6)\,ds$$

$$\le \varepsilon_6\|u\| \int_0^1 J_3(s)\,ds + C_7 \int_0^1 J_3(s)\,ds + \varepsilon_5\|v\| \int_0^1 J_4(s)\,ds$$
$$+ C_6 \int_0^1 J_4(s)\,ds$$
$$= \varepsilon_6\|u\|M_3 + C_7M_3 + \varepsilon_5\|v\|M_4 + C_6M_4$$
$$\le \frac{1}{4}\|u\| + \frac{1}{4}\|v\| + C_9 = \frac{1}{4}\|(u,v)\|_Y + C_9, \quad \forall t \in [0,1],$$
$$C_9 = C_7M_3 + C_6M_4,$$

and so

$$\|\mathcal{Q}(u,v)\|_Y = \|\mathcal{Q}_1(u,v)\| + \|\mathcal{Q}_2(u,v)\| \le \frac{1}{2}\|(u,v)\|_Y + C_{10}, \quad C_{10} = C_8 + C_9.$$

Then there exists a sufficiently large $R_2 \ge \max\{4C_{10}, 1\}$ such that

$$\|\mathcal{Q}(u,v)\|_Y \le \frac{3}{4}\|(u,v)\|_Y, \quad \forall (u,v) \in P', \quad \|(u,v)\|_Y \ge R_2.$$

Hence, $\|\mathcal{Q}(u,v)\|_Y < \|(u,v)\|_Y$ for all $(u,v) \in \partial B_{R_2} \cap P'$, and from Theorem 1.3.1 we have

$$i(\mathcal{Q}, B_{R_2} \cap P', P') = 1. \tag{5.14}$$

On the other hand, from (I6) we conclude that for $\varepsilon_7 = \frac{1}{\gamma_2(m_3+m_4)}$ and $\varepsilon_8 = \frac{1}{\gamma_1(m_1+m_2)}$, there exists $r_2 \in (0,1)$ such that

$$f(t,u) \ge \varepsilon_7 u^{\hat{p}}, \quad g(t,u) \ge \varepsilon_8 u^{\hat{q}}, \quad \forall (t,u) \in [c, 1-c] \times [0, r_2], \tag{5.15}$$

where $m_i = \int_c^{1-c} J_i(s)\,ds$, $i = 1, \ldots, 4$.

From (5.15) and Lemma 5.1.5, we deduce that for any $(u,v) \in \bar{B}_{r_2} \cap P'$

$$\mathcal{Q}_1(u,v)(t) \ge \int_c^{1-c} G_1(t,s)f(s,v(s))\,ds + \int_c^{1-c} G_2(t,s)g(s,u(s))\,ds$$
$$\ge \varepsilon_7 \int_c^{1-c} G_1(t,s)(v(s))^{\hat{p}}\,ds + \varepsilon_8 \int_c^{1-c} G_2(t,s)(u(s))^{\hat{q}}\,ds$$
$$\ge \varepsilon_7 \int_c^{1-c} G_1(t,s)v(s)\,ds + \varepsilon_8 \int_c^{1-c} G_2(t,s)u(s)\,ds =: L_1(u,v)(t),$$
$$\forall t \in [0,1],$$
$$\mathcal{Q}_2(u,v)(t) \ge \int_c^{1-c} G_3(t,s)g(s,u(s))\,ds + \int_c^{1-c} G_4(t,s)f(s,v(s))\,ds$$
$$\ge \varepsilon_8 \int_c^{1-c} G_3(t,s)(u(s))^{\hat{q}}\,ds + \varepsilon_7 \int_c^{1-c} G_4(t,s)(v(s))^{\hat{p}}\,ds$$
$$\ge \varepsilon_8 \int_c^{1-c} G_3(t,s)u(s)\,ds + \varepsilon_7 \int_c^{1-c} G_4(t,s)v(s)\,ds =: L_2(u,v)(t),$$
$$\forall t \in [0,1].$$

Hence,

$$\mathcal{Q}(u,v) \geq L(u,v), \quad \forall\, (u,v) \in \partial B_{r_2} \cap P', \tag{5.16}$$

where the linear operator $L : P' \to P'$ is defined by $L(u,v) = (L_1(u,v), L_2(u,v))$.

For $(\tilde{u}_0, \tilde{v}_0) \in P' \setminus \{(0,0)\}$ defined by

$$\tilde{u}_0(t) = \int_c^{1-c} G_1(t,s)\, ds + \int_c^{1-c} G_2(t,s)\, ds, \quad t \in [0,1],$$

$$\tilde{v}_0(t) = \int_c^{1-c} G_3(t,s)\, ds + \int_c^{1-c} G_4(t,s)\, ds, \quad t \in [0,1],$$

we have $L(\tilde{u}_0, \tilde{v}_0) = (L_1(\tilde{u}_0, \tilde{v}_0), L_2(\tilde{u}_0, \tilde{v}_0))$, with

$$\begin{aligned}
L_1(\tilde{u}_0, \tilde{v}_0)(t) &= \varepsilon_7 \int_c^{1-c} G_1(t,s) \left(\int_c^{1-c} G_3(s,\tau)\, d\tau + \int_c^{1-c} G_4(s,\tau)\, d\tau \right) ds \\
&\quad + \varepsilon_8 \int_c^{1-c} G_2(t,s) \left(\int_c^{1-c} G_1(s,\tau)\, d\tau + \int_c^{1-c} G_2(s,\tau)\, d\tau \right) ds \\
&\geq \varepsilon_7 \int_c^{1-c} G_1(t,s) \left(\int_c^{1-c} \gamma_2 J_3(\tau)\, d\tau + \int_c^{1-c} \gamma_2 J_4(\tau)\, d\tau \right) ds \\
&\quad + \varepsilon_8 \int_c^{1-c} G_2(t,s) \left(\int_c^{1-c} \gamma_1 J_1(\tau)\, d\tau + \int_c^{1-c} \gamma_1 J_2(\tau)\, d\tau \right) ds \\
&= \varepsilon_7 \gamma_2 (m_3 + m_4) \int_c^{1-c} G_1(t,s)\, ds + \varepsilon_8 \gamma_1 (m_1 + m_2) \int_c^{1-c} G_2(t,s)\, ds \\
&= \int_c^{1-c} G_1(t,s)\, ds + \int_c^{1-c} G_2(t,s)\, ds = \tilde{u}_0(t), \quad \forall\, t \in [0,1], \\
L_2(\tilde{u}_0, \tilde{v}_0)(t) &= \varepsilon_8 \int_c^{1-c} G_3(t,s) \left(\int_c^{1-c} G_1(s,\tau)\, d\tau + \int_c^{1-c} G_2(s,\tau)\, d\tau \right) ds \\
&\quad + \varepsilon_7 \int_c^{1-c} G_4(t,s) \left(\int_c^{1-c} G_3(s,\tau)\, d\tau + \int_c^{1-c} G_4(s,\tau)\, d\tau \right) ds \\
&\geq \varepsilon_8 \int_c^{1-c} G_3(t,s) \left(\int_c^{1-c} \gamma_1 J_1(\tau)\, d\tau + \int_c^{1-c} \gamma_1 J_2(\tau)\, d\tau \right) ds \\
&\quad + \varepsilon_7 \int_c^{1-c} G_4(t,s) \left(\int_c^{1-c} \gamma_2 J_3(\tau)\, d\tau + \int_c^{1-c} \gamma_2 J_4(\tau)\, d\tau \right) ds \\
&= \varepsilon_8 \gamma_1 (m_1 + m_2) \int_c^{1-c} G_3(t,s)\, ds + \varepsilon_7 \gamma_2 (m_3 + m_4) \int_c^{1-c} G_4(t,s)\, ds \\
&= \int_c^{1-c} G_3(t,s)\, ds + \int_c^{1-c} G_4(t,s)\, ds = \tilde{v}_0(t), \quad \forall\, t \in [0,1].
\end{aligned}$$

So

$$L(\tilde{u}_0, \tilde{v}_0) \geq (\tilde{u}_0, \tilde{v}_0). \tag{5.17}$$

We may suppose that $\mathcal{Q}$ has no fixed point on $\partial B_{r_2} \cap P'$ (otherwise the proof is finished). From (5.16), (5.17), and Theorem 1.3.3, we conclude that

$$i(\mathcal{Q}, B_{r_2} \cap P', P') = 0. \tag{5.18}$$

Therefore, from (5.14) and (5.18), we have

$$i(\mathcal{Q}, (B_{R_2} \setminus \bar{B}_{r_2}) \cap P', P') = i(\mathcal{Q}, B_{R_2} \cap P', P') - i(\mathcal{Q}, B_{r_2} \cap P', P') = 1.$$

Then $\mathcal{Q}$ has at least one fixed point $(u, v) \in (B_{R_2} \setminus \bar{B}_{r_2}) \cap P'$—that is, $r_2 < \|(u, v)\|_Y < R_2$. Thus, problem (S′)–(BC) has at least one positive solution $(u, v) \in P'$. This completes the proof of Theorem 5.2.2. □

Theorem 5.2.3. *Assume that (I1)–(I3) and (I6) hold. If the functions f and g also satisfy the following condition (I7), then problem* (S′)–(BC) *has at least two positive solutions $(u_1(t), v_1(t)), (u_2(t), v_2(t)), t \in [0, 1]$:*

(I7) *For each $t \in [0, 1]$, $f(t, u)$ and $g(t, u)$ are nondecreasing with respect to u, and there exists a constant $N > 0$ such that*

$$f(t, N) < \frac{N}{4m_0}, \quad g(t, N) < \frac{N}{4m_0}, \quad \forall t \in [0, 1],$$

where $m_0 = \max\{M_i,\ i = 1, \ldots, 4\}$ and $M_i = \int_0^1 J_i(s)\,ds,\ i = 1, \ldots, 4$.

Proof. By using (I7), for any $(u, v) \in \partial B_N \cap P'$, we obtain

$$\begin{aligned}
\mathcal{Q}_1(u, v)(t) &\le \int_0^1 G_1(t, s) f(s, N)\,ds + \int_0^1 G_2(t, s) g(s, N)\,ds \\
&\le \int_0^1 J_1(s) f(s, N)\,ds + \int_0^1 J_2(s) g(s, N)\,ds \\
&< \frac{N}{4m_0}\int_0^1 J_1(s)\,ds + \frac{N}{4m_0}\int_0^1 J_2(s)\,ds = \frac{NM_1}{4m_0} + \frac{NM_2}{4m_0} \le \frac{N}{2}, \\
&\quad \forall t \in [0, 1], \\
\mathcal{Q}_2(u, v)(t) &\le \int_0^1 G_3(t, s) g(s, N)\,ds + \int_0^1 G_4(t, s) f(s, N)\,ds \\
&\le \int_0^1 J_3(s) g(s, N)\,ds + \int_0^1 J_4(s) f(s, N)\,ds \\
&< \frac{N}{4m_0}\int_0^1 J_3(s)\,ds + \frac{N}{4m_0}\int_0^1 J_4(s)\,ds = \frac{NM_3}{4m_0} + \frac{NM_4}{4m_0} \le \frac{N}{2}, \\
&\quad \forall t \in [0, 1].
\end{aligned}$$

Then we deduce

$$\|\mathcal{Q}(u, v)\|_Y = \|\mathcal{Q}_1(u, v)\| + \|\mathcal{Q}_2(u, v)\| < N = \|(u, v)\|_Y, \quad \forall (u, v) \in \partial B_N \cap P'.$$

By Theorem 1.3.1, we conclude that

$$i(\mathcal{Q}, B_N \cap P', P') = 1. \tag{5.19}$$

On the other hand, from (I3), (I6), and the proofs of Theorems 5.2.1 and 5.2.2, we know that there exists a sufficiently large $R_1 > N$ and a sufficiently small $r_2 \in (0, N)$ such that

$$i(\mathcal{Q}, B_{R_1} \cap P', P') = 0, \quad i(\mathcal{Q}, B_{r_2} \cap P', P') = 0. \tag{5.20}$$

From relations (5.19) and (5.20), we obtain

$$i(\mathcal{Q}, (B_{R_1} \setminus \bar{B}_N) \cap P', P') = i(\mathcal{Q}, B_{R_1} \cap P', P') - i(\mathcal{Q}, B_N \cap P', P') = -1,$$
$$i(\mathcal{Q}, (B_N \setminus \bar{B}_{r_2}) \cap P', P') = i(\mathcal{Q}, B_N \cap P', P') - i(\mathcal{Q}, B_{r_2} \cap P', P') = 1.$$

Then $\mathcal{Q}$ has at least one fixed point $(u_1, v_1) \in (B_{R_1} \setminus \bar{B}_N) \cap P'$ and at least one fixed point $(u_2, v_2) \in (B_N \setminus \bar{B}_{r_2}) \cap P'$. Therefore, problem (S′)–(BC) has two distinct positive solutions (u_1, v_1), (u_2, v_2). The proof of Theorem 5.2.3 is completed. □

We present now an example which illustrates our results above.

Example 5.2.1. Let $\alpha = 7/3$ $(n=3)$, $\beta = 5/2$ $(m = 3)$, $H(t) = \begin{cases} 0, & t \in [0, 1/3), \\ 1/2, & t \in [1/3, 2/3), \\ 3/2, & t \in [2/3, 1], \end{cases}$ and $K(t) = t^2$ for all $t \in [0, 1]$. Then, $\int_0^1 v(s)\,\mathrm{d}H(s) = \frac{1}{2} v\left(\frac{1}{3}\right) + v\left(\frac{2}{3}\right)$ and $\int_0^1 u(s)\,\mathrm{d}K(s) = 2\int_0^1 su(s)\,\mathrm{d}s$.

We consider the system of fractional differential equations

$$\begin{cases} D_{0+}^{7/3} u(t) + f(t, v(t)) = 0, & t \in (0, 1), \\ D_{0+}^{5/2} v(t) + g(t, u(t)) = 0, & t \in (0, 1), \end{cases} \tag{S$_2$}$$

with the boundary conditions

$$\begin{cases} u(0) = u'(0) = 0, \quad u(1) = \dfrac{1}{2} v\left(\dfrac{1}{3}\right) + v\left(\dfrac{2}{3}\right), \\ v(0) = v'(0) = 0, \quad v(1) = 2\displaystyle\int_0^1 su(s)\,\mathrm{d}s. \end{cases} \tag{BC$_2$}$$

Here we consider

$$f(t, u) = a_0(u^{\alpha_0} + u^{\beta_0}), \quad g(t, u) = b_0(u^{\gamma_0} + u^{\delta_0}), \quad \forall t \in [0, 1], \quad u \in [0, \infty),$$

where $\alpha_0 > 1$, $0 < \beta_0 < 1$, $\gamma_0 > 1$, $0 < \delta_0 < 1$, and $a_0, b_0 > 0$.

Then we obtain $\Delta = \frac{10\sqrt{3} - 1 - 4\sqrt{2}}{10\sqrt{3}} \approx 0.61566634 > 0$. We also deduce

$$g_1(t, s) = \frac{1}{\Gamma(7/3)} \begin{cases} t^{4/3}(1 - s)^{4/3} - (t - s)^{4/3}, & 0 \le s \le t \le 1, \\ t^{4/3}(1 - s)^{4/3}, & 0 \le t \le s \le 1, \end{cases}$$

$$g_2(t, s) = \frac{4}{3\sqrt{\pi}} \begin{cases} t^{3/2}(1 - s)^{3/2} - (t - s)^{3/2}, & 0 \le s \le t \le 1, \\ t^{3/2}(1 - s)^{3/2}, & 0 \le t \le s \le 1, \end{cases}$$

$\theta_1(s) = \frac{1}{4-6s+4s^2-s^3}$, and $\theta_2(s) = \frac{1}{3-3s+s^2}$ for all $s \in [0,1]$. For the functions J_i, $i = 1, \dots, 4$, we obtain

$$J_1(s) = \frac{1}{\Gamma(7/3)} \left\{ \frac{s(1-s)^{4/3}}{(4-6s+4s^2-s^3)^{1/3}} + \frac{1+4\sqrt{2}}{\Delta\sqrt{3}} \left[\frac{1}{10}(1-s)^{4/3} - \frac{1}{10}(1-s)^{10/3} - \frac{1}{7}s(1-s)^{7/3} \right] \right\}, \quad s \in [0,1],$$

$$J_2(s) = \begin{cases} \frac{2}{9\sqrt{3\pi}\Delta} \left[(1-s)^{3/2} - (1-3s)^{3/2} + 2(2-2s)^{3/2} - 2(2-3s)^{3/2} \right], \\ 0 \le s < \frac{1}{3}, \\ \frac{2}{9\sqrt{3\pi}\Delta} \left[(1-s)^{3/2} + 2(2-2s)^{3/2} - 2(2-3s)^{3/2} \right], \quad \frac{1}{3} \le s < \frac{2}{3}, \\ \frac{2}{9\sqrt{3\pi}\Delta} \left[(1-s)^{3/2} + 2(2-2s)^{3/2} \right], \quad \frac{2}{3} \le s \le 1, \end{cases}$$

$$J_3(s) = \begin{cases} \frac{4}{3\sqrt{\pi}} \left\{ \frac{s(1-s)^{3/2}}{(3-3s+s^2)^{1/2}} + \frac{1}{10\sqrt{3}\Delta} \left[(1-s)^{3/2} - (1-3s)^{3/2} \right.\right. \\ \qquad \left.\left. +2(2-2s)^{3/2} - 2(2-3s)^{3/2} \right] \right\}, \quad 0 \le s < \frac{1}{3}, \\ \frac{4}{3\sqrt{\pi}} \left\{ \frac{s(1-s)^{3/2}}{(3-3s+s^2)^{1/2}} + \frac{1}{10\sqrt{3}\Delta} \left[(1-s)^{3/2} + 2(2-2s)^{3/2} \right.\right. \\ \qquad \left.\left. -2(2-3s)^{3/2} \right] \right\}, \quad \frac{1}{3} \le s < \frac{2}{3}, \\ \frac{4}{3\sqrt{\pi}} \left\{ \frac{s(1-s)^{3/2}}{(3-3s+s^2)^{1/2}} + \frac{1}{10\sqrt{3}\Delta} \left[(1-s)^{3/2} + 2(2-2s)^{3/2} \right] \right\}, \\ \frac{2}{3} \le s \le 1, \end{cases}$$

$$J_4(s) = \frac{6}{\Gamma(7/3)\Delta} \left[\frac{1}{10}(1-s)^{4/3} - \frac{1}{10}(1-s)^{10/3} - \frac{1}{7}s(1-s)^{7/3} \right], \quad s \in [0,1].$$

We also deduce $M_1 = \int_0^1 J_1(s)\,ds \approx 0.13794382$, $M_2 = \int_0^1 J_2(s)\,ds \approx 0.12003161$, $M_3 = \int_0^1 J_3(s)\,ds \approx 0.13555753$, and $M_4 = \int_0^1 J_4(s)\,ds \approx 0.08095184$.

In addition, we have $m_0 = \max\{M_i, i = 1, \dots, 4\} = M_1$. The functions $f(t,u)$ and $g(t,u)$ are nondecreasing with respect to u for any $t \in [0,1]$, and for $\hat{p} = \hat{q} = 1$ and $c \in (0, 1/2)$ assumptions (I3) and (I6) are satisfied; we obtain $\tilde{f}_\infty^{\mathrm{i}} = \infty$, $\tilde{g}_\infty^{\mathrm{i}} = \infty$, $\tilde{f}_0^{\mathrm{i}} = \infty$, and $\tilde{g}_0^{\mathrm{i}} = \infty$. We take $N = 1$ and then $f(t,N) = 2a_0$, $g(t,N) = 2b_0$ for all $t \in [0,1]$. If $a_0 < \frac{1}{8m_0}$ and $b_0 < \frac{1}{8m_0}$, then assumption (I7) is satisfied. For example, if $a_0 \le 0.9$ and $b_0 \le 0.9$, then by Theorem 5.2.3, we deduce that problem (S_2)–(BC_2) has at least two positive solutions.

5.2.2 Singular nonlinearities

We consider the system of nonlinear ordinary fractional differential equations

$$\begin{cases} D_{0+}^{\alpha}u(t) + f(t, v(t)) = 0, & t \in (0, 1), \\ D_{0+}^{\beta}v(t) + g(t, u(t)) = 0, & t \in (0, 1), \end{cases} \tag{S'}$$

with the coupled integral boundary conditions

$$\begin{cases} u(0) = u'(0) = \cdots = u^{(n-2)}(0) = 0, & u(1) = \displaystyle\int_0^1 v(s)\,\mathrm{d}H(s), \\ v(0) = v'(0) = \cdots = v^{(m-2)}(0) = 0, & v(1) = \displaystyle\int_0^1 u(s)\,\mathrm{d}K(s), \end{cases} \tag{BC}$$

where $n - 1 < \alpha \leq n$, $m - 1 < \beta \leq m$, $n, m \in \mathbb{N}$, $n, m \geq 3$, D_{0+}^{α} and D_{0+}^{β} denote the Riemann–Liouville derivatives of orders α and β, respectively, and the integrals from (BC) are Riemann–Stieltjes integrals.

We investigate the existence of positive solutions for problem (S′)–(BC) under various assumptions on functions f and g which may be singular at $t = 0$ and/or $t = 1$. By a positive solution of problem (S′)–(BC) we mean a pair of functions $(u, v) \in C([0, 1]) \times C([0, 1])$ satisfying (S′) and (BC) with $u(t) \geq 0$, $v(t) \geq 0$ for all $t \in [0, 1]$ and $(u, v) \neq (0, 0)$.

We present the basic assumptions that we shall use in the sequel:

(L1) $H, K : [0, 1] \to \mathbb{R}$ are nondecreasing functions, and $\Delta = 1 - \left(\int_0^1 \tau^{\alpha-1}\,\mathrm{d}K(\tau)\right) \times \left(\int_0^1 \tau^{\beta-1}\,\mathrm{d}H(\tau)\right) > 0$.

(L2) The functions $f, g \in C((0, 1) \times \mathbb{R}_+, \mathbb{R}_+)$, and there exist the functions $p_i \in C((0, 1), \mathbb{R}_+)$ and $q_i \in C(\mathbb{R}_+, \mathbb{R}_+)$ with $p_i \not\equiv 0$ (there exists $t_i \in (0, 1)$ such that $p_i(t_i) > 0$), $i = 1, 2$ and $\int_0^1 (1 - s)^{\alpha-1} p_1(s)\,\mathrm{d}s < \infty$, $\int_0^1 (1 - s)^{\beta-1} p_2(s)\,\mathrm{d}s < \infty$ such that

$$f(t, x) \leq p_1(t)q_1(x), \quad g(t, x) \leq p_2(t)q_2(x), \quad \forall t \in (0, 1), \quad x \in [0, \infty).$$

Under assumption (L1), we have all auxiliary results in Lemmas 5.1.1–5.1.6 from Section 5.1.2.

By using the functions G_i, $i = 1, \ldots, 4$ from Section 5.1.2 (Lemma 5.1.2), we can write our problem (S′)–(BC) equivalently as the following nonlinear system of integral equations:

$$\begin{cases} u(t) = \displaystyle\int_0^1 G_1(t, s)f(s, v(s))\,\mathrm{d}s + \int_0^1 G_2(t, s)g(s, u(s))\,\mathrm{d}s, & t \in [0, 1], \\ v(t) = \displaystyle\int_0^1 G_3(t, s)g(s, u(s))\,\mathrm{d}s + \int_0^1 G_4(t, s)f(s, v(s))\,\mathrm{d}s, & t \in [0, 1]. \end{cases}$$

We consider again the Banach space $X = C([0, 1])$ with the supremum norm $\|\cdot\|$ and the Banach space $Y = X \times X$ with the norm $\|(u, v)\|_Y = \|u\| + \|v\|$. We define the cone $P' \subset Y$ by

$$P' = \{(u, v) \in Y, \quad u(t) \geq 0, \quad v(t) \geq 0, \quad \forall t \in [0, 1]\}.$$

We introduce the operators $\tilde{Q}_1, \tilde{Q}_2 : Y \to X$ and $\tilde{Q} : Y \to Y$ defined by

$$\tilde{Q}_1(u, v)(t) = \int_0^1 G_1(t, s)f(s, v(s))\,ds + \int_0^1 G_2(t, s)g(s, u(s))\,ds, \quad 0 \leq t \leq 1,$$

$$\tilde{Q}_2(u, v)(t) = \int_0^1 G_3(t, s)g(s, u(s))\,ds + \int_0^1 G_4(t, s)f(s, v(s))\,ds, \quad 0 \leq t \leq 1,$$

and $\tilde{Q}(u, v) = (\tilde{Q}_1(u, v), \tilde{Q}_2(u, v)), (u, v) \in Y$.

Lemma 5.2.1. *Assume that (L1) and (L2) hold. Then $\tilde{Q} : P' \to P'$ is completely continuous.*

Proof. We denote $\tilde{\alpha} = \int_0^1 J_1(s)p_1(s)\,ds$, $\tilde{\beta} = \int_0^1 J_2(s)p_2(s)\,ds$, $\tilde{\gamma} = \int_0^1 J_3(s)p_2(s)\,ds$, and $\tilde{\delta} = \int_0^1 J_4(s)p_1(s)\,ds$. Using (L2) and Lemma 5.1.5, we deduce that $\tilde{\alpha}$, $\tilde{\beta}$, $\tilde{\gamma}$, $\tilde{\delta} > 0$ and

$$\begin{aligned}
\tilde{\alpha} &\leq \int_0^1 \left[g_1(\theta_1(s), s) + \frac{1}{\Delta} \left(\int_0^1 \tau^{\beta-1} dH(\tau) \right) \left(\int_0^1 g_1(\theta_1(s), s)\,dK(\tau) \right) \right] p_1(s)\,ds \\
&= \left[1 + \frac{1}{\Delta}(K(1) - K(0)) \int_0^1 \tau^{\beta-1} dH(\tau) \right] \int_0^1 g_1(\theta_1(s), s)p_1(s)\,ds \\
&\leq \left[1 + \frac{1}{\Delta}(K(1) - K(0)) \int_0^1 \tau^{\beta-1} dH(\tau) \right] \int_0^1 \frac{1}{\Gamma(\alpha)}(1 - s)^{\alpha-1} p_1(s)\,ds < \infty, \\
\tilde{\beta} &\leq \frac{1}{\Delta} \int_0^1 \left(\int_0^1 g_2(\theta_2(s), s)\,dH(\tau) \right) p_2(s)\,ds \\
&\leq \frac{1}{\Delta\Gamma(\beta)}(H(1) - H(0)) \int_0^1 (1 - s)^{\beta-1} p_2(s)\,ds < \infty.
\end{aligned}$$

In a similar manner, we have

$$\begin{aligned}
\tilde{\gamma} &\leq \left[1 + \frac{1}{\Delta}(H(1) - H(0)) \int_0^1 \tau^{\alpha-1} dK(\tau) \right] \int_0^1 \frac{1}{\Gamma(\beta)}(1 - s)^{\beta-1} p_2(s)\,ds < \infty, \\
\tilde{\delta} &\leq \frac{1}{\Delta\Gamma(\alpha)}(K(1) - K(0)) \int_0^1 (1 - s)^{\alpha-1} p_1(s)\,ds < \infty.
\end{aligned}$$

By Lemma 5.1.4, we also conclude that $\tilde{Q}$ maps P' into P'.

We shall prove that $\tilde{Q}$ maps bounded sets into relatively compact sets. Suppose $D \subset P'$ is an arbitrary bounded set. Then there exists $\tilde{M}_1 > 0$ such that $\|(u, v)\|_Y \leq \tilde{M}_1$ for all $(u, v) \in D$. Then $\|u\| \leq \tilde{M}_1$ and $\|v\| \leq \tilde{M}_1$ for all $(u, v) \in D$. By the continuity of q_1 and q_2, there exists $\tilde{M}_2 > 0$ such that $\tilde{M}_2 = \max\left\{\sup_{x \in [0, \tilde{M}_1]} q_1(x), \sup_{x \in [0, \tilde{M}_1]} q_2(x)\right\}$. By using Lemma 5.1.5, for any $(u, v) \in D$ and $t \in [0, 1]$, we obtain

$$\tilde{Q}_1(u, v)(t) = \int_0^1 G_1(t, s)f(s, v(s))\,ds + \int_0^1 G_2(t, s)g(s, u(s))\,ds$$

$$
\begin{aligned}
&\le \int_0^1 J_1(s)f(s, v(s))\,ds + \int_0^1 J_2(s)g(s, u(s))\,ds \\
&\le \int_0^1 J_1(s)p_1(s)q_1(v(s))\,ds + \int_0^1 J_2(s)p_2(s)q_2(u(s))\,ds \\
&\le \tilde{M}_2 \int_0^1 J_1(s)p_1(s)\,ds + \tilde{M}_2 \int_0^1 J_2(s)p_2(s)\,ds = \tilde{M}_2(\tilde{\alpha} + \tilde{\beta}),
\end{aligned}
$$

$$
\begin{aligned}
\tilde{Q}_2(u, v)(t) &= \int_0^1 G_3(t, s)g(s, u(s))\,ds + \int_0^1 G_4(t, s)f(s, v(s))\,ds \\
&\le \int_0^1 J_3(s)g(s, u(s))\,ds + \int_0^1 J_4(s)f(s, v(s))\,ds \\
&\le \int_0^1 J_3(s)p_2(s)q_2(u(s))\,ds + \int_0^1 J_4(s)p_1(s)q_1(v(s))\,ds \\
&\le \tilde{M}_2 \int_0^1 J_3(s)p_2(s)\,ds + \tilde{M}_2 \int_0^1 J_4(s)p_1(s)\,ds = \tilde{M}_2(\tilde{\gamma} + \tilde{\delta}).
\end{aligned}
$$

Therefore, $\|\tilde{Q}_1(u, v)\| \le \tilde{M}_2(\tilde{\alpha}+\tilde{\beta})$ and $\|\tilde{Q}_2(u, v)\| \le \tilde{M}_2(\tilde{\gamma}+\tilde{\delta})$ for all $(u, v) \in D$, and so $\tilde{Q}_1(D)$ and $\tilde{Q}_2(D)$ are bounded.

In what follows, we shall prove that $\tilde{\mathcal{Q}}(D)$ is equicontinuous. By using Lemma 5.1.2, we have for any $(u, v) \in D$ and $t \in [0, 1]$

$$
\begin{aligned}
\tilde{Q}_1(u, v)(t) &= \int_0^1 \left[g_1(t, s) + \frac{t^{\alpha-1}}{\Delta} \left(\int_0^1 \tau^{\beta-1} dH(\tau) \right) \left(\int_0^1 g_1(\tau, s)\,dK(\tau) \right) \right] \\
&\quad \times f(s, v(s))\,ds + \int_0^1 \left[\frac{t^{\alpha-1}}{\Delta} \int_0^1 g_2(\tau, s)\,dH(\tau) \right] g(s, u(s))\,ds \\
&= \int_0^t \frac{1}{\Gamma(\alpha)} \left[t^{\alpha-1}(1-s)^{\alpha-1} - (t-s)^{\alpha-1} \right] f(s, v(s))\,ds \\
&\quad + \int_t^1 \frac{1}{\Gamma(\alpha)} t^{\alpha-1}(1-s)^{\alpha-1} f(s, v(s))\,ds \\
&\quad + \frac{t^{\alpha-1}}{\Delta} \left(\int_0^1 \tau^{\beta-1} dH(\tau) \right) \left(\int_0^1 \left(\int_0^1 g_1(\tau, s)\,dK(\tau) \right) f(s, v(s))\,ds \right) \\
&\quad + \frac{t^{\alpha-1}}{\Delta} \int_0^1 \left(\int_0^1 g_2(\tau, s)\,dH(\tau) \right) g(s, u(s))\,ds.
\end{aligned}
$$

Therefore, for any $(u, v) \in D$ and $t \in (0, 1)$ we conclude

$$
\begin{aligned}
(\tilde{Q}_1(u, v))'(t) &= \frac{1}{\Gamma(\alpha)} \int_0^t [(\alpha-1)t^{\alpha-2}(1-s)^{\alpha-1} - (\alpha-1)(t-s)^{\alpha-2}] f(s, v(s))\,ds \\
&\quad + \frac{1}{\Gamma(\alpha)} \int_t^1 (\alpha-1)t^{\alpha-2}(1-s)^{\alpha-1} f(s, v(s))\,ds
\end{aligned}
$$

$$+\frac{(\alpha-1)t^{\alpha-2}}{\Delta}\left(\int_0^1 \tau^{\beta-1}\mathrm{d}H(\tau)\right)\left(\int_0^1\left(\int_0^1 g_1(\tau,s)\,\mathrm{d}K(\tau)\right)\right.$$
$$\left.\times f(s,v(s))\,\mathrm{d}s\right)+\frac{(\alpha-1)t^{\alpha-2}}{\Delta}\int_0^1\left(\int_0^1 g_2(\tau,s)\,\mathrm{d}H(\tau)\right)g(s,u(s))\,\mathrm{d}s.$$

So, for any $(u,v)\in D$ and $t\in(0,1)$ we deduce

$$\begin{aligned}
|(\tilde{Q}_1(u,v))'(t)| &\le \frac{1}{\Gamma(\alpha-1)}\int_0^t [t^{\alpha-2}(1-s)^{\alpha-1}+(t-s)^{\alpha-2}]p_1(s)q_1(v(s))\,\mathrm{d}s\\
&\quad+\frac{1}{\Gamma(\alpha-1)}\int_t^1 t^{\alpha-2}(1-s)^{\alpha-1}p_1(s)q_1(v(s))\,\mathrm{d}s\\
&\quad+\frac{(\alpha-1)t^{\alpha-2}}{\Delta}\left(\int_0^1 \tau^{\beta-1}\mathrm{d}H(\tau)\right)\left(\int_0^1\left(\int_0^1 g_1(\tau,s)\,\mathrm{d}K(\tau)\right)\right.\\
&\quad\left.\times\, p_1(s)q_1(v(s))\,\mathrm{d}s\right)+\frac{(\alpha-1)t^{\alpha-2}}{\Delta}\\
&\quad\times\int_0^1\left(\int_0^1 g_2(\tau,s)\,\mathrm{d}H(\tau)\right)p_2(s)q_2(u(s))\,\mathrm{d}s\\
&\le \tilde{M}_2\left[\frac{1}{\Gamma(\alpha-1)}\int_0^t [t^{\alpha-2}(1-s)^{\alpha-1}+(t-s)^{\alpha-2}]p_1(s)\,\mathrm{d}s\right.\\
&\quad+\frac{1}{\Gamma(\alpha-1)}\int_t^1 t^{\alpha-2}(1-s)^{\alpha-1}p_1(s)\,\mathrm{d}s\\
&\quad+\frac{(\alpha-1)t^{\alpha-2}}{\Delta}\left(\int_0^1 \tau^{\beta-1}\mathrm{d}H(\tau)\right)\left(\int_0^1\left(\int_0^1 g_1(\tau,s)\,\mathrm{d}K(\tau)\right)\right.\\
&\quad\left.\left.\times\, p_1(s)\,\mathrm{d}s\right)+\frac{(\alpha-1)t^{\alpha-2}}{\Delta}\int_0^1\left(\int_0^1 g_2(\tau,s)\,\mathrm{d}H(\tau)\right)p_2(s)\,\mathrm{d}s\right].
\end{aligned}\tag{5.21}$$

We denote

$$\begin{aligned}
h(t) &= \frac{1}{\Gamma(\alpha-1)}\int_0^t [t^{\alpha-2}(1-s)^{\alpha-1}+(t-s)^{\alpha-2}]p_1(s)\,\mathrm{d}s\\
&\quad+\frac{1}{\Gamma(\alpha-1)}\int_t^1 t^{\alpha-2}(1-s)^{\alpha-1}p_1(s)\,\mathrm{d}s,\quad t\in(0,1),\\
\mu(t) &= h(t)+\frac{(\alpha-1)t^{\alpha-2}}{\Delta}\left(\int_0^1 \tau^{\beta-1}\mathrm{d}H(\tau)\right)\left(\int_0^1\left(\int_0^1 g_1(\tau,s)\,\mathrm{d}K(\tau)\right)p_1(s)\,\mathrm{d}s\right)\\
&\quad+\frac{(\alpha-1)t^{\alpha-2}}{\Delta}\int_0^1\left(\int_0^1 g_2(\tau,s)\,\mathrm{d}H(\tau)\right)p_2(s)\,\mathrm{d}s,\quad t\in(0,1).
\end{aligned}$$

For the integral of the function h, by exchanging the order of integration, we obtain after some computations

$$\int_0^1 h(t)\,\mathrm{d}t = \frac{2}{\Gamma(\alpha)} \int_0^1 (1-s)^{\alpha-1} p_1(s)\,\mathrm{d}s < \infty.$$

For the integral of the function μ, we have

$$\begin{aligned}
\int_0^1 \mu(t)\,\mathrm{d}t &= \int_0^1 h(t)\,\mathrm{d}t + \frac{\alpha-1}{\Delta}\left(\int_0^1 t^{\alpha-2}\,\mathrm{d}t\right)\left(\int_0^1 \tau^{\beta-1}\mathrm{d}H(\tau)\right) \\
&\quad \times \left(\int_0^1 \left(\int_0^1 g_1(\tau,s)\,\mathrm{d}K(\tau)\right) p_1(s)\,\mathrm{d}s\right) \\
&\quad + \frac{\alpha-1}{\Delta}\left(\int_0^1 t^{\alpha-2}\mathrm{d}t\right)\left(\int_0^1 \left(\int_0^1 g_2(\tau,s)\,\mathrm{d}H(\tau)\right) p_2(s)\,\mathrm{d}s\right) \\
&\le \frac{2}{\Gamma(\alpha)} \int_0^1 (1-s)^{\alpha-1} p_1(s)\,\mathrm{d}s + \frac{1}{\Delta}\left(\int_0^1 \tau^{\beta-1}\mathrm{d}H(\tau)\right) \\
&\quad \times \left(\int_0^1 \left(\int_0^1 g_1(\theta_1(s),s)\,\mathrm{d}K(\tau)\right) p_1(s)\,\mathrm{d}s\right) \\
&\quad + \frac{1}{\Delta}\left(\int_0^1 \left(\int_0^1 g_2(\theta_2(s),s)\,\mathrm{d}H(\tau)\right) p_2(s)\,\mathrm{d}s\right) \\
&= \frac{2}{\Gamma(\alpha)} \int_0^1 (1-s)^{\alpha-1} p_1(s)\,\mathrm{d}s + \frac{1}{\Delta}\left(\int_0^1 \tau^{\beta-1}\mathrm{d}H(\tau)\right) \\
&\quad \times \left(\int_0^1 g_1(\theta_1(s),s) p_1(s)\,\mathrm{d}s\right)\left(\int_0^1 \mathrm{d}K(\tau)\right) \\
&\quad + \frac{1}{\Delta}\left(\int_0^1 g_2(\theta_2(s),s) p_2(s)\,\mathrm{d}s\right)\left(\int_0^1 \mathrm{d}H(\tau)\right) \\
&\le \frac{2}{\Gamma(\alpha)} \int_0^1 (1-s)^{\alpha-1} p_1(s)\,\mathrm{d}s + \frac{K(1)-K(0)}{\Delta\Gamma(\alpha)}\left(\int_0^1 \tau^{\beta-1}\mathrm{d}H(\tau)\right) \\
&\quad \times \int_0^1 (1-s)^{\alpha-1} p_1(s)\,\mathrm{d}s + \frac{H(1)-H(0)}{\Delta\Gamma(\beta)} \int_0^1 (1-s)^{\beta-1} p_2(s)\,\mathrm{d}s.
\end{aligned}$$

Therefore, we obtain

$$\begin{aligned}
\int_0^1 \mu(t)\,\mathrm{d}t &\le \frac{1}{\Gamma(\alpha)}\left(2 + \frac{K(1)-K(0)}{\Delta}\int_0^1 \tau^{\beta-1}\mathrm{d}H(\tau)\right)\int_0^1 (1-s)^{\alpha-1} p_1(s)\,\mathrm{d}s \\
&\quad + \frac{H(1)-H(0)}{\Delta\Gamma(\beta)} \int_0^1 (1-s)^{\beta-1} p_2(s)\,\mathrm{d}s < \infty. \qquad (5.22)
\end{aligned}$$

We deduce that $\mu \in L^1(0,1)$. Thus, for any $t_1, t_2 \in [0,1]$ with $t_1 \le t_2$ and $(u,v) \in D$, by (5.21) and (5.22), we conclude

$$|\tilde{Q}_1(u,v)(t_1) - \tilde{Q}_1(u,v)(t_2)| = \left|\int_{t_1}^{t_2} (\tilde{Q}_1(u,v))'(t)\,\mathrm{d}t\right| \le \tilde{M}_2 \int_{t_1}^{t_2} \mu(t)\,\mathrm{d}t. \quad (5.23)$$

From (5.22), (5.23), and the absolute continuity of the integral function, we find that $\tilde{Q}_1(D)$ is equicontinuous. In a similar manner we deduce that $\tilde{Q}_2(D)$ is also equicontinuous. By the Ascoli-Arzèla theorem, we conclude that $\tilde{Q}_1(D)$ and $\tilde{Q}_2(D)$ are relatively compact sets, and so $\tilde{\mathcal{Q}}(D)$ is also relatively compact; therefore, $\tilde{\mathcal{Q}}$ is a compact operator. Besides, we can easily show that $\tilde{\mathcal{Q}}$ is continuous on P'; hence, $\tilde{\mathcal{Q}} : P' \to P'$ is completely continuous. □

For $c \in (0, 1/2)$, we define the cone

$$P_0 = \left\{(u, v) \in P', \quad \inf_{t\in[c,1-c]} u(t) \ge \gamma_1 \|u\|, \quad \inf_{t\in[c,1-c]} v(t) \ge \gamma_2 \|v\|\right\}.$$

Under assumptions (L1) and (L2) and Lemma 5.1.6, we have $\tilde{\mathcal{Q}}(P') \subset P_0$, and so $\tilde{\mathcal{Q}}|_{P_0} : P_0 \to P_0$ (denoted again by $\tilde{\mathcal{Q}}$) is also a completely continuous operator.

Theorem 5.2.4. *Assume that (L1) and (L2) hold. If the functions f and g also satisfy the following conditions (L3) and (L4), then problem* (S′)–(BC) *has at least one positive solution $(u(t), v(t)), t \in [0, 1]$:*

(L3) (1) $q_{1\infty} = \lim_{x\to\infty} \dfrac{q_1(x)}{x} = 0$ *and* (2) $q_{2\infty} = \lim_{x\to\infty} \dfrac{q_2(x)}{x} = 0.$

(L4) *There exist $c \in (0, 1/2)$, $a \in (0, 1]$, and $b \in (0, 1]$ such that*

$$(1)\ \hat{f}_0^i = \lim_{x\to 0^+} \inf_{t\in[c,1-c]} \frac{f(t,x)}{x^a} = \infty \quad \textit{and} \quad (2)\ \hat{g}_0^i = \lim_{x\to 0^+} \inf_{t\in[c,1-c]} \frac{g(t,x)}{x^b} = \infty.$$

Proof. We consider the cone P_0 with c given in (L4). From (L3) we deduce that for $\varepsilon_9 \in \left(0, \frac{1}{\tilde{\alpha}+\tilde{\delta}}\right)$ and $\varepsilon_{10} \in \left(0, \frac{1}{\tilde{\beta}+\tilde{\gamma}}\right)$ there exist $C_{11}, C_{12} > 0$ such that

$$q_1(x) \le \varepsilon_9 x + C_{11}, \quad q_2(x) \le \varepsilon_{10} x + C_{12}, \quad \forall x \in [0, \infty). \tag{5.24}$$

By using Lemma 5.1.5, (5.24), and (L2), for any $(u, v) \in P_0$ we conclude

$$\begin{aligned}
\tilde{Q}_1(u, v)(t) &\le \int_0^1 J_1(s)p_1(s)q_1(v(s))\,ds + \int_0^1 J_2(s)p_2(s)q_2(u(s))\,ds\\
&\le \int_0^1 J_1(s)p_1(s)(\varepsilon_9 v(s)+C_{11})\,ds+\int_0^1 J_2(s)p_2(s)(\varepsilon_{10}u(s)+C_{12})\,ds\\
&\le (\varepsilon_9\|v\| + C_{11})\tilde{\alpha} + (\varepsilon_{10}\|u\| + C_{12})\tilde{\beta}, \quad \forall t \in [0, 1],\\
\tilde{Q}_2(u, v)(t) &\le \int_0^1 J_3(s)p_2(s)q_2(u(s))\,ds + \int_0^1 J_4(s)p_1(s)q_1(v(s))\,ds\\
&\le \int_0^1 J_3(s)p_2(s)(\varepsilon_{10}u(s)+C_{12})\,ds+\int_0^1 J_4(s)p_1(s)(\varepsilon_9 v(s)+C_{11})\,ds\\
&\le (\varepsilon_{10}\|u\| + C_{12})\tilde{\gamma} + (\varepsilon_9\|v\| + C_{11})\tilde{\delta}, \quad \forall t \in [0, 1].
\end{aligned}$$

Therefore,

$$\begin{aligned}
\|\tilde{Q}_1(u, v)\| &\le \varepsilon_{10}\tilde{\beta}\|u\| + \varepsilon_9\tilde{\alpha}\|v\| + C_{11}\tilde{\alpha} + C_{12}\tilde{\beta},\\
\|\tilde{Q}_2(u, v)\| &\le \varepsilon_{10}\tilde{\gamma}\|u\| + \varepsilon_9\tilde{\delta}\|v\|+C_{12}\tilde{\gamma}+C_{11}\tilde{\delta},
\end{aligned}$$

and so

$$\begin{aligned}\|\tilde{\mathcal{Q}}(u,v)\|_Y &\le \varepsilon_{10}(\tilde{\beta}+\tilde{\gamma})\|u\| + \varepsilon_9(\tilde{\alpha}+\tilde{\delta})\|v\| + C_{11}(\tilde{\alpha}+\tilde{\delta}) + C_{12}(\tilde{\beta}+\tilde{\gamma})\\ &\le \max\{\varepsilon_{10}(\tilde{\beta}+\tilde{\gamma}), \varepsilon_9(\tilde{\alpha}+\tilde{\delta})\}\|(u,v)\|_Y + C_{11}(\tilde{\alpha}+\tilde{\delta}) + C_{12}(\tilde{\beta}+\tilde{\gamma}).\end{aligned}$$

We can choose large $R_3 > 1$ such that

$$\|\tilde{\mathcal{Q}}(u,v)\|_Y \le \|(u,v)\|_Y, \quad \forall\,(u,v) \in \partial B_{R_3} \cap P_0. \tag{5.25}$$

From (L4), we deduce that for $\varepsilon_{11} = \frac{1}{\gamma_1\gamma_2 m_1} > 0$ and $\varepsilon_{12} = \frac{1}{\gamma_1^2 m_2} > 0$ there exists $r_3 \in (0,1]$ such that

$$f(t,x) \ge \varepsilon_{11}x^a, \quad g(t,x) \ge \varepsilon_{12}x^b, \quad \forall\,(t,x) \in [c,1-c]\times[0,r_3]. \tag{5.26}$$

Then by using (5.26), for any $(u,v) \in \partial B_{r_3} \cap P_0$ and $t \in [c,1-c]$, we have

$$\begin{aligned}\tilde{Q}_1(u,v)(t) &\ge \int_c^{1-c} G_1(t,s)f(s,v(s))\,ds + \int_c^{1-c} G_2(t,s)g(s,u(s))\,ds\\ &\ge \gamma_1\int_c^{1-c} J_1(s)\varepsilon_{11}(v(s))^a\,ds + \gamma_1\int_c^{1-c} J_2(s)\varepsilon_{12}(u(s))^b\,ds\\ &\ge \gamma_1\left(\min_{s\in[c,1-c]}(v(s))^a\right)\varepsilon_{11}m_1 + \gamma_1\left(\min_{s\in[c,1-c]}(u(s))^b\right)\varepsilon_{12}m_2\\ &\ge \gamma_1\varepsilon_{11}m_1\min_{s\in[c,1-c]}v(s) + \gamma_1\varepsilon_{12}m_2\min_{s\in[c,1-c]}u(s)\\ &\ge \gamma_1\gamma_2\varepsilon_{11}m_1\|v\| + \gamma_1^2\varepsilon_{12}m_2\|u\|\\ &= \|v\| + \|u\| = \|(u,v)\|_Y.\end{aligned}$$

Therefore, $\|\tilde{Q}_1(u,v)\| \ge \|(u,v)\|_Y$ for all $(u,v) \in \partial B_{r_3} \cap P_0$, and so

$$\|\tilde{\mathcal{Q}}(u,v)\|_Y \ge \|\tilde{Q}_1(u,v)\| \ge \|(u,v)\|_Y, \quad \forall\,(u,v) \in \partial B_{r_3} \cap P_0. \tag{5.27}$$

By (5.25), (5.27), Lemma 5.2.1, and Theorem 1.1.1, we deduce that $\tilde{\mathcal{Q}}$ has at least one fixed point $(u,v) \in (\bar{B}_{R_3}\setminus B_{r_3})\cap P_0$—that is, $r_3 \le \|(u,v)\|_Y \le R_3$. This completes the proof of Theorem 5.2.4. □

Theorem 5.2.5. *Assume that (L1) and (L2) hold. If the functions f and g also satisfy the following conditions (L5) and (L6), then problem* (S′)–(BC) *has at least one positive solution $(u(t),v(t)), t \in [0,1]$:*

(L5) *There exist $\hat{a} \ge 1$ and $\hat{b} \ge 1$ such that*

$$(1)\quad q_{10} = \lim_{x\to 0^+}\frac{q_1(x)}{x^{\hat{a}}} = 0 \quad \textit{and} \quad (2)\quad q_{20} = \lim_{x\to 0^+}\frac{q_2(x)}{x^{\hat{b}}} = 0.$$

(L6) *There exists $c \in \left(0,\frac{1}{2}\right)$ such that*

$$(1)\quad \hat{f}^i_\infty = \lim_{x\to\infty}\inf_{t\in[c,1-c]}\frac{f(t,x)}{x} = \infty \quad \textit{and} \quad (2)\quad \hat{g}^i_\infty = \lim_{x\to\infty}\inf_{t\in[c,1-c]}\frac{g(t,x)}{x} = \infty.$$

Proof. We consider again the cone P_0 with c given in (L6). From (L5), we deduce that for $\varepsilon_{13} = \min\left\{\frac{1}{2\tilde{\alpha}}, \frac{1}{2\tilde{\delta}}\right\} > 0$ and $\varepsilon_{14} = \min\left\{\frac{1}{2\tilde{\beta}}, \frac{1}{2\tilde{\gamma}}\right\} > 0$ there exists $r_4 \in (0, 1)$ such that

$$q_1(x) \le \varepsilon_{13} x^{\hat{a}}, \quad q_2(x) \le \varepsilon_{14} x^{\hat{b}}, \quad \forall x \in [0, r_4]. \tag{5.28}$$

Then, by (5.28) and Lemma 5.1.5, for any $(u, v) \in \partial B_{r_4} \cap P_0$ and $t \in [0, 1]$ we obtain

$$\begin{aligned}
\tilde{Q}_1(u, v)(t) &\le \int_0^1 J_1(s)p_1(s)q_1(v(s))\,ds + \int_0^1 J_2(s)p_2(s)q_2(u(s))\,ds \\
&\le \varepsilon_{13} \int_0^1 J_1(s)p_1(s)(v(s))^{\hat{a}}\,ds + \varepsilon_{14} \int_0^1 J_2(s)p_2(s)(u(s))^{\hat{b}}\,ds \\
&\le \varepsilon_{13}\|v\|^{\hat{a}}\tilde{\alpha} + \varepsilon_{14}\|u\|^{\hat{b}}\tilde{\beta} \le \varepsilon_{13}\tilde{\alpha}\|v\| + \varepsilon_{14}\tilde{\beta}\|u\| \le \frac{1}{2}\|(u, v)\|_Y, \\
\tilde{Q}_2(u, v)(t) &\le \int_0^1 J_3(s)p_2(s)q_2(u(s))\,ds + \int_0^1 J_4(s)p_1(s)q_1(v(s))\,ds \\
&\le \varepsilon_{14} \int_0^1 J_3(s)p_2(s)(u(s))^{\hat{b}}\,ds + \varepsilon_{13} \int_0^1 J_4(s)p_1(s)(v(s))^{\hat{a}}\,ds \\
&\le \varepsilon_{14}\|u\|^{\hat{b}}\tilde{\gamma} + \varepsilon_{13}\|v\|^{\hat{a}}\tilde{\delta} \le \varepsilon_{14}\tilde{\gamma}\|u\| + \varepsilon_{13}\tilde{\delta}\|v\| \le \frac{1}{2}\|(u, v)\|_Y.
\end{aligned}$$

Therefore, we deduce $\|\tilde{Q}_1(u, v)\| \le \frac{1}{2}\|(u, v)\|_Y$, $\|\tilde{Q}_2(u, v)\| \le \frac{1}{2}\|(u, v)\|_Y$ for all $(u, v) \in \partial B_{r_4} \cap P_0$, and so

$$\|\tilde{Q}(u, v)\|_Y \le \|(u, v)\|_Y, \quad \forall\, (u, v) \in \partial B_{r_4} \cap P_0. \tag{5.29}$$

From (L6), for $\varepsilon_{15} = \frac{2}{\gamma_1\gamma_2 m_1}$ and $\varepsilon_{16} = \frac{2}{\gamma_1^2 m_2}$ there exist $C_{13}, C_{14} > 0$ such that

$$f(t, x) \ge \varepsilon_{15}x - C_{13}, \quad g(t, x) \ge \varepsilon_{16}x - C_{14}, \quad \forall\, (t, x) \in [c, 1-c] \times [0, \infty). \tag{5.30}$$

Then, by using (5.30), for any $(u, v) \in P_0$ and $t \in [c, 1 - c]$ we have

$$\begin{aligned}
\tilde{Q}_1(u, v)(t) &\ge \int_c^{1-c} G_1(t, s)f(s, v(s))\,ds + \int_c^{1-c} G_2(t, s)g(s, u(s))\,ds \\
&\ge \gamma_1 \int_c^{1-c} J_1(s)f(s, v(s))\,ds + \gamma_1 \int_c^{1-c} J_2(s)g(s, u(s))\,ds \\
&\ge \gamma_1 \int_c^{1-c} J_1(s)(\varepsilon_{15}v(s) - C_{13})\,ds + \gamma_1 \int_c^{1-c} J_2(s)(\varepsilon_{16}u(s) - C_{14})\,ds \\
&\ge \gamma_1 m_1 \varepsilon_{15} \min_{s\in[c,1-c]} v(s) - \gamma_1 m_1 C_{13} + \gamma_1 m_2 \varepsilon_{16} \min_{s\in[c,1-c]} u(s) - \gamma_1 m_2 C_{14} \\
&\ge \gamma_1 m_1 \varepsilon_{15}\gamma_2\|v\| + \gamma_1 m_2 \varepsilon_{16}\gamma_1\|u\| - \gamma_1 m_1 C_{13} - \gamma_1 m_2 C_{14} \\
&= 2\|(u, v)\|_Y - C_{15}, \quad C_{15} = \gamma_1(m_1 C_{13} + m_2 C_{14}).
\end{aligned}$$

Hence, we obtain $\|\tilde{Q}_1(u,v)\| \geq 2\|(u,v)\|_Y - C_{15}$ for all $(u,v) \in P_0$. We can choose $R_4 \geq \max\{C_{15}, 1\}$, and then we deduce

$$\|\tilde{Q}(u,v)\|_Y \geq \|\tilde{Q}_1(u,v)\| \geq \|(u,v)\|_Y, \quad \forall\, (u,v) \in \partial B_{R_4} \cap P_0. \tag{5.31}$$

By (5.29), (5.31), Lemma 5.2.1, and Theorem 1.1.1, we conclude that $\tilde{Q}$ has a fixed point $(u,v) \in (\bar{B}_{R_4} \setminus B_{r_4}) \cap P_0$—that is, $r_4 \leq \|(u,v)\|_Y \leq R_4$. The proof of Theorem 5.2.5 is completed. □

We present now an example for the above results.

Example 5.2.2. We consider problem (S$_2$)–(BC$_2$) from Example 5.2.1, where $f(t,x) = \frac{x^{a_0}}{t^{\zeta_1}(1-t)^{\rho_1}}$ and $g(t,x) = \frac{x^{b_0}}{t^{\zeta_2}(1-t)^{\rho_2}}$ for all $t \in (0,1)$, $x \in [0,\infty)$, with $a_0, b_0 > 1$ and $\zeta_1, \rho_1, \zeta_2, \rho_2 \in (0,1)$. Here $f(t,x) = p_1(t)q_1(x)$ and $g(t,x) = p_2(t)q_2(x)$, where $p_1(t) = \frac{1}{t^{\zeta_1}(1-t)^{\rho_1}}$ and $p_2(t) = \frac{1}{t^{\zeta_2}(1-t)^{\rho_2}}$ for all $t \in (0,1)$, and $q_1(x) = x^{a_0}$ and $q_2(x) = x^{b_0}$ for all $x \in [0,\infty)$. We have $0 < \int_0^1 (1-s)^{\alpha-1} p_1(s)\,ds < \infty$ and $0 < \int_0^1 (1-s)^{\beta-1} p_2(s)\,ds < \infty$.

In (L5), for $\hat{a} = \hat{b} = 1$ we obtain $q_{10} = 0$ and $q_{20} = 0$. In (L6), for $c \in \left(0, \frac{1}{2}\right)$ we have $\hat{f}^{\mathrm{i}}_{\infty} = \infty$ and $\hat{g}^{\mathrm{i}}_{\infty} = \infty$. Then, by Theorem 5.2.5, we deduce that problem (S$_2$)–(BC$_2$) has at least one positive solution.

Remark 5.2.1. The results presented in this section were published in Henderson et al. (2015a).

5.3 Coupled boundary conditions with additional positive constants

In this section, we shall investigate the existence and nonexistence of positive solutions for a system of nonlinear ordinary fractional differential equations with coupled integral boundary conditions in which some positive constants appear.

5.3.1 Presentation of the problem

We consider the system of nonlinear ordinary fractional differential equations

$$\begin{cases} D_{0+}^{\alpha} u(t) + a(t) f(v(t)) = 0, & t \in (0,1), \\ D_{0+}^{\beta} v(t) + b(t) g(u(t)) = 0, & t \in (0,1), \end{cases} \tag{S^0}$$

with the coupled integral boundary conditions

$$\begin{cases} u(0) = u'(0) = \cdots = u^{(n-2)}(0) = 0, & u(1) = \displaystyle\int_0^1 v(s)\,dH(s) + a_0, \\ v(0) = v'(0) = \cdots = v^{(m-2)}(0) = 0, & v(1) = \displaystyle\int_0^1 u(s)\,dK(s) + b_0, \end{cases} \tag{BC^0}$$

where $n-1<\alpha\le n$, $m-1<\beta\le m$, $n,m\in\mathbb{N}$, $n,m\ge 3$, D_{0+}^{α} and D_{0+}^{β} denote the Riemann–Liouville derivatives of orders α and β, respectively, the integrals from (BC^0) are Riemann–Stieltjes integrals, and a_0 and b_0 are positive constants.

By using the Schauder fixed point theorem (Theorem 1.6.1), we shall prove the existence of positive solutions of problem (S^0)–(BC^0). By a positive solution of (S^0)–(BC^0) we mean a pair of functions $(u,v)\in C([0,1];\mathbb{R}_+)\times C([0,1];\mathbb{R}_+)$ satisfying (S^0) and (BC^0) with $u(t)>0$, $v(t)>0$ for all $t\in(0,1]$. We shall also give sufficient conditions for the nonexistence of positive solutions for this problem.

We present the assumptions that we shall use in the sequel:

(J1) $H,K:[0,1]\to\mathbb{R}$ are nondecreasing functions, and $\Delta=1-\left(\int_0^1\tau^{\alpha-1}\,dK(\tau)\right)\times\left(\int_0^1\tau^{\beta-1}\,dH(\tau)\right)>0$.

(J2) The functions $a,b:[0,1]\to[0,\infty)$ are continuous, and there exist $t_1,t_2\in(0,1)$ such that $a(t_1)>0$, $b(t_2)>0$.

(J3) $f,g:[0,\infty)\to[0,\infty)$ are continuous functions, and there exists $c_0>0$ such that $f(u)<\frac{c_0}{L}$, $g(u)<\frac{c_0}{L}$ for all $u\in[0,c_0]$, where $L=\max\{\int_0^1 a(s)J_1(s)\,ds+\int_0^1 b(s)J_2(s)\,ds,\ \int_0^1 b(s)J_3(s)\,ds+\int_0^1 a(s)J_4(s)\,ds\}$, and $J_i,i=1,\dots,4$ are defined in Section 5.1.2.

(J4) $f,g:[0,\infty)\to[0,\infty)$ are continuous functions and satisfy the conditions $\lim_{u\to\infty}\frac{f(u)}{u}=\infty$, $\lim_{u\to\infty}\frac{g(u)}{u}=\infty$.

Under assumption (J1), we have all auxiliary results in Lemmas 5.1.1–5.1.6 from Section 5.1.2. Besides, by (J2) we deduce that $\int_0^1 a(s)J_1(s)\,ds>0$, $\int_0^1 b(s)J_2(s)\,ds>0$, $\int_0^1 b(s)J_3(s)\,ds>0$, and $\int_0^1 a(s)J_4(s)\,ds>0$—that is, the constant L from (J3) is positive.

5.3.2 Main results

Our first theorem is the following existence result for problem (S^0)–(BC^0):

Theorem 5.3.1. *Assume that assumptions (J1)–(J3) hold. Then problem* (S^0)–(BC^0) *has at least one positive solution for* $a_0>0$ *and* $b_0>0$ *sufficiently small.*

Proof. We consider the system of ordinary fractional differential equations

$$\begin{cases} D_{0+}^{\alpha}h(t)=0, & t\in(0,1),\\ D_{0+}^{\beta}k(t)=0, & t\in(0,1), \end{cases}\tag{5.32}$$

with the coupled integral boundary conditions

$$\begin{cases} h(0)=h'(0)=\cdots=h^{(n-2)}(0)=0, & h(1)=\displaystyle\int_0^1 k(s)\,dH(s)+a_0,\\ k(0)=k'(0)=\cdots=k^{(m-2)}(0)=0, & k(1)=\displaystyle\int_0^1 h(s)\,dK(s)+b_0, \end{cases}\tag{5.33}$$

with $a_0>0$ and $b_0>0$.

Problem (5.32)–(5.33) has the solution

$$h(t) = \frac{t^{\alpha-1}}{\Delta}\left(a_0 + b_0 \int_0^1 s^{\beta-1} \mathrm{d}H(s)\right), \quad t \in [0,1],$$

$$k(t) = \frac{t^{\beta-1}}{\Delta}\left(b_0 + a_0 \int_0^1 s^{\alpha-1} \mathrm{d}K(s)\right), \quad t \in [0,1], \tag{5.34}$$

where Δ is defined in (J1). By assumption (J1), we obtain $h(t) > 0$ and $k(t) > 0$ for all $t \in (0,1]$.

We define the functions $x(t)$ and $y(t)$, $t \in [0,1]$ by

$$x(t) = u(t) - h(t), \quad y(t) = v(t) - k(t), \quad \forall t \in [0,1],$$

where (u, v) is a solution of (S^0)–(BC^0). Then (S^0)–(BC^0) can be equivalently written as

$$\begin{cases} D_{0+}^{\alpha} x(t) + a(t) f(y(t) + k(t)) = 0, & t \in (0,1), \\ D_{0+}^{\beta} y(t) + b(t) g(x(t) + h(t)) = 0, & t \in (0,1), \end{cases} \tag{5.35}$$

with the boundary conditions

$$\begin{cases} x(0) = x'(0) = \cdots = x^{(n-2)}(0) = 0, \quad x(1) = \displaystyle\int_0^1 y(s)\, \mathrm{d}H(s), \\ y(0) = y'(0) = \cdots = y^{(m-2)}(0) = 0, \quad y(1) = \displaystyle\int_0^1 x(s)\, \mathrm{d}K(s). \end{cases} \tag{5.36}$$

Using the Green's functions G_i, $i = 1, \dots, 4$ from Section 5.1.2, we find a pair (x, y) is a solution of problem (5.35)–(5.36) if and only if (x, y) is a solution for the nonlinear integral equations

$$\begin{cases} x(t) = \displaystyle\int_0^1 G_1(t,s) a(s) f(y(s) + k(s))\, \mathrm{d}s + \int_0^1 G_2(t,s) b(s) g(x(s) + h(s))\, \mathrm{d}s, & t \in [0,1], \\ y(t) = \displaystyle\int_0^1 G_3(t,s) b(s) g(x(s) + h(s))\, \mathrm{d}s + \int_0^1 G_4(t,s) a(s) f(y(s) + k(s))\, \mathrm{d}s, & t \in [0,1], \end{cases} \tag{5.37}$$

where $h(t)$ and $k(t)$, $t \in [0,1]$ are given in (5.34).

We consider the Banach space $X = C([0,1])$ with the supremum norm $\|\cdot\|$ and the space $Y = X \times X$ with the norm $\|(x,y)\|_Y = \|x\| + \|y\|$, and we define the set

$$E = \{x \in C([0,1]), \quad 0 \le x(t) \le c_0, \quad \forall t \in [0,1]\} \subset X.$$

We also define the operators $S_1, S_2 : E \times E \to X$ and $\mathcal{S} : E \times E \to Y$ by

$$S_1(x,y)(t) = \int_0^1 G_1(t,s) a(s) f(y(s) + k(s))\, \mathrm{d}s + \int_0^1 G_2(t,s) b(s) g(x(s) + h(s))\, \mathrm{d}s, \quad t \in [0,1],$$

$$S_2(x,y)(t) = \int_0^1 G_3(t,s) b(s) g(x(s) + h(s))\, \mathrm{d}s + \int_0^1 G_4(t,s) a(s) f(y(s) + k(s))\, \mathrm{d}s, \quad t \in [0,1],$$

and $\mathcal{S}(x,y) = (S_1(x,y), S_2(x,y)), (x,y) \in E \times E$.

For sufficiently small $a_0 > 0$ and $b_0 > 0$, by (J3), we deduce

$$f(y(t)+k(t)) \le \frac{c_0}{L}, \quad g(x(t)+h(t)) \le \frac{c_0}{L}, \quad \forall t \in [0,1], \quad \forall x,\, y \in E.$$

Then, by using Lemma 5.1.4, we obtain $S_1(x,y)(t) \ge 0$, $S_2(x,y)(t) \ge 0$ for all $t \in [0,1]$ and $(x,y) \in E \times E$. By Lemma 5.1.5, for all $(x,y) \in E \times E$, we have

$$\begin{aligned} S_1(x,y)(t) &\le \int_0^1 J_1(s)a(s)f(y(s)+k(s))\,ds + \int_0^1 J_2(s)b(s)g(x(s)+h(s))\,ds \\ &\le \frac{c_0}{L}\left(\int_0^1 a(s)J_1(s)\,ds + \int_0^1 b(s)J_2(s)\,ds\right) \le c_0, \quad \forall t \in [0,1], \end{aligned}$$

and

$$\begin{aligned} S_2(x,y)(t) &\le \int_0^1 J_3(s)b(s)g(x(s)+h(s))\,ds + \int_0^1 J_4(s)a(s)f(y(s)+k(s))\,ds \\ &\le \frac{c_0}{L}\left(\int_0^1 b(s)J_3(s)\,ds + \int_0^1 a(s)J_4(s)\,ds\right) \le c_0, \quad \forall t \in [0,1]. \end{aligned}$$

Therefore, $\mathcal{S}(E \times E) \subset E \times E$.

Using standard arguments, we deduce that $\mathcal{S}$ is completely continuous. By Theorem 1.6.1, we conclude that $\mathcal{S}$ has a fixed point $(x,y) \in E \times E$, which represents a solution for problem (5.35)–(5.36). This shows that our problem (S^0)–(BC^0) has a positive solution (u,v) with $u = x+h$, $v = y+k$ for sufficiently small $a_0 > 0$ and $b_0 > 0$. □

In what follows, we present sufficient conditions for the nonexistence of positive solutions of (S^0)–(BC^0).

Theorem 5.3.2. *Assume that assumptions (J1), (J2), and (J4) hold. Then problem* (S^0)–(BC^0) *has no positive solution for a_0 and b_0 sufficiently large.*

Proof. We suppose that (u,v) is a positive solution of (S^0)–(BC^0). Then (x,y) with $x = u-h$, $y = v-k$ is a solution for problem (5.35)–(5.36), where (h,k) is the solution of problem (5.32)–(5.33) (given by (5.34)). By (J2), there exists $c \in (0,1/2)$ such that $t_1, t_2 \in (c, 1-c)$, and then $\int_c^{1-c} a(s)J_1(s)\,ds > 0$, $\int_c^{1-c} b(s)J_2(s)\,ds > 0$, $\int_c^{1-c} b(s)J_3(s)\,ds > 0$, and $\int_c^{1-c} a(s)J_4(s)\,ds > 0$. Now by using Lemma 5.1.4, we have $x(t) \ge 0$, $y(t) \ge 0$ for all $t \in [0,1]$, and by Lemma 5.1.6 we obtain $\inf_{t\in[c,1-c]} x(t) \ge \gamma_1\|x\|$ and $\inf_{t\in[c,1-c]} y(t) \ge \gamma_2\|y\|$.

Using now (5.34), we deduce that $\inf_{t\in[c,1-c]} h(t) = \gamma_1\|h\|$ and $\inf_{t\in[c,1-c]} k(t) = \gamma_2\|k\|$. Therefore, we obtain $\inf_{t\in[c,1-c]}(x(t)+h(t)) \ge \gamma_1\|x+h\|$ and $\inf_{t\in[c,1-c]}(y(t)+k(t)) \ge \gamma_2\|y+k\|$.

We now consider $R = \left(\gamma_1^2 \int_c^{1-c} b(s)J_2(s)\,ds\right)^{-1} > 0$. By using (J4), for R defined above, we conclude that there exists $M > 0$ such that $f(u) > 2Ru$, $g(u) > 2Ru$ for all $u \ge M$. We consider $a_0 > 0$ and $b_0 > 0$ sufficiently large such that $\inf_{t\in[c,1-c]}(x(t)+h(t)) \ge M$ and $\inf_{t\in[c,1-c]}(y(t)+k(t)) \ge M$. By (J2), (5.35), (5.36), and the above inequalities, we deduce that $\|x\| > 0$ and $\|y\| > 0$.

Now by using Lemma 5.1.5 and the above considerations, we have

$$\begin{aligned}
x(c) &= \int_0^1 G_1(c,s)a(s)f(y(s)+k(s))\,\mathrm{d}s + \int_0^1 G_2(c,s)b(s)g(x(s)+h(s))\,\mathrm{d}s \\
&\geq \gamma_1 \int_0^1 J_2(s)b(s)g(x(s)+h(s))\,\mathrm{d}s \geq \gamma_1 \int_c^{1-c} J_2(s)b(s)g(x(s)+h(s))\,\mathrm{d}s \\
&\geq 2R\gamma_1 \int_c^{1-c} J_2(s)b(s)(x(s)+h(s))\,\mathrm{d}s \\
&\geq 2R\gamma_1 \int_c^{1-c} J_2(s)b(s) \inf_{\tau\in[c,1-c]} (x(\tau)+h(\tau))\,\mathrm{d}s \\
&\geq 2R\gamma_1^2 \int_c^{1-c} J_2(s)b(s)\|x+h\|\,\mathrm{d}s = 2\|x+h\| \geq 2\|x\|.
\end{aligned}$$

Therefore, we obtain $\|x\| \leq \frac{1}{2}x(c) \leq \frac{1}{2}\|x\|$, which is a contradiction, because $\|x\| > 0$. Then, for a_0 and b_0 sufficiently large, our problem (S^0)–(BC^0) has no positive solution. □

5.3.3 An example

Example 5.3.1. We consider $a(t) = 1$, $b(t) = 1$ for all $t \in [0,1]$, $\alpha = 7/3$ $(n = 3)$, $\beta = 5/2$ $(m = 3)$, $H(t) = t^2$, for all $t \in [0,1]$, and $K(t) = \begin{cases} 0, & t \in [0,1/3), \\ 1, & t \in [1/3,2/3), \\ 3/2, & t \in [2/3,1]. \end{cases}$

Then, $\int_0^1 v(s)\,\mathrm{d}H(s) = 2\int_0^1 sv(s)\,\mathrm{d}s$ and $\int_0^1 u(s)\,\mathrm{d}K(s) = u\left(\frac{1}{3}\right) + \frac{1}{2}u\left(\frac{2}{3}\right)$. We also consider the functions $f, g : [0,\infty) \to [0,\infty)$, $f(x) = \tilde{a}x^2$, $g(x) = \tilde{b}x^3$ for all $x \in [0,\infty)$, with $\tilde{a}, \tilde{b} > 0$. We have $\lim_{x\to\infty} f(x)/x = \lim_{x\to\infty} g(x)/x = \infty$.

Therefore, we consider the system of fractional differential equations

$$\begin{cases} D_{0+}^{7/3}u(t) + \tilde{a}v^2(t) = 0, & t \in (0,1), \\ D_{0+}^{5/2}v(t) + \tilde{b}u^3(t) = 0, & t \in (0,1), \end{cases} \qquad (\mathrm{S}_1^0)$$

with the boundary conditions

$$\begin{cases} u(0) = u'(0) = 0, & u(1) = 2\displaystyle\int_0^1 sv(s)\,\mathrm{d}s + a_0, \\ v(0) = v'(0) = 0, & v(1) = u\left(\dfrac{1}{3}\right) + \dfrac{1}{2}u\left(\dfrac{2}{3}\right) + b_0. \end{cases} \qquad (\mathrm{BC}_1^0)$$

By using Δ from Section 5.1.4, we deduce that assumptions (J1), (J2), and (J4) are satisfied. In addition, by using the functions J_i, $i = 1,\ldots,4$ from Section 5.1.4, we obtain $A = \int_0^1 J_1(s)\,\mathrm{d}s \approx 0.15972386$, $B = \int_0^1 J_2(s)\,\mathrm{d}s \approx 0.05446581$, $C = \int_0^1 J_3(s)\,\mathrm{d}s \approx 0.09198682$, $D = \int_0^1 J_4(s)\,\mathrm{d}s \approx 0.12885992$, and then

$L = \max\{A+B, C+D\} \approx 0.22084674$. We choose $c_0 = 1$, and if we select $\tilde{a} < \frac{1}{L}$, $\tilde{b} < \frac{1}{L}$, then we conclude that $f(x) < \frac{1}{L}$ and $g(x) < \frac{1}{L}$ for all $x \in [0,1]$. For example, if $\tilde{a} \le 4.52$ and $\tilde{b} \le 4.52$, then the above conditions for f and g are satisfied. So, assumption (J3) is also satisfied. By Theorems 5.3.1 and 5.3.2, we deduce that problem (S_1^0)–(BC_1^0) has at least one positive solution for sufficiently small $a_0 > 0$ and $b_0 > 0$, and no positive solution for sufficiently large a_0 and b_0.

5.4 A system of semipositone coupled fractional boundary value problems

In this section, we investigate the existence of positive solutions for a system of nonlinear ordinary fractional differential equations with sign-changing nonlinearities subject to coupled integral boundary conditions.

5.4.1 Presentation of the problem

We consider the system of nonlinear ordinary fractional differential equations

$$\begin{cases} D_{0+}^{\alpha} u(t) + \lambda f(t, u(t), v(t)) = 0, & t \in (0,1), \\ D_{0+}^{\beta} v(t) + \mu g(t, u(t), v(t)) = 0, & t \in (0,1), \end{cases} \tag{$\tilde{S}$}$$

with the coupled integral boundary conditions

$$\begin{cases} u(0) = u'(0) = \cdots = u^{(n-2)}(0) = 0, & u(1) = \displaystyle\int_0^1 v(s)\,\mathrm{d}H(s), \\ v(0) = v'(0) = \cdots = v^{(m-2)}(0) = 0, & v(1) = \displaystyle\int_0^1 u(s)\,\mathrm{d}K(s), \end{cases} \tag{$\widetilde{BC}$}$$

where $n-1 < \alpha \le n$, $m-1 < \beta \le m$, $n, m \in \mathbb{N}$, $n, m \ge 3$, D_{0+}^{α} and D_{0+}^{β} denote the Riemann–Liouville derivatives of orders α and β, respectively, the integrals from (BC) are Riemann–Stieltjes integrals, and f and g are sign-changing continuous functions (i.e., we have a so-called system of semipositone boundary value problems). These functions may be nonsingular or singular at $t = 0$ and/or $t = 1$.

We present intervals for parameters λ and μ such that problem $(\tilde{S})$–$(\widetilde{BC})$ has at least one positive solution. By a positive solution of problem $(\tilde{S})$–$(\widetilde{BC})$, we mean a pair of functions $(u, v) \in C([0,1]; \mathbb{R}_+) \times C([0,1]; \mathbb{R}_+)$ satisfying $(\tilde{S})$ and $(\widetilde{BC})$ with $u(t) > 0$, $v(t) > 0$ for all $t \in (0,1)$. For the case when f and g are nonnegative, the existence of positive solutions for the above problem ($u(t) \ge 0$, $v(t) \ge 0$ for all $t \in [0,1]$, $(u,v) \ne (0,0)$) was studied in Section 5.1 by using the Guo–Krasnosel'skii fixed point theorem. The positive solutions of system $(\widetilde{S})$ with $\lambda = \mu = 1$ and with $f(t,u,v)$ and $g(t,u,v)$ replaced by $\tilde{f}(t,v)$ and $\tilde{g}(t,u)$, respectively, ($\tilde{f}$ and $\tilde{g}$ are nonnegative functions) with the boundary conditions $(\widetilde{BC})$ were investigated in Section 5.2 (the nonsingular and singular cases) by applying some theorems from the fixed point index theory and the Guo–Krasnosel'skii fixed point theorem.

5.4.2 Auxiliary results

In this section, we present some auxiliary results related to the following system of fractional differential equations

$$\begin{cases} D_{0+}^{\alpha}u(t) + \tilde{x}(t) = 0, \quad t \in (0,1), \\ D_{0+}^{\beta}v(t) + \tilde{y}(t) = 0, \quad t \in (0,1), \end{cases} \tag{5.38}$$

with the coupled integral boundary conditions

$$\begin{cases} u(0) = u'(0) = \cdots = u^{(n-2)}(0) = 0, \quad u(1) = \int_0^1 v(s)\,dH(s), \\ v(0) = v'(0) = \cdots = v^{(m-2)}(0) = 0, \quad v(1) = \int_0^1 u(s)\,dK(s), \end{cases} \tag{5.39}$$

where $n-1 < \alpha \le n$, $m-1 < \beta \le m$, $n, m \in \mathbb{N}$, $n, m \ge 3$, and $H, K : [0,1] \to \mathbb{R}$ are functions of bounded variation.

Lemma 5.4.1. *If $H, K : [0,1] \to \mathbb{R}$ are functions of bounded variation, $\Delta = 1 - \left(\int_0^1 \tau^{\alpha-1}\,dK(\tau)\right)\left(\int_0^1 \tau^{\beta-1}\,dH(\tau)\right) \neq 0$, and $\tilde{x}, \tilde{y} \in C(0,1) \cap L^1(0,1)$, then the unique solution of problem* (5.38)–(5.39) *is given by*

$$\begin{cases} u(t) = \int_0^1 G_1(t,s)\tilde{x}(s)\,ds + \int_0^1 G_2(t,s)\tilde{y}(s)\,ds, \\ v(t) = \int_0^1 G_3(t,s)\tilde{y}(s)\,ds + \int_0^1 G_4(t,s)\tilde{x}(s)\,ds, \quad t \in [0,1], \end{cases} \tag{5.40}$$

where

$$\begin{cases} G_1(t,s) = g_1(t,s) + \dfrac{t^{\alpha-1}}{\Delta}\left(\int_0^1 \tau^{\beta-1}dH(\tau)\right)\left(\int_0^1 g_1(\tau,s)\,dK(\tau)\right), \\ G_2(t,s) = \dfrac{t^{\alpha-1}}{\Delta}\int_0^1 g_2(\tau,s)\,dH(\tau), \\ G_3(t,s) = g_2(t,s) + \dfrac{t^{\beta-1}}{\Delta}\left(\int_0^1 \tau^{\alpha-1}dK(\tau)\right)\left(\int_0^1 g_2(\tau,s)\,dH(\tau)\right), \\ G_4(t,s) = \dfrac{t^{\beta-1}}{\Delta}\int_0^1 g_1(\tau,s)\,dK(\tau), \quad \forall t, s \in [0,1], \end{cases} \tag{5.41}$$

and

$$\begin{cases} g_1(t,s) = \dfrac{1}{\Gamma(\alpha)}\begin{cases} t^{\alpha-1}(1-s)^{\alpha-1} - (t-s)^{\alpha-1}, & 0 \le s \le t \le 1, \\ t^{\alpha-1}(1-s)^{\alpha-1}, & 0 \le t \le s \le 1, \end{cases} \\ g_2(t,s) = \dfrac{1}{\Gamma(\beta)}\begin{cases} t^{\beta-1}(1-s)^{\beta-1} - (t-s)^{\beta-1}, & 0 \le s \le t \le 1, \\ t^{\beta-1}(1-s)^{\beta-1}, & 0 \le t \le s \le 1. \end{cases} \end{cases} \tag{5.42}$$

For the proof of the above lemma, see Lemmas 5.1.1 and 5.1.2.

Lemma 5.4.2. *The functions g_1 and g_2 given by* (5.42) *have the following properties:*

(a) $g_1, g_2 : [0,1] \times [0,1] \to \mathbb{R}_+$ *are continuous functions, and* $g_1(t,s) > 0$, $g_2(t,s) > 0$ *for all* $(t,s) \in (0,1) \times (0,1)$.

(b) $g_1(t,s) \le h_1(s)$ *and* $g_2(t,s) \le h_2(s)$ *for all* $(t,s) \in [0,1] \times [0,1]$, *where* $h_1(s) = \frac{s(1-s)^{\alpha-1}}{\Gamma(\alpha-1)}$ *and* $h_2(s) = \frac{s(1-s)^{\beta-1}}{\Gamma(\beta-1)}$ *for all* $s \in [0,1]$.

(c) $g_1(t,s) \ge k_1(t)h_1(s)$ *and* $g_2(t,s) \ge k_2(t)h_2(s)$ *for all* $(t,s) \in [0,1] \times [0,1]$, *where*

$$k_1(t) = \min\left\{\frac{(1-t)t^{\alpha-2}}{\alpha-1}, \frac{t^{\alpha-1}}{\alpha-1}\right\} = \begin{cases} \dfrac{t^{\alpha-1}}{\alpha-1}, & 0 \le t \le \frac{1}{2}, \\ \dfrac{(1-t)t^{\alpha-2}}{\alpha-1}, & \frac{1}{2} \le t \le 1, \end{cases}$$

$$k_2(t) = \min\left\{\frac{(1-t)t^{\beta-2}}{\beta-1}, \frac{t^{\beta-1}}{\beta-1}\right\} = \begin{cases} \dfrac{t^{\beta-1}}{\beta-1}, & 0 \le t \le \frac{1}{2}, \\ \dfrac{(1-t)t^{\beta-2}}{\beta-1}, & \frac{1}{2} \le t \le 1. \end{cases}$$

(d) *For any* $(t,s) \in [0,1] \times [0,1]$, *we have*

$$g_1(t,s) \le \frac{(1-t)t^{\alpha-1}}{\Gamma(\alpha-1)} \le \frac{t^{\alpha-1}}{\Gamma(\alpha-1)}, \quad g_2(t,s) \le \frac{(1-t)t^{\beta-1}}{\Gamma(\beta-1)} \le \frac{t^{\beta-1}}{\Gamma(\beta-1)}.$$

For the proof of Lemma 5.4.2 (a) and (b), see Yuan (2010), for the proof of Lemma 5.4.2 (c), see Lemma 4.4.2, and the proof of Lemma 5.4.2 (d) is based on the relations $g_1(t,s) = g_1(1-s, 1-t)$ and $g_2(t,s) = g_2(1-s, 1-t)$ and relations (b) above.

Lemma 5.4.3. *If H, K : $[0,1] \to \mathbb{R}$ are nondecreasing functions, and $\Delta > 0$, then G_i, $i = 1, \ldots, 4$ given by* (5.41) *are continuous functions on* $[0,1] \times [0,1]$ *and satisfy* $G_i(t,s) \ge 0$ *for all* $(t,s) \in [0,1] \times [0,1]$, $i = 1, \ldots, 4$. *Moreover, if* $\tilde{x}, \tilde{y} \in C(0,1) \cap L^1(0,1)$ *satisfy* $\tilde{x}(t) \ge 0$, $\tilde{y}(t) \ge 0$ *for all* $t \in (0,1)$, *then the solution* (u,v) *of problem* (5.38)–(5.39) *(given by* (5.40)*) satisfies* $u(t) \ge 0$, $v(t) \ge 0$ *for all* $t \in [0,1]$.

For the proof of the above lemma, see Lemma 5.1.4.

Lemma 5.4.4. *Assume that $H, K : [0,1] \to \mathbb{R}$ are nondecreasing functions, $\Delta > 0$, $\int_0^1 \tau^{\alpha-1}(1-\tau)\,\mathrm{d}K(\tau) > 0$, and $\int_0^1 \tau^{\beta-1}(1-\tau)\,\mathrm{d}H(\tau) > 0$. Then the functions $G_i, i = 1, \ldots, 4$ satisfy the following inequalities:*

(a_1) $G_1(t,s) \le \sigma_1 h_1(s)$, $\forall (t,s) \in [0,1] \times [0,1]$, *where*

$$\sigma_1 = 1 + \frac{1}{\Delta}(K(1) - K(0)) \int_0^1 \tau^{\beta-1}\,\mathrm{d}H(\tau) > 0.$$

(a_2) $G_1(t,s) \le \delta_1 t^{\alpha-1}$, $\forall (t,s) \in [0,1] \times [0,1]$, *where*

$$\delta_1 = \frac{1}{\Gamma(\alpha-1)}\left[1 + \frac{1}{\Delta}\left(\int_0^1 \tau^{\beta-1}\,\mathrm{d}H(\tau)\right)\left(\int_0^1 (1-\tau)\tau^{\alpha-1}\,\mathrm{d}K(\tau)\right)\right] > 0.$$

(a_3) $G_1(t,s) \ge \varrho_1 t^{\alpha-1} h_1(s)$, $(t,s) \in [0,1] \times [0,1]$, *where*

$$\varrho_1 = \frac{1}{\Delta}\left(\int_0^1 \tau^{\beta-1}\,\mathrm{d}H(\tau)\right)\left(\int_0^1 k_1(\tau)\,\mathrm{d}K(\tau)\right) > 0.$$

(b_1) $G_2(t,s) \le \sigma_2 h_2(s)$, $\forall\,(t,s) \in [0,1] \times [0,1]$, *where* $\sigma_2 = \frac{1}{\Delta}(H(1) - H(0)) > 0$.

(b_2) $G_2(t,s) \le \delta_2 t^{\alpha-1}$, $\forall\,(t,s) \in [0,1]\times[0,1]$, *where* $\delta_2 = \frac{1}{\Delta\Gamma(\beta-1)} \int_0^1 (1-\tau)\tau^{\beta-1}\,\mathrm{d}H(\tau) > 0$.

(b_3) $G_2(t,s) \ge \varrho_2 t^{\alpha-1} h_2(s)$, $\forall\,(t,s) \in [0,1] \times [0,1]$, *where* $\varrho_2 = \frac{1}{\Delta} \int_0^1 k_2(\tau)\,\mathrm{d}H(\tau) > 0$.

(c_1) $G_3(t,s) \le \sigma_3 h_2(s)$, $\forall\,(t,s) \in [0,1] \times [0,1]$, *where*

$$\sigma_3 = 1 + \frac{1}{\Delta}(H(1) - H(0)) \int_0^1 \tau^{\alpha-1}\,\mathrm{d}K(\tau) > 0.$$

(c_2) $G_3(t,s) \le \delta_3 t^{\beta-1}$, $\forall\,(t,s) \in [0,1] \times [0,1]$, *where*

$$\delta_3 = \frac{1}{\Gamma(\beta-1)}\left[1 + \frac{1}{\Delta}\left(\int_0^1 \tau^{\alpha-1}\,\mathrm{d}K(\tau)\right)\left(\int_0^1 (1-\tau)\tau^{\beta-1}\,\mathrm{d}H(\tau)\right)\right] > 0.$$

(c_3) $G_3(t,s) \ge \varrho_3 t^{\beta-1} h_2(s)$, $\forall\,(t,s) \in [0,1] \times [0,1]$, *where*

$$\varrho_3 = \frac{1}{\Delta}\left(\int_0^1 \tau^{\alpha-1}\,\mathrm{d}K(\tau)\right)\left(\int_0^1 k_2(\tau)\,\mathrm{d}H(\tau)\right) > 0.$$

(d_1) $G_4(t,s) \le \sigma_4 h_1(s)$, $\forall\,(t,s) \in [0,1] \times [0,1]$, *where* $\sigma_4 = \frac{1}{\Delta}(K(1) - K(0)) > 0$.

(d_2) $G_4(t,s) \le \delta_4 t^{\beta-1}$, $\forall\,(t,s) \in [0,1] \times [0,1]$, *where* $\delta_4 = \frac{1}{\Delta\Gamma(\alpha-1)} \int_0^1 (1-\tau)\tau^{\alpha-1}\,\mathrm{d}K(\tau) > 0$.

(d_3) $G_4(t,s) \ge \varrho_4 t^{\beta-1} h_1(s)$, $\forall\,(t,s) \in [0,1] \times [0,1]$, *where* $\varrho_4 = \frac{1}{\Delta} \int_0^1 k_1(\tau)\,\mathrm{d}K(\tau) > 0$.

Proof. From the assumptions of this lemma, we obtain

$$\int_0^1 \tau^{\alpha-1}\,\mathrm{d}K(\tau) \ge \int_0^1 \tau^{\alpha-1}(1-\tau)\,\mathrm{d}K(\tau) > 0,$$

$$\int_0^1 (1-\tau)\tau^{\alpha-2}\,\mathrm{d}K(\tau) \ge \int_0^1 (1-\tau)\tau^{\alpha-1}\,\mathrm{d}K(\tau) > 0,$$

$$\int_0^1 k_1(\tau)\,\mathrm{d}K(\tau) \ge \frac{1}{\alpha-1}\int_0^1 \tau^{\alpha-1}(1-\tau)\,\mathrm{d}K(\tau) > 0,$$

$$\int_0^1 \tau^{\beta-1}\,\mathrm{d}H(\tau) \ge \int_0^1 \tau^{\beta-1}(1-\tau)\,\mathrm{d}H(\tau) > 0,$$

$$\int_0^1 (1-\tau)\tau^{\beta-2}\,\mathrm{d}H(\tau) \ge \int_0^1 (1-\tau)\tau^{\beta-1}\,\mathrm{d}H(\tau) > 0,$$

$$\int_0^1 k_2(\tau)\,\mathrm{d}H(\tau) \ge \frac{1}{\beta-1}\int_0^1 \tau^{\beta-1}(1-\tau)\,\mathrm{d}H(\tau) > 0,$$

$$K(1) - K(0) = \int_0^1 \mathrm{d}K(\tau) \ge \int_0^1 \tau^{\alpha-1}(1-\tau)\,\mathrm{d}K(\tau) > 0,$$

$$H(1) - H(0) = \int_0^1 \mathrm{d}H(\tau) \ge \int_0^1 \tau^{\beta-1}(1-\tau)\,\mathrm{d}H(\tau) > 0.$$

By using Lemma 5.4.2, we deduce the following relations for all $(t,s) \in [0,1] \times [0,1]$:

(a$_1$)

$$G_1(t,s) = g_1(t,s) + \frac{t^{\alpha-1}}{\Delta}\left(\int_0^1 \tau^{\beta-1}\,\mathrm{d}H(\tau)\right)\left(\int_0^1 g_1(\tau,s)\,\mathrm{d}K(\tau)\right)$$
$$\le h_1(s) + \frac{1}{\Delta}\left(\int_0^1 \tau^{\beta-1}\,\mathrm{d}H(\tau)\right)\left(\int_0^1 h_1(s)\,\mathrm{d}K(\tau)\right)$$
$$= h_1(s)\left[1 + \frac{1}{\Delta}(K(1)-K(0))\int_0^1 \tau^{\beta-1}\,\mathrm{d}H(\tau)\right] = \sigma_1 h_1(s).$$

(a$_2$)

$$G_1(t,s) \le \frac{t^{\alpha-1}}{\Gamma(\alpha-1)} + \frac{t^{\alpha-1}}{\Delta}\left(\int_0^1 \tau^{\beta-1}\,\mathrm{d}H(\tau)\right)\left(\int_0^1 \frac{(1-\tau)\tau^{\alpha-1}}{\Gamma(\alpha-1)}\mathrm{d}K(\tau)\right)$$
$$= t^{\alpha-1}\frac{1}{\Gamma(\alpha-1)}\left[1 + \frac{1}{\Delta}\left(\int_0^1 \tau^{\beta-1}\,\mathrm{d}H(\tau)\right)\left(\int_0^1 (1-\tau)\tau^{\alpha-1}\,\mathrm{d}K(\tau)\right)\right] = \delta_1 t^{\alpha-1}.$$

(a$_3$)

$$G_1(t,s) \ge \frac{t^{\alpha-1}}{\Delta}\left(\int_0^1 \tau^{\beta-1}\,\mathrm{d}H(\tau)\right)\left(\int_0^1 k_1(\tau)h_1(s)\,\mathrm{d}K(\tau)\right)$$
$$= t^{\alpha-1}h_1(s)\frac{1}{\Delta}\left(\int_0^1 \tau^{\beta-1}\,\mathrm{d}H(\tau)\right)\left(\int_0^1 k_1(\tau)\,\mathrm{d}K(\tau)\right) = \varrho_1 t^{\alpha-1}h_1(s).$$

(b$_1$)

$$G_2(t,s) = \frac{t^{\alpha-1}}{\Delta}\int_0^1 g_2(\tau,s)\,\mathrm{d}H(\tau) \le \frac{1}{\Delta}\int_0^1 h_2(s)\,\mathrm{d}H(\tau)$$
$$= \frac{1}{\Delta}(H(1)-H(0))h_2(s) = \sigma_2 h_2(s).$$

(b$_2$)

$$G_2(t,s) \le \frac{t^{\alpha-1}}{\Delta}\int_0^1 \frac{(1-\tau)\tau^{\beta-1}}{\Gamma(\beta-1)}\mathrm{d}H(\tau) = \delta_2 t^{\alpha-1}.$$

(b$_3$)

$$G_2(t,s) \ge \frac{t^{\alpha-1}}{\Delta}\int_0^1 k_2(\tau)h_2(s)\,\mathrm{d}H(\tau) = \frac{t^{\alpha-1}}{\Delta}h_2(s)\int_0^1 k_2(\tau)\,\mathrm{d}H(\tau) = \varrho_2 t^{\alpha-1}h_2(s).$$

(c$_1$)

$$G_3(t,s) = g_2(t,s) + \frac{t^{\beta-1}}{\Delta}\left(\int_0^1 \tau^{\alpha-1}\,\mathrm{d}K(\tau)\right)\left(\int_0^1 g_2(\tau,s)\,\mathrm{d}H(\tau)\right)$$
$$\le h_2(s) + \frac{1}{\Delta}\left(\int_0^1 \tau^{\alpha-1}\,\mathrm{d}K(\tau)\right)\left(\int_0^1 h_2(s)\,\mathrm{d}H(\tau)\right)$$
$$= h_2(s)\left[1 + \frac{1}{\Delta}(H(1)-H(0))\int_0^1 \tau^{\alpha-1}\,\mathrm{d}K(\tau)\right) = \sigma_3 h_2(s).$$

(c$_2$)

$$G_3(t,s) \le \frac{(1-t)t^{\beta-1}}{\Gamma(\beta-1)} + \frac{t^{\beta-1}}{\Delta}\left(\int_0^1 \tau^{\alpha-1}\,\mathrm{d}K(\tau)\right)\left(\int_0^1 \frac{(1-\tau)\tau^{\beta-1}}{\Gamma(\beta-1)}\mathrm{d}H(\tau)\right)$$
$$\le \frac{t^{\beta-1}}{\Gamma(\beta-1)}\left[1+\frac{1}{\Delta}\left(\int_0^1 \tau^{\alpha-1}\mathrm{d}K(\tau)\right)\left(\int_0^1 (1-\tau)\tau^{\beta-1}\mathrm{d}H(\tau)\right)\right] = \delta_3 t^{\beta-1}.$$

(c$_3$)

$$G_3(t,s) \ge \frac{t^{\beta-1}}{\Delta}\left(\int_0^1 \tau^{\alpha-1}\mathrm{d}K(\tau)\right)\left(\int_0^1 k_2(\tau)h_2(s)\,\mathrm{d}H(\tau)\right)$$
$$= t^{\beta-1}h_2(s)\frac{1}{\Delta}\left(\int_0^1 \tau^{\alpha-1}\mathrm{d}K(\tau)\right)\left(\int_0^1 k_2(\tau)\,\mathrm{d}H(\tau)\right) = \varrho_3 t^{\beta-1}h_2(s).$$

(d$_1$)

$$G_4(t,s) = \frac{t^{\beta-1}}{\Delta}\int_0^1 g_1(\tau,s)\,\mathrm{d}K(\tau) \le \frac{1}{\Delta}\int_0^1 h_1(s)\,\mathrm{d}K(\tau)$$
$$= h_1(s)\frac{1}{\Delta}(K(1)-K(0)) = \sigma_4 h_1(s).$$

(d$_2$)

$$G_4(t,s) \le \frac{t^{\beta-1}}{\Delta}\int_0^1 \frac{(1-\tau)\tau^{\alpha-1}}{\Gamma(\alpha-1)}\mathrm{d}K(\tau) = \delta_4 t^{\beta-1}.$$

(d$_3$)

$$G_4(t,s) \ge \frac{t^{\beta-1}}{\Delta}\int_0^1 k_1(\tau)h_1(s)\,\mathrm{d}K(\tau) = t^{\beta-1}h_1(s)\frac{1}{\Delta}\int_0^1 k_1(s)\,\mathrm{d}K(\tau) = \varrho_4 t^{\beta-1}h_1(s).$$

□

Lemma 5.4.5. *Assume that $H, K : [0,1] \to \mathbb{R}$ are nondecreasing functions, $\Delta > 0$, $\int_0^1 \tau^{\alpha-1}(1-\tau)\,\mathrm{d}K(\tau) > 0$, $\int_0^1 \tau^{\beta-1}(1-\tau)\,\mathrm{d}H(\tau) > 0$, and $\tilde{x}, \tilde{y} \in C(0,1) \cap L^1(0,1)$, $\tilde{x}(t) \ge 0$, $\tilde{y}(t) \ge 0$ for all $t \in (0,1)$. Then the solution $(u(t), v(t))$, $t \in [0,1]$ of problem (5.38)–(5.39) satisfies the inequalities $u(t) \ge \tilde{\gamma}_1 t^{\alpha-1}u(t')$ and $v(t) \ge \tilde{\gamma}_2 t^{\beta-1}v(t')$ for all $t, t' \in [0,1]$, where $\tilde{\gamma}_1 = \min\left\{\frac{\varrho_1}{\sigma_1}, \frac{\varrho_2}{\sigma_2}\right\} > 0$ and $\tilde{\gamma}_2 = \min\left\{\frac{\varrho_3}{\sigma_3}, \frac{\varrho_4}{\sigma_4}\right\} > 0$.*

Proof. By using Lemma 5.4.4, we obtain

$$u(t) = \int_0^1 G_1(t,s)\tilde{x}(s)\,\mathrm{d}s + \int_0^1 G_2(t,s)\tilde{y}(s)\,\mathrm{d}s$$
$$\ge \int_0^1 \varrho_1 t^{\alpha-1}h_1(s)\tilde{x}(s)\,\mathrm{d}s + \int_0^1 \varrho_2 t^{\alpha-1}h_2(s)\tilde{y}(s)\,\mathrm{d}s$$
$$= t^{\alpha-1}\left(\varrho_1\int_0^1 h_1(s)\tilde{x}(s)\,\mathrm{d}s + \varrho_2\int_0^1 h_2(s)\tilde{y}(s)\,\mathrm{d}s\right)$$
$$\ge t^{\alpha-1}\left(\frac{\varrho_1}{\sigma_1}\int_0^1 G_1(t',s)\tilde{x}(s)\,\mathrm{d}s + \frac{\varrho_2}{\sigma_2}\int_0^1 G_2(t',s)\tilde{y}(s)\,\mathrm{d}s\right)$$

$$\geq t^{\alpha-1} \min\left\{\frac{\varrho_1}{\sigma_1}, \frac{\varrho_2}{\sigma_2}\right\} \left(\int_0^1 G_1(t', s)\tilde{x}(s)\, ds + \int_0^1 G_2(t', s)\tilde{y}(s)\, ds\right)$$

$$= \tilde{\gamma}_1 t^{\alpha-1} u(t'), \quad \forall t, t' \in [0, 1], \quad \text{where} \quad \tilde{\gamma}_1 = \min\left\{\frac{\varrho_1}{\sigma_1}, \frac{\varrho_2}{\sigma_2}\right\} > 0,$$

and

$$v(t) = \int_0^1 G_3(t, s)\tilde{y}(s)\, ds + \int_0^1 G_4(t, s)\tilde{x}(s)\, ds$$

$$\geq \int_0^1 \varrho_3 t^{\beta-1} h_2(s)\tilde{y}(s)\, ds + \int_0^1 \varrho_4 t^{\beta-1} h_1(s)\tilde{x}(s)\, ds$$

$$= t^{\beta-1}\left(\varrho_3 \int_0^1 h_2(s)\tilde{y}(s)\, ds + \varrho_4 \int_0^1 h_1(s)\tilde{x}(s)\, ds\right)$$

$$\geq t^{\beta-1}\left(\frac{\varrho_3}{\sigma_3}\int_0^1 G_3(t', s)\tilde{y}(s)\, ds + \frac{\varrho_4}{\sigma_4}\int_0^1 G_4(t', s)\tilde{x}(s)\, ds\right)$$

$$\geq t^{\beta-1} \min\left\{\frac{\varrho_3}{\sigma_3}, \frac{\varrho_4}{\sigma_4}\right\} \left(\int_0^1 G_3(t', s)\tilde{y}(s)\, ds + \int_0^1 G_4(t', s)\tilde{x}(s)\, ds\right)$$

$$= \tilde{\gamma}_2 t^{\beta-1} v(t'), \quad \forall t, t' \in [0, 1], \quad \text{where} \quad \tilde{\gamma}_2 = \min\left\{\frac{\varrho_3}{\sigma_3}, \frac{\varrho_4}{\sigma_4}\right\} > 0.$$

□

5.4.3 Main results

In this section, we investigate the existence and multiplicity of positive solutions for our problem $(\widetilde{\mathrm{S}})$–$(\widetilde{\mathrm{BC}})$. We present now the assumptions that we shall use in the sequel:

$(\widetilde{\mathrm{H1}})$ $H, K : [0, 1] \to \mathbb{R}$ are nondecreasing functions, $\Delta = 1 - \left(\int_0^1 \tau^{\alpha-1}\, dK(\tau)\right) \times \left(\int_0^1 \tau^{\beta-1}\, dH(\tau)\right) > 0$, and $\int_0^1 \tau^{\alpha-1}(1-\tau)\, dK(\tau) > 0$, $\int_0^1 \tau^{\beta-1}(1-\tau)\, dH(\tau) > 0$.

$(\widetilde{\mathrm{H2}})$ The functions $f, g \in C([0, 1] \times [0, \infty) \times [0, \infty), (-\infty, +\infty))$, and there exist functions $p_1, p_2 \in C([0, 1], [0, \infty))$ such that $f(t, u, v) \geq -p_1(t)$ and $g(t, u, v) \geq -p_2(t)$ for any $t \in [0, 1]$ and $u, v \in [0, \infty)$.

$(\widetilde{\mathrm{H3}})$ $f(t, 0, 0) > 0$ and $g(t, 0, 0) > 0$ for all $t \in [0, 1]$.

$(\widetilde{\mathrm{H4}})$ The functions $f, g \in C((0, 1) \times [0, \infty) \times [0, \infty), (-\infty, +\infty))$, f, g may be singular at $t = 0$ and/or $t = 1$, and there exist functions $p_1, p_2 \in C((0, 1), [0, \infty))$, $\alpha_1, \alpha_2 \in C((0, 1), [0, \infty))$, and $\beta_1, \beta_2 \in C([0, 1] \times [0, \infty) \times [0, \infty), [0, \infty))$ such that

$$-p_1(t) \leq f(t, u, v) \leq \alpha_1(t)\beta_1(t, u, v), \quad -p_2(t) \leq g(t, u, v) \leq \alpha_2(t)\beta_2(t, u, v)$$

for all $t \in (0, 1)$ and $u, v \in [0, \infty)$, with $0 < \int_0^1 p_i(s)\, ds < \infty$ and $\int_0^1 \alpha_i(s)\, ds < \infty$, $i = 1, 2$.

$(\widetilde{\mathrm{H5}})$ There exists $c \in (0, 1/2)$ such that

$$f_\infty = \lim_{u+v\to\infty} \min_{t\in[c,1-c]} \frac{f(t, u, v)}{u+v} = \infty \quad \text{or} \quad g_\infty = \lim_{u+v\to\infty} \min_{t\in[c,1-c]} \frac{g(t, u, v)}{u+v} = \infty.$$

$(\widetilde{\mathrm{H6}})$ $\beta_{i\infty} = \lim_{u+v\to\infty} \max_{t\in[0,1]} \dfrac{\beta_i(t, u, v)}{u+v} = 0, \quad i = 1, 2.$

We consider the system of nonlinear fractional differential equations

$$\begin{cases} D_{0+}^{\alpha}x(t) + \lambda(f(t,[x(t)-q_1(t)]^*,[y(t)-q_2(t)]^*) + p_1(t)) = 0, & 0 < t < 1, \\ D_{0+}^{\beta}y(t) + \mu(g(t,[x(t)-q_1(t)]^*,[y(t)-q_2(t)]^*) + p_2(t)) = 0, & 0 < t < 1, \end{cases} \tag{5.43}$$

with the integral boundary conditions

$$\begin{cases} x(0) = x'(0) = \cdots = x^{(n-2)}(0) = 0, \quad x(1) = \int_0^1 y(s)\,\mathrm{d}H(s), \\ y(0) = y'(0) = \cdots = y^{(m-2)}(0) = 0, \quad y(1) = \int_0^1 x(s)\,\mathrm{d}K(s), \end{cases} \tag{5.44}$$

where $z(t)^* = z(t)$ if $z(t) \geq 0$, and $z(t)^* = 0$ if $z(t) < 0$. Here, (q_1, q_2) with

$$q_1(t) = \lambda \int_0^1 G_1(t,s)p_1(s)\,\mathrm{d}s + \mu \int_0^1 G_2(t,s)p_2(s)\,\mathrm{d}s, \quad t \in [0,1],$$

$$q_2(t) = \mu \int_0^1 G_3(t,s)p_2(s)\,\mathrm{d}s + \lambda \int_0^1 G_4(t,s)p_1(s)\,\mathrm{d}s, \quad t \in [0,1]$$

is the solution of the system of fractional differential equations

$$\begin{cases} D_{0+}^{\alpha}q_1(t) + \lambda p_1(t) = 0, & 0 < t < 1, \\ D_{0+}^{\beta}q_2(t) + \mu p_2(t) = 0, & 0 < t < 1 \end{cases} \tag{5.45}$$

with the integral boundary conditions

$$\begin{cases} q_1(0) = q_1'(0) = \cdots = q_1^{(n-2)}(0) = 0, \quad q_1(1) = \int_0^1 q_2(s)\,\mathrm{d}H(s), \\ q_2(0) = q_2'(0) = \cdots = q_2^{(m-2)}(0) = 0, \quad q_2(1) = \int_0^1 q_1(s)\,\mathrm{d}K(s). \end{cases} \tag{5.46}$$

Under assumptions $(\widetilde{\mathrm{H1}})$ and $(\widetilde{\mathrm{H2}})$, or $(\widetilde{\mathrm{H1}})$ and $(\widetilde{\mathrm{H4}})$, we have $q_1(t) \geq 0$, $q_2(t) \geq 0$ for all $t \in [0,1]$.

We shall prove that there exists a solution (x, y) for the boundary value problem (5.43)–(5.44) with $x(t) \geq q_1(t)$ and $y(t) \geq q_2(t)$ on $[0,1]$ and $x(t) > q_1(t)$ and $y(t) > q_2(t)$ on $(0,1)$. In this case, (u, v) with $u(t) = x(t) - q_1(t)$ and $v(t) = y(t) - q_2(t)$, $t \in [0,1]$ represents a positive solution of the boundary value problem $(\widetilde{\mathrm{S}})$–$(\widetilde{\mathrm{BC}})$. By (5.43)–(5.46), we have

$$\begin{aligned} D_{0+}^{\alpha}u(t) &= D_{0+}^{\alpha}x(t) - D_{0+}^{\alpha}q_1(t) = -\lambda f(t,[x(t)-q_1(t)]^*,[y(t)-q_2(t)]^*) \\ &\quad - \lambda p_1(t) + \lambda p_1(t) = -\lambda f(t,u(t),v(t)), \quad \forall t \in (0,1), \\ D_{0+}^{\beta}v(t) &= D_{0+}^{\beta}y(t) - D_{0+}^{\beta}q_2(t) = -\mu g(t,[x(t)-q_1(t)]^*,[y(t)-q_2(t)]^*) \\ &\quad - \mu p_2(t) + \mu p_2(t) = -\mu g(t,u(t),v(t)), \quad \forall t \in (0,1), \end{aligned}$$

and

$$u(0) = x(0) - q_1(0) = 0, \dots, u^{(n-2)}(0) = x^{(n-2)}(0) - q_1^{(n-2)}(0) = 0,$$
$$v(0) = y(0) - q_2(0) = 0, \dots, v^{(m-2)}(0) = y^{(m-2)}(0) - q_2^{(m-2)}(0) = 0,$$
$$u(1) = x(1) - q_1(1) = \int_0^1 y(s)\, dH(s) - \int_0^1 q_2(s)\, dH(s) = \int_0^1 v(s)\, dH(s),$$
$$v(1) = y(1) - q_2(1) = \int_0^1 x(s)\, dK(s) - \int_0^1 q_1(s)\, dK(s) = \int_0^1 u(s)\, dK(s).$$

Therefore, in what follows, we shall investigate the boundary value problem (5.43)–(5.44).

By using Lemma 5.4.1 (relations (5.40)), we find problem (5.43)–(5.44) is equivalent to the system

$$\begin{cases} x(t) = \lambda \int_0^1 G_1(t,s)(f(s,[x(s)-q_1(s)]^*,[y(s)-q_2(s)]^*) + p_1(s))\, ds \\ \qquad + \mu \int_0^1 G_2(t,s)(g(s,[x(s)-q_1(s)]^*,[y(s)-q_2(s)]^*) + p_2(s))\, ds, \quad t \in [0,1], \\ y(t) = \mu \int_0^1 G_3(t,s)(g(s,[x(s)-q_1(s)]^*,[y(s)-q_2(s)]^*) + p_2(s))\, ds \\ \qquad + \lambda \int_0^1 G_4(t,s)(f(s,[x(s)-q_1(s)]^*,[y(s)-q_2(s)]^*) + p_1(s))\, ds, \quad t \in [0,1]. \end{cases}$$

We consider the Banach space $X = C([0,1])$ with the supremum norm $\|\cdot\|$ and the Banach space $Y = X \times X$ with the norm $\|(u,v)\|_Y = \|u\| + \|v\|$. We define the cones

$$P_1 = \{x \in X, \quad x(t) \ge \tilde{\gamma}_1 t^{\alpha-1}\|x\|, \quad \forall t \in [0,1]\}, \quad P_2 = \{y \in X, \quad y(t) \ge \tilde{\gamma}_2 t^{\beta-1}\|y\|, \ \forall t \in [0,1]\},$$

where $\tilde{\gamma}_1$ and $\tilde{\gamma}_2$ are defined in Section 5.4.2 (Lemma 5.4.5), and $P = P_1 \times P_2 \subset Y$.

For $\lambda, \mu > 0$, we introduce the operators $\tilde{\mathcal{T}}_1, \tilde{\mathcal{T}}_2 : Y \to X$, and $\tilde{\mathcal{T}} : Y \to Y$ defined by $\tilde{\mathcal{T}}(x,y) = (\tilde{\mathcal{T}}_1(x,y), \tilde{\mathcal{T}}_2(x,y))$, $(x,y) \in Y$ with

$$\tilde{\mathcal{T}}_1(x,y)(t) = \lambda \int_0^1 G_1(t,s)(f(s,[x(s)-q_1(s)]^*,[y(s)-q_2(s)]^*) + p_1(s))\, ds$$
$$+ \mu \int_0^1 G_2(t,s)(g(s,[x(s)-q_1(s)]^*,[y(s)-q_2(s)]^*) + p_2(s))\, ds,$$
$$t \in [0,1],$$
$$\tilde{\mathcal{T}}_2(x,y)(t) = \mu \int_0^1 G_3(t,s)(g(s,[x(s)-q_1(s)]^*,[y(s)-q_2(s)]^*) + p_2(s))\, ds$$
$$+ \lambda \int_0^1 G_4(t,s)(f(s,[x(s)-q_1(s)]^*,[y(s)-q_2(s)]^*) + p_1(s))\, ds,$$
$$t \in [0,1].$$

It is clear that $(x,y) \in P$ is a solution of problem (5.43)–(5.44) if and only if (x,y) is a fixed point of the operator $\tilde{\mathcal{T}}$.

Lemma 5.4.6. *If* $(\widetilde{\mathrm{H}1})$ *and* $(\widetilde{\mathrm{H}2})$, *or* $(\widetilde{\mathrm{H}1})$ *and* $(\widetilde{\mathrm{H}4})$ *hold, then the operator* $\tilde{\mathcal{T}}: P \to P$ *is a completely continuous operator.*

Proof. The operators $\tilde{T}_1$ and $\tilde{T}_2$ are well defined. To prove this, let $(x, y) \in P$ be fixed with $\|(x, y)\|_Y = \tilde{L}$. Then we have

$$[x(s) - q_1(s)]^* \le x(s) \le \|x\| \le \|(x, y)\|_Y = \tilde{L}, \quad \forall s \in [0, 1],$$
$$[y(s) - q_2(s)]^* \le y(s) \le \|y\| \le \|(x, y)\|_Y = \tilde{L}, \quad \forall s \in [0, 1].$$

If $(\widetilde{\mathrm{H}1})$ and $(\widetilde{\mathrm{H}2})$ hold, we obtain

$$\begin{aligned}
\tilde{T}_1(x, y)(t) &\le \lambda\sigma_1 \int_0^1 h_1(s)(f(s, [x(s) - q_1(s)]^*, [y(s) - q_2(s)]^*) + p_1(s))\, ds \\
&\quad + \mu\sigma_2 \int_0^1 h_2(s)(g(s, [x(s) - q_1(s)]^*, [y(s) - q_2(s)]^*) + p_2(s))\, ds \\
&\le 2M\left(\lambda\sigma_1 \int_0^1 h_1(s)\, ds + \mu\sigma_2 \int_0^1 h_2(s)\, ds\right) < \infty, \quad \forall t \in [0, 1],
\end{aligned}$$

$$\begin{aligned}
\tilde{T}_2(x, y)(t) &\le \mu\sigma_3 \int_0^1 h_2(s)(g(s, [x(s) - q_1(s)]^*, [y(s) - q_2(s)]^*) + p_2(s))\, ds \\
&\quad + \lambda\sigma_4 \int_0^1 h_1(s)(f(s, [x(s) - q_1(s)]^*, [y(s) - q_2(s)]^*) + p_1(s))\, ds \\
&\le 2M\left(\mu\sigma_3 \int_0^1 h_2(s)\, ds + \lambda\sigma_4 \int_0^1 h_1(s)\, ds\right) < \infty, \quad \forall t \in [0, 1],
\end{aligned}$$

where

$$M = \max\left\{\max_{t\in[0,1],\, u,v\in[0,\tilde{L}]} f(t, u, v), \max_{t\in[0,1],\, u,v\in[0,\tilde{L}]} g(t, u, v), \max_{t\in[0,1]} p_1(t), \max_{t\in[0,1]} p_2(t)\right\}.$$

If $(\widetilde{\mathrm{H}1})$ and $(\widetilde{\mathrm{H}4})$ hold, we deduce for all $t \in [0, 1]$

$$\begin{aligned}
\tilde{T}_1(x, y)(t) &\le \lambda\sigma_1 \int_0^1 h_1(s)(f(x, [x(s) - q_1(s)]^*, [y(s) - q_2(s)]^*) + p_1(s))\, ds \\
&\quad + \mu\sigma_2 \int_0^1 h_2(s)(g(s, [x(s) - q_1(s)]^*, [y(s) - q_2(s)]^*) + p_2(s))\, ds \\
&\le \lambda\sigma_1 \int_0^1 h_1(s)[\alpha_1(s)\beta_1(s, [x(s) - q_1(s)]^*, [y(s) - q_2(s)]^*) + p_1(s)]\, ds \\
&\quad + \mu\sigma_2 \int_0^1 h_2(s)[\alpha_2(s)\beta_2(s, [x(s) - q_1(s)]^*, [y(s) - q_2(s)]^*) + p_2(s)]\, ds \\
&\le \tilde{M}\left(\lambda\sigma_1 \int_0^1 h_1(s)(\alpha_1(s) + p_1(s))\, ds + \mu\sigma_2 \int_0^1 h_2(s)(\alpha_2(s) + p_2(s))\, ds\right) < \infty,
\end{aligned}$$

$$\begin{aligned}
\tilde{T}_2(x,y)(t) &\le \mu\sigma_3\int_0^1 h_2(s)(g(x,[x(s)-q_1(s)]^*,[y(s)-q_2(s)]^*)+p_2(s))\,ds \\
&\quad +\lambda\sigma_4\int_0^1 h_1(s)(f(s,[x(s)-q_1(s)]^*,[y(s)-q_2(s)]^*)+p_1(s))\,ds \\
&\le \mu\sigma_3\int_0^1 h_2(s)[\alpha_2(s)\beta_2(s,[x(s)-q_1(s)]^*,[y(s)-q_2(s)]^*)+p_2(s)]\,ds \\
&\quad +\lambda\sigma_4\int_0^1 h_1(s)[\alpha_1(s)\beta_1(s,[x(s)-q_1(s)]^*,[y(s)-q_2(s)]^*)+p_1(s)]\,ds \\
&\le \tilde{M}\left(\mu\sigma_3\int_0^1 h_2(s)(\alpha_2(s)+p_2(s))\,ds+\lambda\sigma_4\int_0^1 h_1(s)(\alpha_1(s)+p_1(s))\,ds\right)<\infty,
\end{aligned}$$

where $\tilde{M}=\max\left\{\max_{t\in[0,1],\,u,v\in[0,\tilde{L}]}\beta_1(t,u,v),\ \max_{t\in[0,1],\,u,v\in[0,\tilde{L}]}\beta_2(t,u,v),1\right\}$.

Besides, by Lemma 5.4.5, we conclude that

$$\tilde{T}_1(x,y)(t)\ge\tilde{\gamma}_1t^{\alpha-1}\|\tilde{T}_1(x,y)\|,\quad \tilde{T}_2(x,y)(t)\ge\tilde{\gamma}_2t^{\beta-1}\|\tilde{T}_2(x,y)\|,\quad \forall t\in[0,1],$$

and so $\tilde{T}_1(x,y),\ \tilde{T}_2(x,y)\in P$.

By using standard arguments, we deduce that the operator $\tilde{\mathcal{T}}:P\to P$ is a completely continuous operator. □

Theorem 5.4.1. *Assume that* $(\widetilde{\mathrm{H}1})$–$(\widetilde{\mathrm{H}3})$ *hold. Then there exist constants* $\lambda_0>0$ *and* $\mu_0>0$ *such that for any* $\lambda\in(0,\lambda_0]$ *and* $\mu\in(0,\mu_0]$ *the boundary value problem* $(\tilde{\mathrm{S}})$–$(\widetilde{\mathrm{BC}})$ *has at least one positive solution.*

Proof. Let $\delta\in(0,1)$ be fixed. From $(\widetilde{\mathrm{H}2})$ and $(\widetilde{\mathrm{H}3})$, there exists $R_0\in(0,1]$ such that

$$f(t,u,v)\ge\delta f(t,0,0),\quad g(t,u,v)\ge\delta g(t,0,0),\quad \forall t\in[0,1],\quad u,\,v\in[0,R_0]. \tag{5.47}$$

We define

$$\begin{aligned}
\bar{f}(R_0)&=\max_{t\in[0,1],\,u,v\in[0,R_0]}\{f(t,u,v)+p_1(t)\}\ge\max_{t\in[0,1]}\{\delta f(t,0,0)+p_1(t)\}>0,\\
\bar{g}(R_0)&=\max_{t\in[0,1],\,u,v\in[0,R_0]}\{g(t,u,v)+p_2(t)\}\ge\max_{t\in[0,1]}\{\delta g(t,0,0)+p_2(t)\}>0,\\
c_1&=\sigma_1\int_0^1 h_1(s)\,ds,\quad c_2=\sigma_2\int_0^1 h_2(s)\,ds,\quad c_3=\sigma_3\int_0^1 h_2(s)\,ds,\\
c_4&=\sigma_4\int_0^1 h_1(s)\,ds,\\
\lambda_0&=\max\left\{\frac{R_0}{8c_1\bar{f}(R_0)},\frac{R_0}{8c_4\bar{f}(R_0)}\right\},\quad \mu_0=\max\left\{\frac{R_0}{8c_2\bar{g}(R_0)},\frac{R_0}{8c_3\bar{g}(R_0)}\right\}.
\end{aligned}$$

We shall show that for any $\lambda\in(0,\lambda_0]$ and $\mu\in(0,\mu_0]$ problem (5.43)–(5.44) has at least one positive solution.

So, let $\lambda\in(0,\lambda_0]$ and $\mu\in(0,\mu_0]$ be arbitrary, but fixed for the moment. We define the set $U=\{(x,y)\in P,\ \|(u,v)\|_Y<R_0\}$. We suppose that there exist $(x,y)\in\partial U$ ($\|(x,y)\|_Y=R_0$ or $\|x\|+\|y\|=R_0$) and $\nu\in(0,1)$ such that $(x,y)=\nu\tilde{\mathcal{T}}(x,y)$ or $x=\nu\tilde{T}_1(x,y)$, $y=\nu\tilde{T}_2(x,y)$.

We deduce that

$$[x(t)-q_1(t)]^* = x(t)-q_1(t) \le x(t) \le R_0, \quad \text{if} \quad x(t)-q_1(t) \ge 0,$$
$$[x(t)-q_1(t)]^* = 0, \quad \text{for} \quad x(t)-q_1(t) < 0, \quad \forall t \in [0,1],$$
$$[y(t)-q_2(t)]^* = y(t)-q_2(t) \le y(t) \le R_0, \quad \text{if} \quad y(t)-q_2(t) \ge 0,$$
$$[y(t)-q_2(t)]^* = 0, \quad \text{for} \quad y(t)-q_2(t) < 0, \quad \forall t \in [0,1].$$

Then, by Lemma 5.4.4, for all $t \in [0,1]$ we obtain

$$\begin{aligned}
x(t) &= \nu \tilde{T}_1(x,y)(t) \le \tilde{T}_1(x,y)(t) \\
&= \lambda \int_0^1 G_1(t,s)(f(s,[x(s)-q_1(s)]^*,[y(s)-q_2(s)]^*) + p_1(s))\,ds \\
&\quad + \mu \int_0^1 G_2(t,s)(g(s,[x(s)-q_1(s)]^*,[y(s)-q_2(s)]^*) + p_2(s))\,ds \\
&\le \lambda\sigma_1 \int_0^1 h_1(s)\bar{f}(R_0)\,ds + \mu\sigma_2 \int_0^1 h_2(s)\bar{g}(R_0)\,ds \\
&\le \lambda_0 c_1 \bar{f}(R_0) + \mu_0 c_2 \bar{g}(R_0) \le \frac{R_0}{8} + \frac{R_0}{8} = \frac{R_0}{4}, \\
y(t) &= \nu \tilde{T}_2(x,y)(t) \le \tilde{T}_2(x,y)(t) \\
&= \mu \int_0^1 G_3(t,s)(g(s,[x(s)-q_1(s)]^*,[y(s)-q_2(s)]^*) + p_2(s))\,ds \\
&\quad + \lambda \int_0^1 G_4(t,s)(f(s,[x(s)-q_1(s)]^*,[y(s)-q_2(s)]^*) + p_1(s))\,ds \\
&\le \mu\sigma_3 \int_0^1 h_2(s)\bar{g}(R_0)\,ds + \lambda\sigma_4 \int_0^1 h_1(s)\bar{f}(R_0)\,ds \\
&\le \mu_0 c_3 \bar{g}(R_0) + \lambda_0 c_4 \bar{f}(R_0) \le \frac{R_0}{8} + \frac{R_0}{8} = \frac{R_0}{4}.
\end{aligned}$$

Hence, $\|x\| \le \frac{R_0}{4}$ and $\|y\| \le \frac{R_0}{4}$. Then $R_0 = \|(x,y)\|_Y = \|x\| + \|y\| \le \frac{R_0}{4} + \frac{R_0}{4} = \frac{R_0}{2}$, which is a contradiction.

Therefore, by Lemma 5.4.6 and Theorem 2.5.1, we deduce that $\tilde{T}$ has a fixed point $(x_0, y_0) \in \bar{U} \cap P$. That is, $(x_0,y_0) = \tilde{T}(x_0,y_0)$ or $x_0 = \tilde{T}_1(x_0,y_0)$, $y_0 = \tilde{T}_2(x_0,y_0)$, and $\|x_0\| + \|y_0\| \le R_0$ with $x_0(t) \ge \tilde{\gamma}_1 t^{\alpha-1}\|x_0\|$ and $y_0(t) \ge \tilde{\gamma}_2 t^{\beta-1}\|y_0\|$ for all $t \in [0,1]$. Moreover, by (5.47), we conclude

$$\begin{aligned}
x_0(t) &= \tilde{T}_1(x_0,y_0)(t) \ge \lambda \int_0^1 G_1(t,s)(\delta f(t,0,0) + p_1(s))\,ds \\
&\quad + \mu \int_0^1 G_2(t,s)(\delta g(t,0,0) + p_2(s))\,ds \\
&\ge \lambda \int_0^1 G_1(t,s)p_1(s)\,ds + \mu \int_0^1 G_2(t,s)p_2(s)\,ds = q_1(t), \quad \forall t \in [0,1],
\end{aligned}$$

$$x_0(t) > \lambda \int_0^1 G_1(t,s)p_1(s)\,\mathrm{d}s + \mu \int_0^1 G_2(t,s)p_2(s)\,\mathrm{d}s = q_1(t), \quad \forall t \in (0,1),$$

$$\begin{aligned} y_0(t) &= \tilde{T}_2(x_0,y_0)(t) \geq \mu \int_0^1 G_3(t,s)(\delta g(t,0,0) + p_2(s))\,\mathrm{d}s \\ &\quad + \lambda \int_0^1 G_4(t,s)(\delta f(t,0,0) + p_1(s))\,\mathrm{d}s \\ &\geq \mu \int_0^1 G_3(t,s)p_2(s)\,\mathrm{d}s + \lambda \int_0^1 G_4(t,s)p_1(s)\,\mathrm{d}s = q_2(t), \quad \forall t \in [0,1], \\ y_0(t) &> \mu \int_0^1 G_3(t,s)p_2(s)\,\mathrm{d}s + \lambda \int_0^1 G_4(t,s)p_1(s)\,\mathrm{d}s = q_2(t), \quad \forall t \in (0,1). \end{aligned}$$

Therefore, $x_0(t) \geq q_1(t)$ and $y_0(t) \geq q_2(t)$ for all $t \in [0,1]$, and $x_0(t) > q_1(t)$ and $y_0(t) > q_2(t)$ for all $t \in (0,1)$. Let $u_0(t) = x_0(t) - q_1(t)$ and $v_0(t) = y_0(t) - q_2(t)$ for all $t \in [0,1]$. Then $u_0(t) \geq 0$ and $v_0(t) \geq 0$ for all $t \in [0,1]$, and $u_0(t) > 0$ and $v_0(t) > 0$ for all $t \in (0,1)$. Therefore, (u_0, v_0) is a positive solution of $(\tilde{\mathrm{S}})$–$(\widetilde{\mathrm{BC}})$. □

Theorem 5.4.2. *Assume that* $(\widetilde{\mathrm{H1}})$, $(\widetilde{\mathrm{H4}})$, *and* $(\widetilde{\mathrm{H5}})$ *hold. Then there exist* $\lambda^* > 0$ *and* $\mu^* > 0$ *such that for any* $\lambda \in (0,\lambda^*]$ *and* $\mu \in (0,\mu^*]$, *the boundary value problem* $(\tilde{\mathrm{S}})$–$(\widetilde{\mathrm{BC}})$ *has at least one positive solution.*

Proof. We choose a positive number

$$\begin{aligned} R_1 > \max\Bigg\{ &1, \frac{2}{\tilde{\gamma}_1} \int_0^1 (\delta_1 p_1(s) + \delta_2 p_2(s))\,\mathrm{d}s, \frac{2}{\tilde{\gamma}_2} \int_0^1 (\delta_3 p_2(s) + \delta_4 p_1(s))\,\mathrm{d}s, \\ &\frac{2}{\tilde{\gamma}_1\tilde{\gamma}_2} \left(\int_0^1 s^{\beta-1}\mathrm{d}H(s) \right)^{-1} \int_0^1 (\delta_1 p_1(s) + \delta_2 p_2(s))\,\mathrm{d}s, \\ &\frac{2}{\tilde{\gamma}_1\tilde{\gamma}_2} \left(\int_0^1 s^{\alpha-1}\mathrm{d}K(s) \right)^{-1} \int_0^1 (\delta_3 p_2(s) + \delta_4 p_1(s))\,\mathrm{d}s \Bigg\}, \end{aligned}$$

and define the set $\Omega_1 = \{(x,y) \in P, \|(x,y)\|_Y < R_1\}$.

We introduce

$$\lambda^* = \min\left\{1, \frac{R_1}{4\sigma_1 M_1} \left(\int_0^1 h_1(s)(\alpha_1(s) + p_1(s))\,\mathrm{d}s \right)^{-1}, \frac{R_1}{4\sigma_4 M_1} \left(\int_0^1 h_1(s)(\alpha_1(s) + p_1(s))\,\mathrm{d}s \right)^{-1} \right\},$$

$$\mu^* = \min\left\{1, \frac{R_1}{4\sigma_2 M_2} \left(\int_0^1 h_2(s)(\alpha_2(s) + p_2(s))\,\mathrm{d}s \right)^{-1}, \frac{R_1}{4\sigma_3 M_2} \left(\int_0^1 h_2(s)(\alpha_2(s) + p_2(s))\,\mathrm{d}s \right)^{-1} \right\},$$

with $M_1 = \max\left\{ \max\limits_{\substack{t\in[0,1] \\ u,v\geq 0,\, u+v\leq R_1}} \beta_1(t,u,v), 1 \right\}$ and $M_2 = \max\left\{ \max\limits_{\substack{t\in[0,1] \\ u,v\geq 0,\, u+v\leq R_1}} \beta_2(t,u,v), 1 \right\}$.

Let $\lambda \in (0,\lambda^*]$ and $\mu \in (0,\mu^*]$. Then for any $(x,y) \in P \cap \partial\Omega_1$ and $s \in [0,1]$, we have

$$[x(s) - q_1(s)]^* \leq x(s) \leq \|x\| \leq R_1, \quad [y(s) - q_2(s)]^* \leq y(s) \leq \|y\| \leq R_1.$$

Then, by using Lemma 5.4.4, for any $(x,y) \in P \cap \partial\Omega_1$ we obtain

$$\begin{aligned}
\|\tilde{T}_1(x,y)\| &\le \lambda\sigma_1 \int_0^1 h_1(s)[\alpha_1(s)\beta_1(s,[x(s)-q_1(s)]^*,[y(s)-q_2(s)]^*)+p_1(s)]\,ds \\
&\quad + \mu\sigma_2 \int_0^1 h_2(s)[\alpha_2(s)\beta_2(s,[x(s)-q_1(s)]^*,[y(s)-q_2(s)]^*)+p_2(s)]\,ds \\
&\le \lambda^*\sigma_1 M_1 \int_0^1 h_1(s)(\alpha_1(s)+p_1(s))\,ds + \mu^*\sigma_2 M_2 \int_0^1 h_2(s)(\alpha_2(s)+p_2(s))\,ds \\
&\le \frac{R_1}{4}+\frac{R_1}{4} = \frac{R_1}{2} = \frac{\|(x,y)\|_Y}{2}, \\
\|\tilde{T}_2(x,y)\| &\le \mu\sigma_3 \int_0^1 h_2(s)[\alpha_2(s)\beta_2(s,[x(s)-q_1(s)]^*,[y(s)-q_2(s)]^*)+p_2(s)]\,ds \\
&\quad + \lambda\sigma_4 \int_0^1 h_1(s)[\alpha_1(s)\beta_1(s,[x(s)-q_1(s)]^*,[y(s)-q_2(s)]^*)+p_1(s)]\,ds \\
&\le \mu^*\sigma_3 M_2 \int_0^1 h_2(s)(\alpha_2(s)+p_2(s))\,ds + \lambda^*\sigma_4 M_1 \int_0^1 h_1(s)(\alpha_1(s)+p_1(s))\,ds \\
&\le \frac{R_1}{4}+\frac{R_1}{4} = \frac{R_1}{2} = \frac{\|(x,y)\|_Y}{2}.
\end{aligned}$$

Therefore,

$$\|\tilde{T}(x,y)\|_Y = \|\tilde{T}_1(x,y)\| + \|\tilde{T}_2(x,y)\| \le \|(x,y)\|_Y, \quad \forall (x,y) \in P \cap \partial\Omega_1. \tag{5.48}$$

On the other hand, we choose a constant $L > 0$ such that

$$\begin{aligned}
&\lambda L \varrho_1 \tilde{\gamma}_1 c^{2(\alpha-1)} \int_c^{1-c} h_1(s)\,ds \ge 4, \quad \lambda L \varrho_4 \tilde{\gamma}_2 c^{2(\beta-1)} \int_c^{1-c} h_1(s)\,ds \ge 4, \\
&\mu L \varrho_2 \tilde{\gamma}_1 c^{2(\alpha-1)} \int_c^{1-c} h_2(s)\,ds \ge 4, \quad \mu L \varrho_3 \tilde{\gamma}_2 c^{2(\beta-1)} \int_c^{1-c} h_2(s)\,ds \ge 4.
\end{aligned}$$

From $(\widetilde{H5})$, we deduce that there exists a constant $M_0 > 0$ such that

$$\begin{aligned}
&f(t,u,v) \ge L(u+v) \quad \text{or} \quad g(t,u,v) \ge L(u+v), \quad \forall t \in [c,1-c], \\
&u,v \in [0,\infty), \quad u+v \ge M_0.
\end{aligned} \tag{5.49}$$

Now we define

$$\begin{aligned}
R_2 = \max\Bigg\{&2R_1, \frac{4M_0}{\tilde{\gamma}_1 c^{\alpha-1}}, \frac{4M_0}{\tilde{\gamma}_2 c^{\beta-1}}, \frac{4}{\tilde{\gamma}_1}\int_0^1 (\delta_1 p_1(s)+\delta_2 p_2(s))\,ds, \\
&\frac{4}{\tilde{\gamma}_2}\int_0^1 (\delta_3 p_2(s)+\delta_4 p_1(s))\,ds\Bigg\} > 0,
\end{aligned}$$

and let $\Omega_2 = \{(x,y) \in P, \ \|(x,y)\|_Y < R_2\}$.

We suppose that $f_\infty = \infty$—that is, $f(t,u,v) \le L(u+v)$ for all $t \in [c,1-c]$ and $u,v \in [0,\infty)$, $u+v \ge M_0$. Then for any $(x,y) \in P \cap \partial\Omega_2$, we have $\|(x,y)\|_Y = R_2$ or $\|x\| + \|y\| = R_2$. We deduce that $\|x\| \ge \frac{R_2}{2}$ or $\|y\| \ge \frac{R_2}{2}$.

We suppose that $\|x\| \ge \frac{R_2}{2}$. Then, by Lemma 5.4.4, for any $(x,y) \in P \cap \partial\Omega_2$ we obtain

$$
\begin{aligned}
x(t) - q_1(t) &= x(t) - \lambda \int_0^1 G_1(t,s)p_1(s)\,\mathrm{d}s - \mu \int_0^1 G_2(t,s)p_2(s)\,\mathrm{d}s \\
&\geq x(t) - t^{\alpha-1}\left(\delta_1 \int_0^1 p_1(s)\,\mathrm{d}s + \delta_2 \int_0^1 p_2(s)\,\mathrm{d}s\right) \\
&\geq x(t) - \frac{x(t)}{\tilde{\gamma}_1 \|x\|} \int_0^1 (\delta_1 p_1(s) + \delta_2 p_2(s))\,\mathrm{d}s \\
&= x(t)\left[1 - \frac{1}{\tilde{\gamma}_1 \|x\|} \int_0^1 (\delta_1 p_1(s) + \delta_2 p_2(s))\,\mathrm{d}s\right] \\
&\geq x(t)\left[1 - \frac{2}{\tilde{\gamma}_1 R_2} \int_0^1 (\delta_1 p_1(s) + \delta_2 p_2(s))\,\mathrm{d}s\right] \geq \frac{1}{2}x(t) \geq 0, \quad \forall t \in [0,1].
\end{aligned}
$$

Therefore, we conclude

$$
\begin{aligned}
[x(t) - q_1(t)]^* &= x(t) - q_1(t) \geq \frac{1}{2}x(t) \geq \frac{1}{2}\tilde{\gamma}_1 t^{\alpha-1}\|x\| \\
&\geq \frac{1}{4}\tilde{\gamma}_1 t^{\alpha-1} R_2 \geq \frac{1}{4}\tilde{\gamma}_1 c^{\alpha-1} R_2 \geq M_0, \quad \forall t \in [c, 1-c].
\end{aligned}
$$

Hence,

$$
[x(t)-q_1(t)]^*+[y(t)-q_2(t)]^* \geq [x(t)-q_1(t)]^* = x(t)-q_1(t) \geq M_0, \quad \forall t \in [c,1-c]. \tag{5.50}
$$

Then, for any $(x,y) \in P \cap \partial\Omega_2$ and $t \in [c, 1-c]$, by (5.49) and (5.50), we deduce

$$
\begin{aligned}
f(t, [x(t) - q_1(t)]^*, [y(t) - q_2(t)]^*) &\geq L([x(t) - q_1(t)]^* + [y(t) - q_2(t)]^*) \\
&\geq L[x(t) - q_1(t)]^* \geq \frac{L}{2}x(t), \quad \forall t \in [c, 1-c].
\end{aligned}
$$

By using Lemma 5.4.4, for any $(x,y) \in P \cap \partial\Omega_2$, $t \in [c, 1-c]$ we obtain

$$
\begin{aligned}
\tilde{T}_1(x,y)(t) &\geq \lambda \int_0^1 G_1(t,s)(f(s,[x(s) - q_1(s)]^*, [y(s) - q_2(s)]^*) + p_1(s))\,\mathrm{d}s \\
&\geq \lambda \int_c^{1-c} G_1(t,s)(f(s,[x(s) - q_1(s)]^*, [y(s) - q_2(s)]^*) + p_1(s))\,\mathrm{d}s \\
&\geq \lambda \int_c^{1-c} G_1(t,s)L([x(s) - q_1(s)]^*)\,\mathrm{d}s \\
&\geq \lambda \int_c^{1-c} G_1(t,s)\frac{1}{4}L\tilde{\gamma}_1 c^{\alpha-1} R_2\,\mathrm{d}s \\
&\geq \lambda \int_c^{1-c} \varrho_1 t^{\alpha-1} h_1(s)\frac{1}{4}L\tilde{\gamma}_1 c^{\alpha-1} R_2\,\mathrm{d}s \\
&\geq \lambda c^{2(\alpha-1)}\frac{1}{4}\varrho_1 L\tilde{\gamma}_1 R_2 \int_c^{1-c} h_1(s)\,\mathrm{d}s \geq R_2.
\end{aligned}
$$

Then $\|\tilde{T}_1(x,y)\| \geq \|(x,y)\|_Y$ and

$$\|\tilde{\mathcal{T}}(x,y)\|_Y \geq \|(x,y)\|_Y, \quad \forall\,(x,y) \in P \cap \partial\Omega_2. \tag{5.51}$$

If $\|y\| \geq \frac{R_2}{2}$, then by Lemma 5.4.4, for any $(x,y) \in P \cap \partial\Omega_2$ we conclude

$$\begin{aligned}
y(t) - q_2(t) &= y(t) - \mu \int_0^1 G_3(t,s)p_2(s)\,ds - \lambda \int_0^1 G_4(t,s)p_1(s)\,ds \\
&\geq y(t) - t^{\beta-1}\left(\delta_3 \int_0^1 p_2(s)\,ds + \delta_4 \int_0^1 p_1(s)\,ds\right) \\
&\geq y(t) - \frac{y(t)}{\tilde{\gamma}_2\|y\|} \int_0^1 (\delta_3 p_2(s) + \delta_4 p_1(s))\,ds \\
&= y(t)\left[1 - \frac{1}{\tilde{\gamma}_2\|y\|} \int_0^1 (\delta_3 p_2(s) + \delta_4 p_1(s))\,ds\right] \\
&\geq y(t)\left[1 - \frac{2}{\tilde{\gamma}_2 R_2} \int_0^1 (\delta_3 p_2(s) + \delta_4 p_1(s))\,ds\right] \\
&\geq \frac{1}{2}y(t) \geq 0, \quad \forall t \in [0,1].
\end{aligned}$$

Therefore, we deduce

$$\begin{aligned}
[y(t) - q_2(t)]^* &= y(t) - q_2(t) \geq \frac{1}{2}y(t) \geq \frac{1}{2}\tilde{\gamma}_2 t^{\beta-1}\|y\| \\
&\geq \frac{1}{4}\tilde{\gamma}_2 t^{\beta-1}R_2 \geq \frac{1}{4}\tilde{\gamma}_2 c^{\beta-1}R_2 \geq M_0, \quad \forall t \in [c, 1-c].
\end{aligned}$$

Hence,

$$[x(t) - q_1(t)]^* + [y(t) - q_2(t)]^* \geq [y(t) - q_2(t)]^* = y(t) - q_2(t) \geq M_0, \quad \forall t \in [c, 1-c]. \tag{5.52}$$

Then, for any $(x,y) \in P \cap \partial\Omega_2$ and $t \in [c, 1-c]$, by (5.49) and (5.52), we obtain

$$\begin{aligned}
f(t, [x(t) - q_1(t)]^*, [y(t) - q_2(t)]^*) &\geq L([x(t) - q_1(t)]^* + [y(t) - q_2(t)]^*) \\
&\geq L[y(t) - q_2(t)]^* \geq \frac{L}{2}y(t), \quad \forall t \in [c, 1-c].
\end{aligned}$$

By using Lemma 5.4.4, for any $(x,y) \in P \cap \partial\Omega_2$ and $t \in [c, 1-c]$, we obtain

$$\begin{aligned}
\tilde{T}_2(x,y)(t) &\geq \lambda \int_0^1 G_4(t,s)(f(s, [x(s) - q_1(s)]^*, [y(s) - q_2(s)]^*) + p_1(s))\,ds \\
&\geq \lambda \int_c^{1-c} G_4(t,s)(f(s, [x(s) - q_1(s)]^*, [y(s) - q_2(s)]^*) + p_1(s))\,ds \\
&\geq \lambda \int_c^{1-c} G_4(t,s)L([y(s) - q_2(s)]^*)\,ds \\
&\geq \lambda \int_c^{1-c} G_4(t,s)\frac{1}{4}L\tilde{\gamma}_2 c^{\beta-1}R_2\,ds
\end{aligned}$$

$$\geq \lambda \int_c^{1-c} \varrho_4 t^{\beta-1} h_1(s) \frac{1}{4} L\tilde{\gamma}_2 c^{\beta-1} R_2 \, ds$$
$$\geq \lambda c^{2(\beta-1)} \frac{1}{4} \varrho_4 L\tilde{\gamma}_2 R_2 \int_c^{1-c} h_1(s) \, ds \geq R_2.$$

Then $\|\tilde{T}_2(x,y)\| \geq \|(x,y)\|_Y$, and we obtain again relation (5.51).

We suppose now that $g_\infty = \infty$—that is, $g(t,u,v) \geq L(u+v)$ for all $t \in [c, 1-c]$ and $u, v \in [0,\infty)$, $u+v \geq M_0$. Then for any $(x,y) \in P \cap \partial\Omega_2$, we have $\|(x,y)\|_Y = R_2$. Hence, $\|x\| \geq \frac{R_2}{2}$ or $\|y\| \geq \frac{R_2}{2}$.

If $\|x\| \geq \frac{R_2}{2}$, then for any $(x,y) \in P \cap \partial\Omega_2$ we deduce in a manner similar to that above that $x(t) - q_1(t) \geq \frac{1}{2}x(t)$ for all $t \in [0,1]$ and

$$\tilde{T}_1(x,y)(t) \geq \mu \int_0^1 G_2(t,s)(g(s,[x(s)-q_1(s)]^*,[y(s)-q_2(s)]^*) + p_2(s)) \, ds$$
$$\geq \mu \int_c^{1-c} G_2(t,s)(g(s,[x(s)-q_1(s)]^*,[y(s)-q_2(s)]^*) + p_2(s)) \, ds$$
$$\geq \mu \int_c^{1-c} G_2(t,s) L([x(s)-q_1(s)]^*) \, ds$$
$$\geq \mu \int_c^{1-c} G_2(t,s) \frac{1}{4} L\tilde{\gamma}_1 c^{\alpha-1} R_2 \, ds$$
$$\geq \mu \int_c^{1-c} \varrho_2 t^{\alpha-1} h_2(s) \frac{1}{4} L\tilde{\gamma}_1 c^{\alpha-1} R_2 \, ds$$
$$\geq \mu c^{2(\alpha-1)} \frac{1}{4} \varrho_2 L\tilde{\gamma}_1 R_2 \int_c^{1-c} h_2(s) \, ds \geq R_2, \quad \forall t \in [c, 1-c].$$

Hence, we obtain relation (5.51).

If $\|y\| \geq \frac{R_2}{2}$, then for any $(x,y) \in P \cap \partial\Omega_2$ we deduce in a manner similar to that above that $y(t) - q_2(t) \geq \frac{1}{2}y(t)$ for all $t \in [0,1]$ and

$$\tilde{T}_2(x,y)(t) \geq \mu \int_0^1 G_3(t,s)(g(s,[x(s)-q_1(s)]^*,[y(s)-q_2(s)]^*) + p_2(s)) \, ds$$
$$\geq \mu \int_c^{1-c} G_3(t,s)(g(s,[x(s)-q_1(s)]^*,[y(s)-q_2(s)]^*) + p_2(s)) \, ds$$
$$\geq \mu \int_c^{1-c} G_3(t,s) L([y(s)-q_2(s)]^*) \, ds$$
$$\geq \mu \int_c^{1-c} G_3(t,s) \frac{1}{4} L\tilde{\gamma}_2 c^{\beta-1} R_2 \, ds$$
$$\geq \mu \int_c^{1-c} \varrho_3 t^{\beta-1} h_2(s) \frac{1}{4} L\tilde{\gamma}_2 c^{\beta-1} R_2 \, ds$$
$$\geq \mu c^{2(\beta-1)} \frac{1}{4} \varrho_3 L\tilde{\gamma}_2 R_2 \int_c^{1-c} h_2(s) \, ds \geq R_2, \quad \forall t \in [c, 1-c].$$

Hence, we obtain again relation (5.51).

Therefore, by Lemma 5.4.6, Theorem 1.1.1, and relations (5.48) and (5.51), we conclude that $\tilde{\mathcal{T}}$ has a fixed point $(x_1, y_1) \in P \cap (\bar{\Omega}_2 \setminus \Omega_1)$—that is, $R_1 \le \|(x_1, y_1)\|_Y \le R_2$. Since $\|(x_1, y_1)\|_Y \ge R_1$, then $\|x_1\| \ge \frac{R_1}{2}$ or $\|y_1\| \ge \frac{R_1}{2}$.

We suppose first that $\|x_1\| \ge \frac{R_1}{2}$. Then we deduce

$$\begin{aligned}
x_1(t) - q_1(t) &= x_1(t) - \lambda \int_0^1 G_1(t,s) p_1(s)\,ds - \mu \int_0^1 G_2(t,s) p_2(s)\,ds \\
&\ge x_1(t) - t^{\alpha-1}\left(\delta_1 \int_0^1 p_1(s)\,ds + \delta_2 \int_0^1 p_2(s)\,ds\right) \\
&\ge x_1(t) - \frac{x_1(t)}{\tilde{\gamma}_1 \|x_1\|} \int_0^1 (\delta_1 p_1(s) + \delta_2 p_2(s))\,ds \\
&\ge \left[1 - \frac{2}{\tilde{\gamma}_1 R_1} \int_0^1 (\delta_1 p_1(s) + \delta_2 p_2(s))\,ds\right] x_1(t) \\
&\ge \left[1 - \frac{2}{\tilde{\gamma}_1 R_1} \int_0^1 (\delta_1 p_1(s) + \delta_2 p_2(s))\,ds\right] \tilde{\gamma}_1 t^{\alpha-1} \|x_1\| \\
&\ge \frac{R_1}{2}\left[1 - \frac{2}{\tilde{\gamma}_1 R_1} \int_0^1 (\delta_1 p_1(s) + \delta_2 p_2(s))\,ds\right] \tilde{\gamma}_1 t^{\alpha-1} \\
&= \Lambda_1 t^{\alpha-1}, \quad \forall t \in [0,1],
\end{aligned}$$

and so $x_1(t) \ge q_1(t) + \Lambda_1 t^{\alpha-1}$ for all $t \in [0,1]$, where $\Lambda_1 = \frac{\tilde{\gamma}_1 R_1}{2} - \int_0^1 (\delta_1 p_1(s) + \delta_2 p_2(s))\,ds > 0$.

Then, $y_1(1) = \int_0^1 x_1(s)\,dK(s) \ge \Lambda_1 \int_0^1 s^{\alpha-1}\,dK(s) > 0$ and

$$\|y_1\| \ge y_1(1) = \int_0^1 x_1(s)\,dK(s) \ge \int_0^1 \tilde{\gamma}_1 s^{\alpha-1} \|x_1\|\,dK(s) \ge \frac{\tilde{\gamma}_1 R_1}{2} \int_0^1 s^{\alpha-1} dK(s) > 0.$$

Therefore, we obtain

$$\begin{aligned}
y_1(t) - q_2(t) &= y_1(t) - \mu \int_0^1 G_3(t,s) p_2(s)\,ds - \lambda \int_0^1 G_4(t,s) p_1(s)\,ds \\
&\ge y_1(t) - t^{\beta-1} \int_0^1 (\delta_3 p_2(s) + \delta_4 p_1(s))\,ds \\
&\ge y_1(t) - \frac{y_1(t)}{\tilde{\gamma}_2 \|y_1\|} \int_0^1 (\delta_3 p_2(s) + \delta_4 p_1(s))\,ds \\
&\ge y_1(t)\left[1 - \frac{2}{\tilde{\gamma}_1 \tilde{\gamma}_2 R_1}\left(\int_0^1 s^{\alpha-1} dK(s)\right)^{-1} \int_0^1 (\delta_3 p_2(s) + \delta_4 p_1(s))\,ds\right] \\
&\ge \tilde{\gamma}_2 t^{\beta-1} \|y_1\| \left[1 - \frac{2}{\tilde{\gamma}_1 \tilde{\gamma}_2 R_1}\left(\int_0^1 s^{\alpha-1} dK(s)\right)^{-1} \int_0^1 (\delta_3 p_2(s) + \delta_4 p_1(s))\,ds\right] \\
&\ge \frac{\tilde{\gamma}_1 \tilde{\gamma}_2 R_1}{2} t^{\beta-1} \int_0^1 s^{\alpha-1} dK(s) \left[1 - \frac{2}{\tilde{\gamma}_1 \tilde{\gamma}_2 R_1}\left(\int_0^1 s^{\alpha-1} dK(s)\right)^{-1}\right.
\end{aligned}$$

$$\times \int_0^1 (\delta_3 p_2(s) + \delta_4 p_1(s))\,\mathrm{d}s\Bigg]$$
$$= \Lambda_2 t^{\beta-1}, \quad \forall t \in [0,1],$$

where $\Lambda_2 = \frac{\tilde{\gamma}_1\tilde{\gamma}_2 R_1}{2}\int_0^1 s^{\alpha-1}\mathrm{d}K(s) - \int_0^1 (\delta_3 p_2(s) + \delta_4 p_1(s))\,\mathrm{d}s > 0$.

Hence, $y_1(t) \geq q_2(t) + \Lambda_2 t^{\beta-1}$ for all $t \in [0,1]$.

If $\|y_1\| \geq \frac{R_1}{2}$, then

$$\begin{aligned}
y_1(t) - q_2(t) &= y_1(t) - \mu\int_0^1 G_3(t,s)p_2(s)\,\mathrm{d}s - \lambda\int_0^1 G_4(t,s)p_1(s)\,\mathrm{d}s \\
&\geq y_1(t) - t^{\beta-1}\left(\delta_3\int_0^1 p_2(s)\,\mathrm{d}s + \delta_4\int_0^1 p_1(s)\,\mathrm{d}s\right) \\
&\geq y_1(t) - \frac{y_1(t)}{\tilde{\gamma}_2\|y_1\|}\int_0^1 (\delta_3 p_2(s) + \delta_4 p_1(s))\,\mathrm{d}s \\
&\geq \left[1 - \frac{2}{\tilde{\gamma}_2 R_1}\int_0^1 (\delta_3 p_2(s) + \delta_4 p_1(s))\,\mathrm{d}s\right] y_1(t) \\
&\geq \left[1 - \frac{2}{\tilde{\gamma}_2 R_1}\int_0^1 (\delta_3 p_2(s) + \delta_4 p_1(s))\,\mathrm{d}s\right] \tilde{\gamma}_2 t^{\beta-1}\|y_1\| \\
&\geq \frac{R_1}{2}\left[1 - \frac{2}{\tilde{\gamma}_2 R_1}\int_0^1 (\delta_3 p_2(s) + \delta_4 p_1(s))\,\mathrm{d}s\right] \tilde{\gamma}_2 t^{\beta-1} \\
&= \Lambda_3 t^{\beta-1}, \quad \forall t \in [0,1],
\end{aligned}$$

and so $y_1(t) \geq q_2(t) + \Lambda_3 t^{\beta-1}$ for all $t \in [0,1]$, where $\Lambda_3 = \frac{\tilde{\gamma}_2 R_1}{2} - \int_0^1 (\delta_3 p_2(s) + \delta_4 p_1(s))\,\mathrm{d}s > 0$.

Then, $x_1(1) = \int_0^1 y_1(s)\,\mathrm{d}H(s) \geq \Lambda_3\int_0^1 s^{\beta-1}\,\mathrm{d}H(s) > 0$ and

$$\|x_1\| \geq x_1(1) = \int_0^1 y_1(s)\,\mathrm{d}H(s) \geq \int_0^1 \tilde{\gamma}_2 s^{\beta-1}\|y_1\|\,\mathrm{d}H(s) \geq \frac{\tilde{\gamma}_2 R_1}{2}\int_0^1 s^{\beta-1}\mathrm{d}H(s) > 0.$$

Therefore, we obtain

$$\begin{aligned}
x_1(t) - q_1(t) &= x_1(t) - \lambda\int_0^1 G_1(t,s)p_1(s)\,\mathrm{d}s - \mu\int_0^1 G_2(t,s)p_2(s)\,\mathrm{d}s \\
&\geq x_1(t) - t^{\alpha-1}\int_0^1 (\delta_1 p_1(s) + \delta_2 p_2(s))\,\mathrm{d}s \\
&\geq x_1(t) - \frac{x_1(t)}{\tilde{\gamma}_1\|x_1\|}\int_0^1 (\delta_1 p_1(s) + \delta_2 p_2(s))\,\mathrm{d}s \\
&\geq x_1(t)\left[1 - \frac{2}{\tilde{\gamma}_1\tilde{\gamma}_2 R_1}\left(\int_0^1 s^{\beta-1}\mathrm{d}H(s)\right)^{-1}\int_0^1 (\delta_1 p_1(s) + \delta_2 p_2(s))\,\mathrm{d}s\right] \\
&\geq \tilde{\gamma}_1 t^{\alpha-1}\|x_1\|\left[1 - \frac{2}{\tilde{\gamma}_1\tilde{\gamma}_2 R_1}\left(\int_0^1 s^{\beta-1}\mathrm{d}H(s)\right)^{-1}\int_0^1 (\delta_1 p_1(s) + \delta_2 p_2(s))\,\mathrm{d}s\right]
\end{aligned}$$

$$
\begin{aligned}
&\geq \frac{\tilde{\gamma}_1\tilde{\gamma}_2 R_1}{2} t^{\alpha-1} \int_0^1 s^{\beta-1} \mathrm{d}H(s) \left[1 - \frac{2}{\tilde{\gamma}_1\tilde{\gamma}_2 R_1} \left(\int_0^1 s^{\beta-1} \mathrm{d}H(s) \right)^{-1} \right. \\
&\quad \left. \times \int_0^1 (\delta_1 p_1(s) + \delta_2 p_2(s))\, \mathrm{d}s \right] \\
&= \Lambda_4 t^{\alpha-1}, \quad \forall t \in [0,1],
\end{aligned}
$$

where $\Lambda_4 = \frac{\tilde{\gamma}_1\tilde{\gamma}_2 R_1}{2} \int_0^1 s^{\beta-1} \mathrm{d}H(s) - \int_0^1 (\delta_1 p_1(s) + \delta_2 p_2(s))\, \mathrm{d}s > 0$.

Hence, $x_1(t) \geq q_1(t) + \Lambda_4 t^{\alpha-1}$ for all $t \in [0,1]$.

Let $u_1(t) = x_1(t) - q_1(t)$ and $v_1(t) = y_1(t) - q_2(t)$ for all $t \in [0,1]$. Then (u_1, v_1) is a positive solution of $(\tilde{\mathrm{S}})$–$(\widetilde{\mathrm{BC}})$ with $u_1(t) \geq \Lambda_5 t^{\alpha-1}$ and $v_1(t) \geq \Lambda_6 t^{\beta-1}$ for all $t \in [0,1]$, where $\Lambda_5 = \min\{\Lambda_1, \Lambda_4\}$ and $\Lambda_6 = \min\{\Lambda_2, \Lambda_3\}$. This completes the proof of Theorem 5.4.2. □

Theorem 5.4.3. *Assume that* $(\widetilde{\mathrm{H1}})$, $(\widetilde{\mathrm{H3}})$, $(\widetilde{\mathrm{H5}})$, *and the following assumption* $(\widetilde{\mathrm{H4'}})$ *hold:*

$(\widetilde{\mathrm{H4'}})$ *The functions* $f, g \in C([0,1] \times [0,\infty) \times [0,\infty), (-\infty, +\infty))$, *and there exist functions* $p_1, p_2, \alpha_1, \alpha_2 \in C([0,1], [0,\infty))$, $\beta_1, \beta_2 \in C([0,1] \times [0,\infty) \times [0,\infty), [0,\infty))$ *such that*

$$
-p_1(t) \leq f(t,u,v) \leq \alpha_1(t)\beta_1(t,u,v), \quad -p_2(t) \leq g(t,u,v) \leq \alpha_2(t)\beta_2(t,u,v),
$$

for all $t \in [0,1]$ *and* $u, v \in [0,\infty)$, *with* $\int_0^1 p_i(s)\, \mathrm{d}s > 0$, $i = 1,2$.

Then the boundary value problem $(\tilde{\mathrm{S}})$–$(\widetilde{\mathrm{BC}})$ *has at least two positive solutions for* $\lambda > 0$ *and* $\mu > 0$ *sufficiently small.*

Proof. Because assumption $(\widetilde{\mathrm{H4'}})$ implies assumptions $(\widetilde{\mathrm{H2}})$ and $(\widetilde{\mathrm{H4}})$, we can apply Theorems 5.4.1 and 5.4.2. Therefore, we deduce that for $0 < \lambda \leq \min\{\lambda_0, \lambda^*\}$ and $0 < \mu \leq \min\{\mu_0, \mu^*\}$, problem $(\tilde{\mathrm{S}})$–$(\widetilde{\mathrm{BC}})$ has at least two positive solutions (u_0, v_0) and (u_1, v_1) with $\|(u_0 + q_1, v_0 + q_2)\|_Y \leq 1$ and $\|(u_1 + q_1, v_1 + q_2)\|_Y > 1$, where (q_1, q_2) is the solution of problem (5.45)–(5.46). □

Theorem 5.4.4. *Assume that* $\lambda = \mu$, *and that* $(\widetilde{\mathrm{H1}})$, $(\widetilde{\mathrm{H4}})$, *and* $(\widetilde{\mathrm{H6}})$ *hold. In addition, if the following assumption* $(\widetilde{\mathrm{H7}})$ *holds, then there exists* $\lambda_* > 0$ *such that for any* $\lambda \geq \lambda_*$, *problem* $(\tilde{\mathrm{S}})$–$(\widetilde{\mathrm{BC}})$ *(with* $\lambda = \mu$*) has at least one positive solution:*

$(\widetilde{\mathrm{H7}})$ *There exists* $c \in (0, 1/2)$ *such that*

$$
f_\infty^{\mathrm{i}} = \liminf_{\substack{u+v\to\infty \\ u,v\geq 0}} \min_{t\in[c,1-c]} f(t,u,v) > L_0 \quad or \quad g_\infty^{\mathrm{i}} = \liminf_{\substack{u+v\to\infty \\ u,v\geq 0}} \min_{t\in[c,1-c]} g(t,u,v) > L_0,
$$

where

$$
\begin{aligned}
L_0 = \max\Bigg\{ &\frac{4}{\tilde{\gamma}_1} \int_0^1 (\delta_1 p_1(s) + \delta_2 p_2(s))\, \mathrm{d}s, \frac{4}{\tilde{\gamma}_2} \int_0^1 (\delta_3 p_2(s) + \delta_4 p_1(s))\, \mathrm{d}s, \\
&\frac{4}{\tilde{\gamma}_1\tilde{\gamma}_2} \left(\int_0^1 s^{\alpha-1} \mathrm{d}K(s) \right)^{-1} \int_0^1 (\delta_3 p_2(s) + \delta_4 p_1(s))\, \mathrm{d}s,
\end{aligned}
$$

$$\left.\frac{4}{\tilde{\gamma}_1\tilde{\gamma}_2}\left(\int_0^1 s^{\beta-1}\mathrm{d}H(s)\right)^{-1}\int_0^1(\delta_1p_1(s)+\delta_2p_2(s))\,\mathrm{d}s\right\}$$
$$\times\left(\min\left\{c^{\alpha-1}\varrho_1\int_c^{1-c}h_1(s)\,\mathrm{d}s,\,c^{\alpha-1}\varrho_2\int_c^{1-c}h_2(s)\,\mathrm{d}s\right\}\right)^{-1}.$$

Proof. By $(\widetilde{\mathrm{H}7})$ we conclude that there exists $M_3>0$ such that

$$f(t,u,v)\geq L_0 \quad\text{or}\quad g(t,u,v)\geq L_0,\quad \forall t\in[c,1-c],\quad u,v\in[0,\infty),\quad u+v\geq M_3.$$

We define

$$\lambda_*=\max\left\{\frac{M_3}{c^{\alpha-1}}\left(\int_0^1(\delta_1p_1(s)+\delta_2p_2(s))\,\mathrm{d}s\right)^{-1},\frac{M_3}{c^{\beta-1}}\left(\int_0^1(\delta_3p_2(s)+\delta_4p_1(s))\,\mathrm{d}s\right)^{-1}\right\}.$$

We assume now $\lambda\geq\lambda_*$. Let

$$R_3=\max\left\{\frac{4\lambda}{\tilde{\gamma}_1}\int_0^1(\delta_1p_1(s)+\delta_2p_2(s))\,\mathrm{d}s,\frac{4\lambda}{\tilde{\gamma}_2}\int_0^1(\delta_3p_2(s)+\delta_4p_1(s))\,\mathrm{d}s,\right.$$
$$\frac{4\lambda}{\tilde{\gamma}_1\tilde{\gamma}_2}\left(\int_0^1 s^{\alpha-1}\mathrm{d}K(s)\right)^{-1}\int_0^1(\delta_3p_2(s)+\delta_4p_1(s))\,\mathrm{d}s,$$
$$\left.\frac{4\lambda}{\tilde{\gamma}_1\tilde{\gamma}_2}\left(\int_0^1 s^{\beta-1}\mathrm{d}H(s)\right)^{-1}\int_0^1(\delta_1p_1(s)+\delta_2p_2(s))\,\mathrm{d}s\right\},$$

and $\Omega_3=\{(x,y)\in P,\ \|(x,y)\|_Y<R_3\}$.

We suppose first that $f_\infty^{\mathrm{i}}>L_0$—that is, $f(t,u,v)\geq L_0$ for all $t\in[c,1-c]$ and $u,v\in[0,\infty)$, $u+v\geq M_3$. Let $(x,y)\in P\cap\partial\Omega_3$. Then $\|(x,y)\|_Y=R_3$, so $\|x\|\geq R_3/2$ or $\|y\|\geq R_3/2$. We assume that $\|x\|\geq R_3/2$. Then by using Lemma 5.4.4, for all $t\in[0,1]$ we deduce

$$\begin{aligned}x(t)-q_1(t)&\geq\tilde{\gamma}_1t^{\alpha-1}\|x\|-\lambda t^{\alpha-1}\delta_1\int_0^1 p_1(s)\,\mathrm{d}s-\lambda t^{\alpha-1}\delta_2\int_0^1 p_2(s)\,\mathrm{d}s\\
&\geq t^{\alpha-1}\left[\frac{\tilde{\gamma}_1R_3}{2}-\lambda\int_0^1(\delta_1p_1(s)+\delta_2p_2(s))\,\mathrm{d}s\right]\\
&\geq t^{\alpha-1}\left[2\lambda\int_0^1(\delta_1p_1(s)+\delta_2p_2(s))\,\mathrm{d}s-\lambda\int_0^1(\delta_1p_1(s)+\delta_2p_2(s))\,\mathrm{d}s\right]\\
&=t^{\alpha-1}\lambda\int_0^1(\delta_1p_1(s)+\delta_2p_2(s))\,\mathrm{d}s\geq t^{\alpha-1}\lambda_*\int_0^1(\delta_1p_1(s)+\delta_2p_2(s))\,\mathrm{d}s\\
&\geq\frac{M_3}{c^{\alpha-1}}t^{\alpha-1}\geq 0.\end{aligned}$$

Therefore, for any $(x,y)\in P\cap\partial\Omega_3$ and $t\in[c,1-c]$, we have

$$[x(t)-q_1(t)]^*+[y(t)-q_2(t)]^*\geq[x(t)-q_1(t)]^*=x(t)-q_1(t)\geq\frac{M_3}{c^{\alpha-1}}t^{\alpha-1}\geq M_3. \tag{5.53}$$

Hence, by Lemma 5.4.4, for any $(x, y) \in P \cap \partial\Omega_3$ and $t \in [c, 1-c]$ we conclude

$$\begin{aligned}\tilde{T}_1(x, y)(t) &\geq \lambda \int_0^1 G_1(t, s)[f(s, [x(s) - q_1(s)]^*, [y(s) - q_2(s)]^*) + p_1(s)]\,ds \\ &\geq \lambda \varrho_1 t^{\alpha-1} \int_c^{1-c} h_1(s) f(s, [x(s) - q_1(s)]^*, [y(s) - q_2(s)]^*)\,ds \\ &\geq \lambda L_0 \varrho_1 t^{\alpha-1} \int_c^{1-c} h_1(s)\,ds \geq \lambda L_0 \varrho_1 c^{\alpha-1} \int_c^{1-c} h_1(s)\,ds \geq R_3 = \|(x, y)\|_Y.\end{aligned}$$

Therefore, we obtain $\|\tilde{T}_1(x, y)\| \geq R_3$ for all $(x, y) \in P \cap \partial\Omega_3$, and so

$$\|\tilde{\mathcal{T}}(x, y)\|_Y \geq R_3 = \|(x, y)\|_Y, \quad \forall (x, y) \in P \cap \partial\Omega_3. \tag{5.54}$$

If $\|y\| \geq R_3/2$, then for all $t \in [0, 1]$ we deduce

$$\begin{aligned}y(t) - q_2(t) &\geq \tilde{\gamma}_2 t^{\beta-1}\|y\| - \lambda t^{\beta-1}\delta_3 \int_0^1 p_2(s)\,ds - \lambda t^{\beta-1}\delta_4 \int_0^1 p_1(s)\,ds \\ &\geq t^{\beta-1}\left[\frac{\tilde{\gamma}_2 R_3}{2} - \lambda \int_0^1 (\delta_3 p_2(s) + \delta_4 p_1(s))\,ds\right] \\ &\geq t^{\beta-1}\left[2\lambda \int_0^1 (\delta_3 p_2(s) + \delta_4 p_1(s))\,ds - \lambda \int_0^1 (\delta_3 p_2(s) + \delta_4 p_1(s))\,ds\right] \\ &= t^{\beta-1}\lambda \int_0^1 (\delta_3 p_2(s) + \delta_4 p_1(s))\,ds \geq t^{\beta-1}\lambda_* \int_0^1 (\delta_3 p_2(s) + \delta_4 p_1(s))\,ds \\ &\geq \frac{M_3}{c^{\beta-1}} t^{\beta-1} \geq 0.\end{aligned}$$

Therefore, for any $(x, y) \in P \cap \partial\Omega_3$ and $t \in [c, 1-c]$ we have

$$[x(t)-q_1(t)]^* + [y(t)-q_2(t)]^* \geq [y(t)-q_2(t)]^* = y(t)-q_2(t) \geq \frac{M_3}{c^{\beta-1}} t^{\beta-1} \geq M_3. \tag{5.55}$$

Hence, for any $(x, y) \in P \cap \partial\Omega_3$ and $t \in [c, 1-c]$, we obtain in a manner similar to that above $\tilde{T}_1(x, y)(t) \geq R_3 = \|(x, y)\|_Y$. Hence, $\|\tilde{T}_1(x, y)\| \geq R_3$ for all $(x, y) \in P \cap \partial\Omega_3$, and we deduce again relation (5.54).

We suppose now that $g_\infty^{\mathrm{i}} > L_0$—that is, $g(t, u, v) \geq L_0$ for all $t \in [c, 1-c]$ and $u, v \in [0, \infty)$, $u + v \geq M_3$. Let $(x, y) \in P \cap \partial\Omega_3$. Then $\|(x, y)\|_Y = R_3$, so $\|x\| \geq R_3/2$ or $\|y\| \geq R_3/2$. If $\|x\| \geq R_3/2$, then we obtain in a manner similar to that in the first case above ($f_\infty^{\mathrm{i}} > L_0$) $x(t) - q_1(t) \geq \frac{M_3}{c^{\alpha-1}} t^{\alpha-1} \geq 0$ for all $t \in [0, 1]$.

Therefore, for any $(x, y) \in P \cap \partial\Omega_3$ and $t \in [c, 1-c]$, we deduce inequalities (5.53).

Hence, by Lemma 5.4.4, for any $(x, y) \in P \cap \partial\Omega_3$ and $t \in [c, 1-c]$ we conclude

$$\begin{aligned}\tilde{T}_1(x, y)(t) &\geq \lambda \int_0^1 G_2(t, s)[g(s, [x(s) - q_1(s)]^*, [y(s) - q_2(s)]^*) + p_2(s)]\,ds \\ &\geq \lambda \varrho_2 t^{\alpha-1} \int_c^{1-c} h_2(s) g(s, [x(s) - q_1(s)]^*, [y(s) - q_2(s)]^*)\,ds\end{aligned}$$

$$\geq \lambda L_0 \varrho_2 t^{\alpha-1} \int_c^{1-c} h_2(s)\,\mathrm{d}s \geq \lambda L_0 \varrho_2 c^{\alpha-1} \int_c^{1-c} h_2(s)\,\mathrm{d}s \geq R_3$$
$$= \|(x,y)\|_Y.$$

Therefore, we obtain $\|\tilde{T}_1(x,y)\| \geq R_3$, and so $\|\tilde{\mathcal{T}}(x,y)\|_Y \geq R_3 = \|(x,y)\|_Y$ for all $(x,y) \in P \cap \partial\Omega_3$—that is, we have relation (5.54).

If $\|y\| \geq R_3/2$, then we conclude in a manner similar to that in the first case above ($f_\infty^{\mathrm{i}} > L_0$) that $y(t) - q_2(t) \geq \frac{M_3}{c^{\beta-1}} t^{\beta-1} \geq 0$ for all $t \in [0,1]$.

Therefore, for any $(x,y) \in P \cap \partial\Omega_3$ and $t \in [c, 1-c]$, we deduce inequalities (5.55).

Hence, for any $(x,y) \in P \cap \partial\Omega_3$ and $t \in [c, 1-c]$ we obtain in a manner similar to that above $\tilde{T}_1(x,y)(t) \geq R_3 = \|(x,y)\|_Y$. Hence, $\|\tilde{T}_1(x,y)\| \geq R_3$ and $\|\tilde{\mathcal{T}}(x,y)\|_Y \geq \|(x,y)\|_Y$ for all $(x,y) \in P \cap \partial\Omega_3$—that is, we have relation (5.54).

On the other hand, we consider the positive number

$$\varepsilon = \min\left\{ \frac{1}{8\lambda\sigma_1}\left(\int_0^1 h_1(s)\alpha_1(s)\,\mathrm{d}s\right)^{-1}, \frac{1}{8\lambda\sigma_2}\left(\int_0^1 h_2(s)\alpha_2(s)\,\mathrm{d}s\right)^{-1}, \right.$$
$$\left. \frac{1}{8\lambda\sigma_3}\left(\int_0^1 h_2(s)\alpha_2(s)\,\mathrm{d}s\right)^{-1}, \frac{1}{8\lambda\sigma_4}\left(\int_0^1 h_1(s)\alpha_1(s)\,\mathrm{d}s\right)^{-1} \right\}.$$

Then, by ($\widetilde{\mathrm{H}6}$), we deduce that there exists $M_4 > 0$ such that

$$\beta_i(t,u,v) \leq \varepsilon(u+v), \quad \forall t \in [0,1], \quad u, v \in [0,\infty), \quad u+v \geq M_4, \quad i = 1,2.$$

Therefore, we obtain

$$\beta_i(t,u,v) \leq M_5 + \varepsilon(u+v), \quad \forall t \in [0,1], \quad u, v \in [0,\infty), \quad i = 1,2,$$

where $M_5 = \max_{i=1,2}\left\{ \max_{t\in[0,1],\, u,v\geq 0,\, u+v\leq M_4} \beta_i(t,u,v) \right\}$.

We define now

$$R_4 = \max\left\{ 2R_3, 8\lambda\sigma_1 \max\{M_5, 1\} \int_0^1 h_1(s)(\alpha_1(s) + p_1(s))\,\mathrm{d}s, \right.$$
$$8\lambda\sigma_2 \max\{M_5, 1\} \int_0^1 h_2(s)(\alpha_2(s) + p_2(s))\,\mathrm{d}s,$$
$$8\lambda\sigma_3 \max\{M_5, 1\} \int_0^1 h_2(s)(\alpha_2(s) + p_2(s))\,\mathrm{d}s,$$
$$\left. 8\lambda\sigma_4 \max\{M_5, 1\} \int_0^1 h_1(s)(\alpha_1(s) + p_1(s))\,\mathrm{d}s \right\},$$

and let $\Omega_4 = \{(x,y) \in P, \|(x,y)\|_Y < R_4\}$.

By using Lemma 5.4.4, for any $(x,y) \in P \cap \partial\Omega_4$, we have

$$\tilde{T}_1(x,y)(t) \leq \lambda \int_0^1 \sigma_1 h_1(s)[\alpha_1(s)\beta_1(s, [x(s) - q_1(s)]^*, [y(s) - q_2(s)]^*) + p_1(s)]\,\mathrm{d}s$$
$$+ \lambda \int_0^1 \sigma_2 h_2(s)[\alpha_2(s)\beta_2(s, [x(s) - q_1(s)]^*, [y(s) - q_2(s)]^*) + p_2(s)]\,\mathrm{d}s$$

$$\leq \lambda\sigma_1 \int_0^1 h_1(s)[\alpha_1(s)(M_5 + \varepsilon([x(s) - q_1(s)]^* + [y(s) - q_2(s)]^*) + p_1(s)]\,ds$$

$$+ \lambda\sigma_2 \int_0^1 h_2(s)[\alpha_2(s)(M_5 + \varepsilon([x(s) - q_1(s)]^* + [y(s) - q_2(s)]^*) + p_2(s)]\,ds$$

$$\leq \lambda\sigma_1 \max\{M_5, 1\} \int_0^1 h_1(s)(\alpha_1(s) + p_1(s))\,ds + \lambda\sigma_1\varepsilon R_4 \int_0^1 h_1(s)\alpha_1(s)\,ds$$

$$+ \lambda\sigma_2 \max\{M_5, 1\} \int_0^1 h_2(s)(\alpha_2(s) + p_2(s))\,ds + \lambda\sigma_2\varepsilon R_4 \int_0^1 h_2(s)\alpha_2(s)\,ds$$

$$\leq \frac{R_4}{8} + \frac{R_4}{8} + \frac{R_4}{8} + \frac{R_4}{8} = \frac{R_4}{2} = \frac{\|(x,y)\|_Y}{2}, \quad \forall t \in [0,1],$$

and so $\|\tilde{T}_1(x,y)\| \leq \frac{\|(x,y)\|_Y}{2}$ for all $(x,y) \in P \cap \partial\Omega_4$.

In a similar manner, we obtain

$$\tilde{T}_2(x,y)(t) \leq \lambda \int_0^1 \sigma_3 h_2(s)[\alpha_2(s)\beta_2(s, [x(s) - q_1(s)]^*, [y(s) - q_2(s)]^*) + p_2(s)]\,ds$$

$$+ \lambda \int_0^1 \sigma_4 h_1(s)[\alpha_1(s)\beta_1(s, [x(s) - q_1(s)]^*, [y(s) - q_2(s)]^*) + p_1(s)]\,ds$$

$$\leq \lambda\sigma_3 \int_0^1 h_2(s)[\alpha_2(s)(M_5 + \varepsilon([x(s) - q_1(s)]^* + [y(s) - q_2(s)]^*) + p_2(s)]\,ds$$

$$+ \lambda\sigma_4 \int_0^1 h_1(s)[\alpha_1(s)(M_5 + \varepsilon([x(s) - q_1(s)]^* + [y(s) - q_2(s)]^*) + p_1(s)]\,ds$$

$$\leq \lambda\sigma_3 \max\{M_5, 1\} \int_0^1 h_2(s)(\alpha_2(s) + p_2(s))\,ds + \lambda\sigma_3\varepsilon R_4 \int_0^1 h_2(s)\alpha_2(s)\,ds$$

$$+ \lambda\sigma_4 \max\{M_5, 1\} \int_0^1 h_1(s)(\alpha_1(s) + p_1(s))\,ds + \lambda\sigma_4\varepsilon R_4 \int_0^1 h_1(s)\alpha_1(s)\,ds$$

$$\leq \frac{R_4}{8} + \frac{R_4}{8} + \frac{R_4}{8} + \frac{R_4}{8} = \frac{R_4}{2} = \frac{\|(x,y)\|_Y}{2}, \quad \forall t \in [0,1],$$

and so $\|\tilde{T}_2(x,y)\| \leq \frac{\|(x,y)\|_Y}{2}$ for all $(x,y) \in P \cap \partial\Omega_4$.

Therefore, we deduce

$$\|\tilde{\mathcal{T}}(x,y)\|_Y \leq \|(x,y)\|_Y, \quad \forall (x,y) \in P \cap \partial\Omega_4. \tag{5.56}$$

Hence by Lemma 5.4.6, Theorem 1.1.1, and relations (5.54) and (5.56), we conclude that $\tilde{\mathcal{T}}$ has a fixed point $(x_1, y_1) \in P \cap (\bar{\Omega}_4 \setminus \Omega_3)$. Since $\|(x_1, y_1)\| \geq R_3$, then $\|x_1\| \geq R_3/2$ or $\|y_1\| \geq R_3/2$.

We suppose that $\|x_1\| \geq R_3/2$. Then $x_1(t) - q_1(t) \geq \frac{M_3}{c^{\alpha-1}} t^{\alpha-1}$ for all $t \in [0,1]$. Besides,

$$y_1(1) = \int_0^1 x_1(s)\,dK(s) \geq \tilde{\gamma}_1 \|x_1\| \int_0^1 s^{\alpha-1}\,dK(s) \geq \frac{\tilde{\gamma}_1 R_3}{2} \int_0^1 s^{\alpha-1}\,dK(s) > 0,$$

and then

$$\|y_1\| \geq y_1(1) = \int_0^1 x_1(s)\,dK(s) \geq \frac{\tilde{\gamma}_1 R_3}{2} \int_0^1 s^{\alpha-1}\,dK(s) > 0.$$

Therefore, we deduce that for all $t \in [0,1]$

$$\begin{aligned}
y_1(t) - q_2(t) &= y_1(t) - \lambda \int_0^1 G_3(t,s)p_2(s)\,\mathrm{d}s - \lambda \int_0^1 G_4(t,s)p_1(s)\,\mathrm{d}s \\
&\geq y_1(t) - \lambda\delta_3 \int_0^1 t^{\beta-1}p_2(s)\,\mathrm{d}s - \lambda\delta_4 \int_0^1 t^{\beta-1}p_1(s)\,\mathrm{d}s \\
&\geq \tilde{\gamma}_2 t^{\beta-1}\|y_1\| - \lambda t^{\beta-1} \int_0^1 (\delta_3 p_2(s) + \delta_4 p_1(s))\,\mathrm{d}s \\
&\geq \frac{\tilde{\gamma}_1\tilde{\gamma}_2 R_3}{2} t^{\beta-1} \int_0^1 s^{\alpha-1}\,\mathrm{d}K(s) - \lambda t^{\beta-1} \int_0^1 (\delta_3 p_2(s) + \delta_4 p_1(s))\,\mathrm{d}s \\
&\geq \lambda t^{\beta-1} \int_0^1 (\delta_3 p_2(s) + \delta_4 p_1(s))\,\mathrm{d}s \\
&\geq \lambda_* t^{\beta-1} \int_0^1 (\delta_3 p_2(s) + \delta_4 p_1(s))\,\mathrm{d}s \geq \frac{M_3}{c^{\beta-1}} t^{\beta-1}.
\end{aligned}$$

If $\|y_1\| \geq R_3/2$, then $y_1(t) - q_2(t) \geq \frac{M_3}{c^{\beta-1}} t^{\beta-1}$ for all $t \in [0,1]$. Besides,

$$x_1(1) = \int_0^1 y_1(s)\,\mathrm{d}H(s) \geq \tilde{\gamma}_2\|y_1\| \int_0^1 s^{\beta-1}\,\mathrm{d}H(s) \geq \frac{\tilde{\gamma}_2 R_3}{2} \int_0^1 s^{\beta-1}\,\mathrm{d}H(s) > 0,$$

and then

$$\|x_1\| \geq x_1(1) = \int_0^1 y_1(s)\,\mathrm{d}H(s) \geq \frac{\tilde{\gamma}_2 R_3}{2} \int_0^1 s^{\beta-1}\,\mathrm{d}H(s) > 0.$$

Therefore, we conclude that for all $t \in [0,1]$

$$\begin{aligned}
x_1(t) - q_1(t) &= x_1(t) - \lambda \int_0^1 G_1(t,s)p_1(s)\,\mathrm{d}s - \lambda \int_0^1 G_2(t,s)p_2(s)\,\mathrm{d}s \\
&\geq x_1(t) - \lambda\delta_1 \int_0^1 t^{\alpha-1}p_1(s)\,\mathrm{d}s - \lambda\delta_2 \int_0^1 t^{\alpha-1}p_2(s)\,\mathrm{d}s \\
&\geq \tilde{\gamma}_1 t^{\alpha-1}\|x_1\| - \lambda t^{\alpha-1} \int_0^1 (\delta_1 p_1(s) + \delta_2 p_2(s))\,\mathrm{d}s \\
&\geq \frac{\tilde{\gamma}_1\tilde{\gamma}_2 R_3}{2} t^{\alpha-1} \int_0^1 s^{\beta-1}\,\mathrm{d}H(s) - \lambda t^{\alpha-1} \int_0^1 (\delta_1 p_1(s) + \delta_2 p_2(s))\,\mathrm{d}s \\
&\geq \lambda t^{\alpha-1} \int_0^1 (\delta_1 p_1(s) + \delta_2 p_2(s))\,\mathrm{d}s \\
&\geq \lambda_* t^{\alpha-1} \int_0^1 (\delta_1 p_1(s) + \delta_2 p_2(s))\,\mathrm{d}s \geq \frac{M_3}{c^{\alpha-1}} t^{\alpha-1}.
\end{aligned}$$

Let $u_1(t) = x_1(t) - q_1(t)$ and $v_1(t) = y_1(t) - q_2(t)$ for all $t \in [0,1]$. Then $u_1(t) \geq \tilde{\Lambda}_1 t^{\alpha-1}$ and $v_1(t) \geq \tilde{\Lambda}_2 t^{\beta-1}$ for all $t \in [0,1]$, where $\tilde{\Lambda}_1 = \frac{M_3}{c^{\alpha-1}}$ and $\tilde{\Lambda}_2 = \frac{M_3}{c^{\beta-1}}$. Hence, we deduce that (u_1, v_1) is a positive solution of $(\tilde{\mathrm{S}})$–$(\widetilde{\mathrm{BC}})$, which completes the proof of Theorem 5.4.4. □

In a manner similar to that in which we proved Theorem 5.4.4, we obtain the following theorems:

Theorem 5.4.5. *Assume that* $\lambda = \mu$, *and that* $(\widetilde{H1})$, $(\widetilde{H4})$, *and* $(\widetilde{H6})$ *hold. In addition, if the following assumption* $(\widetilde{H7'})$ *holds, then there exists* $\lambda'_* > 0$ *such that for any* $\lambda \geq \lambda'_*$, *problem* $(\tilde{S})$–$(\widetilde{BC})$ *(with* $\lambda = \mu$*) has at least one positive solution:*

$(\widetilde{H7'})$ *There exists* $c \in (0, 1/2)$ *such that*

$$f_\infty^{\rm i} = \liminf_{\substack{u+v\to\infty\\ u,v\geq 0}} \min_{t\in[c,1-c]} f(t,u,v) > \tilde{L}_0 \quad or \quad g_\infty^{\rm i} = \liminf_{\substack{u+v\to\infty\\ u,v\geq 0}} \min_{t\in[c,1-c]} g(t,u,v) > \tilde{L}_0,$$

where

$$\tilde{L}_0 = \max\left\{\frac{4}{\tilde{\gamma}_1}\int_0^1 (\delta_1 p_1(s) + \delta_2 p_2(s))\,{\rm d}s, \frac{4}{\tilde{\gamma}_2}\int_0^1 (\delta_3 p_2(s) + \delta_4 p_1(s))\,{\rm d}s,\right.$$
$$\frac{4}{\tilde{\gamma}_1\tilde{\gamma}_2}\left(\int_0^1 s^{\alpha-1}{\rm d}K(s)\right)^{-1}\int_0^1 (\delta_3 p_2(s) + \delta_4 p_1(s))\,{\rm d}s,$$
$$\left.\frac{4}{\tilde{\gamma}_1\tilde{\gamma}_2}\left(\int_0^1 s^{\beta-1}{\rm d}H(s)\right)^{-1}\int_0^1 (\delta_1 p_1(s) + \delta_2 p_2(s))\,{\rm d}s\right\}$$
$$\times\left(\min\left\{c^{\beta-1}\varrho_3\int_c^{1-c} h_2(s)\,{\rm d}s, c^{\beta-1}\varrho_4\int_c^{1-c} h_1(s)\,{\rm d}s\right\}\right)^{-1}.$$

Theorem 5.4.6. *Assume that* $\lambda = \mu$, *and that* $(\widetilde{H1})$, $(\widetilde{H4})$, *and* $(\widetilde{H6})$ *hold. In addition, if the following assumption* $(\widetilde{H8})$ *holds, then there exists* $\tilde{\lambda}_* > 0$ *such that for any* $\lambda \geq \tilde{\lambda}_*$, *problem* $(\tilde{S})$–$(\widetilde{BC})$ *(with* $\lambda = \mu$*) has at least one positive solution:*

$(\widetilde{H8})$ *There exists* $c \in (0, 1/2)$ *such that*

$$\hat{f}_\infty = \lim_{\substack{u+v\to\infty\\ u,v\geq 0}} \min_{t\in[c,1-c]} f(t,u,v) = \infty \quad or \quad \hat{g}_\infty = \lim_{\substack{u+v\to\infty\\ u,v\geq 0}} \min_{t\in[c,1-c]} g(t,u,v) = \infty.$$

5.4.4 Examples

Let $\alpha = 5/2$ ($n = 3$), $\beta = 7/3$ ($m = 3$), $H(t) = t^2$, and $K(t) = t^3$. Then $\int_0^1 u(s)\,{\rm d}K(s) = 3\int_0^1 s^2 u(s)\,{\rm d}s$ and $\int_0^1 v(s)\,{\rm d}H(s) = 2\int_0^1 sv(s)\,{\rm d}s$.

We consider the system of fractional differential equations

$$\begin{cases} D_{0+}^{5/2}u(t) + \lambda f(t,u(t),v(t)) = 0, & t\in(0,1),\\ D_{0+}^{7/3}v(t) + \mu g(t,u(t),v(t)) = 0, & t\in(0,1), \end{cases} \qquad (\widetilde{S_0})$$

with the boundary conditions

$$\begin{cases} u(0) = u'(0) = 0, & u(1) = 2\displaystyle\int_0^1 sv(s)\,{\rm d}s,\\ v(0) = v'(0) = 0, & v(1) = 3\displaystyle\int_0^1 s^2u(s)\,{\rm d}s. \end{cases} \qquad (\widetilde{BC_0})$$

Then, we obtain $\Delta = 1-\left(\int_0^1 s^{\alpha-1}\mathrm{d}K(s)\right)\left(\int_0^1 s^{\beta-1}\mathrm{d}H(s)\right) = \frac{3}{5} > 0$, $\int_0^1 \tau^{\alpha-1}(1-\tau)\,\mathrm{d}K(\tau) = \frac{4}{33} > 0$, and $\int_0^1 \tau^{\beta-1}(1-\tau)\,\mathrm{d}H(\tau) = \frac{9}{65} > 0$. The functions H and K are nondecreasing, and so assumption $(\widetilde{\mathrm{H}1})$ is satisfied. Besides, we deduce

$$g_1(t,s) = \frac{4}{3\sqrt{\pi}}\begin{cases} t^{3/2}(1-s)^{3/2} - (t-s)^{3/2}, & 0 \le s \le t \le 1, \\ t^{3/2}(1-s)^{3/2}, & 0 \le t \le s \le 1, \end{cases}$$

$$g_2(t,s) = \frac{1}{\Gamma(7/3)}\begin{cases} t^{4/3}(1-s)^{4/3} - (t-s)^{4/3}, & 0 \le s \le t \le 1, \\ t^{4/3}(1-s)^{4/3}, & 0 \le t \le s \le 1, \end{cases}$$

$$G_1(t,s) = g_1(t,s)+3t^{3/2}\int_0^1 \tau^2 g_1(\tau,s)\,\mathrm{d}\tau, \quad G_2(t,s) = \frac{10}{3}t^{3/2}\int_0^1 \tau g_2(\tau,s)\,\mathrm{d}\tau,$$

$$G_3(t,s) = g_2(t,s)+\frac{20}{9}t^{4/3}\int_0^1 \tau g_2(\tau,s)\,\mathrm{d}\tau, \quad G_4(t,s) = 5t^{4/3}\int_0^1 \tau^2 g_1(\tau,s)\,\mathrm{d}\tau.$$

We also obtain $h_1(s) = \frac{2}{\sqrt{\pi}}s(1-s)^{3/2}$, $h_2(s) = \frac{1}{\Gamma(4/3)}s(1-s)^{4/3}$,

$$k_1(t) = \begin{cases} \frac{2}{3}t^{3/2}, & 0 \le t \le 1/2, \\ \frac{2}{3}(1-t)t^{1/2}, & 1/2 \le t \le 1, \end{cases} \qquad k_2(t) = \begin{cases} \frac{3}{4}t^{4/3}, & 0 \le t \le 1/2, \\ \frac{3}{4}(1-t)t^{1/3}, & 1/2 \le t \le 1. \end{cases}$$

In addition, we have $\sigma_1 = 2$, $\delta_1 = \frac{74}{33\sqrt{\pi}}$, $\varrho_1 = \frac{8\sqrt{2}-1}{63\sqrt{2}}$, $\sigma_2 = \frac{5}{3}$, $\delta_2 = \frac{3}{13\Gamma(4/3)}$, $\varrho_2 = \frac{36\sqrt[3]{2}-9}{112\sqrt[3]{2}}$, $\sigma_3 = \frac{19}{9}$, $\delta_3 = \frac{15}{13\Gamma(4/3)}$, $\varrho_3 = \frac{12\sqrt[3]{2}-3}{56\sqrt[3]{2}}$, $\sigma_4 = \frac{5}{3}$, $\delta_4 = \frac{40}{99\sqrt{\pi}}$, $\varrho_4 = \frac{40\sqrt{2}-5}{189\sqrt{2}}$, $\tilde{\gamma}_1 = \frac{8\sqrt{2}-1}{126\sqrt{2}} \approx 0.0578801$, and $\tilde{\gamma}_2 = \frac{9(12\sqrt[3]{2}-3)}{1064\sqrt[3]{2}} \approx 0.08136286$.

Example 5.4.1. We consider the functions

$$f(t,u,v) = \frac{(u+v)^2}{\sqrt{t(1-t)}} + \ln t, \quad g(t,u,v) = \frac{2+\sin(u+v)}{\sqrt{t(1-t)}} + \ln(1-t),$$

for all $t \in (0,1)$ and $u, v \in [0,\infty)$. We have $p_1(t) = -\ln t$, $p_2(t) = -\ln(1-t)$, and $\alpha_1(t) = \alpha_2(t) = \frac{1}{\sqrt{t(1-t)}}$ for all $t \in (0,1)$, $\beta_1(t,u,v) = (u+v)^2$ and $\beta_2(t,u,v) = 2+\sin(u+v)$ for all $t \in [0,1]$ and $u, v \in [0,\infty)$, $\int_0^1 p_1(t)\,\mathrm{d}t = 1$, $\int_0^1 p_2(t)\,\mathrm{d}t = 1$, and $\int_0^1 \alpha_i(t)\,\mathrm{d}t = \pi$, $i = 1,2$. Therefore, assumption $(\widetilde{\mathrm{H}4})$ is satisfied. In addition, for $c \in (0,1/2)$ fixed, assumption $(\widetilde{\mathrm{H}5})$ is also satisfied ($f_\infty = \infty$).

After some computations, we deduce $\int_0^1 (\delta_1 p_1(s) + \delta_2 p_2(s)\,\mathrm{d}s \approx 1.52357852$, $\int_0^1 (\delta_3 p_2(s) + \delta_4 p_1(s))\,\mathrm{d}s \approx 1.520086$, $\int_0^1 h_1(s)(\alpha_1(s)+p_1(s))\,\mathrm{d}s \approx 0.42548534$, and $\int_0^1 h_2(s)(\alpha_2(s)+p_2(s))\,\mathrm{d}s \approx 0.44092924$. We choose $R_1 = 1080$, which satisfies the condition from the beginning of the proof of Theorem 5.4.2. Then $M_1 = R_1^2$, $M_2 = 3$, $\lambda^* \approx 2.7202 \cdot 10^{-4}$, and $\mu^* = 1$. By Theorem 5.4.2, we conclude that $(\widetilde{\mathrm{S}_0})$–$(\widetilde{\mathrm{BC}_0})$ has at least one positive solution for any $\lambda \in (0,\lambda^*]$ and $\mu \in (0,\mu^*]$.

Example 5.4.2. We consider the functions

$$f(t,u,v) = (u+v)^2 + \cos u, \quad g(t,u,v) = (u+v)^{1/2} + \cos v,$$

for all $t \in [0,1]$ and $u,\ v \in [0,\infty)$. We have $p_1(t) = p_2(t) = 1$ for all $t \in [0,1]$, and then assumption $(\widetilde{H2})$ is satisfied. Besides, assumption $(\widetilde{H3})$ is also satisfied because $f(t,0,0) = 1$ and $g(t,0,0) = 1$ for all $t \in [0,1]$.

Let $\delta = \frac{1}{2} < 1$ and $R_0 = 1$. Then

$$f(t,u,v) \geq \delta f(t,0,0) = \frac{1}{2}, \quad g(t,u,v) \geq \delta g(t,0,0) = \frac{1}{2}, \quad \forall t \in [0,1], \quad u,\ v \in [0,1].$$

In addition,

$$\bar{f}(R_0) = \bar{f}(1) = \max_{t\in[0,1],\ u,v\in[0,1]} \{f(t,u,v) + p_1(t)\} \approx 5.5403023,$$

$$\bar{g}(R_0) = \bar{g}(1) = \max_{t\in[0,1],\ u,v\in[0,1]} \{g(t,u,v) + p_2(t)\} \approx 3.10479256.$$

We also obtain $c_1 \approx 0.25791523$, $c_2 \approx 0.23996711$, $c_3 \approx 0.30395834$, $c_4 \approx 0.21492936$, and then $\lambda_0 = \max\left\{\frac{R_0}{8c_1\bar{f}(R_0)}, \frac{R_0}{8c_4\bar{f}(R_0)}\right\} \approx 0.10497377$, and $\mu_0 = \max\left\{\frac{R_0}{8c_2\bar{g}(R_0)}, \frac{R_0}{8c_3\bar{g}(R_0)}\right\} \approx 0.1677744$.

By Theorem 5.4.1, for any $\lambda \in (0,\lambda_0]$ and $\mu \in (0,\mu_0]$, we deduce that problem $(\widetilde{S}_0)$–$(\widetilde{BC}_0)$ has at least one positive solution.

Because assumption $(\widetilde{H4'})$ is satisfied ($\alpha_1(t) = \alpha_2(t) = 1$, $\beta_1(t,u,v) = (u+v)^2+1$, $\beta_2(t,u,v) = (u+v)^{1/2}+1$ for all $t \in [0,1]$ and $u, v \in [0,\infty)$), and assumption $(\widetilde{H5})$ is also satisfied ($f_\infty = \infty$), by Theorem 5.4.3 we conclude that problem $(\widetilde{S}_0)$–$(\widetilde{BC}_0)$ has at least two positive solutions for λ and μ sufficiently small.

Example 5.4.3. We consider $\lambda = \mu$ and the functions

$$f(t,u,v) = \frac{(u+v)^a}{\sqrt[3]{t^2(1-t)}} - \frac{1}{\sqrt{t}}, \quad g(t,u,v) = \frac{\ln(1+u+v)}{\sqrt[3]{t(1-t)^2}} - \frac{1}{\sqrt{1-t}},$$

for all $t \in (0,1)$ and $u,\ v \in [0,\infty)$, where $a \in (0,1)$.

Here we have $p_1(t) = \frac{1}{\sqrt{t}}$, $p_2(t) = \frac{1}{\sqrt{1-t}}$, $\alpha_1(t) = \frac{1}{\sqrt[3]{t^2(1-t)}}$, and $\alpha_2(t) = \frac{1}{\sqrt[3]{t(1-t)^2}}$ for all $t \in (0,1)$, and $\beta_1(t,u,v) = (u+v)^a$ and $\beta_2(t,u,v) = \ln(1+u+v)$ for all $t \in [0,1]$ and $u, v \in [0,\infty)$. For $c \in (0,1/2)$ fixed, assumptions $(\widetilde{H4})$, $(\widetilde{H6})$, and $(\widetilde{H8})$ are satisfied ($\beta_{i\infty} = 0$ for $i = 1,2$, and $\hat{f}_\infty = \infty$).

Then, by Theorem 5.4.6, we deduce that there exists $\tilde{\lambda}_* > 0$ such that for any $\lambda \geq \tilde{\lambda}_*$ problem $(\widetilde{S}_0)$–$(\widetilde{BC}_0)$ (with $\lambda = \mu$) has at least one positive solution.

Remark 5.4.1. The results presented in this section were published in Henderson and Luca (submitted).

Bibliography

Agarwal, R.P., Andrade, B., Cuevas, C., 2010a. Weighted pseudo-almost periodic solutions of a class of semilinear fractional differential equations. Nonlinear Anal. Real World Appl. 11, 3532–3554.

Agarwal, R.P., Zhou, Y., He, Y., 2010b. Existence of fractional neutral functional differential equations. Comput. Math. Appl. 59, 1095–1100.

Agarwal, R.P., Meehan, M., O'Regan, D., 2001. Fixed Point Theory and Applications. Cambridge University Press, UK.

Aghajani, A., Jalilian, Y., Trujillo, J.J., 2012. On the existence of solutions of fractional integro-differential equations. Fract. Calc. Appl. Anal. 15, 44–69.

Ahmad, B., Alsaedi, A., Alghamdi, B.S., 2008. Analytic approximation of solutions of the forced Duffing equation with integral boundary conditions. Nonlinear Anal. Real World Appl. 9, 1727–1740.

Ahmad, B., Ntouyas, S.K., 2012a. Nonlinear fractional differential equations and inclusions of arbitrary order and multi-strip boundary conditions. Electron. J. Differ. Equ. 2012 (98), 1–22.

Ahmad, B., Ntouyas, S.K., 2012b. A note on fractional differential equations with fractional separated boundary conditions. Abstr. Appl. Anal., 1–11. Article ID 818703.

Amann, H., 1976. Fixed point equations and nonlinear eigenvalue problems in ordered Banach spaces. SIAM Rev. 18, 620–709.

Amann, H., 1986. Parabolic evolution equations with nonlinear boundary conditions. In: Browder, F. (Ed.), Nonlinear Functional Analysis and its Applications. Proceedings of Symposia in Pure Mathematics, vol. 45. American Mathematical Society, Providence, RI, pp. 17–27.

Amann, H., 1988. Parabolic evolution equations and nonlinear boundary conditions. J. Differ. Equ. 72, 201–269.

Aronson, D.G., 1978. A comparison method for stability analysis of nonlinear parabolic problems. SIAM Rev. 20, 245–264.

Atici, F.M., Guseinov, G.S., 2002. On Green's functions and positive solutions for boundary value problems on time scales. J. Comput. Appl. Math. 141, 75–99.

Bai, Z., 2010. On positive solutions of a nonlocal fractional boundary value problem. Nonlinear Anal. 72, 916–924.

Balachandran, K., Trujillo, J.J., 2010. The nonlocal Cauchy problem for nonlinear fractional integrodifferential equations in Banach spaces. Nonlinear Anal. 72, 4587–4593.

Baleanu, D., Diethelm, K., Scalas, E., Trujillo, J.J., 2012. Fractional calculus models and numerical methods. In: Series on Complexity, Nonlinearity and Chaos. World Scientific, Boston.

Baleanu, D., Mustafa, O.G., Agarwal, R.P., 2010. An existence result for a superlinear fractional differential equation. Appl. Math. Lett. 23, 1129–1132.

Boucherif, A., 2009. Second-order boundary value problems with integral boundary conditions. Nonlinear Anal. 70, 364–371.

Boucherif, A., Henderson, J., 2006. Positive solutions of second order boundary value problems with changing signs Caratheodory nonlinearites. Electron. J. Qual. Theory Differ. Equ. 2006 (7), 1–14.

Cac, N.P., Fink, A.M., Gatica, J.A., 1997. Nonnegative solutions of quasilinear elliptic boundary value problems with nonnegative coefficients. J. Math. Anal. Appl. 206, 1–9.

Cannon, J.R., 1964. The solution of the heat equation subject to the specification of energy. Q. Appl. Math. 22, 155–160.

Chegis, R.Y., 1984. Numerical solution of the heat conduction problem with an integral condition. Litovsk. Mat. Sb. 24, 209–215.

Cui, Y., Sun, J., 2012. On existence of positive solutions of coupled integral boundary value problems for a nonlinear singular superlinear differential system. Electron. J. Qual. Theory Differ. Equ. 2012 (41), 1–13.

Das, S., 2008. Functional Fractional Calculus for System Identification and Control. Springer, New York.

de Figueiredo, D.G., Lions, P.L., Nussbaum, R.D., 1982. A priori estimates and existence of positive solutions of semilinear elliptic equations. J. Math. Pures Appl. 61, 41–63.

Deng, K., 1995. Global existence and blow-up for a system of heat equations with nonlinear boundary conditions. Math. Methods Appl. Sci. 18, 307–315.

Deng, K., 1996. Blow-up rates for parabolic systems. Z. Angew. Math. Phys. 47, 132–143.

El-Shahed, M., Nieto, J.J., 2010. Nontrivial solutions for a nonlinear multi-point boundary value problem of fractional order. Comput. Math. Appl. 59, 3438–3443.

Eloe, P.W., Ahmad, B., 2005. Positive solutions of a nonlinear nth order boundary value problem with nonlocal conditions. Appl. Math. Lett. 18, 521–527.

Eloe, P.W., Henderson, J., 1997. Positive solutions for $(n-1, 1)$ conjugate boundary value problems. Nonlinear Anal. 28, 1669–1680.

Goodrich, C.S., 2012. Nonlocal systems of BVPs with asymptotically superlinear boundary conditions. Comment. Math. Univ. Carol. 53, 79–97.

Graef, J.R., Kong, L., Kong, Q., Wang, M., 2012. Uniqueness of positive solutions of fractional boundary value problems with non-homogeneous integral boundary conditions. Fract. Calc. Appl. Anal. 15, 509–528.

Guo, D., Lakshmikantham, V., 1988a. Nonlinear Problems in Abstract Cones. Academic Press, New York.

Guo, D.J., Lakshmikantham, V., 1988b. Multiple solutions of two-point boundary value problems of ordinary differential equations in Banach spaces. J. Math. Anal. Appl. 129, 211–222.

Hao, X., Liu, L., Wu, Y., 2012. Positive solutions for second order differential systems with nonlocal conditions. Fixed Point Theory 13, 507–516.

Henderson, J., Luca, R., 2011. Positive solutions for a system of higher-order multi-point boundary value problems. Comput. Math. Appl. 62, 3920–3932.

Henderson, J., Luca, R., 2012a. Existence and multiplicity for positive solutions of a multi-point boundary value problem. Appl. Math. Comput. 218, 10572–10585.

Henderson, J., Luca, R., 2012b. Existence of positive solutions for a system of higher-order multi-point boundary value problems. Appl. Math. Comput. 219, 3709–3720.

Henderson, J., Luca, R., 2012c. On a system of higher-order multi-point boundary value problems. Electron. J. Qual. Theory Differ. Equ. 2012 (49), 1–14.

Henderson, J., Luca, R., 2012d. On a system of second-order multi-point boundary value problems. Appl. Math. Lett. 25, 2089–2094.

Henderson, J., Luca, R., 2012e. Positive solutions for a system of second-order multi-point boundary value problems. Appl. Math. Comput. 218, 6083–6094.

Henderson, J., Luca, R., 2012f. Positive solutions for a system of second-order multi-point discrete boundary value problems. J. Differ. Equ. Appl. 18, 1575–1592.

Henderson, J., Luca, R., 2013a. Existence and multiplicity for positive solutions of a second-order multi-point discrete boundary value problem. J. Differ. Equ. Appl. 19, 418–438.

Henderson, J., Luca, R., 2013b. Existence and multiplicity for positive solutions of a system of higher-order multi-point boundary value problems. Nonlinear Differ. Equ. Appl. 20, 1035–1054.

Henderson, J., Luca, R., 2013c. Existence and multiplicity of positive solutions for a system of higher-order multi-point boundary value problems. Adv. Dyn. Syst. Appl. 8, 233–245.

Henderson, J., Luca, R., 2013d. Existence of positive solutions for a system of second-order multi-point discrete boundary value problems. J. Differ. Equ. Appl. 19, 1889–1906.

Henderson, J., Luca, R., 2013e. On a multi-point discrete boundary value problem. J. Differ. Equ. Appl. 19, 690–699.

Henderson, J., Luca, R., 2013f. Positive solutions for a system of nonlocal fractional boundary value problems. Fract. Calc. Appl. Anal. 16, 985–1008.

Henderson, J., Luca, R., 2013g. Positive solutions for a system of second-order nonlinear multi-point eigenvalue problems. Appl. Math. Comput. 223, 197–208.

Henderson, J., Luca, R., 2013h. Positive solutions for singular systems of higher-order multi-point boundary value problems. Math. Model. Anal. 18, 309–324.

Henderson, J., Luca, R., 2013i. Positive solutions for singular systems of multi-point boundary value problems. Math. Methods Appl. Sci. 36, 814–828.

Henderson, J., Luca, R., 2013j. Positive solutions for systems of second-order integral boundary value problems. Electron. J. Qual. Theory Differ. Equ. 2013 (70), 1–21.

Henderson, J., Luca, R., 2014a. Existence and multiplicity of positive solutions for a system of fractional boundary value problems. Bound. Value Probl. 60, 1–17.

Henderson, J., Luca, R., 2014b. On a second-order nonlinear discrete multi-point eigenvalue problem. J. Differ. Equ. Appl. 20, 1005–1018.

Henderson, J., Luca, R., 2014c. Positive solutions for a system of fractional differential equations with coupled integral boundary conditions. Appl. Math. Comput. 249, 182–197.

Henderson, J., Luca, R., 2014d. Positive solutions for systems of multi-point nonlinear boundary value problems. Commun. Appl. Nonlinear Anal. 21, 1–12.

Henderson, J., Luca, R., 2014e. Positive solutions for systems of nonlinear second-order multi-point boundary value problems. Math. Methods Appl. Sci. 37, 2502–2516.

Henderson, J., Luca, R., 2014f. Positive solutions for systems of singular higher-order multi-point boundary value problems. Br. J. Math. Comput. Sci. 4, 460–473.

Henderson, J., Luca, R., submitted. Positive solutions for a system of semipositone coupled fractional boundary value problems.

Henderson, J., Luca, R., Tudorache, A., 2014. Multiple positive solutions for a multi-point discrete boundary value problem. Commun. Fac. Sci. Univ. Ank. Ser. A1: Math. Stat. 63 (2), 59–70.

Henderson, J., Luca, R., Tudorache, A., 2015a. On a system of fractional differential equations with coupled integral boundary conditions. Fract. Calc. Appl. Anal. 18, 361–386.

Henderson, J., Luca, R., Tudorache, A., 2015b. Positive solutions for a fractional boundary value problem. Nonlinear Stud. 22, 1–13.

Henderson, J., Ntouyas, S.K., 2007. Positive solutions for systems of *n*th order three-point nonlocal boundary value problems. Electron. J. Qual. Theory Differ. Equ. 2007 (18), 1–12.

Henderson, J., Ntouyas, S.K., 2008a. Positive solutions for systems of nonlinear boundary value problems. Nonlinear Stud. 15, 51–60.

Henderson, J., Ntouyas, S.K., 2008b. Positive solutions for systems of three-point nonlinear boundary value problems. Aust. J. Math. Anal. Appl. 5, 1–9.

Henderson, J., Ntouyas, S.K., Purnaras, I.K., 2008a. Positive solutions for systems of generalized three-point nonlinear boundary value problems. Comment. Math. Univ. Carol. 49, 79–91.

Henderson, J., Ntouyas, S.K., Purnaras, I.K., 2008b. Positive solutions for systems of m-point nonlinear boundary value problems. Math. Model. Anal. 13, 357–370.

Henderson, J., Ntouyas, S.K., Purnaras, I.K., 2008c. Positive solutions for systems of three-point nonlinear discrete boundary value problems. Neural Parallel Sci. Comput. 16, 209–224.

Henderson, J., Ntouyas, S.K., Purnaras, I.K., 2009. Positive solutions for systems of nonlinear discrete boundary value problems. J. Differ. Equ. Appl. 15, 895–912.

Il'in, V.A., Moiseev, E.I., 1987a. Nonlocal boundary value problem of the second kind for a Sturm–Liouville operator. Differ. Equ. 23, 979–987.

Il'in, V.A., Moiseev, E.I., 1987b. Nonlocal boundary value problems of the first kind for a Sturm–Liouville operator in its differential and finite difference aspects. Differ. Equ. 23, 803–810.

Infante, G., Minhos, F.M., Pietramala, P., 2012. Non-negative solutions of systems of ODEs with coupled boundary conditions. Commun. Nonlinear Sci. Numer. Simul. 17, 4952–4960.

Infante, G., Pietramala, P., 2009a. Eigenvalues and non-negative solutions of a system with nonlocal BCs. Nonlinear Stud. 16, 187–196.

Infante, G., Pietramala, P., 2009b. Existence and multiplicity of non-negative solutions for systems of perturbed Hammerstein integral equations. Nonlinear Anal. 71, 1301–1310.

Ionkin, N.I., 1977. Solution of a boundary-value problem in heat conduction with a nonclassical boundary condition. Differ. Equ. 13, 204–211.

Jankowski, T., 2013. Positive solutions to second-order differential equations with dependence on the first-order derivative and nonlocal boundary conditions. Bound. Value Probl. 8, 1–20.

Ji, Y., Guo, Y., 2009. The existence of countably many positive solutions for some nonlinear nth order m-point boundary value problems. J. Comput. Appl. Math. 232, 187–200.

Ji, Y., Guo, Y., Yu, C., 2009. Positive solutions to $(n-1,n)$ m-point boundary value problems with dependence on the first order derivative. Appl. Math. Mech. (Engl. Ed.) 30, 527–536.

Jia, M., Wang, P., 2012. Multiple positive solutions for integro-differential equations with integral boundary conditions and sign changing nonlinearities. Electron. J. Differ. Equ. 2012 (31), 1–13.

Jiang, D., Yuan, C., 2010. The positive properties of the Green function for Dirichlet-type boundary value problems of nonlinear fractional differential equations and its application. Nonlinear Anal. 72, 710–719.

Joseph, D.D., Sparrow, E.M., 1970. Nonlinear diffusion induced by nonlinear sources. Q. Appl. Math. 28, 327–342.

Kang, P., Wei, Z., 2009. Three positive solutions of singular nonlocal boundary value problems for systems of nonlinear second-order ordinary differential equations. Nonlinear Anal. 70, 444–451.

Karakostas, G.L., Tsamatos, P.C., 2002. Multiple positive solutions of some Fredholm integral equations arisen from nonlocal boundary-value problems. Electron. J. Differ. Equ. 2002 (30), 1–17.

Keller, H.B., Cohen, D.S., 1967. Some positone problems suggested by nonlinear heat generation. J. Math. Mech. 16, 1361–1376.

Kelley, W.G., Peterson, A.C., 2001. Difference Equations. An Introduction with Applications, second ed. Academic Press, San Diego, CA.

Kilbas, A.A., Srivastava, H.M., Trujillo, J.J., 2006. Theory and applications of fractional differential equations. In: North-Holland Mathematics Studies, vol. 204. Elsevier Science B.V., Amsterdam.

Lakshmikantham, V., Trigiante, D., 1988. Theory of difference equations. Numerical methods and applications. In: Mathematics in Science and Engineering, vol. 181. Academic Press, Inc., Boston, MA.

Lan, K.Q., 2011. Positive solutions of systems of Hammerstein integral equations. Commun. Appl. Anal. 15, 521–528.

Lan, K.Q., Lin, W., 2011. Multiple positive solutions of systems of Hammerstein integral equations with applications to fractional differential equations. J. Lond. Math. Soc. 83, 449–469.

Li, W.T., Sun, H.R., 2006. Positive solutions for second-order m-point boundary value problems on times scales. Acta Math. Sin. Engl. Ser. 22, 1797–1804.

Liang, S., Zhang, J., 2009. Positive solutions for boundary value problems of nonlinear fractional differential equation. Nonlinear Anal. 71, 5545–5550.

Lin, Z., Xie, C., 1998. The blow-up rate for a system of heat equations with nonlinear boundary conditions. Nonlinear Anal. 34, 767–778.

Liu, B., Liu, L., Wu, Y., 2007. Positive solutions for singular systems of three-point boundary value problems. Comput. Math. Appl. 53, 1429–1438.

Luca, R., 2009. Positive solutions for $m+1$-point discrete boundary value problems. Lib. Math. 29, 65–82.

Luca, R., 2010. Existence of positive solutions for a class of higher-order m-point boundary value problems. Electron. J. Qual. Theory Differ. Equ. 2010 (74), 1–15.

Luca, R., 2011. Positive solutions for a second-order m-point boundary value problem. Dyn. Contin. Discrete Impuls. Syst. Ser. A: Math. Anal. 18, 161–176.

Luca, R., 2012a. Existence of positive solutions for a second-order $m+1$-point discrete boundary value problem. J. Differ. Equ. Appl. 18, 865–877.

Luca, R., 2012b. Positive solutions for a higher-order m-point boundary value problem. Mediterr. J. Math. 9, 381–394.

Luca, R., Tudorache, A., 2014. Positive solutions to a system of semipositone fractional boundary value problems. Adv. Differ. Equ. 179, 1–11.

Luca, R., Tudorache, A., 2015. Existence of positive solutions to a system of higher-order semipositone integral boundary value problems. Commun. Appl. Anal. 19, 589–604.

Ma, R., An, Y., 2009. Global structure of positive solutions for nonlocal boundary value problems involving integral conditions. Nonlinear Anal. 71, 4364–4376.

Ma, R., Thompson, B., 2004. Positive solutions for nonlinear m-point eigenvalue problems. J. Math. Anal. Appl. 297, 24–37.

Moshinsky, M., 1950. Sobre los problemas de condiciones a la frontiera en una dimension de caracteristicas discontinuas. Bol. Soc. Mat. Mex. 7, 1–25.

Pedersen, M., Lin, Z., 2001. Blow-up analysis for a system of heat equations coupled through a nonlinear boundary condition. Appl. Math. Lett. 14, 171–176.

Podlubny, I., 1999. Fractional Differential Equations. Academic Press, San Diego.

Sabatier, J., Agrawal, O.P., Machado, J.A.T. (Eds.), 2007. Advances in Fractional Calculus: Theoretical Developments and Applications in Physics and Engineering. Springer, Dordrecht.

Samarskii, A.A., 1980. Some problems of the theory of differential equations. Differ. Uravn. 16, 1925–1935.

Samko, S.G., Kilbas, A.A., Marichev, O.I., 1993. Fractional Integrals and Derivatives. Theory and Applications. Gordon and Breach, Yverdon.

Song, W., Gao, W., 2011. Positive solutions for a second-order system with integral boundary conditions. Electron. J. Differ. Equ. 2011 (13), 1–9.

Su, H., Wei, Z., Zhang, X., Liu, J., 2007. Positive solutions of n-order and m-order multi-point singular boundary value system. Appl. Math. Comput. 188, 1234–1243.

Webb, J.R.L., Infante, G., 2008. Positive solutions of nonlocal boundary value problems involving integral conditions. Nonlinear Differ. Equ. Appl. 15, 45–67.

Xie, D., Bai, C., Liu, Y., Wang, C., 2008. Positive solutions for nonlinear semipositone nth-order boundary value problems. Electron. J. Qual. Theory Differ. Equ. 2008 (7), 1–12.

Yang, Z., 2005. Positive solutions to a system of second-order nonlocal boundary value problems. Nonlinear Anal. 62, 1251–1265.

Yang, Z., 2006. Positive solutions of a second-order integral boundary value problem. J. Math. Anal. Appl. 321, 751–765.

Yang, Z., O'Regan, D., 2005. Positive solvability of systems of nonlinear Hammerstein integral equations. J. Math. Anal. Appl. 311, 600–614.

Yang, Z., Zhang, Z., 2012. Positive solutions for a system of nonlinear singular Hammerstein integral equations via nonnegative matrices and applications. Positivity 16, 783–800.

Yuan, C., 2010. Multiple positive solutions for $(n-1, 1)$-type semipositone conjugate boundary value problems of nonlinear fractional differential equations. Electron. J. Qual. Theory Differ. Equ. 2010 (36), 1–12.

Yuan, C., Jiang, D., O'Regan, D., Agarwal, R.P., 2012. Multiple positive solutions to systems of nonlinear semipositone fractional differential equations with coupled boundary conditions. Electron. J. Qual. Theory Differ. Equ. 2012 (13), 1–17.

Zhou, Y., Xu, Y., 2006. Positive solutions of three-point boundary value problems for systems of nonlinear second order ordinary differential equations. J. Math. Anal. Appl. 320, 578–590.

Index

Printed in the United States
By Bookmasters